TIME DELAY SYSTEMS
(TDS 2001)

A Proceedings volume from the 3rd IFAC Workshop,
Santa Fe, New Mexico, USA, 8 – 10 December 2001

Edited by

K. GU
Department of Mechanical & Industrial Engineering,
Southern Illinois Univeristy at Edwardsville, Illinois, USA

C.T. ABDALLAH
Department of Electrical & Computer Engineering,
University of New Mexico, USA

and

S.-I. NICULESCU
Université de Technologie de Compiègne, France

Published for the

INTERNATIONAL FEDERATION OF AUTOMATIC CONTROL

by

PERGAMON
An Imprint of Elsevier Science

ELSEVIER SCIENCE Ltd
The Boulevard, Langford Lane
Kidlington, Oxford OX5 1GB, UK

Elsevier Science Internet Homepage
http://www.elsevier.com

Consult the Elsevier Homepage for full catalogue information on all books, journals and electronic products and services.

IFAC Publications Internet Homepage
http://www.elsevier.com/locate/ifac

Consult the IFAC Publications Homepage for full details on the preparation of IFAC meeting papers, published/forthcoming IFAC books, and information about the IFAC Journals and affiliated journals.

First edition 2002

Library of Congress Cataloging in Publication Data

A catalogue record for this book is available from the Library of Congress

British Library Cataloguing in Publication Data

A catalogue record for this book is available from the British Library

ISBN 0-08-044004 5
ISSN 1474-6670

Printed in Great Britain

To Contact the Publisher

Elsevier Science welcomes enquiries concerning publishing proposals: books, journal special issues, conference proceedings, etc. All formats and media can be considered. Should you have a publishing proposal you wish to discuss, please contact, without obligation, the publisher responsible for Elsevier's industrial and control engineering publishing programme:

Dr Martin Ruck
Publishing Editor
Elsevier Science Ltd
The Boulevard, Langford Lane
Kidlington, Oxford
OX5 1GB, UK

Phone: +44 1865 843230
Fax: +44 1865 843920
E.mail: m.ruck@elsevier.co.uk

General enquiries, including placing orders, should be directed to Elsevier's Regional Sales Offices – please access the Elsevier homepage for full contact details (homepage details at the top of this page).

3rd IFAC WORKSHOP ON TIME DELAY SYSTEMS 2001

Sponsored by
International Federation of Automatic Control (IFAC)
Technical Committee on Linear Systems

Co-sponsored by
International Federation of Automatic Control (IFAC)
Technical Committee on Nonlinear Systems

Supported by
EECE Department, University of New Mexico
Ibero-American Science and Technology Education Consortium (ISTEC)

International Programme Committee (IPC)
S.-I. Niculescu (France, Chairman)
K. Gu (USA, Program Editor)

M. Ariola (Italy)
B. A. Leon de la Barra (Chile)
A. Bellen (Italy)
R. Byrne (USA)
J. Chiasson (USA)
J. Chen (USA)
R. Datko (USA)
J.-M. Dion (France)
L. Dugard (France)
M. Fliess (France)
W. Haddad (USA)
J. K. Hale (USA)
M. Jamshidi (USA)
V. Kapila (USA)
V. Kharitonov (Mexico)
V.B. Kolmanovskii (Russia)
J.-F. Lafay (France)
J. Louisell (USA)
A. Matasov (Russia)
Z.J. Palmor (Israel)
L. Pandolfi (Italy)
A.M. Perdon (Italy)
V. Rasvan (Romania)
J.-P. Richard (France)
M. Sorine (France)
C.E. de Souza (Brazil)
K. Uchida (Japan)
S.M. Verduyn Lunel (The Netherlands)
E.I. Verriest (USA)
G. Weiss (UK)

National Organizing Committee (NOC)
C.T. Abdallah (Chairman)

K. Gu (Program Editor)
M. Jamshidi (Local Arrangement)
Qing-Long Han

FOREWORD

These Proceedings contain the papers presented at the 3rd IFAC International Workshop on Time Delay Systems, Santa Fe, New Mexico, USA, 8 - 10 December 2001.

The first workshop in this series took place in Grenoble, France, 1998 and the second workshop was held in Ancona, Italy, 2000. This third workshop drops the "linear" adjective as its scope has expanded to include nonlinear systems. This workshop also contains various "new" applications of delay systems including those of communication networks and computing. The effects of delay are currently being felt in many research areas and this workshop was held in an attempt to bring together researchers from various technical fields in order to advance our understanding of delay phenomena in dynamical systems.

The workshop organizers and their collaborators would like to thank the National Science Foundation (NSF), USA, and the Centre National de Recherches Scientifiques (CNRS), France for supporting their research and the organization of the workshop. This workshop, in large measure, culminates the activities of the NSF-CNRS project. The organizers would also like to thank the Electrical and Computer Engineering Department of the University of New Mexico and the Ibero-American Science and Technology Education Consortium (ISTEC) for their generous support.

Chaouki T. Abdallah, *NOC Chair*
Keqin Gu, *Program Editor*
Silviu-Iulian Niculescu, *IPC Chair*

CONTENTS

ROBUST STABILIZATION

FEEDBACK CONTROL DESIGN

PLENARY PAPER

APPLICATION IN ENGINEERING AND MANUFACTURING

STABILITY AND CONTROL OF TIME-DELAY SYSTEMS

STABILITY AND TRACKING

STABILITY ANALYSIS

APPLICATION IN NETWORK AND COMPUTING

NUMERICAL COMPUTATION AND IMPLEMENTATION

SPECTRUM ASSIGNMENT AND SLIDING MODE CONTROL

NONLINEAR AND LEARNING CONTROL

MODELING, IDENTIFICATION AND PASSIVITY

www.elsevier.com/locate/ifac

H_∞ CONTROL OF DEAD-TIME SYSTEMS BASED ON A TRANSFORMATION [1]

Qing-Chang Zhong [*,2]

[*] *Faculty of Mechanical Engineering*
Technion-Israel Institute of Technology
Technion City, Haifa, 32000
Israel

Abstract: This paper presents a transformation that makes all the robust control problems of dead-time systems able to be solved similarly as in the finite dimensional situations. With trade-off of the performance, some advantages obtained are : (i) The controller has a quite simple and transparent structure; (ii) There are no any *additional* hidden modes in the Smith predictor and, hence, there is no *additional* hidden possibility to destabilize the system; (iii) it can be applied to systems with long dead-time without any difficulties. *Copyright © 2001 IFAC*

Keywords: Dead-time compensator, robust control, Smith predictor, performance index

1. INTRODUCTION

Dead-time systems are systems in which the action of control inputs takes a certain time before it affects the measured outputs. The typical dead-time systems are those with input delays. It is well known that it is very difficult to control such systems. Smith predictor (Smith, 1957) is the first effective way to control such systems, and then, many modified Smith predictors were presented (Palmor, 1996) . In recent years, the research interests are moving to the optimal control of dead-time systems, especially in H_∞ control, i.e., to find a controller to internally stabilize the system (if so, called an admissible controller) and to minimize the H_∞-norm of the transfer matrix from the external input signals, such as noises, disturbances and reference signals, to the output signals, such as controlled signals and tracking errors.

With the help of the modified Smith predictor (abbreviated as MSP) , many robust control problems for dead-time systems, such as robust stability, tracking and model-matching and input sensitivity minimization, can be solved as in the finite dimensional situations (Meinsma and Zwart, 2000; Nobuyama, 1992). However, the sensitivity minimization, the mixed sensitivity minimization and/or the standard H_∞ control problems cannot be solved in this way. The newly notable results in this area are (Mirkin, 2000; Meinsma and Zwart, 2000; Tadmor, 2000; Nagpal and Ravi, 1997). The results in (Mirkin, 2000; Meinsma and Zwart, 2000) were presented in the form of a modified Smith predictor. This paper focuses on the H_∞ control of dead-time systems of using a Smith predictor-type controller. For other approaches to H_∞ control of dead-time systems, we refer to (Mirkin and Tadmor, 2001) and the references therein.

The results in (Mirkin, 2000; Meinsma and Zwart, 2000) are quite elegant and the ideas are

[1] This research was supported by THE ISRAEL SCIENCE FOUNDATION founded by The Israel Academy of Sciences and Humanities.
[2] *Corresponding author*: *Dr.* Qingchang Zhong *is currently a research associate at Imperial College, London, UK. Tel*: 44-20-7594 6295, *Fax*: 44-20-7594 6282, *Email*: zhongqc@ic.ac.uk, *URL*: http://come.to/zhongqc

very tricky, but the controllers are too involved. Specifically, the Smith predictor is quite complex and, even more, relates to the performance level γ. Hence, there exist some problems to apply the method to systems with long dead-time, as pointed out by the authors in(Meinsma and Zwart, 2000). Another disadvantage is that the predictor always includes *additional* unstable hidden modes, even with stable plants, because the hidden modes are the eigenvalues of a Hamiltonian matrix. It has been pointed out in (Assche *et al.*, 1999; Manitius and Olbrot, 1979) that such hidden modes are not safe and tend to destabilize the system when implemented. Hence, the practical significance of these results may be limited.

This paper presents a transformation on the closed-loop transfer matrix. With this transformation, all the robust control problem can be solved as in the finite dimensional situations. The controller obtained has a quite simple and transparent structure. The resulted Smith predictor only depends on the real plant, here P_{22}, and is independent of the performance level γ and of the performance evaluation mechanism. There exist no any additional hidden modes and, hence, there is no hidden possibility to destabilize the system. This method can be applied to systems with long dead-time without any difficulties. The cost to pay for these advantages is some performance degradation. Qualitatively, this transformation is reasonable. However, it is very difficult to quantify the performance degradation in general cases. When the dead-time h becomes 0, the transformation becomes null. The transformation is quite ease to be applied because it just, virtually, subtracts an finite impulse response (FIR) block from the original system. The FIR block also relates to the original system, in fact, the performance evaluation mechanism. In single-input-single-output (SISO) case, it can be interpreted as just to care the performance when $t > h$ rather than the whole time range because the system behavior during $0 < t < h$ is uncontrollable and should not be considered in the performance index.

Notation Assume

$$G(s) = \left[\begin{array}{c|c} A & B \\ \hline C & D \end{array}\right] \tag{1}$$

is a rational transfer matrix. Two operators (Mirkin, 2000): *truncation* operator $\tau_h\{G\}$ and *completion* operator $\pi_h\{e^{-sh}G\}$, are defined as

$$\begin{aligned}\tau_h\{G\} &= \left[\begin{array}{c|c} A & B \\ \hline C & D \end{array}\right] - e^{-sh}\left[\begin{array}{c|c} A & e^{Ah}B \\ \hline C & 0 \end{array}\right] \\ &= G(s) - e^{-sh}\tilde{G}(s)\end{aligned} \tag{2}$$

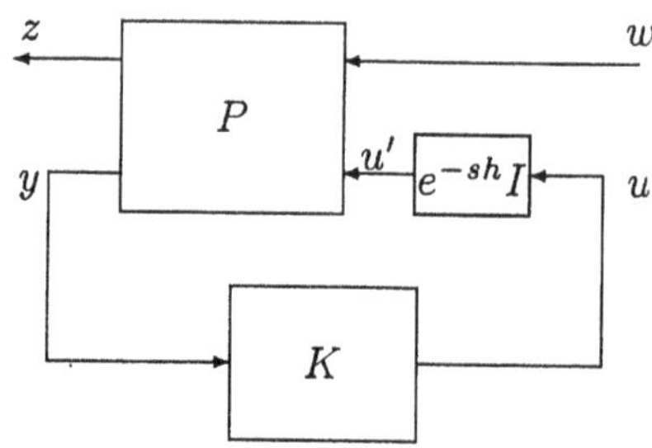

Fig. 1. General control setup of dead-time system

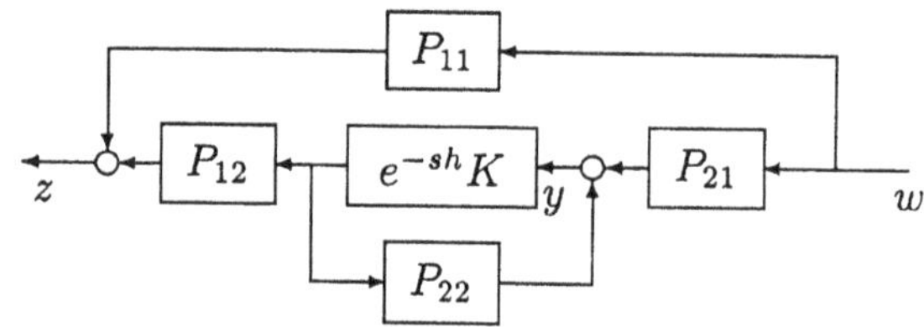

Fig. 2. Equivalent structure

$$\begin{aligned}\pi_h\{e^{-sh}G\} &= \left[\begin{array}{c|c} A & B \\ \hline Ce^{-Ah} & 0 \end{array}\right] - e^{-sh}\left[\begin{array}{c|c} A & B \\ \hline C & D \end{array}\right] \\ &= \hat{G}(s) - e^{-sh}G(s)\end{aligned} \tag{3}$$

These two operators map the transfer matrix $G(s)$ into FIR blocks. *Truncation* operator $\tau_h\{G\}$ truncates the impulse response of $G(s)$ at $t = h$; and *completion* operator $\pi_h\{e^{-sh}G\}$ extends the impulse response of $e^{-sh}G(s)$ to $t = 0 \sim h$. They all have finite impulse response with the support in $[0, h]$.

2. THE TRANSFORMATION

The general control setup for dead-time systems with a input or output delay is shown in Figure 1, where

$$P(s) = \begin{bmatrix} P_{11}(s) & P_{12}(s) \\ P_{21}(s) & P_{22}(s) \end{bmatrix}. \tag{4}$$

The closed-loop transfer matrix from w to z can be formulated as

$$T_{zw}(s) = P_{11} + e^{-sh}P_{12}K(I - e^{-sh}P_{22}K)^{-1}P_{21} \tag{5}$$

This means there exists instant response without delay through the path P_{11}. A clearer equivalent structure is shown in Figure 2. It's not difficult to recognize that, during the period $t = 0 \sim h$ after w is applied, the output z is not controllable and is only determined by P_{11}. It seems that one should *not* consider this period when evaluating the system performance.

Define the following FIR block

$$\Delta_1(s) = \tau_h\{P_{11}\} = P_{11}(s) - \tilde{P}_{11}(s)e^{-sh} \tag{6}$$

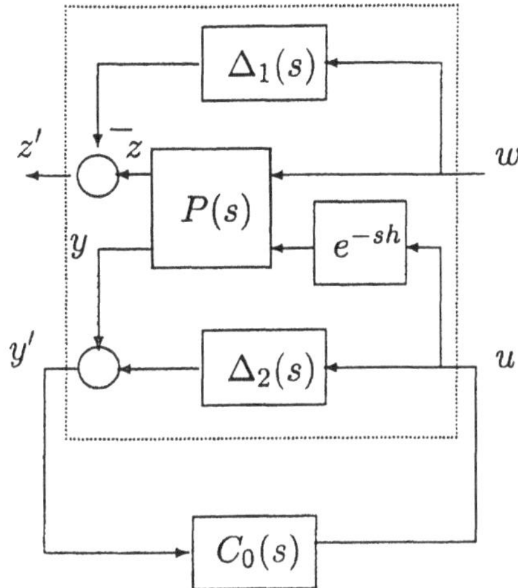

Fig. 3. Graphic interpretation of the transformation

which is exactly the uncontrollable part in $T_{zw}(s)$. Subtract it from the feed-forward path P_{11}, as shown in Figure 3, then,

$$T_{z'w}(s)=e^{-sh}\{\tilde{P}_{11}+P_{12}K(I-e^{-sh}P_{22}K)^{-1}P_{21}\} \quad (7)$$

or equivalently,

$$T_{zw}(s) = \Delta_1(s) + T_{z'w}(s) \quad (8)$$

This fact is firstly recognized in (Mirkin and Raskin, 1999). It is this FIR block that makes the control problems so complex and difficult to be solved. It has been shown (Mirkin, 2000) that this FIR block determines the achievable minimal performance level γ_{min}. As one can see, this block results from the performance evaluation mechanism. At some extent, it is not a part of the real control system, but an artificial part consisting of the weighting functions to evaluate the system performance. The idea in this paper is to optimize $T_{z'w}(s)$ rather than $T_{zw}(s)$, in other words, to change the performance evaluation mechanism.

In SISO case, the interpretation of the transformation in time domain can be given in Figure 4. $g(t)$ is the impulse response of $P_{11}(s)$ and $\tilde{g}(t)$ is the impulse response of $\tilde{P}_{11}(s)e^{-sh}$ which is truncated from $g(t)$ to $t > h$. Hence, with this transformation, only the system behavior when $t > h$ is considered and the behavior during $0 < t < h$ is not considered because it is not *controllable*.

3. H_∞ CONTROL OF SYSTEMS WITH A SINGLE DELAY

Assume that the realization of the rational part of the generalized process in Figure 1 is taken to be of the form

$$P(s) = \left[\begin{array}{c|cc} A & B_1 & B_2 \\ \hline C_1 & D_{11} & D_{12} \\ C_2 & D_{21} & D_{22} \end{array}\right] \quad (9)$$

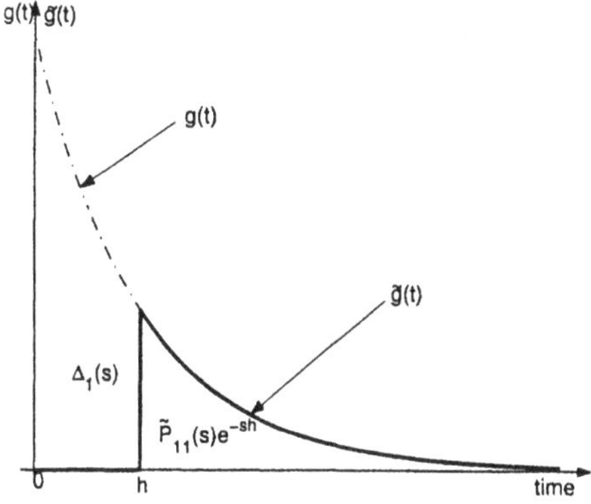

Fig. 4. Interpretation of the transformation in time domain

and the following standard assumptions hold.

(A1) (A, B_2) is stabilizable and (C_2, A) is detectable;

(A2) $\begin{bmatrix} A - j\omega I & B_2 \\ C_1 & D_{12} \end{bmatrix}$ has full column rank $\forall \omega \in R$;

(A3) $\begin{bmatrix} A - j\omega I & B_1 \\ C_2 & D_{21} \end{bmatrix}$ has full column rank $\forall \omega \in R$;

(A4) $D_{12}^* D_{12} = I$ and $D_{21} D_{21}^* = I$.

Assumption (A4) is made just to simplify the exposition. In fact, only the non-singularity of the matrices $D_{12}^* D_{12}$ and $D_{21} D_{21}^*$ is required.

Consider the Smith predictor-type controller as in Figure 3, in which the predictor is designed as

$$\Delta_2(s) = \left[\begin{array}{c|c} A & B_2 \\ \hline C_2 e^{-Ah} & 0 \end{array}\right] - e^{-sh}\left[\begin{array}{c|c} A & B_2 \\ \hline C_2 & D_{22} \end{array}\right]$$
$$= \pi_h\{e^{-sh}P_{22}\} \quad (10)$$

then the system can be re-formulated as

$$\begin{bmatrix} z'' \\ y' \end{bmatrix} = \tilde{P}(s) \begin{bmatrix} w \\ u \end{bmatrix} \quad (11)$$

with $e^{-sh}z'' \doteq z'$, where

$$\tilde{P}(s) = \left[\begin{array}{c|cc} A & e^{Ah}B_1 & B_2 \\ \hline C_1 & 0 & D_{12} \\ C_2 e^{-Ah} & D_{21} & 0 \end{array}\right] \quad (12)$$

is free of dead-time. Hence, the closed-loop transfer matrix of the transformed system is,

$$T_{z'w}(s)=e^{-sh}T_{z''w}(s)=e^{-sh}\mathcal{F}_l(\tilde{P}(s),C_0(s)) \quad (13)$$

and the H_∞ control problem

$$\|T_{z'w}(s)\|_\infty < \gamma \quad (14)$$

is converted to

$$\left\|\mathcal{F}_l(\tilde{P}(s), C_0(s))\right\|_\infty < \gamma \quad (15)$$

This is a finite dimensional problem and can be solved with the well-known results. Since this is for general setup of systems with an input delay, all the H_∞ control problems, such as robust stability, tracking and model-matching, input sensitivity minimization, output sensitivity minimization, mixed sensitivity minimization and the standard H_∞ control problem etc., can be solved similarly as in the finite dimensional situation.

The H_∞ solution involves the following two Hamiltonian matrices

$$H_h = \begin{bmatrix} A & \gamma^{-2}e^{Ah}B_1B_1^*e^{A^*h} \\ -C_1^*C_1 & -A^* \end{bmatrix} - \begin{bmatrix} B_2 \\ -C_1^*D_{12} \end{bmatrix} \begin{bmatrix} D_{12}^*C_1 & B_2^* \end{bmatrix} \quad (16)$$

$$J_h = \begin{bmatrix} A^* & \gamma^{-2}C_1^*C_1 \\ -e^{Ah}B_1B_1^*e^{A^*h} & -A \end{bmatrix} - \begin{bmatrix} e^{-A^*h}C_2^* \\ -e^{Ah}B_1D_{21}^* \end{bmatrix} \begin{bmatrix} D_{21}B_1^*e^{A^*h} & C_2e^{-Ah} \end{bmatrix} (17)$$

and the solution is given by the following theorem.

Theorem 1. There exists an admissible main controller such that $\|T_{z'w}(s)\|_\infty < \gamma$ in Figure 3 iff the following three conditions hold:

(i) $H_h \in dom(Ric)$ and $X = Ric(H_h) \geq 0$;

(ii) $J_h \in dom(Ric)$ and $Y = Ric(J_h) \geq 0$;

(iii) $\rho(XY) < \gamma^2$.

Moreover, when the conditions hold, one such main controller is

$$C_0(s) = \left[\begin{array}{c|c} A_h & -L_h \\ \hline F_hZ_h & 0 \end{array}\right] \quad (18)$$

where

$$\begin{aligned} A_h &= A + L_hC_2e^{-Ah} + \gamma^{-2}YC_1^*C_1 \\ &\quad + (B_2 + \gamma^{-2}YC_1^*D_{12})\, F_hZ_h, \\ F_h &= -(B_2^*X + D_{12}^*C_1), \\ L_h &= -(Ye^{-A^*h}C_2^* + e^{Ah}B_1D_{21}^*), \\ Z_h &= (I - \gamma^{-2}YX)^{-1}. \end{aligned} \quad (19)$$

Furthermore, if define

$$F_{h2} = -\left(C_2e^{-Ah} + \gamma^{-2}D_{21}B_1^*e^{A^*h}X\right), (20)$$

and

$$L_{h2} = -(B_2 + \gamma^{-2}YC_1^*D_{12}), \quad (21)$$

then the set of all admissible main controllers such that $\|T_{z'w}(s)\|_\infty < \gamma$ can be parameterized as

$$C_0(s) = \mathcal{F}_l(M(s), Q(s)) \quad (22)$$

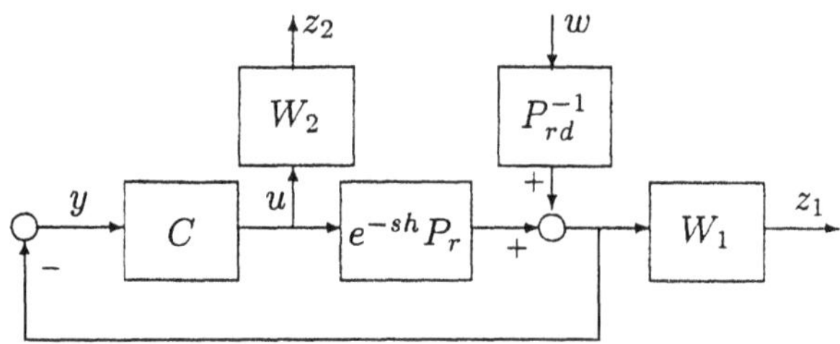

Fig. 5. Setup for Mixed sensitivity minimization

where

$$M(s) = \left[\begin{array}{c|cc} A_h & -L_h & -L_{h2} \\ \hline F_hZ_h & 0 & I \\ F_{h2}Z_h & I & 0 \end{array}\right] \quad (23)$$

and $Q(s) \in H_\infty$, $\|Q(s)\|_\infty < \gamma$.

PROOF. First of all, check if $\tilde{P}(s)$ meets the standard assumptions (A1-A4).

(A1) (C_2e^{-Ah}, A) is detectable for $A + e^{Ah}LC_2e^{-Ah} \sim A + LC_2$ and (C_2, A) is detectable;

(A3) $\begin{bmatrix} A - j\omega I & e^{Ah}B_1 \\ C_2e^{-Ah} & D_{21} \end{bmatrix}$ has full column rank $\forall\omega \in R$ because $\begin{bmatrix} A - j\omega I & e^{Ah}B_1 \\ C_2e^{-Ah} & D_{21} \end{bmatrix} \sim \begin{bmatrix} A - j\omega I & B_1 \\ C_2 & D_{21} \end{bmatrix}$ has full column rank $\forall\omega \in R$.

The assumptions (A2) and (A4) remain unchanged. Hence, $\tilde{P}(s)$ meets all the standard assumptions. Theorem 5.1 in (Green *et al.*, 1990) can be directly used to solve the problem. Substitute $\tilde{P}(s)$ into that Theorem, then the above result is obtained with ease.

Remark 2. It should be pointed out that, under the transformation, either $D_{11} \neq 0$ or $D_{22} \neq 0$ does not make the problem more complex. This seems more reasonable. In fact, when $h = 0$, $\Delta_2(s) = -D_{22}$ is the common controller transformation to make $D_{22} = 0$.

4. EXAMPLE

Consider the example studied in (Meinsma and Zwart, 2000), see Figure 5, where

$$P_r(s) = \frac{1}{s-1} \text{ and } P_{rd}(s) = \frac{s-1}{s+1} \quad (24)$$

$$W_1(s) = \frac{2(s+1)}{10s+1} \text{ and } W_2(s) = \frac{s+1.1}{5(s+1)} \quad (25)$$

This problem can be arranged as a standard problem with

Table 1. Performance Comparison

h	Meinsma's $\|T_{zw}(s)\|_\infty$	Proposed $\|T_{z'w}(s)\|_\infty$	Proposed $\|T_{zw}(s)\|_\infty$
0.2	*0.68*	0.58	*0.88*
2	*5.22*	3.86	*8.34*
5	109.5*	78.63	*183.7*

*Note: For $h > 3.2$, some matrices are close to singular or badly scaled and the results may be inaccurate.

$$P(s) = \left[\begin{array}{c|c} \frac{2(s+1)^2}{(10s+1)(s-1)} & \frac{2(s+1)}{(10s+1)(s-1)} \\ 0 & \frac{0.2(s+1.1)}{s+1} \\ \hline -\frac{s+1}{s-1} & -\frac{1}{s-1} \end{array}\right] \qquad (26)$$

where $W_2(s)$ should be delayed as $W_2 e^{-hs}$ and the result will not be affected.

Here, the predictor is designed as

$$\Delta_2(s) = -\frac{e^{-h} - e^{-sh}}{s-1} \qquad (27)$$

However, the one obtained in (Meinsma and Zwart, 2000) when $h = 0.2sec$ is

$$\Delta_2(s) = -\frac{10.75s^2 + 0.245s - 0.3298}{(s-1)(12.03s^2+1)} + \frac{13.16s^2 - 0.1316}{(s-1)(12.03s^2+1)} e^{-sh} \qquad (28)$$

It is much more complex. In addition to the pole of the real plant $s_0 = 1$, there are two additional hidden modes $s_{1,2} = \pm 0.2883j$. The achievable performance is listed in Table 1 for different delays h. The performance of the proposed method degrades not so much to pay for other advantages.

The frequency responses of both cases are shown in Figure 6. Although the H_∞-norm achieved in this paper is a little bit larger than that achieved by the method of Meinsma and Zwart (noted as M-Z in the figures), the proposed method obtains better performance in a quite broad frequency band ranging from $1Hz$ to about $10,000Hz$. From the engineering point of view, this solution is much better.

5. CONCLUSIONS

This is not a solution to the original H_∞ control problem of time delay systems. However, it considerably simplifies the problem with reasonable performance degradation and has many advantages. Especially, the practical significance is obvious. The quantitative analysis of the performance degradation would be further studied.

ACKNOWLEDGEMENTS

The author would like to thank Dr. Mirkin and Prof. K. M. Zhou for their helpful suggestions.

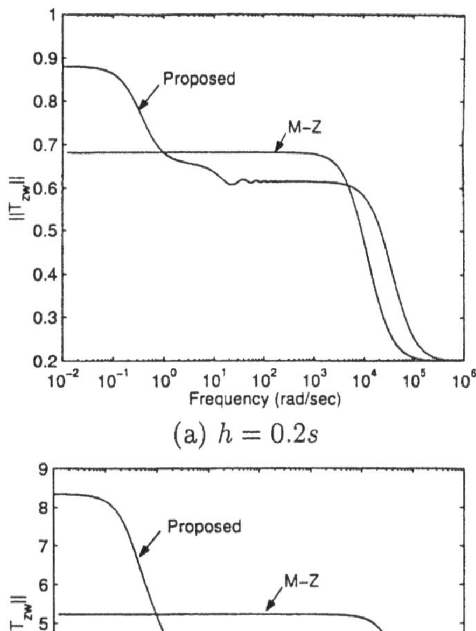

(a) $h = 0.2s$

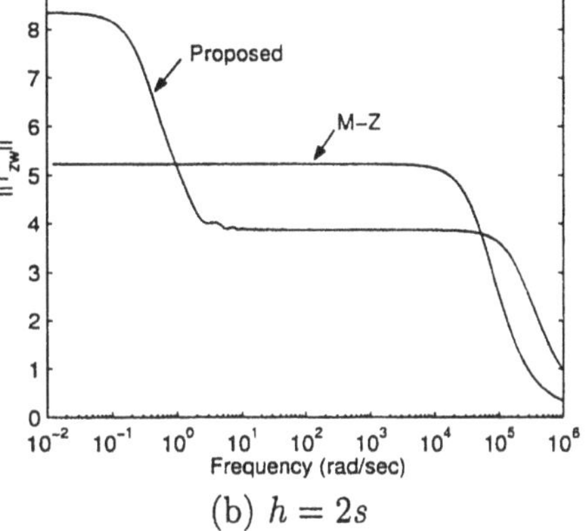

(b) $h = 2s$

Fig. 6. Comparison of $\|T_{zw}(j\omega)\|$

REFERENCES

Assche, V. Van, M. Dambrine and J.F. Lafay (1999). Some problems arising in the implementation of distributed-delay control laws. In: *Proc. of the 38th IEEE Conference on Decision & control.* Phoenix, Arizona, USA. pp. 4668–4672.

Green, M., K. Glover, D. Limebeer and J. Doyle (1990). A J-spectral factorization approach to H_∞ control. *SIAM J. Control Optim.* **28**(6), 1350–1371.

Manitius, A.Z. and A. W. Olbrot (1979). Finite spectrum assignment problem for systems with delays. *IEEE Trans. Automat. Control* **24**(4), 541–553.

Meinsma, G. and H. Zwart (2000). On $\mathcal{H}_\infty$ control for dead-time systems. *IEEE Trans. Automat. Control* **45**(2), 272–285.

Mirkin, L. (2000). On the extraction of dead-time controllers from delay-free parametrizations. In: *Proc. of 2nd IFAC Workshop on Linear Time Delay Systems.* Ancona, Italy. pp. 157–162.

Mirkin, L. and G. Tadmor (2001). H_∞ control of systems with I/O delay: A review of some problem-oriented methods. to appear in IMA J. Math. Control & Information.

Mirkin, L. and N. Raskin (1999). State-space parametrization of all stabilizing dead-time controllers. In: *Proc. of the 38th Conf. on*

Decision & Control. Phoenix, Arizona, USA. pp. 221–226.

Nagpal, K. M. and R. Ravi (1997). H_∞ control and estimation problems with delayed measurements: State-space solutions. *SIAM J. Control Optim.* **35**(4), 1217–1243.

Nobuyama, E. (1992). Robust stabilization of time-delay systems via reduction to delay-free model matching problems. In: *Proc. of the 31th IEEE Conference on Decision & control.* Vol. 1. Tucson, Arizona. pp. 357–358.

Palmor, Z. J. (1996). Time-delay compensation — Smith predictor and its modifications. In: *The Control Handbook* (S. Levine, Ed.). pp. 224–237. CRC Press.

Smith, O. J. M. (1957). Closer control of loops with dead time. *Chem. Eng. Progress* **53**(5), 217–219.

Tadmor, G. (2000). The standard H_∞ problem in systems with a single input delay. *IEEE Trans. Automat. Control* **45**(3), 382 –397.

www.elsevier.com/locate/ifac

Another Look at Finite Horizon H^∞ Control Problems for Systems with Input Delays

K. Uchida†, M. Fujita‡ and K. Ikeda†

†*Department of Electrical, Electronics and Computer Engineering, Waseda University, Tokyo169-8555, Japan, {kuchida, Ikeda}@uchi.elec.waseda.ac.jp*
‡*Department of Electrical and Electronic Engineering, Kanazawa University, Kanazawa 920-8667, Japan, fujita@t.kanazawa-u.ac.jp*

Abstract. We discuss a finite horizon H^∞ control problem for systems with input delays. Clarifying a relationship between two H^∞ control problems in input delay case and in measurement delay case, we derive a solution in input delay case based on the known result for the H^∞ control problem in measurement delay case, and show that the solution has the same predictor-observer structure as the solution in measurement delay case has. Using this structural information on the solution, we also present a direct proof of the solution to the finite horizon H^∞ control problem for systems with input delays, which is based only on completion of squares.

Keywords: Time delay systems; Input delay; Finite horizon; H^∞ control; Predictor

1. Introduction

In control system designs, input delay appears rather often and is considered as a small but cumbersome obstacle. For systems with input delays, the H^∞ control problem was actively investigated in parallel with development of H^∞ control theory. The problem was solved by Kojima and Ishijima (1994) in state-space form, the parameterization of all the solutions was given by Tadmor (1995), and a particular (predictor-observer) structure of the solutions has been recently been pointed out by Mirkin (2000). We can find detailed reviews of this area in (Tadmor, 2000, Mirkin, 2000) and the references insides. In this paper, we revisit the H^∞ control problem for systems with input delays in the framework of finite horizon. The first objective is to discuss further the predictor-observer structure of the solution, which is pointed out by Milkin (2000), from a novel viewpoint. Being suggested by the first discussion, secondly, we try to develop an elementary approach to the problem, which is completely different from the abstract approach based on evolution equations taken in (Kojima and Ishijima, 1994, Tadmor, 1995) and requires only completion of squares.

More specifically, the content of this paper is stated and organized as follows. In Section 2, we formulate the H^∞ control problem for systems with input delays together with two related H^∞ control problems. In Section 3, we first clarify a relationship between our problem and an H^∞ control problem for systems with measurement delays. Next, we derive a solution based on the known result for the H^∞ control problem in measurement delay case, and show that the solution has the same predictor-observer structure as the solution in measurement delay case has. In Section 4, using this structural information on the solution, we present a direct proof of the solution to the finite horizon H^∞ control problem for systems with input delays, which is based only on completion of squares.

Notations: $L^2(a,b;R^k)$ is the space of square integrable functions of k-dimension defined on the time interval $[a,b]$. When $a=t_0$ and $b=t_1$, the L^2-norm of f in $L^2(a,b;R^k)$ is denoted as $\|f\|_2^2$. $\|x\|$ denotes the Euclidean norm of x in R^k. For symmetric matrices X and Y, $X \geq Y$ $(X > Y)$ implies that $X-Y$ is positive semidefinite (positive definite). I is the identity matrix of appropriate dimension. $(\ )'$ denotes the transpose of vector or matrix. $\rho(X)$ denotes the spectral radius of matrix X.

2. System Description and Problem Statement

Consider the linear time-varying system with the time-delay $h>0$ in the control input, which is defined on the interval $[t_0, t_1]$ and described by

$$\frac{d}{dt}x(t) = A(t)x(t) + B(t)u(t-h) + D(t)v(t),$$
$$y(t) = C(t)x(t) + w(t), \qquad (1)$$
$$g(t) = \begin{bmatrix} z(t) \\ u(t) \end{bmatrix}, \qquad z(t) = F(t)x(t)$$

where $x(t)$ the n -dimensional internal-variable; $u(t)$ is the r -dimensional control input; $y(t)$ is the m -dimensional measurement output; $g(t)$ is the $(q+r)$ -dimensional controlled output; $d(t) = (v(t)', w(t)')'$ is the $(p+m)$ -dimensional disturbance; the initial condition $(x(t_0), u_{t_0})$ (where $u_t = \{u(t+\beta), -h \le \beta \le 0\}$) in $R^n \times L^2(-h,0;R^r)$, is given by a constant matrix $N > 0$ and an n -dimensional parameter ξ as

$$x(t_0) = N\xi, \qquad u_{t_0} = 0. \qquad (2)$$

$A(t), B(t), C(t), D(t)$ and $F(t)$ are matrices of appropriate dimensions whose elements are continuous functions of time. For the system (1) with the initial condition (2), the admissible control $u(t) = \Phi_{ID}(t, y)$ is given by a causal function of the measurement data specifically to be the form

$$u(t) = \Phi_{ID}(t, \{y(s), t_0 \le s \le t\}), \qquad t_0 \le t \le t_1 - h. \qquad (3)$$

Problem ID *(H^∞ Control Problem with Input Delay):* Given the system described by (1) and (2) and a constant number $\gamma > 0$, the problem is to find an admissible control (3) which satisfies the inequality

$$\|g\|_2^2 < \gamma^2(\|d\|_2^2 + \xi' N\xi) \qquad (u_{t_1} = 0) \qquad \text{(ID)}$$

for all $d = (v', w')'$ in $L^2(t_0, t_1; R^{p+m})$ and all ξ in R^n such that $(d, \xi) \ne 0$.

Remark : The terminal penalty $\xi' N\xi$ implies that the controls are determined so as to attenuate the effect of the uncertain initial internal-variable $x(t_0)$ which is known to be in $\mathrm{Im}(N)$; if $N = 0$, the initial internal-variable is completely (to be zero), and if N is nonsingular (positive definite), the initial internal-variable is completely unknown. For simplicity, in this paper, we discuss only the nonsingular case. The extension to the singular case can be done by a perturbation technique (Uchida and Fujita, 1992).

The H^∞ control problem for systems with input delays was solved by Kojima and Ishijima (1994), the parameterization of all the solutions was given by Tadmor (1995), and a particular (predictor-observer) structure of the solutions has been recently been pointed out by Mirkin (2000). Problem ID is an extension of the problem discussed in these literatures in the points that the system with a finite horizon is time varying and the criterion includes a terminal penalty, and could be solved by extending the arguments of (Kojima and Ishijima, 1994, Tadmor, 1995). In this paper, instead of pursuing this line, we will develop another approach and provide a new characterization of the solutions, which is inspired by the observation in (Mirkin, 2000).

We first consider an auxiliary problem to Problem ID. The system is defined on $[t_0, t_1]$ and described by

$$\frac{d}{dt}x(t) = A(t)x(t) + B(t)u(t-h) + D(t)v(t),$$
$$y(t) = C(t)x(t) + w(t), \qquad (4)$$
$$g(t) = \begin{bmatrix} z(t) \\ u(t-h) \end{bmatrix}, \quad z(t) = F(t)x(t)$$

with the initial condition

$$x(t_0) = N\xi, \qquad (5)$$

and the admissible control $u(t) = \Phi_{AID}(t, y)$ is given by a causal function of the measurement data specifically to be the form

$$u(t) = \begin{cases} \Phi_{AID}(t, \{y(s), t_0 \le s \le t\}), & t_0 \le t < t_1 - h \\ \Phi_{AID}(t), & t_0 - h \le t \le t_0. \end{cases} \qquad (6)$$

Problem AID *(Auxiliary H^∞ Control Problem with Input Delay):* Given the system described by (4) and (5) and a constant number $\gamma > 0$, the problem is to find an admissible control (6) which satisfies the inequality

$$\|g\|_2^2 < \gamma^2(\|d\|_2^2 + \xi' N\xi) \qquad \text{(AID)}$$

for all $d = (v', w')'$ in $L^2(t_0, t_1; R^{p+m})$ and all ξ in R^n such that $(d, \xi) \ne 0$.

The difference between Problem ID and Problem AID is found only in the role of $u_{t_0} = \{u(t_0 + \beta), -h \le \beta \le 0\}$, that is, u_{t_0} is fixed (to be zero function) as a part of the initial condition in Problem ID, while u_{t_0} is a part of the control input to be determined in Problem AID. Although Problem AID itself is an H^∞ control problem applicable to some control designs, we will use Problem AID to bridge a gap between Problem ID and another H^∞ control problem introduced in the following.

We consider next an H^∞ problem which does not

have input delays but has measurement delays. The system is defined on $[t_0, t_1]$ and described by

$$\frac{d}{dt}x(t) = A(t)x(t) + B(t)u(t) + D(t)v(t),$$
$$y(t) = C(t)x(t) + w(t), \quad (7)$$
$$g(t) = \begin{bmatrix} z(t) \\ u(t) \end{bmatrix}, \qquad z(t) = F(t)x(t)$$

with the initial condition

$$x(t_0) = N\xi, \quad (8)$$

and the admissible control $u(t) = \Phi_{MD}(t, y)$ is given by a causal function of the delayed measurement data specifically to be the form

$$u(t) = \begin{cases} \Phi_{MD}(t, \{y(s), t_0 \le s \le t-h\}), & t_0 + h \le t \le t_1 \\ \Phi_{MD}(t), & t_0 \le t \le t_0 + h. \end{cases} \quad (9)$$

Problem MD *(H^∞ Control Problem with Measurement Delay):* Given the system described by (7) and (8) and a constant number $\gamma > 0$, the problem is to find an admissible control (9) which satisfies the inequality

$$\|g\|_2^2 < \gamma^2(\|d\|_2^2 + \xi' N\xi) \quad \text{(MD)}$$

for all $d = (v', w')'$ in $L^2(t_0, t_1; R^{p+m})$ and all ξ in R^n such that $(d, \xi) \neq 0$.

The H^∞ control problem for systems with measurement delays was also solved completely in (Basar and Bernhard, 1991, Nagpal and Ravy, 1997), and, as is expected from existence of information delays in constructing control inputs, the solution has a natural predictor-observer structure.

Our approach to Problem ID, which we will take in the following sections, is summarized as follows. We establish first some relationships between Problem ID and Problem MD via Problem AID, and try to solve Problem ID based on the relationships and the solution of Problem MD, so that the solution of Problem ID has the same predictor-observer structure as the solution of Problem MD has.

3. Structure and Characterization of Solution

To find relations among three Problems ID, AID and MD, we observe the detail of the term of the controlled output in each criterion. In (ID), if we take an admissible control given by (3), we have

$$\|g\|_2^2 = \int_{t_0}^{t_1} \|z(t)\|^2 dt + \int_{t_0+h}^{t_1} \|\Phi_{ID}(t-h, \{y(s), t_0 \le s \le t-h\})\|^2 dt. \quad (10)$$

In (AID), if we take an admissible control given by (6), we have

$$\|g\|_2^2 = \int_{t_0}^{t_1} \|z(t)\|^2 dt + \int_{t_0+h}^{t_1} \|\Phi_{AID}(t-h, \{y(s), t_0 \le s \le t-h\})\|^2 dt + \int_{t_0}^{t_0+h} \|\Phi_{AID}(t-h)\|^2 dt. \quad (11)$$

In (MD), if we take an admissible control given by (9), we have

$$\|g\|_2^2 = \int_{t_0}^{t_1} \|z(t)\|^2 dt + \int_{t_0+h}^{t_1} \|\Phi_{MD}(t, \{y(s), t_0 \le s \le t-h\})\|^2 dt + \int_{t_0}^{t_0+h} \|\Phi_{AID}(t)\|^2 dt. \quad (12)$$

The following result is an immediate conclusion from the descriptions of three Problems (ID), (AID) and (MD) and the expressions of (10), (11) and (12).

Proposition 1: a) If $u(t) = \Phi_{ID}(t, y)$ defined by (3) is a solution to Problem ID, the control $u(t) = \Phi_{ID}(t, y)$ together with $u_{t_0} = 0$ is a solution to Problem AID. Conversely, if a control $u(t) = \Phi_{AID}(t, y)$ defined by (6) is a solution to Problem AID and satisfies $u_{t_0} = 0$, the control $u(t) = \Phi_{AID}(t, y)$ is a solution to Problem ID.

b) If a control $u(t) = \Phi_{AID}(t, y)$ given by (6) is a solution to Problem AID, the delayed control $u(t) = \Phi_{AID}(t-h, y)$ is a solution to Problem MD. Conversely, if $u(t) = \Phi_{MD}(t, y)$ defined by (9) is a solution to Problem MD, the advanced control $u(t) = \Phi_{MD}(t+h, y)$ is a solution to Problem AID.

c) If $u(t) = \Phi_{MD}(t, y)$ defined by (9) is a solution to Problem MD and satisfies $u_{t_0+h} = 0$, the advanced control $u(t) = \Phi_{MD}(t+h, y)$ is a solution to Problem ID. Conversely, if $u(t) = \Phi_{ID}(t, y)$ given by (3) is a solution to Problem ID, the delayed control $u(t) = \Phi_{ID}(t-h, y)$ together with $u_{t_0+h} = 0$ is a solution to Problem MD.

Using the fact c) in Proposition 1 and a solution to Problem MD, we will derive a solution of Problem ID. Now we present the solution to Problem MD, which is a slight modification of the result given by Basar and Bernhard (1991). We need to introduce the following four conditions.

(C1) There exists a solution $M(t)$, $t_0 \le t \le t_1$ to the Riccati differential equation

$$-\frac{d}{dt}M(t) = M(t)A(t) + A(t)'M(t) + F(t)'F(t) - M(t)(B(t)B(t)' - \gamma^{-2}D(t)D(t)')M(t),$$
$$M(t_1) = 0. \tag{13}$$

(C2) There exists a solution $P(t)$, $t_0 \le t \le t_1 - h$ to the Riccati differential equation

$$\frac{d}{dt}P(t) = A(t)P(t) + P(t)A(t)' + D(t)D(t)' - P(t)(C(t)'C(t) - \gamma^{-2}F(t)'F(t))P(t),$$
$$P(t_0) = N. \tag{14}$$

(C3) There exists a solution $Q(t+\beta, t-h)$, $t_0 + h \le t \le t_1$, $-h \le \beta \le 0$ to the Riccati differential equation

$$\frac{\partial}{\partial\beta}Q(t+\beta, t-h) = A(t+\beta)Q(t+\beta, t-h) + Q(t+\beta, t-h)A(t+\beta)' + D(t+\beta)D(t+\beta)' + \gamma^{-2}Q(t+\beta, t-h)F(t+\beta)'F(t+\beta)Q(t+\beta, t-h),$$
$$Q(t-h, t-h) = P(t-h). \tag{15}$$

(C4) $\rho(M(t+\beta)Q(t+\beta, t-h)) < \gamma^2$, $t_0 + h \le t \le t_1$, $-h \le \beta \le 0$.

Proposition 2: Assume that the conditions (C1)-(C4) are satisfied. Then, a solution to Problem MD is given by

$$u(t) = \begin{cases} -B(t)'S(t, t-h)\bar{x}(t, t-h), & t_0 + h \le t \le t_1 \\ 0, & t_0 \le t \le t_0 + h \end{cases} \tag{16}$$

where $S(t, t-h)$ is defined by

$$S(t+\beta, t-h) = M(t+\beta)(I - \gamma^{-2}Q(t+\beta, t-h)M(t+\beta))^{-1}, \quad -h \le \beta \le 0 \tag{17}$$

and $\bar{x}(t, t-h)$ is predicted with the "predictor"

$$\frac{\partial}{\partial\beta}\bar{x}(t+\beta, t-h) = (A(t+\beta) + \gamma^{-2}Q(t+\beta, t-h)F(t+\beta)'F(t+\beta) - B(t+\beta)B(t+\beta)'S(t+\beta, t-h))\bar{x}(t+\beta, t-h), \quad -h \le \beta \le 0 \tag{18}$$

from the estimate $\bar{x}(t-h, t-h) = \hat{x}(t-h)$ which is estimated with the "observer"

$$\frac{d}{dt}\hat{x}(t) = (A(t) + \gamma^{-2}P(t)F(t)'F(t) - B(t)B(t)'S(t))\hat{x}(t) + P(t)C(t)'(y(t) - C(t)\hat{x}(t)),$$
$$\hat{x}(t_0) = 0. \tag{19}$$

The proof can be found in the next section. From b) in Proposition 1 and Proposition 2, a solution to Problem AID is given by

$$u(t) = \begin{cases} -B(t+h)'S(t+h, t)\bar{x}(t+h, t), & t_0 \le t \le t_1 - h \\ 0 & t_0 - h \le t \le t_0 \end{cases} \tag{20}$$

which is the advanced form of the control (16). Moreover, since the solution (16) satisfies $u_{t_0+h} = 0$, it follows from c) in Proposition 1 and Proposition 2 that the advanced version of (16) given as

$$u(t) = -B(t+h)'S(t+h, t)\bar{x}(t+h, t), \quad t_0 \le t \le t_1 - h \tag{21}$$

is a solution to Problem ID. Here note that the solutions (20) and (21) have the same predictor-observer structure. That is, in constructing the controls (20) and (21), the estimate $\hat{x}(t)$ is estimated with the observer (19) based on the data $\{y(s), t_0 \le s \le t\}$, and $\bar{x}(t+h, t)$ is predicted with the predictor (18) from the estimate $\bar{x}(t,t) = \hat{x}(t)$. It is also noted that the conditions (C1)-(C4) form the same sufficient condition for existence of solutions to Problems AID and ID. We can summarize these facts, together with necessity of the conditions (C1)-(C4), in the following form.

Theorem: a) There exists a solution to Problem ID if and only if the conditions (C1)-(C4) are satisfied. If the conditions (C1)-(C4) are satisfied, the control (21) is a solution to Problem ID.

b) There exists a solution to Problem AID if and only if the conditions (C1)-(C4) are satisfied. If the conditions (C1)-(C4) are satisfied, the control (20) is a solution to Problem AID.

c) There exists a solution to Problem MD if and only if the conditions (C1)-(C4) are satisfied. If the conditions (C1)-(C4) are satisfied, the control (16) is a solution to Problem MD.

In the next section, we provide a direct proof of this theorem by using an elementary argument based only on completion of squares (Uchida and Fujita, 1990).

4. Proof of Theorem (Completion of Squares)

We prove only b) in Theorem because a) and c) follows from b) and Proposition 1. Before starting the proof, we present a preliminary result.

Lemma 1: Let $Q(t+\beta,t-h)$, $t_0+h\le t\le t_1$, $-h\le\beta\le 0$ be a solution to the Riccati differential equation (15) in (C3) with the initial condition given in (C2). The solution satisfies the Riccati differential equations

$$\frac{d}{dt}Q(t,t-h)=A(t)Q(t,t-h)+Q(t,t-h)A(t)'$$
$$+D(t)D(t)'+\gamma^{-2}Q(t,t-h)F(t)'F(t)Q(t,t-h)$$
$$-\Psi(t,t-h)P(t-h)C(t-h)'C(t-h)P(t-h)\Psi(t,t-h)',$$
$$\frac{\partial}{\partial\beta}Q(t_0+h+\beta,t_0)=A(t_0+h+\beta)Q(t_0+h+\beta,t_0)$$
$$+Q(t_0+h+\beta,t_0)A(t_0+h+\beta)'+D(t_0+h+\beta)D(t_0+h+\beta)'$$
$$+\gamma^{-2}Q(t_0+h+\beta,t_0)F(t_0+h+\beta)'F(t_0+h+\beta)Q(t_0+h+\beta,t_0),$$
$$Q(t_0,t_0)=N. \tag{22}$$

where $\Psi(\tau,t-h)$ is the transition matrix associated with $A(\tau)+\gamma^{-2}Q(\tau,t-h)F(\tau)'F(\tau)$.

Lemma 2: The condition formed by (C1), (C2), (C3) and (C4) is equivalent to the condition formed by (C14), (C2) and (C3), where (C14) is defined as follows.

(C14) There exists a solution $S(t+\beta,t-h)$, $t_0+h\le t\le t_1$, $-h\le\beta\le 0$ to the Riccati differential equations

$$-\frac{d}{dt}S(t,t-h)=S(t,t-h)\Gamma(t,t-h)+\Gamma(t,t-h)'S(t,t-h)$$
$$+F(t)'F(t)-S(t,t-h)(B(t)B(t)'-\gamma^{-2}\Psi(t,t-h)P(t-h)\times$$
$$\times C(t-h)'C(t-h)P(t-h)\Psi(t,t-h)')S(t,t-h),$$
$$S(t_1,t_1-h)=0,$$
$$\frac{\partial}{\partial\beta}S(t+\beta,t-h)=S(t+\beta,t-h)\Gamma(t+\beta,t-h)$$
$$+\Gamma(t+\beta,t-h)'S(t+\beta,t-h)+F(t+\beta)'F(t+\beta)$$
$$-S(t+\beta,t-h)B(t+\beta)B(t+\beta)'S(t+\beta,t-h),$$
$$\Gamma(t+\beta,t-h)=A(t+\beta)+\gamma^{-2}Q(t+\beta,t-h)F(t+\beta)'F(t+\beta). \tag{23}$$

Proof of Sufficiency of b) in Theorem: Assume that the conditions (C1), (C2), (C3) and (C4) are satisfied so that (C14) is also satisfied, and consider the functionals

$$V_1(t+\beta,t-h)=\bar{x}(t+\beta,t-h)'S(t+\beta,t-h)\bar{x}(t+\beta,t-h)$$
$$V_2(t+\beta,t-h)=(x(t+\beta)-\bar{x}(t+\beta,t-h))'\gamma^2Q(t+\beta,t-h)^{-1}\times$$
$$\times(x(t+\beta)-\bar{x}(t+\beta,t-h))$$

where $Q(t-h,t-h)=P(t-h)$, and further assume that, for a fixed admissible control $u(t)$, $x(t+\beta)$ is generated by (3) and (4) and $\bar{x}(t+\beta,t-h)$ given by

$$\bar{x}(t+\beta,t-h)=\Psi(t+\beta,t-h)\hat{x}(t-h)+\int_{t-h}^{t+\beta}\Psi(t+\beta,s)B(s)u(s-h)ds. \tag{24}$$

$$\frac{d}{dt}\hat{x}(t-h)=(A(t-h)+\gamma^{-2}P(t-h)F(t-h)'F(t-h))\hat{x}(t-h)$$
$$+B(t-h)u(t-h)+P(t-h)C(t-h)'(y(t-h)-C(t-h)\hat{x}(t-h)),$$
$$\hat{x}(t_0)=0. \tag{25}$$

Substituting the definitions (24) and (25) together with the formulas (14), (22) and (23) into the following identities

$$\int_{t_0+h}^{t_1}\{\frac{d}{dt}V_1(t,t-h)+\frac{\partial}{\partial\beta}V_2(t+\beta,t-h)\big|_{\beta=0}$$
$$-\frac{\partial}{\partial\beta}V_2(t+\beta,t-h)\big|_{\beta=-h}+\frac{d}{dt}V_2(t-h,t-h)\}dt$$
$$=V_1(t_1,t_1-h)+V_2(t_1,t_1-h)-V_1(t_0+h,t_0)-V_2(t_0+h,t_0)$$

in the interval $[t_0+h,t_1]$ and

$$\int_{-h}^{0}\{\frac{\partial}{\partial\beta}V_1(t_0+h+\beta,t_0)+\frac{\partial}{\partial\beta}V_2(t_0+h+\beta,t_0)\}d\beta$$
$$=V_1(t_0+h,t_0)+V_2(t_0+h,t_0)-V_1(t_0,t_0)-V_2(t_0,t_0),$$

in the interval $[t_0,t_0+h]$, and rearranging terms, we obtain

$$\|g\|_2^2-\gamma^2(\|d\|_2^2+\xi'N\xi)=\int_{t_0}^{t_1}\{\|u(t-h)-u_{\min}(t-h)\|^2$$
$$-\gamma^2\|v(t)-v_{\max}(t)\|^2-\gamma^2\|w(t)-w_{\max}(t)\|^2\}dt$$
$$-\gamma^2(x(t_1)-\bar{x}(t_1,t_1-h))'Q(t_1,t_1-h)^{-1}(x(t_1)-\bar{x}(t_1,t_1-h)) \tag{26}$$

where $u_{\min}(t)$, $v_{\max}(t)$ and $w_{\max}(t)$ are defined by

$$u_{\min}(t)=\begin{cases}-B(t+h)'S(t+h,t)\bar{x}(t+h,t), & t_0\le t\le t_1-h\\ -B(t+h)'S(t+h,t_0)\bar{x}(t+h,t_0), & t_0-h\le t\le t_0\end{cases}$$

$$v_{\max}(t)=D(t)'Q(t,t-h)^{-1}(x(t)-\bar{x}(t,t-h))$$

$$w_{\max}(t)=\begin{cases}0, & t_1-h\le t\le t_1\\ -C(t)(x(t)-\hat{x}(t))+\gamma^{-2}C(t)P(t)\times\\ \times\Psi(t+h,t)'S(t+h,t)\bar{x}(t+h,t), & t_0\le t\le t_1-h.\end{cases}$$

From (26), we see that $u(t)=u_{\min}(t)$ assures $\|g\|_2^2-\gamma^2(\|d\|_2^2+\xi'N\xi)\le 0$, and also see that the equality holds only if $(v(t),w(t))=(v_{\max}(t),w_{\max}(t))$ and $x(t_1)=\bar{x}(t_1,t_1-h)$ so that $(d,\xi)=0$. Thus $u(t)=u_{\min}(t)$ is a solution to Problem AID. Furthermore, when $u(t)=u_{\min}(t)$, it follows from (24) and (25) that $\bar{x}(t+h,t)$ is generated also by (18) and (19) and $\bar{x}(t+h,t_0)=0$, $t_0-h\le t\le t_0$. (Note that the above proof together with Proposition 1 proves Proposition 2.)

Proof of Necessity of b) in Theorem: We show that, if there exists a solution to Problem AID, (C2), (C3) and (C14) must be satisfied. Then, necessity of the conditions (C1), (C2), (C3) and (C4) follows from Lemma 2.

Suppose that the condition (C2) does not hold; then, we can find the smallest time $t^* \in [t_0, t_1)$ such that (14) has a solution $P(t),\ t_0 \le t < t^*$ and there exists a nonzero vector α such that $\lim_{t \to t^*} P(t)^{-1}\alpha = 0$ Now using the functional $V_2(t-h, t-h),\ t_0 + h \le t \le T + h$ and a same argument as in the proof of sufficiency, choosing nonzero (d, ξ) such that

$$(v(t), w(t)) = \begin{cases} (0, 0), & T \le t \le t_1 - h \\ (v_{\max}(t), -C(t)(x(t) - \hat{x}(t))), & t_0 \le t < T \end{cases}$$

$$(T < t^*)$$

and ξ guarantees $x(T) - \hat{x}(T) = \alpha$, and taking T as $T \to t^*$, we have

$$\|g\|_2^2 - \gamma^2(\|d\|_2^2 + \xi' N \xi) \ge \int_{t_0}^{t^*} \{\|F(t)\hat{x}(t)\|^2 + \|u(t)\|^2\} dt \ge 0$$

for all admissible controls $u(t)$. This inequality contradicts the existence of a solution. Thus, (C2) must hold.

As to the condition (C3), by using the functionals $V_1(t_0 + h + \beta, t_0)$ and $V_1(t + \beta, t - h)$ and modifying slightly the above argument for (C2), we can show the existence of solutions $Q(t_0 + h + \beta, t_0)$ and $Q(t + \beta, t - h)$ to the Riccati equations (15). Thus, (C3) must hold.

Suppose that the condition (C14) does not hold; then, we can find the largest time $t_* \in (t_0 + h, t_1]$ such that the first equation of (23) has a solution $S(t, t-h),\ t_* < t \le t_1$ and there exists a nonzero vector β such that $\lim_{T \to t_*} S(T, T-h)\beta = \infty$. Now using the functionals $V_1(t + \beta, t - h)$ and $V_2(t + \beta, t - h)$, $T \le t \le t_1$, repeating the same argument as in the proof of sufficiency, and choosing nonzero (d, ξ) such that

$$(v(t), w(t)) = \begin{cases} (v_{\max}(t), w_{\max}(t)), & T < t \le t_1 \\ (0, 0), & t_0 + h \le t \le T \end{cases} \quad (t_* < T)$$

and ξ assure $x(t_1) - \bar{x}(t_1, t_1 - h) = 0$ and $\bar{x}(T, T-h) = \beta$, where $\beta \ne 0$ assures $(d, \xi) \ne 0$, we have

$$\|g\|_2^2 - \gamma^2(\|d\|_2^2 + \xi' N \xi) \ge \int_T^{t_1} \|u(t) - u_{\min}(t)\| dt$$
$$+ (x(T) - \beta)' Q(T, T-h)^{-1}(x(T) - \beta) + \beta' S(T, T-\beta)\beta - \xi' N \xi.$$

Taking T as $T \to t^*$, the right hand side of the above inequality becomes arbitrary large. This contradicts the existence of a solution. Thus, the first equation of (23) has a solution on the whole interval. Using this solution $S(t, t-h)$ as a terminal condition for the second equation of (23) and repeating the same argument, we can show that the second equation of (23) has a solution $S(t + \beta, t - h),\ -h \le \beta \le 0$. Thus, the condition (C14) must hold.

5. Conclusion

We discussed a finite horizon H^∞ control problem for systems with input delays. We derived a solution based on the known result for the H^∞ control problem in measurement delay case, and showed that the solution has the same predictor-observer structure as the solution in measurement delay case has. Using this structural information on the solution, we also presented a direct proof of the solution to the finite horizon H^∞ control problem for systems with input delays, which is based only on completion of squares.

References

Basar, T. and P. Bernhard (1991), *H_∞ Optimal Control and Related Minimax Design Problems: A Dynamic Game Approach*, Birkhauser.

Kojima, A. and S. Ishijima (1994), Robust Controller Design for Delay Systems in the Gap Metric, in *Proc. of American Control Conference 1994*, pp.1939-1944.

Mirkin, L. (2000), On the Extraction of Dead-Time Controllers from Delay-Free Parametrizations, in *Proc. of 2nd IFAC Workshop on Linear Time Delay Systems*, pp.157-162.

Nagpal, K.M. and A.R. Ravi (1997), H_∞ Control and Estimation Problems with Delayed Measurements: State-Space Solutions, *SIAM J. Control Optim.*, **35-4**, pp.1217-1243.

Tadmor, G. (1995), H_∞ Control in Systems with a Single Input Lag, in *Proc. American Control Conference 1995*, pp.321-325.

Tadmor, G. (2000), The Standard H_∞ Problem in Systems with a Single Input Delay, *IEEE Trans. Automat. Control*, **45-3**, pp.382-397.

Uchida, K. and M. Fujita (1990), Controllers Attenuating Disturbances and Initial-Uncertainties for Time-Varying Systems, in *Proc. of 4th Int. Symp. Differential Games Appl.*, pp.188-196.

Uchida, K. and M. Fujita (1992), Finite Horizon H^∞ Control Problems with Terminal Penalties, *IEEE Trans. Automat. Control*, **37-11**, pp.1762-1767.

www.elsevier.com/locate/ifac

ON A GENERALIZATION OF PID REGULATORS FOR DELAY SYSTEMS ⋆

Michel Fliess * **Richard Marquez** ***,**
Hugues Mounier ****

* *Centre de Mathématiques et Leurs Applications, ENS Cachan, 61, avenue du Président Wilson, 94235 Cachan Cedex, France.*
e-mail: `fliess@cmla.ens-cachan.fr`
** *Laboratoire des signaux et systèmes, CNRS - Supélec - Université Paris sud, Plateau de Moulon, 91192 Gif-sur-Yvette, France.*
*** *Departamento de Sistemas de Control, Facultad de Ingeniería, Universidad de Los Andes, Mérida 5101, Venezuela.*
e-mail: `marquez@ula.ve`
**** *Département AXIS, Institut d'Électronique Fondamentale, Bât. 220, Université Paris-Sud, 91405 Orsay, France.*
e-mail: `mounier@ief.u-psud.fr`

Abstract: We derive PID-like regulators utilizing flatness-based predictive control (M. Fliess and R. Marquez, 2000, Continuous-time linear predictive control and flatness: a module-theoretic setting with examples, vol. 73, 606–623), generalized PI controllers (M. Fliess, R. Marquez, E. Delaleau, and H. Sira-Ramirez, 2001, Correcteurs proportionnels-intégraux généralisés, *ESAIM: Control, Optimisation, and Calculus of Variations*) and a kind of predictors which are reminiscent of Smith predictors. Our formalism is analogous to the two-sided Laplace transform.

Keywords: Predictive control, PID control, time delay systems, stabilization and disturbance attenuation.

1. INTRODUCTION

We are here proposing an extension of predictive control, PID regulators, and Smith predictors to a class of linear delay systems of the form $G(s)e^{-Ls}$, where $G(s)$ is a rational transfer function which might be unstable. To give a flavor of the problems we are going to study, we will start with the following important SISO system, a first-order-plus-time-delay process (see, e.g., Aström and Hägglund 1995, Aström and Hägglund 2000, O'Dwyer 2000, Palmor 1996, Smith 1958, Tan *et al.* 1999, Shinskey 1996),

$$sy = ay + b\,e^{-Ls}u + \gamma\frac{e^{-\varphi s}}{s} + y(0) \qquad (1)$$

$a, b, \gamma, L, \varphi \in \mathbb{R}$, $b \neq 0$, L, $\varphi \geq 0$. The quantity b/a represents the process static gain, $1/a$ is the system time constant, and L is a time delay. The constant load disturbance $\gamma\frac{e^{-\varphi s}}{s}$ appears at time $t = \varphi$. The initial condition $y(0)$ will be considered as another perturbation. Note that (1) is unstable

⋆ This work has been partially supported by the European Commission's Training and Mobility of Researchers (TMR) under Contract ERBFMRXT-CT970137. One author (RM) was partially supported by the *Consejo Nacional de Investigaciones Científicas y Tecnológicas (CONICIT), Venezuela.*

if $a \geq 0$ (for $a = 0$, system (1) corresponds to a pure integrator with delay).

A given output reference signal y^* yields the *open loop reference control* given by

$$u^* = \frac{s-a}{b} e^{Ls} y^* \tag{2}$$

in terms of the nominal (unperturbed) system

$$sy^{\text{nom}} = ay^{\text{nom}} + be^{-Ls}u$$

The advance $e^{Ls}y^*$ in (2) is not an impediment since y^* is supposed to be known for all t. Note the variable u^* can be seen as a *predicted input*, (cf. Fliess and Marquez 2000). The error dynamics results

$$sy_e = ay_e + be^{-Ls}u_e + \gamma \frac{e^{-\varphi s}}{s} + y(0) \tag{3}$$

where $u_e = u - u^*$, $y_e = y - y^*$. Now consider the non-causal PI controller

$$u_e = -(k_P + k_I s^{-1})e^{Ls} y_e \tag{4}$$

In our formalism, which is analogous to the two-sided Laplace transform, we have

$$\frac{f}{s} = \int_{-\infty}^{t} f(\tau)d\tau \neq \int_{0}^{t} f(\tau)d\tau$$

See (Fliess et al. 2001, Marquez 2001) for details, also (Fliess et al. 2001). Thus, expression (4) reads

$$\begin{aligned} u_e(t) &= -k_P y_e(t+L) - k_I \int_0^{t+L} y_e(\tau)d\tau \\ &= -k_P y_e(t+L) - k_I \int_0^t y_e(\tau)d\tau - k_I \int_{t-L}^{t} y_e(\sigma+L)d\sigma \end{aligned} \tag{5}$$

We will replace $e^{Ls}y_e$ in (4) by a value y_e^+ stemming from the unperturbed error dynamics $sy_e^{\text{nom}} = ay_e^{\text{nom}} + be^{-Ls}u_e^{\text{nom}}$ as follows.

From $y_e^{\text{nom}} = b\dfrac{e^{-Ls}}{s-a} u_e^{\text{nom}}$ we obtain

$$\begin{aligned} e^{Ls} y_e^{\text{nom}} &= \frac{b}{s-a}\left(e^{-L(s-a)} + 1 - e^{-L(s-a)}\right)u_e^{\text{nom}} \\ &= e^{aL} y_e^{\text{nom}} + b\frac{1-e^{-L(s-a)}}{s-a} u_e^{\text{nom}} \end{aligned} \tag{6}$$

Using expression (6), we define a *predictor* (Manitius and Olbrot 1979)see, e.g., in terms of the actual output and input values (y_e, u_e),

$$y_e^+ = e^{aL} y_e + b\frac{1-e^{-L(s-a)}}{s-a} u_e \tag{7}$$

This predictor reads in the time domain

$$y_e^+(t) = e^{aL} y_e(t) + \int_{t-L}^{t} e^{a(t-\tau)} b\, u_e(\tau)d\tau$$

Thus, instead of (5), we obtain the following causal PI regulator (we replace $y_e(t+L)$ by $y_e^+(t)$):

$$u_e(t) = -k_P y_e^+(t) - k_I \left(\int_0^t y_e(\tau)d\tau + \int_{t-L}^{t} y_e^+(\sigma)d\sigma\right)$$

i.e.

$$u_e = -k_P y_e^+ - k_I \left(\frac{y_e}{s} + \frac{1-e^{-Ls}}{s} y_e^+\right) \tag{8}$$

The closed loop system (3)-(7)-(8) then becomes

$$\left(s^2 + (k_P b - a)s + k_I b\right)y_e = \left[1 + b\left(k_P + k_I \frac{1-e^{-Ls}}{s}\right)\frac{1-e^{-L(s-a)}}{s-a}\right]\left(\gamma e^{-\varphi s} + sy(0)\right)$$

The gains k_P and k_I are choosen such that the polynomial $s^2 - (a + k_P b)s - k_I b$ is Hurwitz. It is obvious, as for the usual PI controller in the undelayed case, that $\lim_{t\to+\infty} y_e(t) = 0$, i.e. $y \to y^*$. Fig. 1 presents a closed-loop response of system (3) illustrating the performance of our control strategy [1]; parameters $L = 5$, $a, b = 0.5$, The closed-loop polynomial is chosen to be $s^2 + 2.236s + 1.25$. We consider a step perturbation at time $t = 50$.

We have thus proposed a PI controller which combines a type of *continuous predictive control*, called FBPC (Fliess and Marquez 2000), which is inspired from the flatness-based control synthesis (Fliess *et al.* 1995), *generalized PI controllers* (GPI) (Fliess et al. 2000, Fliess et al. 2001, Sira-Ramírez et al. 2001), and an *output predictor* (slightly different from the Smith predictor (Smith 1958); see also (Marquez *et al.* 2001)). The GPI controller permits to avoid any derivatives of the measured quantities. The remaining part of this communication is devoted to a generalization for second order systems of the form

$$y = \frac{b_1 s + b_0}{s^2 + a_1 s + a_0} e^{-Ls} u$$

The calculation of the predictor is of course a little more involved. Two case studies are analyzed, one of them exhibiting a non-minimum phase behavior. Some conclusions draw up the paper. We give a summary of a complete theory presented in (Fliess et al. 2001), which is valid for multivariable systems.

2. CONTROL OF SECOND ORDER QUASI-FINITE SYSTEMS

Consider a perturbed second order delay system, very common in process control, given by

$$y = \frac{b_1 s + b_0}{s^2 + a_1 s + a_0} e^{-Ls} u + \varpi \frac{e^{-\varphi s}}{s} \tag{9}$$

where

- $b_1^2 + b_2^2 \neq 0$,

[1] The final implementation of our proposed control law necessitates some particular care (see Engelborghs *et al.* 2001).

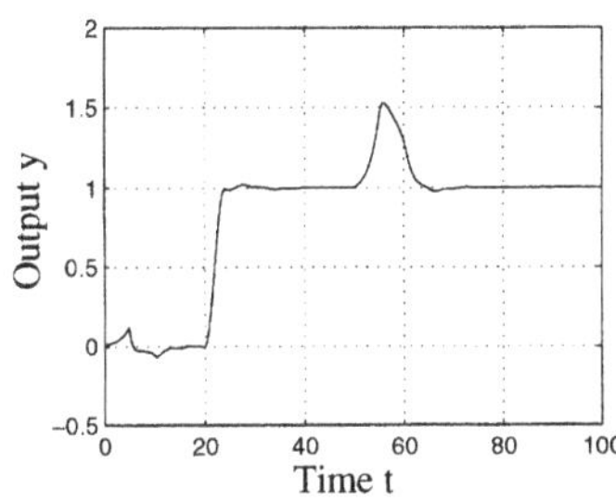

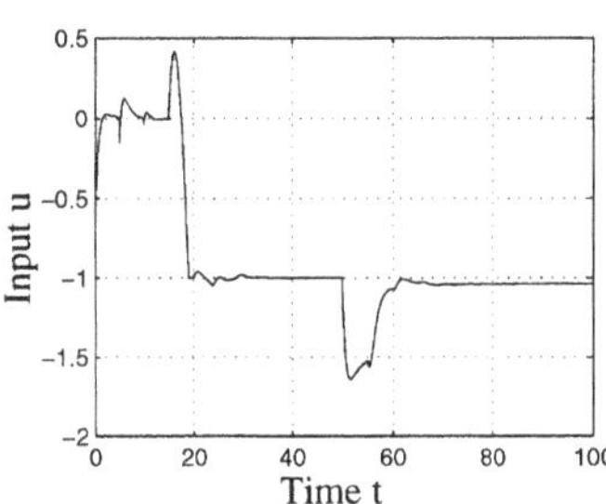

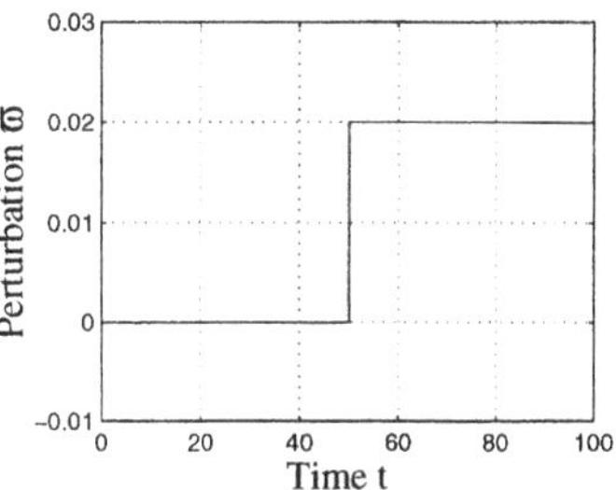

Fig. 1. Closed-loop system response of an unstable first-order system

- $s^2 + a_1 s + a_0$ and $b_1 s + b_0$ are coprime.
- $\varpi \dfrac{e^{-\varphi s}}{s}$ is a load perturbation.

2.1 *Introductory principles*

Replace (9) by

$$(s^2 + a_1 s + a_0)y = (b_1 s + b_0)e^{-Ls}u + \gamma \frac{e^{-\varphi s}}{s} \tag{10}$$

where $\gamma = (s^2 + a_1 s + a_0)\varpi$. The corresponding nominal *quasi-finite* system (see (Fliess et al. 2001)) reads

$$(s^2 + a_1 s + a_0)y^+ = (b_1 s + b_0)u$$

where $y^+ = e^{Ls}y$. The *system predicted trajectories*, (cf. Fliess and Marquez 2000), are then given by

$$\begin{aligned} y^* &= (b_1 s + b_0)z^* \\ u^* &= (s^2 + a_1 s + a_0)e^{Ls}z^* \end{aligned} \tag{11}$$

where z^* is the flat output. The advance of z^* of L units of time is certainly not an impediment because the predicted trajectory is known at least over L units of time in the future. Remember (see, e.g., (Fliess and Marquez 2000)) that trajectory redesign and time scaling permit to handle a large variety of constraints, especially input constraints. To consider perturbation attenuation and feedback stabilization we will be adapting results from (Fliess et al. 2000, Fliess et al. 2001, Sira-Ramírez et al. 2001). Calculations will become more transparent with the following state-variable representation, derived from (10),

$$sx = Ax + Be^{-Ls}u$$

$$y = Cx + \gamma \frac{e^{-\varphi s}}{s}$$

where

$$A = \begin{pmatrix} -a_1 & 1 \\ -a_0 & 0 \end{pmatrix}, \quad B = \begin{pmatrix} b_0 \\ b_1 \end{pmatrix}, \quad C = (1 \ 0)$$

which is controllable and observable. The system error, computed with respect to (11), is given by

$$\begin{aligned} sx_e &= Ax_e + Be^{-Ls}u_e \\ y_e &= Cx_e + \gamma \frac{e^{-\varphi s}}{s} \end{aligned} \tag{12}$$

with $y_e = y - y^*$, $u_e = u - u^*$.

2.2 *Predictors*

In order to extend the output predictor (7), by analogy with system (1), a state predictor results

$$x^+ = e^{LA}x + (1 - e^{-L(sI-A)})(sI - A)^{-1}Bu \tag{13}$$

see, e.g., (Manitius and Olbrot 1979). This yields, in the time domain,

$$x^+(t) = e^{LA}x(t) + \int_{t-L}^{t} e^{A(t-\tau)}Bu(\tau)d\tau$$

Therefore we have

$$y^+ = Cx^+ \tag{14}$$

Thus the output predictor reads

$$y^+ = Ce^{LA}x + C(1 - e^{-L(sI-A)})(sI - A)^{-1}Bu$$

Note that the transfer function relating y^+ and u is given by $C(sI - A)^{-1}B$.

As the state x is not available for measurement, we will replace it by a reconstructed state $\mathtt{REC}(x)$ in the following section.

2.3 *Reconstructors*

A state reconstructor of the type presented in (Fliess et al. 2000, Fliess et al. 2001, Sira-Ramírez et al. 2001) is introduced in this section. From $y_e^{\text{nom}} = Cx_e^{\text{nom}}$ and $sy_e^{\text{nom}} = CAx_e^{\text{nom}} + CBe^{-Ls}u_e^{\text{nom}}$, the following relation is satisfied

$$x_e^{\text{nom}} = \mathcal{O}^{-1} \begin{pmatrix} y_e^{\text{nom}} \\ s\, y_e^{\text{nom}} \end{pmatrix} - \mathcal{O}^{-1} \begin{pmatrix} 0 \\ CB \end{pmatrix} e^{-Ls}u_e^{\text{nom}}$$

where the observability matrix

$$\mathcal{O} = (C, \ CA)^T$$

is invertible. Then, from the expression

$$x_e^{\text{nom}} = A\frac{x_e^{\text{nom}}}{s} + B\frac{e^{-Ls}u_e^{\text{nom}}}{s} \tag{15}$$

we define the *integral reconstructor* (cf. Fliess et al. 2000, Fliess et al. 2001, Sira-Ramírez et al. 2001)

$$\mathtt{REC}(x_e) = A\mathcal{O}^{-1} \begin{pmatrix} \dfrac{y_e}{s} \\ y_e \end{pmatrix} - A\mathcal{O}^{-1}\begin{pmatrix} 0 \\ CB \end{pmatrix} \frac{e^{-Ls}u_e}{s} + B\frac{e^{-Ls}u_e}{s}$$

Note that $\boldsymbol{x}_e$ is identical to $\mathtt{REC}(\boldsymbol{x}_e)$ when there are no perturbations affecting the system.

Our state predictor (13) is redefined as

$$\mathtt{REC}(\boldsymbol{x}^+) = e^{LA}\mathtt{REC}(\boldsymbol{x}_e) + (1 - e^{-L(sI-A)})(sI - A)^{-1}Bu_e \quad (16)$$

As before, the output predictor $\mathtt{REC}(\boldsymbol{y}^+)$ is given by

$$\mathtt{REC}(y^+) = C\,\mathtt{REC}(\boldsymbol{x}^+) \quad (17)$$

Its integral $\frac{\mathtt{REC}(y^+)}{s}$ splits into two terms, see Eq. (8), as follows:

$$\frac{\mathtt{REC}(y^+)}{s} = \frac{y_e}{s} + \frac{1 - e^{-Ls}}{s}\mathtt{REC}(y^+)$$

The second order integral $\frac{\mathtt{REC}(y^+)}{s^2}$ may also be written as

$$\frac{\mathtt{REC}(y^+)}{s^2} = \frac{y_e}{s^2} + \frac{1 - e^{-Ls}}{s}\frac{\mathtt{REC}(y^+)}{s}$$

Both integrals depend explicitly on the output y_e.

2.4 *Generalized PI controller (GPI)*

By analogy with (Fliess et al. 2000, Fliess et al. 2001, Sira-Ramírez et al. 2001), define now a generalized PI-controller for (12):

$$u_e = -K_p\,\mathtt{REC}(\boldsymbol{x}^+) - k_i\frac{\mathtt{REC}(y^+)}{s} - k_{ii}\frac{\mathtt{REC}(y^+)}{s^2}$$

The parameters $K_p = (k_p^1\ k_p^2)$, k_i, and k_{ii} are chosen such that the dynamics given by $s\boldsymbol{\xi} = \breve{A}\boldsymbol{\xi} + \breve{B}u$ with control u and state $\boldsymbol{\xi} = (x_{1e}^{\text{nom},+}, x_{2e}^{\text{nom},+}, \frac{y_e^{\text{nom},+}}{s}, \frac{y_e^{\text{nom},+}}{s^2})$ becomes stable when $u = -K\boldsymbol{\xi}$, with $K = (k_p^1, k_p^2, k_i, k_{ii})$, where

$$\breve{A} = \begin{pmatrix} A & \mathbf{0}_{2\times1} & \mathbf{0}_{2\times1} \\ C & 0 & 0 \\ \mathbf{0}_{1\times2} & 1 & 0 \end{pmatrix}, \qquad \breve{B} = \begin{pmatrix} B \\ 0 \\ 0 \end{pmatrix}$$

The fourth order closed-loop characteristic polynomial equation is then given by $\det(sI_4 - \breve{A} + \breve{B}K) \equiv s^2\det(sI_2 - A + BK_p + \frac{Bk_iC}{s} + \frac{Bk_{ii}C}{s^2})$.

2.5 *Disturbance attenuation and stabilization*

Using

$$y^+ = C\boldsymbol{x}^+ + \gamma\frac{e^{-\varphi s}}{s}$$

the definition of $\mathtt{REC}(\boldsymbol{x}_e)$, and Eq. (16) we obtain

$$\boldsymbol{x}^+ + e^{AL}A\mathcal{O}^{-1}\begin{pmatrix} \gamma\frac{e^{-\varphi s}}{s^2} \\ \gamma\frac{e^{-\varphi s}}{s} \end{pmatrix}$$

where the second term is due to the difference between the nominal and the perturbed case. The final control law is thus given by:

$$u_e = -K_p\left[\boldsymbol{x}^+ + e^{AL}A\mathcal{O}^{-1}\begin{pmatrix} \gamma\frac{e^{-\varphi s}}{s^2} \\ \gamma\frac{e^{-\varphi s}}{s} \end{pmatrix}\right] - k_i\left[\frac{y^+}{s} + \frac{1 - e^{-Ls}}{s}Ce^{AL}A\mathcal{O}^{-1}\begin{pmatrix} \gamma\frac{e^{-\varphi s}}{s^2} \\ \gamma\frac{e^{-\varphi s}}{s} \end{pmatrix}\right] - k_{ii}\left[\frac{y^+}{s^2} + \left(\frac{1 - e^{-Ls}}{s}\right)^2 Ce^{AL}A\mathcal{O}^{-1}\begin{pmatrix} \gamma\frac{e^{-\varphi s}}{s^2} \\ \gamma\frac{e^{-\varphi s}}{s} \end{pmatrix}\right]$$

Therefore, the closed-loop system reads

$$s^2\det\left(sI_2 - A + BK_p + \frac{Bk_iC}{s} + \frac{Bk_{ii}C}{s^2}\right)\boldsymbol{x}^+ = -B\left[K_p - k_i\frac{1 - e^{-Ls}}{s}C - k_{ii}\left(\frac{1 - e^{-Ls}}{s}\right)^2 C\right]\cdot e^{AL}A\mathcal{O}^{-1}\begin{pmatrix} \gamma e^{-\varphi s} \\ \gamma s e^{-\varphi s} \end{pmatrix}$$

The real parts of the zeros of

$$s^2\det\left(sI_2 - A + BK_p + \frac{Bk_iC}{s} + \frac{Bk_{ii}C}{s^2}\right) \quad (18)$$

are chosen to be strictly negative. This guarantees perturbation attenuation and asymptotic stability.

Remark 2.1. *If the perturbation would be a ramp, i.e., of the form* $\varpi\frac{e^{-\varphi s}}{s^2}$, *one would need in the GPI controller an integrator* $\frac{y^+}{s^3}$ *of third order (see Fliess et al. 2000, Fliess et al. 2001, Sira-Ramírez et al. 2001).*

3. CASE STUDIES

Two examples serve to illustrate the applicability of our approach. The first example is typical in process control. The other is academic but cannot be treated by using a Smith Predictor. In all cases we consider long delays and external disturbances of step type.

3.1 *A heat exchanger*

This first example is taken from (Franklin *et al.* 1995). Figure 2 represents a heat exchanger which can be modelled as follows:

$$G(s) = \frac{\kappa e^{-Ls}}{(1 + \tau_1 s)(1 + \tau_2 s)} \quad (19)$$

The delay L can be introduced by the sensor or by the system dynamics (this system can be actually

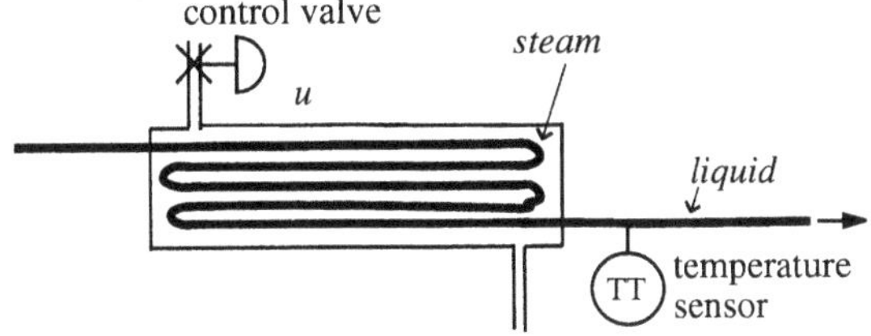

Fig. 2. Schematic of a heat exchanger

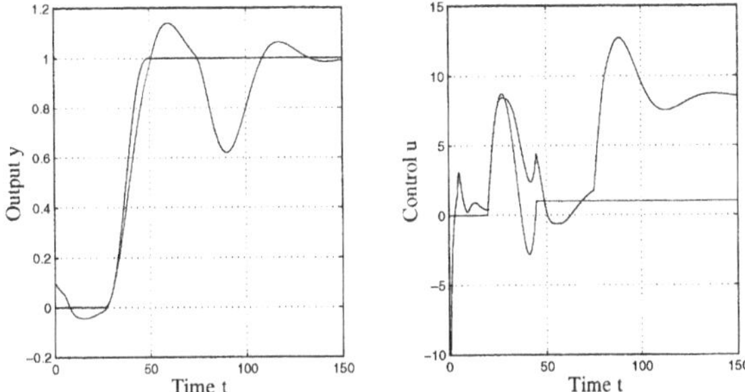

Fig. 3. Response of a heat exchanger: 10% on τ_1: $\tau_1^{real} = 60 + 6 = 66$ sec, 10% on τ_2: $\tau_2^{real} = 10 - 1 = 9$ sec, 15% on L: $L_{real} = L + \Delta L = 5 - 0.75 = 4.25$ sec, 30% on κ: $\kappa_{real} = 1 - 0.3 = 0.7$

modelled by a partial differential equation). Model parameters are given by $\tau_1 = 10$ sec, $\tau_2 = 60$ sec, $L = 5$ sec, $\kappa = 1$. Poles are placed at $(-0.25, -0.25, -0.5, -0.5)$. Controller gain of GPI is given by $K = (389.66\ 830.00\ 112.50\ 9.37)$. Some numerical simulations are given to illustrate the performance of the proposed control strategy.

A numerical simulation, shown in Figure 3, illustrates the performance of the proposed control strategy. We consider parameter uncertainty; the actual values of system parameters are: $\tau_1^{real} = 66$ sec, $\tau_2^{real} = 9$ sec, $L_{real} = 4.25$ sec, $\kappa_{real} = 0.7$.

3.2 *Unstable system*

Consider the following unstable system, (cf. Aström *et al.* 2000)

$$G(s) = \frac{a}{(s+a)(s-1)}$$

Adding a delay L we obtain the system

$$G(s) = \frac{ae^{-Ls}}{(s+a)(s-1)} \tag{20}$$

The model parameters are given by $L = 2$, $a = 1$. In the following numerical simulation, we consider an output transfer for $y = 0$ at $t = 5$ to $y = 1$ at $t = 7$. The system delay is given by $L_{real} = 2 - 0.2 = 1.8$. The performance of our dead-time compensation controller is shown in Figure 4. As expected, the system diverges between 0 and 2 sec because there is no control

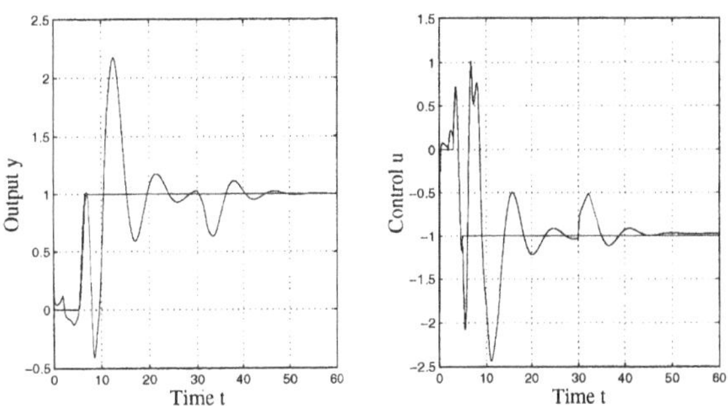

Fig. 4. Response of an unstable system: load change at $t = 30$ sec, 10% on L: $L_{real} = 2 - 0.2 = 1.8$

in this interval. The response to load changes is illustrated by introducing a step load of amplitude 0.1 at $t = 30$ sec. Note that the uncertainty on L degrades the response to set-point changes. The poles of the characteristic closed-loop polynomial are chosen to be $\{-1, -1, -2, -2\}$.

4. CONCLUSION

The full theory with more examples will be presented in (Fliess et al. 2001). It is valid for *quasi-finite delay systems*, i.e., multivariable delay systems with transfer matrices of the form

$$\begin{pmatrix} e^{-L_1 s} & & & \mathbf{0} \\ & e^{-L_2 s} & & \\ & & \ddots & \\ \mathbf{0} & & & e^{-L_p s} \end{pmatrix} G(s) \tag{21}$$

where $G(s)$ is a $p \times m$ rational transfer matrix. Since any transfer matrix description is necessarily ambiguous, we have to utilize equations. This is achieved in an intrinsic manner thanks to a module-theoretic setting (see (Fliess and Mounier 1998)). The following state-variable representation may be given

$$s\begin{pmatrix} x_1^+ \\ \vdots \\ x_n^+ \end{pmatrix} = A\begin{pmatrix} x_1^+ \\ \vdots \\ x_n^+ \end{pmatrix} + B\begin{pmatrix} u_1 \\ \vdots \\ u_m \end{pmatrix}$$

$$\begin{pmatrix} y_1^+ \\ \vdots \\ y_p^+ \end{pmatrix} = C\begin{pmatrix} x_1^+ \\ \vdots \\ x_n^+ \end{pmatrix}$$

where $(y_1^+, \ldots, y_p^+) = (e^{L_1 s}y_1, \ldots, e^{L_p s}y_p)$, $A \in \mathbb{R}^{n\times n}$, $B \in \mathbb{R}^{n\times m}$, $C \in \mathbb{R}^{p\times n}$, $D \in \mathbb{R}^{p\times m}$.

Most monovariable delay systems encountered in practice are indeed of the form (21). This is unfortunately not the case for multivariable systems. We therefore extend in (Fliess et al. 2001) our generalized PID-like controller to systems of the form

$$\begin{pmatrix} \frac{b_{11}e^{-L_{11}s}}{s-a_{11}} & \frac{b_{12}e^{-L_{12}s}}{s-a_{12}} \\ \frac{b_{21}e^{-L_{21}s}}{s-a_{21}} & \frac{b_{22}e^{-L_{22}s}}{s-a_{22}} \end{pmatrix}$$

which have been often studied in the literature (Camacho and Bordons 1999, Richalet 1993, Tan *et al.* 1999). In this case we are only able to attenuate constant load perturbations of the form $\varpi \frac{e^{-\varphi s}}{s}$, $\varpi \in \mathbb{R}$.

The connection with Smith predictors is analyzed in (Fliess et al. 2001) and (Marquez *et al.* 2001).

5. REFERENCES

Aström, K.J. (2000) Limitations on Control Performance, *European Journal on Control*, **6**, 2–20.

Aström, K.J., and T. Hägglund (1995) *PID Controllers: Theory, Design, and Tuning* (Research Triangle Park, NC: Instrument Society of America).

Aström, K.J. and T. Hägglund (2000) The future of PID Control. *IFAC Workshop on Digital Control: Past, Present and Future of PID Control*, Terrasa, Spain, pp. 19–30.

Aström, K.J., H. Panagopoulos and T. Hägglund (1998) Design of PI controllers based on non-convex optimization, *Automatica*, **34**, no. 5, 585–601.

Camacho, E. and C. Bordons (1999) *Model Predictive Control* (London: Springer).

Engelborghs, K., M. Dambrine and D. Roose (2001) Limitations of a class of stabilization methods for delay systems, *IEEE Transactions on Automatic Control*, **46**, 336–339.

Fliess, M., J. Lévine, P. Martin, and P. Rouchon (1995) Flatness and defect of non-linear systems: introductory theory and applications, *International Journal of Control*, **61**, 1327-1361.

Fliess, M. and R. Marquez (2000) Continuous-time linear predictive control and flatness: a module-theoretic setting with examples, *International Journal of Control*, **73**, 606–623.

Fliess, M., R. Marquez and E. Delaleau (2000) State feedbacks without asymptotic observers and generalized PID regulators. In A. Isidori, F. Lamnabhi-Lagarrigue, W. Respondek (eds.), *Nonlinear Control in the Year 2000*, Lecture Notes Control Infor. Sci. (London: Springer) **258**, pp. 367–384.

Fliess, M., R. Marquez, E. Delaleau and H. Sira-Ramírez, Correcteurs proportionnels intégraux généralisés, *ESAIM COCV, to appear.*

Fliess, M., R. Marquez and H. Mounier (2001) An extension of predictive control, PID regulators and Smith predictors to some linear delay systems, *submitted.*

Fliess, M. and H. Mounier (1998) Controllability and observability of linear delay systems: an algebraic approach, *ESAIM COCV*, **3**, 301–314. `http://www.emath.fr/Maths/Cocv/cocv.html`

Franklin, G.F., D.J. Powell and A. Emami-Naeni (1995) *Feedback Control of Dynamic Systems* (Reading, Massachusetts: Addison-Wesley).

Manitius, A.J. and A.W. Olbrot (1979) Finite spectrum assignment problem for systems with delays, *IEEE Transactions on Automatic Control*, **24**, 541–553.

R. Marquez (2001) *À propos de quelques méthodes classiques de commande linéaire: commande prédictive, correcteurs proportionnels-intégraux, prédicteurs de Smith* Ph.D. Thesis, Université Paris XI, Orsay, France, September 2001.

Marquez, R., M. Fliess and H. Mounier (2001) A non-conventional robust PI-controller for the Smith predictor, *submitted.*

O'Dwyer, A. (2000) PI and PID controller tuning rules for time delay processes: a summary, Dublin Institute of Technology, School of Control
Systems and Electrical Engineering, n° AOD-00-01. http://docsee.kst.dit.ie/aodweb+

Palmor, Z.J. (1996) Time-delay compensation: Smith predictor and its modifications. In W. Levine (ed.), *The Control Handbook* (Florida, USA: CRC Press), pp. 224–237.

Richalet, J. (1993) *Pratique de la commande prédictive* (Paris: Hermès).

Sira-Ramírez H., R. Marquez and M. Fliess (2001) On the generalized PID control of linear dynamic systems, *European Control Conference*, ECC'2001.

Smith, O.J.M. (1958) *Feedback Control Systems* (New York: McGraw Hill).

Tan, K.K., Q.-G. Wang, C.C. Hang and T.J. Hägglund (1999) *Advances in PID Control*, (London: Springer).

Shinskey, F.G. (1996) *Process Control Systems: Application, design, and tuning*, 4th edition, (Boston: McGraw Hill).

Copyright © IFAC Time Delay Systems,
New Mexico, USA, 2001

www.elsevier.com/locate/ifac

AN APPROACH TO CONSTRAINED STATE FEEDBACK H^{∞} CONTROL SYNTHESIS

Kojiro Ikeda* Kenko Uchida*

* *Dept. of Electrical Electronics and Computer Engineering, Waseda University*
3-4-1, Okubo, Shinjuku-ku, Tokyo, 169-8555 Japan
E-mail: {ikeda, kuchida}@uchi.elec.waseda.ac.jp

Abstract: In this paper, we propose an approach to H^{∞} controller synthesis problem for linear systems with time-delay, when the control input is constrained. This approach is based on a state reachable set analysis. We deal with a case of memory state feedback controllers.

Keywords: Time-delay, H^{∞} control, Input constraints, State Reachable Sets, Linear Matrix Inequality(LMI)

1. INTRODUCTION

In this paper, we consider H^{∞} control in memory state feedback for the state delay systems, when the control input is constrained. First, we propose an H^{∞} controller synthesis which make the closed loop system asymptotically stable and its L^2 gain less than a specified value. Next, extending the analysis method of state reachable sets for systems no delay proposed in (Watanabe and Fujita, 1998),(Boyd and Valakrisham, 1994), we provide a method to evaluate an admissible range of state. Finally, based on the reachable set analysis, we propose an H^{∞} controller synthesis method when control constraint is imposed.

For recent and related developments in this area, see (Niculescu and L.Dugard, 1996) and (Tarbourieh, 2000), where the memory feedback case presented in this paper is not discussed.

2. SYSTEM DESCRIPTION AND PROBLEM STATEMENT

System Description

Consider a linear system with delay in state. The system is defined over the interval $[0, \infty)$ and described by

$$\begin{aligned}\dot{x}(t) =& A_0 x(t) + A_1 x(t-h) \\ &+ \int_{-h}^{0} A_{01}(\beta) x(t+\beta) d\beta + Bu(t) + Dw(t) \qquad (1)\\ z(t) =& Cx(t)\end{aligned}$$

Here, $w(t)$ is the disturbance vector; $u(t)$ is the control input vector; $z(t)$ is the controlled output vector; and the state at time t of the system is described by $(x(t), x_t)$, here, $x_t = \{x(t+\beta) | -h \leq \beta \leq 0\} \in L^2([-h, 0]; R^n)$. The initial condition is $(x(0), x_0) \in R^n \times L^2([-h, 0]; R^n)$. The number h denotes the length of time delay and $h > 0$. The parameters A_0, A_1, B, D, C are constant matrices and the parameter $A_{01}(\beta)$, is a matrix function whose elements are bounded continuous functions.

In this paper, we consider this constraint condition about disturbance w

$$\begin{aligned} w(t) \in \mathcal{W}, \quad \forall t \in [0, \infty) \qquad (2)\\ \mathcal{W} = \{w | w' W_D w \leq 1\}. \end{aligned}$$

Here, W_D is the given and positive definite matrix.

We consider the feedback controller for the time-delay system as described by

$$u(t) = K_0 x(t) + \int_{-h}^{0} K_{01}(\beta) x(t+\beta) d\beta. \quad (3)$$

Here, K_0 is a constant matrix and $K_{01}(\beta)$ is a matrix function whose elements are in $L^2[-h,0]$.A closed loop system applied the controller (3) to the system (1) is described as

$$\dot{x}(t) = \tilde{A}_0 x(t) + \tilde{A}_1 x(t-h) + \int_{-h}^{0} \tilde{A}_{01}(\beta) x(t+\beta) d\beta$$

where, $\tilde{A}_0 = A_0 + BK_0$, $\tilde{A}_1 = A_1$, $\tilde{A}_{01}(\beta) = A_{01}(\beta) + BK_{01}$.

Now, we prepare a notation for defining a quadratic form of the state. Denote by $\{P, R, S\}$ a triplet of three matrices P, $R(\beta)$ and $S(\alpha,\beta)$ with the same dimensions such that P is a constant matrix, $R(\beta)$ is a matrix function whose elements are in $L^2[-h,0]$ and $S(\alpha,\beta)$ is a matrix function whose elements are in $L^2([-h,0]\times[-h,0])$. A triplet $\{P,R,S\}$ is called symmetric if $P' = P$ and $S'(\alpha,\beta) = S(\alpha,\beta)$. For a given symmetric triplet $\{P,R,S\}$, a quadratic form associated with this triplet is defined as follows:

$$\begin{aligned}(\xi,\zeta)&\{P,R,S\}(\xi,\zeta) \\ &:= \xi' P \xi + 2\xi' \int_{-h}^{0} R(\beta)\zeta(\beta) d\beta \quad (4) \\ &+ \int_{-h}^{0}\int_{-h}^{0} \zeta'(\alpha) S(\alpha,\beta)\zeta(\beta) d\alpha d\beta,\end{aligned}$$

here, (ξ,ζ) satisfies $(\xi,\zeta) \in R^n \times L^2([-h,0];R^n)$. A symmetric triplet $\{P,R,S\}$ is called positive semi-definite if $(\xi,\zeta)'\{P,R,S\}(\xi,\zeta) \geq 0$ for all $(\xi,\ \zeta)$ and, in particular called positive definite if there exists a positive number ϵ such that $(\xi,\zeta)'\{P,R,S\}(\xi,\zeta(\beta)) \geq (\xi,\zeta)'\{\epsilon I\ 0\ 0\}(\xi,\zeta)$ for all ξ, ζ, where I denotes identity matrix. We denote $\{P,R,S\} \geq 0\ (>0)$ when $\{P,R,S\}$ is positive semi-definite (definite). Negative semi-definiteness and negative definiteness are similarly defined.

We describe $L^2([-h,0];R^n)$ as L^2 for the simplicity.

State Reachable Sets

Here, we assume the case that the disturbance of system (1) is constrained by (2). Now, we make the following definitions.

Definition 1. For $\lambda = (\xi,\zeta) \in R^n \times L^2$, if there exists a disturbance w that satisfies (2) and there exists a time $T < \infty$ that satisfies $\xi = x(T), \zeta = x_T$, then λ is called state reachable from $(x(0), x_0)$.

Definition 2. A reachable set $\mathcal{R}(x(0),x_0)$ from $(x(0),x_0)$ is defines as

$$\begin{aligned}&\mathcal{R}(x(0),x_0) \\ &= \{(\xi,\zeta) \in R^n \times L^2 \\ &\quad : (\xi,\zeta) \text{ is state reachable from } (x(0),x_0)\}.\end{aligned}$$

Now, for simplicity we assume about the matrix $\mathcal{W}$ that constrains the disturbance w as

$$W_D = I$$

And we define a set $\mathcal{E}$ as follows.

Definition 3. For any positive definite triplet $\{P,R,S\}$, $\mathcal{E}$ is defined as

$$\begin{aligned}&\mathcal{E}(P,R,S) = \\ &\{\lambda = (\xi,\zeta) \in R^n \times L^2 | \lambda^T \{P,R,S\}\lambda \leq 1\}.\end{aligned}$$

Lemma 1. Assume that there exists a positive definite triplet $\{P,R,S\}$ that satisfies LMI condition LMI-1 for any $\lambda = (\xi,\zeta) \in R^n \times L^2$, $\lambda'\{P,R,S\}\lambda \geq 1$ and for any $w \in \mathcal{W}$, where

(LMI – 1)

$$\int_{-h}^{0}\int_{-h}^{0} \begin{bmatrix} \frac{1}{h}\xi \\ \frac{1}{h}\zeta(-h) \\ \zeta(\alpha) \\ \frac{1}{h}w \end{bmatrix}^T \begin{bmatrix} \Delta_{11} & \Delta_{12} & \Delta_{13} & \Delta_{14} \\ \Delta_{12}' & \Delta_{22} & \Delta_{23} & \Delta_{24} \\ \Delta_{13}' & \Delta_{23}' & \Delta_{33} & \Delta_{34} \\ \Delta_{14}' & \Delta_{24}' & \Delta_{34}' & \Delta_{44} \end{bmatrix} (\alpha,\beta) \begin{bmatrix} \frac{1}{h}\xi \\ \frac{1}{h}\zeta(-h) \\ \zeta(\beta) \\ \frac{1}{h}w \end{bmatrix} d\beta d\alpha \leq 0.$$

Where,

$$\begin{aligned}\Delta_{11} &= A_0'P + PA_0 + R(0)' + R(0) \\ \Delta_{12} &= PA_1 - R(-h) \\ \Delta_{13}(\beta) &= PA_{01}(\beta) + A_0'R(\beta) \\ &\quad - \frac{\partial}{\partial\beta}R(\beta) + S(0,\beta) \\ \Delta_{14} &= PB, \quad \Delta_{22} = 0 \\ \Delta_{23}(\beta) &= A_1'R(\beta) - S(-h,\beta), \quad \Delta_{24} = 0 \\ \Delta_{33}(\alpha,\beta) &= R(\alpha)'A_{01}(\beta) + A_{01}(\alpha)'R(\beta) \\ &\quad - (\frac{\partial}{\partial\beta} + \frac{\partial}{\partial\alpha})S(\alpha,\beta) \\ \Delta_{34}(\alpha) &= R(\alpha)'B, \quad \Delta_{44} = 0\end{aligned}$$

and, if Δ_{ij} is a function of parameter α or β, Δ_{ij}' is defined as follows:

$$\begin{aligned}\Delta_{ij}'(\beta) &= \Delta_{ji}(\alpha), \qquad \Delta_{ij}'(\alpha) = \Delta_{ji}(\beta) \\ \Delta_{ij}'(\alpha,\beta) &= \Delta_{ji}(\beta,\alpha)\end{aligned}$$

Then, the state reachable set $\mathcal{R}(0,0)$ of the time-delay system described by (1) and (2) satisfies

$$\mathcal{R}(0,0) \subset \mathcal{E}(P,R,S)$$

□

Problem Statement

Our objective of this paper is to design a state-feedback controller (3) when the closed loop system is asymptotically stable and the L^2 gain, defined by (5), of the system is less than a scalar γ for any disturbance w which constrained by $\mathcal{W}$, and the input u of the closed loop system is included by $\mathcal{U}$:

$$\mathcal{U} = \{u(t)|u(t)'Uu(t) \leq 1\},$$

where, W_D is the given and positive definite matrix.

Here, an L^2 gain g is defined by

$$g = \sup_{w \in L^2, w \neq 0} \frac{||z||_{L_2}}{||w||_{L_2}}. \quad (5)$$

In this case, we call the L^2-gain as "semi-global L^2-gain", because the disturbance is constrained by $\mathcal{W}$.

3. H^∞ CONTROLLER SYNTHESIS

In this section, we propose a synthesis method of state-feedback controller that makes the closed loop system asymptotically stable and semi-global L^2 gain of the closed system less than γ. First, consider the case that we do not limit the size of the input.

Theorem 1. Assume that there exist non-negative scalar p, matrices and matrix functions W,$Y(\alpha, \beta)$, Z_0 and $Z_{01}(\beta)$ that satisfies conditions LMI-2,

$$(\mathbf{LMI-2})$$

$$\begin{bmatrix} W & W \\ W & Y(\alpha,\beta) \end{bmatrix} > 0, \quad X > 0,$$

$$\begin{bmatrix} \Gamma_{11} & \Gamma_{12} & \Gamma_{13} & \Gamma_{14} \\ \Gamma_{12}' & \Gamma_{22} & \Gamma_{23} & \Gamma_{24} \\ \Gamma_{13}' & \Gamma_{23}' & \Gamma_{33} & \Gamma_{34} \\ \Gamma_{14}' & \Gamma_{24}' & \Gamma_{34}' & \Gamma_{44} \end{bmatrix} (\alpha,\beta) \leq 0$$

Where,

$$\Gamma_{11} = WA_0' + A_0W + Z_0'B + BZ_0 + 2W + pW$$
$$\Gamma_{12} = A_1W - W$$
$$\Gamma_{13}(\beta) = A_{01}(\beta)W + BZ_{01}(\beta) + WA_0' + Z_0B' + Y(0,\beta) + pW$$
$$\Gamma_{14} = BW, \quad \Gamma_{22} = 0$$
$$\Gamma_{23}(\beta) = WA_1 - Y(-h,\beta)$$
$$\Gamma_{24} = 0$$
$$\Gamma_{33}(\alpha,\beta) = A_{01}(\beta)W + BZ_{01}(\beta) + WA_{01}(\beta)' + Z_{01}(\beta)'B' - (\frac{\partial}{\partial\alpha} + \frac{\partial}{\partial\beta})Y(\alpha,\beta) + pY(\alpha,\beta)$$
$$\Gamma_{34} = BW, \quad \Gamma_{44} = -pI$$

$$\begin{bmatrix} \Lambda_{11} & \Lambda_{12} & \Lambda_{13} & \Lambda_{14} & \Lambda_{15} \\ \Lambda_{12}' & \Lambda_{22} & \Lambda_{23} & \Lambda_{24} & \Lambda_{25} \\ \Lambda_{13}' & \Lambda_{23}' & \Lambda_{33} & \Lambda_{34} & \Lambda_{35} \\ \Lambda_{14}' & \Lambda_{24}' & \Lambda_{34}' & \Lambda_{44} & \Lambda_{45} \\ \Lambda_{15}' & \Lambda_{25}' & \Lambda_{35}' & \Lambda_{45}' & \Lambda_{55} \end{bmatrix} (\alpha,\beta) < 0.$$

Where,

$$\Lambda_{11} = WA_0' + A_0W + X + 2W + BZ_0 + Z_0'B'$$
$$\Lambda_{12} = A_1W - W$$
$$\Lambda_{13}(\beta) = A_{01}(\beta)W + BZ_{01}(\beta) + WA_0' + Z_0'B' + Y(0,\beta)$$
$$\Lambda_{14} = WC', \quad \Lambda_{15} = D, \quad \Lambda_{22} = -X$$
$$\Lambda_{23}(\beta) = WA_1' - Y(-h,\beta), \quad \Lambda_{24} = 0$$
$$\Lambda_{25} = 0$$
$$\Lambda_{33}(\alpha,\beta) = A_{01}(\beta)W + WA_{01}(\alpha)' + BZ_{01}(\beta) + Z_{01}(\alpha)'B' - (\frac{\partial}{\partial\alpha} + \frac{\partial}{\partial\beta})Y(\alpha,\beta)$$
$$\Lambda_{34} = 0, \quad \Lambda_{35} = D, \quad \Lambda_{44} = -I$$
$$\Lambda_{45} = 0, \quad \Lambda_{55} = -\gamma^2 I$$

where, $W = W'$, $Y(\alpha,\beta) = Y'(\beta,\alpha)$. Then, the closed loop system with the feedback controller(3) for the system(1) is asymptotically stable and the trajectory of the closed loop system exists in a range $\mathcal{M}$:

$$\mathcal{M} = \{(x(t), x_t)| (x(t), x_t)'\{W^{-1}, W^{-1}, W^{-1}YW^{-1}\}(x(t), x_t) \leq 1\},$$

and the semi-global L^2 gain of the closed loop system is less than $\gamma > 0$. Here, the feedback gain K_0 and $K_{01}(\beta)$ are given by

$$K_0 = Z_0W^{-1}, \quad K_{01}(\beta) = Z_{01}(\beta)W^{-1}$$

□

Here, we note that the state reachable set of the closed loop system where a disturbance w is constrained by $\mathcal{W}$ is evaluated by

$$\mathcal{E}(W^{-1}, W^{-1}, W^{-1}YW^{-1}),$$

so we obtain the next theorem to design a state feedback gain when the input of the closed loop system is in $\mathcal{U}$.

Theorem 2. If there exists a non-negative scalar p and matrices or matrix functions $W, Y(\alpha,\beta), Z_0$ and $Z_{01}(\beta)$ that satisfies conditions LMI-3 and LMI-2 in Theorem2,

$$(\mathbf{LMI-3})$$

$$\begin{bmatrix} W & W & Z_0' \\ W & Y & Z_{01}(\beta)' \\ Z_0 & Z_{01}(\alpha) & U^{-1} \end{bmatrix} > 0$$

then, the closed loop is asymptotically stable and its semi global L^2 gain is less than γ, and its input is included in $\mathcal{U}$. Here, the feedback gains are given by

$$K_0 = Z_0W^{-1}, \quad K_{01}(\beta) = Z_{01}(\beta)W^{-1} \quad (6)$$

Proof.

From (6),

$$u(t)'Uu(t) = (x(t), x_t)'\{W^{-1}Z_0'Z_0W^{-1}, W^{-1}Z_{01}(\beta)'Z_0W^{-1}, W^{-1}Z_{01}(\beta)'Z_{01}(\beta)W^{-1}\}(x(t), x_t).$$

Now, the state reachable set is included in

$$\mathcal{E}(W^{-1}, W^{-1}, W^{-1}YW^{-1}) = \{\lambda \in R^n \times L^2| \lambda'\{W^{-1}, W^{-1}, W^{-1}YW^{-1}\}\lambda \leq 1\},$$

so, if the following condition is satisfied, the input of the system is included in $\mathcal{U}$.

$$\begin{aligned}
&\{W^{-1}, W^{-1}, W^{-1}YW^{-1}\} \\
&- \{W^{-1}Z_0'Z_0W^{-1}, W^{-1}Z_0'Z_{01}(\beta)W^{-1}, W^{-1}Z_{01}(\beta)'Z_{01}(\beta)W^{-1}\} > 0 \\
&\Leftarrow \begin{bmatrix} W^{-1} & W^{-1} \\ W^{-1} & W^{-1}YW^{-1} \end{bmatrix} - \begin{bmatrix} W^{-1}Z_0'Z_0W^{-1} & W^{-1}Z_0'Z_{01}(\beta)W^{-1} \\ * & W^{-1}Z_{01}(\beta)'Z_{01}(\beta)W^{-1} \end{bmatrix} > 0 \\
&\Longleftrightarrow \\
&\begin{bmatrix} W & W \\ W & Y \end{bmatrix} - \begin{bmatrix} Z_0'Z_0 & Z_0'Z_{01}(\beta) \\ * & Z_{01}(\beta)'Z_{01}(\beta) \end{bmatrix} > 0 \\
&\Longleftrightarrow \begin{bmatrix} W & W \\ W & Y \end{bmatrix} - \begin{bmatrix} Z_0' \\ Z_{01}(\beta)' \end{bmatrix} \begin{bmatrix} Z_0' \\ Z_{01}(\beta)' \end{bmatrix}' > 0 \\
&\Longleftrightarrow \textbf{(LMI-3)}
\end{aligned}$$

Q.E.D.

As a special case, we consider the case when $A_{01}(\beta)$ in (1) and $K_{01}(\beta)$ in (3) are zero. In this case, the system has only the point delay and the controller is memoryless controller. The condition of this system which corresponds to that of Theorem 2 is described in the following theorem.

Theorem 3. If there exists a non-negative scalar p and matrices $W,'(\alpha,\beta), Z_0$ that satisfies the conditions:

$$W > 0,\ X > 0,$$

$$\begin{bmatrix} \begin{pmatrix} WA_0' + A_0W \\ +BZ_0 + Z'\_B' + pW \end{pmatrix} & A_1W & D \\ WA_1' & 0 & 0 \\ D' & 0 & -pW_D \end{bmatrix} \leq 0,$$

$$\begin{bmatrix} \begin{pmatrix} WA_0' + A_0 + W \\ +X \\ +BZ_0 + Z_0'B' \end{pmatrix} & WC' & A_1W & D \\ CW & -I & 0 & 0 \\ WA_1' & 0 & -X & 0 \\ D' & 0 & 0 & -\gamma^2 I \end{bmatrix} < 0,$$

$$\begin{bmatrix} W & Z_0' \\ Z_0 & U^{-1} \end{bmatrix} > 0$$

then, the closed loop system is asymptotically stable and its semi-global L^2 gain is less than γ, and its input is included in $\mathcal{U}$. Here, the feedback gain is given by

$$K_0 = Z_0W^{-1}$$

4. ALGORITHM

The condition given by Theorem2 includes bilinear terms, difficult to solve . Here, we propose an algorithm which overcomes such difficulty in solving LMI conditions iteratively.

Algorithm

Step 1 Define the initial values of W and $Y(\alpha,\beta)$ as $\tilde{W} = \epsilon_1 I, \tilde{Y}(\alpha,\beta) = \epsilon_2 I$

Step 2 Solve the following LMIs **(LMI-2')**, **(LMI-3)** and **(LMI-4)** given as

$$\textbf{(LMI-2')}$$

$$\begin{bmatrix} W & W \\ W & Y(\alpha,\beta) \end{bmatrix} > 0, \quad X > 0,$$

$$\begin{bmatrix} \tilde{\Gamma}_{11} & \Gamma_{12} & \tilde{\Gamma}_{13} & \Gamma_{14} \\ \Gamma_{12}' & \Gamma_{22} & \Gamma_{23} & \Gamma_{24} \\ \tilde{\Gamma}_{13}' & \Gamma_{23}' & \tilde{\Gamma}_{33} & \Gamma_{34} \\ \Gamma_{14}' & \Gamma_{24}' & \Gamma_{34}' & \Gamma_{44} \end{bmatrix} (\alpha,\beta) \leq 0$$

where,

$$\tilde{\Gamma}_{11} = WA_0' + A_0W + Z_0'B + BZ_0 + 2W + p\tilde{W}$$

$$\tilde{\Gamma}_{13}(\beta) = A_{01}(\beta)W + BZ_{01}(\beta) + WA_0' + Z_0B' + Y(0,\beta) + p\tilde{W}$$

$$\tilde{\Gamma}_{33}(\alpha,\beta) = A_{01}(\beta)W + BZ_{01}(\beta) + WA_{01}(\beta)' + Z_{01}(\beta)'B' - (\frac{\partial}{\partial\alpha} + \frac{\partial}{\partial\beta})Y(\alpha,\beta) + p\tilde{Y}(\alpha,\beta)$$

$$\begin{bmatrix} \Lambda_{11} & \Lambda_{12} & \Lambda_{13} & \Lambda_{14} & \Lambda_{15} \\ \Lambda_{12}' & \Lambda_{22} & \Lambda_{23} & \Lambda_{24} & \Lambda_{25} \\ \Lambda_{13}' & \Lambda_{23}' & \Lambda_{33} & \Lambda_{34} & \Lambda_{35} \\ \Lambda_{14}' & \Lambda_{24}' & \Lambda_{34}' & \Lambda_{44} & \Lambda_{45} \\ \Lambda_{15}' & \Lambda_{25}' & \Lambda_{35}' & \Lambda_{45}' & \Lambda_{55} \end{bmatrix} (\alpha,\beta) < 0$$

(LMI − 3)

$$\begin{bmatrix} W & W & Z_0' \\ W & Y & Z_{01}(\beta)' \\ Z_0 & Z_{01}(\alpha) & U^{-1} \end{bmatrix} > 0$$

(LMI − 4)

$$\begin{bmatrix} W & W \\ W & Y \end{bmatrix} \leq \begin{bmatrix} \tilde{W} & \tilde{W} \\ \tilde{W} & \tilde{Y} \end{bmatrix}$$

Step 3 If solutions $W, Y(\alpha,\beta)$ are obtained, then the W, $Y(\alpha,\beta)$ are solutions to the problem. If the solutions W, $Y(\alpha,\beta)$ is not satisfactory, put $\tilde{W} = W, \tilde{Y}(\alpha,\beta) = Y(\alpha,\beta)$ and go to Step 2.

Remark 1. The conditions in step 2 are linear, but infinite-dimensional ones, and the computational complexity is very high. Here, using a technique which we show in the next section, the conditions are reduced to the finite dimensional conditions, and becomes feasible ones.

Remark 2. A special case of this algorithm is applicable to the conditions in Theorem 3. Here, we note that when we apply the special case of this algorithm to the conditions in Theorem 3, we can show a convegence property of the algorithm.

5. REDUCTION TO FINITE-DIMENSIONAL LMIS

These LMI conditions, proposed in the algorithm, are infinite-dimensional LMI conditions. These infinite-dimensional LMI conditions are still difficult to solve,and some idea is needed to solve them. Now, we propose a method to reduce infinite-dimensional LMI conditions to finite-dimensional LMI conditions(Ikeda *et al.*, 2001). This method doesn't need descritization.

Now, we assume that, in LMI conditions **(LMI-2')**, **(LMI-3)** and **(LMI-4)**, $A_{01}(\beta), Z_{01}(\beta)$ and $Y(\alpha,\beta)$ is described like:

$$A_{01}(\beta) = A_{01}^0 + \beta A_{01}^1 + \beta^2 A_{01}^2 + \cdots + \beta^l A_{01}^l$$
$$Z_{01}(\beta) = Z_{01}^0 + \beta Z_{01}^1 + \beta^2 Z_{01}^2 + \cdots + \beta^l Z_{01}^l$$
$$Y(\alpha,\beta) = Y_0 + (\alpha+\beta)Y_1 + \cdots + (\alpha^l + \beta^l)Y_l,$$

then, LMI conditions in **(LMI-2')**, **(LMI-3)** and **(LMI-4)** are written as:

$$F_0(M) + f_1(\theta)F_1(M) + f_2(\theta)F_2(M) + \cdots + f_r(\theta)F_r(M) \leq 0$$

Here, θ is a vector like:

$$\theta \in \Theta \equiv \{[\alpha\ \beta]' | \alpha,\beta \in [-h,0]\}$$

and $f_i : R^2 \mapsto R$ is a polynomial function of θ, F_i is an affine matrix function for a symmetric and unknown matrix M. Here, an LMI written like this form is able to be reduced to a finite-dimensional LMI without so-called discritization.

6. NUMERICAL EXAMPLE

In this section, we show a numerical example with a system which has only point delay:

$$\dot{x}(t) = A_0x(t) + A_1x(t-h) + Bu(t) + Dw(t) \quad (7)$$
$$z(t) = Cx(t).$$

and a memoryless controller:

$$u(t) = K_0x(t). \quad (8)$$

In this case, the system has only point delay, then, the matrix inequality conditions given by Theorem 3 are finite dimensional.

The system parameters of (7) are given as:

$$A_0 = \begin{bmatrix} -2.0 & 0.0 \\ 0.0 & -1.5 \end{bmatrix}, A_1 = \begin{bmatrix} -0.2 & 0.0 \\ 0.0 & 0.5 \end{bmatrix},$$
$$B = \begin{bmatrix} 1.0 \\ 0.5 \end{bmatrix}, D = \begin{bmatrix} -0.1 \\ -0.2 \end{bmatrix}.$$

The values of W_D, U which determines the range of w, u respectively and the value of the L^2 gain are chosen as:

$$W_D = 1.0 \times 10^1,\ U = 1.0 \times 10^{-2},\ \gamma = 1.732.$$

With these conditions, we solve the LMI conditions by the special case algorithm introduced the section 4, and we obtain these solutions in the 1st iteration.

$$W = \begin{bmatrix} 0.0536 & 0.0253 \\ 0.0253 & 0.0151 \end{bmatrix}, X = \begin{bmatrix} 3.1548 & 1.5652 \\ 1.5652 & 0.7781 \end{bmatrix},$$
$$Z_0 = \begin{bmatrix} -2.2261 \\ -1.1356 \end{bmatrix}, K_0 = \begin{bmatrix} -32.2513 \\ -21.0170 \end{bmatrix},$$
$$p = 3.5357 \times 10^{-5}.$$

And in the 20th iteration, we obtain these solutions. Here, we really found the decreasing property of W with iterations.

$$W = \begin{bmatrix} 0.0021 & -0.0007 \\ -0.0007 & 0.0021 \end{bmatrix},$$
$$X = \begin{bmatrix} 0.4035 & 0.1994 \\ 0.1994 & 0.1005 \end{bmatrix},$$
$$Z_0 = \begin{bmatrix} -0.3000 \\ -0.1579 \end{bmatrix}, K_0 = \begin{bmatrix} -184.4005 \\ -134.7405 \end{bmatrix},$$
$$p = 0.0306.$$

The responce of the step disturbance ($0.5 \leq t \leq 1.5$) with this feedback gain is described in Fig.1.

We found that $u(t)$ stays in the range $\mathcal{U}$ which is defined by the given matrix U.

7. CONCLUSION

We proposed an H^∞ controller synthesis method for linear time-delay system, when a size of the

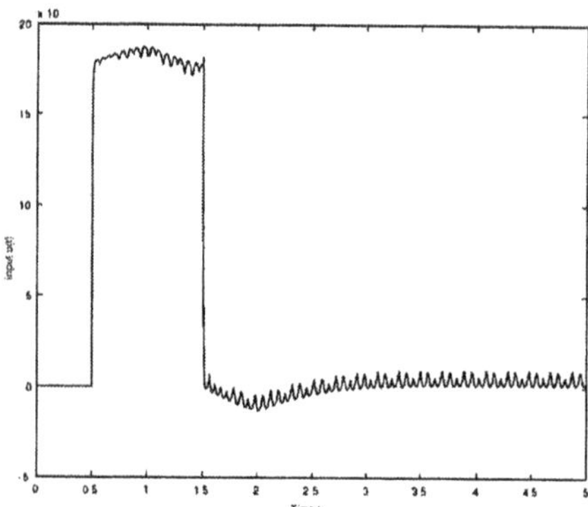

Fig. 1. Simulation result : Input $u(t)$

input is constrained. The controller makes the closed loop is semi-globally asymptotically stable and its L^2 gain less than a specified value.

8. REFERENCES

Boyd, S., L.El Ghaoui-E. Feron and V. Valakrisham (1994). *Linear matrix inequalities in systems and control theory.* Vol. 15. SIAM.

Ikeda, K. and K. Uchida (2000*a*). Analysis of state reachable sets for linear time-delay systems. In: *Proc. of SICE Annual Conference 2000.* p. #0235. (in Japanese).

Ikeda, K. and K. Uchida (2000*b*). Constrained state feedback H^∞ control of time-delay systems. In: *Proc. of 23th Dynamical System Theory Synposium.* pp. 29–32. (in Japanese).

Ikeda, K., T. Azuma and K. Uchida (2000*c*). A construction method of convex polyhedron in infinite number LMI approach for linear time-delay systems. In: *Proc. of 2000 National convention I.E.E. Japan.* pp. 1006–1007. (in Japanese).

Ikeda, K., T. Azuma and K. Uchida (2001). Infinite-dimensional LMI approach to analysis and synthesis for linear time-delay systems. *Kybernetika.* (to appear).

Niculescu, S-I., J-M. Dion and L.Dugard (1996). Robust stabilization for uncertain time-delay systems containing saturating actuators.. *IEEE Trans. A. C.* **41**(5), 742–747.

Tarbourieh, S. (2000). Synthesis of controllers for continuous-time delay systems with saturating controls via LMI's.. *IEEE Trans. A.C.* **45**(1), 105–111.

Watanabe, R., K. Uchida and M. Fujita (1998). Analysis of state reachable sets for quadratic systems. *Trans. SICE* **34**(11), 1590–1595. (in Japanese).

www.elsevier.com/locate/ifac

ANISOCHRONIC STATE FEEDBACK DESIGN COMPENSATING FOR SYSTEM DELAYS

Pavel Zítek, Tomáš Vyhlídal, Jaromír Fišer

Czech Technical University in Prague, Inst. of Instrumentation and Control Engineering,
cAk - Centre for Applied Cybernetics
fax: +4202 31 16414, zitek@fsid.cvut.cz, vyhlidal@fsid.cvut.cz, fiserj@fsid.cvut.cz

Abstract: An original extension of Ackermann formula has been developed to design the state feedback control in systems with delays described by convolution integrals. The used modelling approach called anisochronic promotes a useful potential to select the state variables primarily as available plant outputs, avoiding the usual need for state observers. In accordance with the functional nature of the plant model the state feedback is designed on the basis of convolution integrals too. Its delay distributions are designed to compensate for the plant model delays with the aim to endow the control system with a finite spectrum of eigenvalues in spite of the infinite original spectrum of the plant. Implementation problems of the feedback feasibility have been overcome by means of multivariable extension of Smith predictor scheme. The presented method is demonstrated on an application to real plant - heat transfer system where the plant delays result both from the transport delays and distributed parameters of heat transfer phenomena. *Copyright © 2001 IFAC*

Keywords: Time delay system, state feedback control, Ackermann formula, Smith predictor, delay compensation

1. INTRODUCTION

Application of state feedback scheme, as design approach providing a completely new prescribed system dynamics by means of memory-less feedback loops, is limited inevitably by a condition the number of state variables is finite and not very high for practical implementation. In such kind of systems, the complete pole assignment can be achieved mostly by the help of state observers. The requirement of finite number of state variables means that any kind of system with infinite spectrum of its eigenvalues, including the time delay systems, does not allow to use the state feedback design approach in its original conception. On the other hand the systems with considerable delays in inputs and feedbacks represent a class where a conversion to a more favourable dynamics, compensating for the harmful effects of delays, is needed more than elsewhere. A considerable research effort has been exerted to modify the state feedback and pole placement ideas towards the systems with delays (Malek-Zawarei and Jamshidi, 1987). Various approaches have been developed with this aim, most of them more or less reducing the role of system state. A method based on transcendental transfer function model of the plant is referred to as Finite Spectrum Assignment (Wang Q.G., et al., 1999). This method avoids the use of state vector at all and provides the pole assignment by means of specific delays in the control feedbacks. Its drawback is that only rather simple forms of input delays can be treated by this approach. Anyway, substituting the memory-less feedback used in conventional scheme by a functional one provides a general possibility to assign the poles of control systems with delays.

Seen from the point of view of conventional state space theory the systems either with delays or distributed parameters result in an infinite number of state attributes, which becomes contradictory to state feedback scheme. A well proved alternative of time delay system model consists in distinguishing between the system state and the state variable vector resulting in the anisochronic state description, which is applicable to systems with any kind of after-effects (Zítek, 1997, Zítek and Hlava, 2001). The state of the system is conceived as the last-past-segment of the vector of state variables $\mathbf{x}$ and the system state equation is as follows

$$\frac{d\mathbf{x}(t)}{dt} = \int_0^T d\mathbf{A}(\tau)\,\mathbf{x}(t-\tau) + \int_0^T d\mathbf{B}(\tau)\,\mathbf{u}(t-\tau) + \int_0^T d\mathbf{F}(\tau)\mathbf{d}(t-\tau) \tag{1}$$

The functional matrices $\mathbf{A}(\tau), \mathbf{B}(\tau), \mathbf{F}(\tau)$ are of respective dimensions $(n,n), (n,1), (n,m)$ and defined on a common interval $\tau \in \langle 0, T\rangle$ where τ is a delay variable and T is its maximum range. Vector $\mathbf{x}$ consists of system state variables and the system state at t is its segment $\hat{\mathbf{x}}\langle t-T, t\rangle$. Due to Stieltjes form of the convolution integrals usually stepwise discontinuous delay distribution functions $a_{ij}(\tau), b_{ij}(\tau), f_{ij}(\tau)$ as elements of the functional matrices $\mathbf{A}(\tau), \mathbf{B}(\tau), \mathbf{F}(\tau)$ may be used without the necessity to introduce Dirac distributions. The functions $a_{ij}(\tau), b_{ij}(\tau), f_{ij}(\tau)$ are considered as finite variation functions defined over the interval $\langle 0, T\rangle$.

2. FUNCTIONAL STATE FEEDBACK

Assume a conventional feedback $u = -\mathbf{K}\,\mathbf{x}$ is applied in system (1). Since $\mathbf{x}(t)$ does not represent the system state, such a feedback is not closed from the "state" in fact. As the state of (1) is defined by the segment $\hat{\mathbf{x}}\langle t-T, t\rangle$, the true state feedback has to be drawn from this segment of the vector $\mathbf{x}$. In other words $u(t)$ has to be determined by a functional defined over the interval $\langle t-T, t\rangle$. Since any linear functional relation between $\hat{\mathbf{x}}\langle t-T, t\rangle$ and scalar $u(t)$ can always be expressed as finite segment convolution the functional generalization of state feedback for (1) is supposed as follows

$$u(t) = -\int_0^T d\mathbf{K}(\tau)\mathbf{x}(t-\tau) \tag{2}$$

where $\mathbf{K}(\tau)$ is a $(1, n)$ matrix of delay distribution functions of finite variation again. In order to simplify the system investigation the Laplace transform of the model is needed. Using the rule of convolution transform the transforms of (1) and (2) are as follows

$$s\mathbf{x}(s) - \mathbf{x}(0) = \mathbf{O}(s) + \mathbf{A}(s)\mathbf{x}(s) + \mathbf{B}(s)u(s) + \mathbf{F}(s)\mathbf{d}(s)$$

$$u(s) = -\mathbf{K}(s)\mathbf{x}(s) \tag{3}$$

where

$$\mathbf{A}(s) = \int_0^T \exp(-s\tau)d\mathbf{A}(\tau), \quad \mathbf{B}(s) = \int_0^T \exp(-s\tau)d\mathbf{B}(\tau),$$

$$\mathbf{K}(s) = \int_0^T \exp(-s\tau)d\mathbf{K}(\tau) \tag{4}$$

are s-multiples of ordinary transforms of original matrices and $\mathbf{O}(s)$ is a summary transform of the initial functions of delays. Everywhere in the next considerations zero-valued initial conditions are assumed. Including the functional state feedback (2) into the system (1) under the zero initial conditions results in the transform description as follows

$$s\,\mathbf{x}(s) = [\mathbf{A}(s) - \mathbf{B}(s)\,\mathbf{K}(s)\,]\mathbf{x}(s) + \mathbf{F}(s)\,\mathbf{d}(s) \tag{5}$$

2.1 Spectral controllability

A key issue of applying the functional state feedback (5) is the controllability of the system (1) by the control variable u. The problem is more involved than in finite order systems because of infinite number of system eigenvalues. The conventional controllability concept was modified for the infinite spectrum systems to spectral controllability issue (Gorecki et al, 1989). The system (3) is spectrally controllable if and only if

$$\operatorname{rank}\left[\mathbf{B}(s), \mathbf{A}(s)\mathbf{B}(s), \ldots, \mathbf{A}^{n-1}(s)\mathbf{B}(s)\right] = \operatorname{rank}[\mathbf{R}(s)] = n \tag{6}$$

The controllability matrix $\mathbf{R}(s)$ is functional again due to exponential delay operators. If it satisfies (6) it may be formally inverted, but to consider $\mathbf{R}^{-1}(s)$ as an operator is not feasible because of non-causal terms resulting from the delay inverse.

3. APPLICATION OF ACKERMANN FORMULA

In conventional linear systems Ackermann formula serves as a method to design a state feedback system endowed with the prescribed spectrum of eigenvalues. In spite of infinite spectrum of the system (1) there exists certain possibility to design a functional state feedback (2) which would convert the system

$$s\mathbf{x}(s) = [\mathbf{A}(s) - \mathbf{B}(s)\mathbf{K}(s)]\mathbf{x}(s) \qquad (7)$$

into n-th order only with prescribed location of n its poles. Assume that closing such a state feedback should result in the following characteristic polynomial

$$M(s) = \det[s\mathbf{1} - \mathbf{A}(s) + \mathbf{B}(s)\mathbf{K}(s)] = s^n + \sum_{i=0}^{n-1} \alpha_i s^i \qquad (8)$$

of the system (5) with $\alpha_i, i = 0, ..., n-1$ as prescribed coefficients. This desired polynomial character of $M(s)$, free of delay terms at all, means that $\mathbf{K}(s)$ is able to compensate for all the delays in $\mathbf{A}(s)$, $\mathbf{B}(s)$. This design of $\mathbf{K}(s)$ may be accomplished by adopting the Ackermann formula to the functional operator matrices.

Theorem 1. Spectrally controllable anisochronic system (1) with a full-row-rank controllability matrix $\mathbf{R}(s)$ is supplemented by functional state feedback (2). A polynomial $M(s)$ of n-th degree is prescribed in the form (8). If the following feedback functional matrix $\mathbf{K}(s)$ is introduced

$$\mathbf{K}(s) = [\,0\,, 0\,, ..., 1\,]\,[\mathbf{R}(s)]^{-1}\, M(\mathbf{A}(s)) \qquad (9)$$

then the system equiped with this feedback is converted into a finite spectrum n-th order system with characteristic polynomial $M(s)$.

Proof. The conventional form of Ackermann formula has been proved for coefficient matrices $\mathbf{A}, \mathbf{B}$ and resulting controllability matrix (Ogata, 1990) for designing coefficient state feedback matrix $\mathbf{K}$. The proof is based on the Caley-Hamilton theorem which holds even for functional matrices (Brogan, 1991). Due to this extension also the functional matrix polynomial $M(\mathbf{A}(s))$ and the basic matrix operations used in the proof are valid. For $M(\mathbf{A}(s))$ the following equality may be obtained again

$$M(\mathbf{A}(s)) = [\mathbf{R}(s)] \begin{bmatrix} \Psi_1(\mathbf{K}(s)) \\ \Psi_2(\mathbf{K}(s)) \\ \mathbf{K}(s) \end{bmatrix} \qquad (10)$$

where $\Psi_1, ..., \Psi_{n-1}$ are various linear functions in $\mathbf{K}$, omitted in the final result of (9). The equation (10) is derived in the same way as for the coefficient matrices since any of the matrix operations holds for both coefficient and functional matrices. The existence of $[\mathbf{R}(s)]^{-1}$ corresponds to the controllability condition and left multiplying by $[0, 0, ..., 1]$ provides that $\Psi_1, ..., \Psi_{n-1}$ are excluded from the result of (9).

A significant merit of the anisochronic model (1) is a low number of state variables $\mathbf{x}$ compared with the conventional state space model. Furthermore, the state variables can then be selected as available (measurable) outputs as a rule and therefore the usual problem of estimating unavailable state variables can be avoided. These favourable features of the model (1) will be seen from the application example added.

On the other hand a crucial issue of potentials to apply the formula (9) is the problem of causality, i.e. feasibility. While the matrix $M(\mathbf{A}(s))$ is always causal (since $\mathbf{A}(s)$ is a causal model matrix) the inverse $[\mathbf{R}(s)]^{-1}$ is not causal as a rule because of non-causal (anticipative, predictive) terms unavoidably resulting from the delay inverse. In each column of the matrix $\mathbf{K}(s)$ therefore may appear an anticipative factor $\exp(T_i s)$, $i = 1, ..., n$ and by means of a factorization

$$\mathbf{K}(s) = [k_1^*(s), k_2^*(s), ...]\operatorname{diag}[\exp(T_1 s), ..., \exp(T_n s)] =$$

$$= \mathbf{K}^*(s)\mathbf{E}(s) \qquad (11)$$

a trimmed matrix $\mathbf{K}^*(s)$ is obtained as free of predictive terms at all. The non-causality is thus concentrated into the diagonal matrix $\mathbf{E}(s)$. The parameters $T_1, ..., T_n$ in $\mathbf{E}(s)$ are given by the maximum of the shifts ahead required by the respective columns of $\mathbf{K}(s)$.

4. SMITH CONTROL SCHEME IMPLEMENTATION

A way to implement the result (9) consists in using the principle of Smith predictor scheme. The approach can be characterized by the following points

- Instead of $\mathbf{K}(s)$ only $\mathbf{K}^*(s)$ is intended to be applied as feasible state feedback operator.
- The model (1) is rearranged to a structure where the delays $\exp(-T_i s)$, $i = 1, ..., n$ are linked up to the respective outputs of $x_1, ..., x_n$, and in this way also the predicted outputs $x_i(t + T_i)$, $i = 1, ..., n$ are available from the model.
- The measured plant outputs and both the model outputs simultaneous and predicting are employed in the following Smith control scheme.

The primary prerequisite to applying Smith predictor principle is a restructuring the model (1) in order to obtain $x_i(t + T_i)$, $i = 1, ..., n$ as available model outputs. The following rearrangement is one of the possibilities how to do it.

Theorem 2. Suppose a causal model (1) given by $\mathbf{A}(s), \mathbf{B}(s)$ as a set of n difference-differential equations with the following form of delays

$$\frac{dx_i(t)}{dt} = \sum_{j=1}^{n} \sum_{k=0}^{\kappa} a_{jk} x_j(t - \phi_{jk}) + \sum_{j=1}^{m} \sum_{k=0}^{\kappa} b_{jk} u_j(t - \tau_{jk}), \quad i = 1, \ldots, n \tag{12}$$

For this model a non-causal functional feedback matrix $\mathbf{K}(s)$ has been found from (9) requiring anticipative shifts with maximal values $T_i, i = 1, \ldots, n$ in the respective columns. To separate these predictive terms from the other functionals the factorization (11) has been performed. If each of the equations (12) is shifted ahead by $T_i, i = 1, \ldots, n$ the following modified structure is obtained

$$\frac{dx_i(t + T_i)}{dt} = \sum_{j=1}^{n} \sum_{k=0}^{\kappa} a_{jk} x_j(t + T_i - \phi_{jk}) + \sum_{j=1}^{m} \sum_{k=0}^{\kappa} b_{jk} u_j(t + T_i - \tau_{jk}), \quad i = 1, \ldots, n \tag{13}$$

where the coefficients do not change, only the delays reduce their sizes. The model remains causal and provides the values $x_i(t + T_i), i = 1, \ldots, n$ on its respective integrator outputs while the outputs $x_i(t), i = 1, \ldots, n$ are obtained delayed by $T_i, i = 1, \ldots, n$. Notice that shift parameters in (13) do not represent the actual sizes of delays because of different shifting of the model equations. If e.g. the variable $x_3, j = 3$ in the first equation (i = 1) is considered it is to be delayed by $\phi_{3k} - T_1$, not only by ϕ_{3k} (the subscript means that more than one delay value is supposed in this interaction).

Proof. The causality of the model (1) means that no anticipative shift can emerge neither in its whole structure nor in any of the equations (13) in particular. As a relation between the delay sizes this property can not be lost in arbitrary i-th equation by shifting it by T_i ahead. Therefore any of the equations (13) remains causal itself and the fact that T_i are different in each of the equations brings about the following,

- some of the delays may be cancelled,
- the other reduce their values,
- the differences between T_i influence the final values of effective delays between the output of j-th and the input of i-th integrator,

while the coefficients a_{jk}, b_{jk} remain unaffected by the shifting. In this way the original transform model matrices $\mathbf{A}(s), \mathbf{B}(s)$ are substituted by the modified ones $\mathbf{A}^*(s), \mathbf{B}^*(s)$ providing the following predicted outputs

$$\mathbf{x}^*(s) = \left[s\mathbf{I} - \mathbf{A}^*(s)\right]^{-1} \mathbf{B}^*(s)\, u(s) = \mathbf{E}(s)\mathbf{x}(s) \tag{14}$$

After this model rearrangement the Smith predictor principle may be adopted in the matrix form, (see Fig. 1).

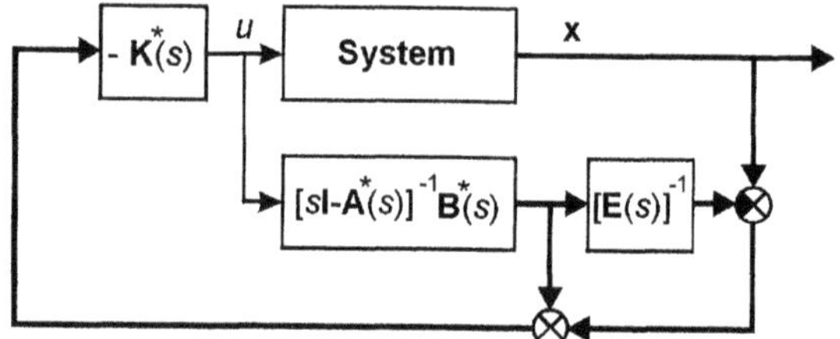

Fig. 1 Smith predictor

Theorem 3. The anisochronic state feedback compensating for the system delays in Smith predictor scheme is according to Fig. 1 as follows

$$u(s) = -\mathbf{K}^*(s)\,\mathbf{x}(s) + \mathbf{K}^*(s)\left[[\mathbf{E}(s)]^{-1} - \mathbf{I}\right]\left[s\mathbf{I} - \mathbf{A}^*(s)\right]^{-1} \mathbf{B}^*(s)\, u(s) \tag{15}$$

where

$$[\mathbf{E}(s)]^{-1} = \text{diag}\left[\exp(-sT_1), \ldots, \exp(-sT_n)\right] \tag{16}$$

is a diagonal matrix of delays and $\mathbf{x}$ are the measured outputs of the plant. The second part of the right-hand side of (15) represents the role of the parallel model of the plant. Apparently from (15), if the model and the real plant behaviour are in a perfect accordance the terms $-\mathbf{K}^*(s)\mathbf{x}(s)$ and $\mathbf{K}^*(s)[\mathbf{E}(s)]^{-1}\left[s\mathbf{I} - \mathbf{A}^*(s)\right]^{-1}\mathbf{B}^*(s)u(s)$ will cancel each other and only the predicting term remains as follows

$$u(s) = -\mathbf{K}^*(s)\left[s\mathbf{I} - \mathbf{A}^*(s)\right]^{-1}\mathbf{B}^*(s)u(s) = -\mathbf{K}^*(s)\mathbf{E}(s)\mathbf{x}(s) \tag{17}$$

This result is identical with the design result (11). However, due to the unavoidable plant-model differences such complete compensation for delays is not achieved in fact.

5. APPLICATION TO A REAL HEAT TRANSFER SYSTEM

A typical area where significant problems of delays are encountered is the control of heating systems. For testing approaches in time delay system control a laboratory-scale set up with heat exchangers, pumps and control devices has been developed at the authors' Department (Zítek and Vyhlídal, 2000). Its scheme is sketched in Fig. 3. Primary hot water is produced by an electric heater and then applied as heating medium in a water-water multiplate heat exchanger. The heating performance of this exchanger is controlled by a mixing valve. Secondary circulating hot water is cooled down in an air cooler the performance of which is controlled by adjusting its propeller velocity. The main system delays are provided as transport delays in piping lines. The mixing valve position u serves as control input. The modelling approach using input as well as state delays is used for the system description. First part of the system, water/water exchanger with measured temperature ϑ_a is described by the equation

$$T_a \frac{d\vartheta_a(t)}{dt} = K_d \vartheta_c(t-\tau_d) - \vartheta_a(t-\phi_a) + \\ + K_u[u(t-\tau_u) + K_e \vartheta_c(t-\tau_d-\tau_e-\tau_u)] \quad (18)$$

and the second part of the system, air/water cooler with measured temperature ϑ_c is described by the equation

$$T_c \frac{d\vartheta_c(t)}{dt} = -\vartheta_c(t-\phi_c) + K_b \vartheta_a(t-\tau_b) \quad (19)$$

Parameters of the model are assessed on the basis of step response of the system. The following parameters of the model consisted of equations (18) and (19) ensure very good agreement between system and model step responses (see Fig. 2).

T_a = 13 s, T_c = 25 s, K_b = 0.78, K_d = 0.5, K_e = 0.65, K_u = 0.19, τ_b = 34 s, τ_d = 6 s, τ_e = 9 s, τ_u = 15 s, ϕ_a = 5.5 s, ϕ_c = 9 s.

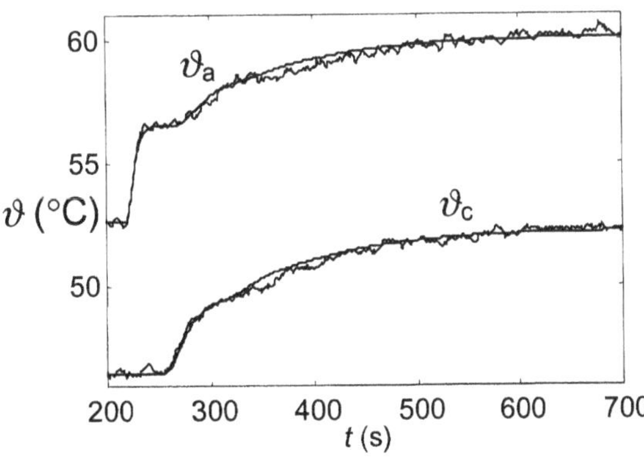

Fig. 2 Step response comparison of the laboratory set-up and its anisochronic model. Model - smooth, plant - influenced by noise

The feedback from state variables ϑ_a and ϑ_c allows to correct only the system dynamics. To achieve a plant control action subject to the desired temperature $\vartheta_{c,set}$ the following integration action is to be added

$$\frac{dI(t)}{dt} = \vartheta_{c,set}(t) - \vartheta_c(t) \quad (20)$$

Final functional state space description of the system results in third order model of the form

$$s\mathbf{x}(s) = \mathbf{A}(s)\mathbf{x}(s) + \mathbf{B}(s)u(s) \\ y(s) = \mathbf{C}(s)\mathbf{x}(s) \quad (21)$$

where $\mathbf{x}(s) = [\vartheta_a(s)\,\vartheta_c(s)\,I(s)]^T$ is the state vector, and $y(s) = \vartheta_c(s)$ is the controlled output. Transform functional matrices are as follows

$$\mathbf{A}(s) = \begin{bmatrix} a_{11}(s) & a_{12}(s) & 0 \\ a_{21}(s) & a_{22}(s) & 0 \\ 0 & -1 & 0 \end{bmatrix}, \mathbf{B}(s) = \begin{bmatrix} b_1(s) \\ 0 \\ 0 \end{bmatrix}, \mathbf{C} = \begin{bmatrix} 0 \\ 1 \\ 0 \end{bmatrix}^T$$

where (22)

$$a_{11}(s) = -\frac{e^{-\phi_a s}}{T_a}, a_{21}(s) = \frac{K_b e^{-s\tau_b}}{T_c}$$

$$a_{12}(s) = \frac{K_d e^{-s\tau_d} + K_u K_e e^{-s(\tau_u+\tau_d+\tau_e)}}{T_a}$$

$$a_{22}(s) = -\frac{e^{-\phi_c s}}{T_c}, \; b_1(s) = \frac{K_u e^{-\tau_u s}}{T_a}$$

State feedback controller is given by the functional matrix $\mathbf{K} = [k_a(s)\,k_c(s)\,k_I(s)]$. If the desired system dynamics are given by a triple real pole $s_{1,2,3} = -\alpha$, the final structure of the feedback vector is according to (9) as follows

$$\mathbf{K}(s) = \begin{bmatrix} e^{s\tau_u} \dfrac{a_{11}(s) + 3\alpha a_{22}(s) e^{s\tau_u}}{b_1} \\ e^{s\tau_g} \dfrac{a_{12}(s)a_{21}(s) + a_{22}(s)^2 + 3\alpha a_{22}(s) + 3\alpha^2}{a_{21}b_1} \\ e^{s\tau_g} \dfrac{-\alpha^3}{a_{21}b_1} \end{bmatrix}^T \quad (23)$$

where $\tau_g = \tau_u + \tau_b$. Because of the positive exponential term in each part of the functional vector, the control algorithm is not feasible. According to equation (14) the predicted outputs

$$\overline{\mathbf{x}}^*(s) = \begin{bmatrix} \vartheta_a(s)e^{s\tau_u} \\ \vartheta_c(s)e^{s\tau_g} \end{bmatrix} = \begin{bmatrix} \vartheta_a^*(s) \\ \vartheta_c^*(s) \end{bmatrix}$$

are obtained from the model given by the modified matrices

$$\overline{\mathbf{A}}^*(s) = \begin{bmatrix} a_{11}(s) & a_{12}^*(s) \\ a_{21}^* & a_{22}(s) \end{bmatrix}, \; \overline{\mathbf{B}}^*(s) = \begin{bmatrix} b_1^* \\ 0 \end{bmatrix} \quad (24)$$

where

$$a_{21}^* = \frac{K_b}{T_c}, \; b_1^* = \frac{K_u}{T_a}$$

$$a_{12}^*(s) = \frac{K_d e^{-s(\tau_b+\tau_d)} + K_u K_e e^{-s(\tau_b+\tau_u+\tau_d+\tau_e)}}{T_a}$$

This model is applied in the feedback scheme (15) as be seen in Fig. 3. The integral action (20) is also modified according to Smith predictor scheme in order to assure the feedback feasibility.

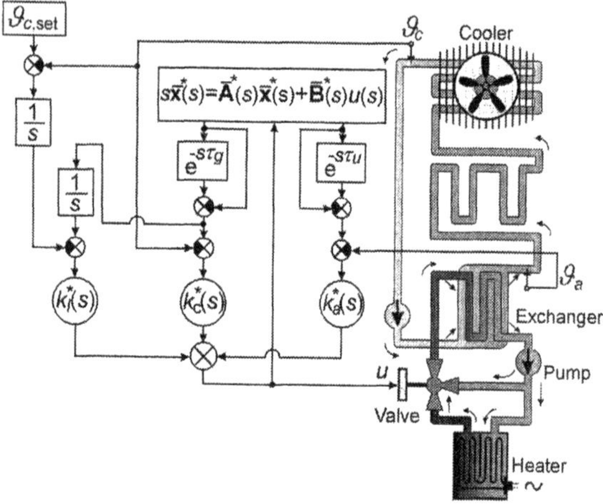

Fig. 3 Scheme of the thermal set-up.

Set-point response of the control system with prescribed dynamics by $s_{1,2,3}$=- α =- 0.06 s^{-1} is shown in Fig. 4.

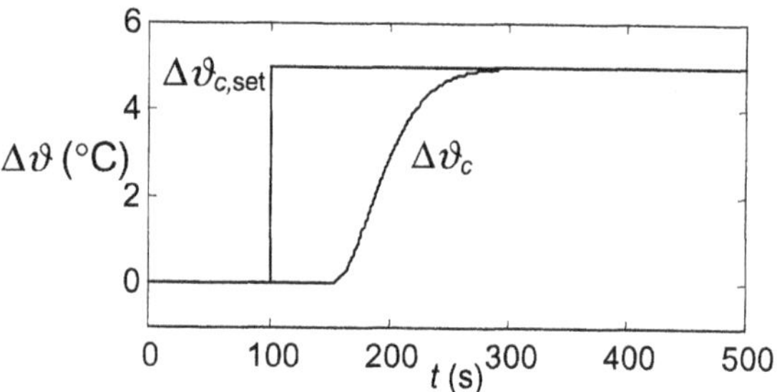

Fig. 4 Set-point response

This setting of α has proved to be proper value for the application.

6. CONCLUSIONS

The anisochronic approach to modelling the plants with delays, latencies or other after-effects promotes the possibility to describe them by means of relatively low number of state variables, almost always available as measurable plant outputs. The used functional description has been proved to be a promising modelling method not only for typical time delay systems but also for infinite spectrum systems in general. Convolution form of delay functionals helps to obtain a model formulation allowing to extend the conventional design methods to the anisochronic systems. The well-known Ackermann formula derived for conventional state space models has been extended to anisochronic plant model (1) and on its basis a functional state feedback can be designed where the feedback delays provide the ability to compensate for all the system delays. On the assumption that the model is perfectly identified with the plant the feedback $u(s) = -\mathbf{K}(s)\,\mathbf{x}(s)$ converts the original system to a conventional one with finite spectrum of prescribed poles. For the sake of the causality condition, the feedback implementation needs to be accomplished via Smith predictor scheme. In comparison with the spectrum assignment method the presented scheme overperforms it with its multivariable potentials of application. As a special point various shifts in particular x_i – branches of the Smith predictor scheme are applied. The sensitivity of the presented scheme to the model-to-plant errors can be effectively tuned by the prescribed dynamics given by the characteristic polynomial $M(s)$. The more moderate are the prescribed dynamics the higher level of robustness is achieved in the system behaviour. A crucial issue of the implementation is the inverse of controllability matrix. In case of more complex form of this matrix the design may result in rather involved form of $K(s)$. In such cases it is supposed that the original functionals are simplified in the terms of interpolation in the required delay distributions.

ACKNOWLEDGEMENT

The presented research was supported by the Ministry of Education of the Czech Rep. under Project LN00B096.

REFERENCES

Brogan, W. L., (1991). *Modern Control Theory.* Prentice-Hall International, Inc., London.

Gorecki, H., S. Fuksa, P. Grabowski and A. Korytowski (1989). *Analysis and Synthesis of Time Delay Systems*, Polish Scient. Publ. Warszawa.

Malek-Zawarei M. and Jamshidi M. (1987). *Time Delay Systems: Analysis, Optimization and Applications*, North Holland, Amsterdam.

Ogata, K. (1990). *Modern Control Engineering.* Prentice-Hall Int., Inc., New Jersey

Wang, Q.G., Lee, T.H. and Tan K.K. (1999). *Finite Spectrum Assignmnent for Time-Delay Systems.* Springer, London.

Zítek, P. (1997). Frequency Domain Synthesis of Hereditary Control Systems via Anisochronic State Space. *International Journal of Control* 66, No. 4, 539-556.

Zítek, P., Hlava, J., (2001). Anisochronic Internal Model Control of Time-Delay Systems. *Control Engineering Practice* 9, 501-516

Zítek, P., Vyhlídal, T. (2000). *State Feedback Control of Time Delays System: Conformal Mapping Aided Design,* In. Proc. of IFAC Symposium on Time Delay Systems, Ancona.

www.elsevier.com/locate/ifac

ON DECOUPLING OF LINEAR TIME DELAY SYSTEMS BY OUTPUT FEEDBACK

Rabah Rabah [*,1] **Michel Malabre** [*]

[*] *Institut de Recherche en Communications et Cybernétique de Nantes, UMR 6597*
1, rue de la Noë, BP 92 101
44321 Nantes Cedex 03, France

Abstract: We consider the row-by-row decoupling problem for linear delay systems by output feedback. The characterization of the solvability of this problem is given in terms of some easily checkable structural conditions. The main contribution is, in particular, to use generalized output feedback laws which may incorporate derivatives of the delayed new reference. *Copyright © 2001 IFAC*

Keywords: Linear Delay Systems, Row-by-row Decoupling, Structure at Infinity, Output feedback.

1. INTRODUCTION

Several authors have considered the row-by-row decoupling problem for delay systems. In (Tzafestas and Paraskevopoulos, 1973) an algebraic solution was given which extends the classical result of Falb and Wolovich (1967). The condition of the non-anticipativity of the state feedback law was studied in (Rekasius and Milzareck, 1977). A more general framework which include delay systems was given by Datta and Hautus (1984). The structural approach was developed in (Sename, *et al.*, 1995) and partial characterization of the solvability of the given problem was discussed. The present authors gave (1999) a more general solution which uses generalized state feedback, i.e. feedback which may include the derivatives of the delayed new reference.

In (Wolovich, 1975), was given a nice characterization of the row-by-row decoupling using output feedback. This approach allows to use only the transfer function matrix in the description of the solution and of the feedback.

[1] Also with École des Mines de Nantes, BP 20722, 4, rue Alfred Kastler, 44307, Nantes Cedex 3

Our purpose is to extend the result of Wolovich to delay systems and to extend our result using generalized feedback. Namely, the aim of the paper is to characterize the row-by-row decoupling problem by generalized output feedback.

1.1 System description

We consider linear time-invariant systems with delays described by:

$$\begin{cases} \dot{x}(t) = A_0x(t) + A_1x(t-1) + B_0u(t) \\ y(t) = C_0x(t) \end{cases} \tag{1}$$

where $x(t) \in \mathbb{R}^n$ is the state, $u(t) \in \mathbb{R}^m$ is the control input, $y(t) \in \mathbb{R}^m$ is the output to be controlled. Without loss of generality, we can assume that B_0 is of full column rank. In order to simplify the notation and some computations, we limit ourselves to systems with a single delay in the state. All results and considerations given here remain valid for systems with several commensurate delays in the state. The transfer function matrix of the system (1) is

$$T(s, e^{-s}) = C_0(sI - A_0 - A_1e^{-s})^{-1}B_0$$

and may be expanded into two different ways, namely as a power series expansion, either in the variable e^{-s} (with coefficients function of s) or in the variable s (with coefficients function of e^{-s}). Both expansions are given using the matrices introduced by Kirillova and Churakova and compared with other tools in (Tsoi, 1978):

$$\begin{aligned} Q_i(j) &= A_0 Q_{i-1}(j) + A_1 Q_{i-1}(j-1), \\ Q_0(0) &= I, \; Q_i(j) = 0, \; i < 0 \text{ or } j < 0. \end{aligned} \quad (2)$$

The first expansion is

$$T(s, e^{-s}) = \sum_{j=0}^{\infty} \sum_{i=0}^{\infty} C_0 Q_i(j) B_0 s^{-i-1} e^{-js}, \quad (3)$$

The other expression, which will be used in this paper, is the following one

$$T(s, e^{-s}) = \sum_{i=0}^{\infty} \sum_{j=0}^{i} C_0 Q_i(j) B_0 e^{-js} s^{-i-1}. \quad (4)$$

These expressions may be obtained by a simple calculation using the relations (2), see (Sename, *et al.*, 1995) and (Tsoi, 1978).

1.2 Problem formulation

We shall consider decoupling of systems like (1). The "open-loop" definition of decoupling is the following: Find a precompensator $K(s, e^{-s})$ and non identically zero scalar transfer functions $h_i(s, e^{-s})$, $i = 1, \ldots, m$, such that

$$T(s, e^{-s}) K(s, e^{-s}) = \operatorname{diag}\{h_1(s, e^{-s}), \ldots, h_m(s, e^{-s})\}.$$

We are interested in output feedback implementations of such decoupling precompensators, when they exist, and we want to connect the properties of $K(s, e^{-s})$ which make it realizable in an output feedback form and the type of, more or less restricted, output feedback laws which may be used.

We have previously shown (Rabah and Malabre, 1996) (see also (Sename, *et al.*, 1995)) that any decoupling solution $K(s, e^{-s})$ belonging to some particular class of precompensators, called *strong biproper* (see Section 2), is equivalent to a *static state feedback control* law of the type: $u(s) = F(e^{-s})x(s) + G(e^{-s})v(s)$, where $F(e^{-s})$ and $G(e^{-s})$ are proper matrices with respect to the variable e^s. In (Rabah and Malabre, 1999) we considered a broader class of decoupling precompensators, called *weak biproper* (see Section 2) and we showed their equivalence with *generalized static state feedback control* laws of the type:

$$u(s) = F(e^{-s})x(s) + G(s, e^{-s})v(s),$$

where

$$\begin{aligned} F(e^{-s}) &= F_0 + F_1 e^{-s} + \cdots, \\ G(s, e^{-s}) &= G_0 + G_1(s) e^{-s} + \cdots, \end{aligned}$$

with possible polynomial matrices $G_i(s), i \geq 1, G_0$ and $F_i, i \in \mathbb{N}$ are constant matrices. This amounts to accepting in the control law some delayed derivatives of the new reference input $v(t)$ and allows to look for more general solutions, assuming that the reference input $v(t)$ is smooth enough for all its involved derivatives to exist.

The aim of the present paper is to extend some above mentioned results to the problem of decoupling by output feedback.

2. PRELIMINARIES

In this section we recall classical results for systems without delays and recent results for decoupling by static state feedback for system with delays. First of all we give the notion of properness for linear systems and its extensions to linear time delay systems.

Definition 2.1. A complex valued function $f(s)$ is called proper if $\lim f(s)$ is finite when $|s| \to \infty$. It is called strictly proper if this limit is 0. It is called biproper if this limit is invertible.

For rational functions this notion may be described by the degrees of the numerator and denominator. For systems with delay, the transfer function contains e^{-s} which is not rational, we need then other notions of properness.

Definition 2.2. A complex valued function $f(s)$ is called weak proper if $\lim f(s)$ is finite when $s \in \mathbb{R}$ tends to ∞. It is called strictly weak proper if this limit is 0. A matrix $B(s)$ is weak biproper if it is weak proper and if this limit is invertible. Weak proper is replaced by strong proper if the same occurs when $s \in \mathbb{C}$ and $\Re e(s) \to \infty$.

It is obvious that strong properness implies weak properness. If the function is analytical at infinity all notions coincide, because the limits at infinity are the same. The strong properness is the natural extension of the classical notion, but does not give a well defined structure at infinity. The weak properness allows to give a good canonical form at infinity (see (Rabah and Malabre, 1999) and references given there).

In the sequel we shall use the different notions

of properness given in Definitions 2.1 and 2.2: proper, weak proper and strong proper.

Theorem 2.3. (Wolovich, 1975) Let us consider the linear system

$$\begin{cases} \dot{x}(t) = Ax(t) + Bu(t) \\ y(t) = Cx(t) \end{cases} \tag{5}$$

with square transfer function matrix $T(s) = C(sI - A)^{-1}B$. The following propositions are equivalent:

(1) There exists a biproper precompensator $K(s)$ and non zero scalar transfer functions $h_i(s)$, $i = 1, \ldots, m$, such that

$$T(s)K(s) = \mathrm{diag}\,\{h_1(s), \ldots, h_m(s)\}.$$

(2) There exist a feedback law $u = Hy + Gv$ and non zero scalar transfer functions $h_i(s), i = 1, \ldots, m$ such that

$$C(sI - A - BHC)^{-1}BG = \mathrm{diag}\,\{h_1(s), \ldots, h_m(s)\}$$

(3) (a) The so-called Falb-Wolovich matrix

$$D = \begin{bmatrix} c_1A^{n_1-1}B \\ \vdots \\ c_mA^{n_m-1}B \end{bmatrix},$$

is invertible. The integer n_i, $i = 1, \ldots, m$ is the order of the zero at infinity of each row subsystem: $c_iA^{n_i-1}B \neq 0$ and $c_iA^jB = 0$ for $j < n_i - 1$.

(b) The off-diagonal elements of $DT(s)^{-1}$ are scalars (independent of s).

Let us note that if the state space representation is not known, then the matrix D may be defined by

$$T(s) = \Delta(s)^{-1}(D + W(s)),$$

where $\Delta(s) = \mathrm{diag}\,\{s^{n_1}, \ldots, s^{n_m}\}$, n_i the order of the zero at infinity of each row i. For the delay system we need the following characterization of the decoupling problem by (generalized) state feedback.

Theorem 2.4. (Rabah and Malabre, 1999) The following propositions are equivalent:

(1) The row-by-row decoupling problem for the delay system (1) is solvable by a *weak biproper* precompensator $K(s, e^{-s})$:

$$T(s, e^{-s})K(s, e^{-s}) = \mathrm{diag}\,\{h_1(s, e^{-s}), \ldots, h_m(s, e^{-s})\}.$$

(2) The generalized Falb-Wolovich matrix:

$$D_0 = \begin{bmatrix} c_1Q_{n_1-1}(k_1)B_0 \\ \vdots \\ c_mQ_{n_m-1}(k_m)B_0 \end{bmatrix},$$

is invertible, where for each row i the integers n_i and k_i are such that: $c_iQ_{n_i-1}(k_i)B_0 \neq 0$ and $c_iQ_l(j)B_0 = 0$ for $l < n_i - 1$ and $j < k_i$.

(3) The decoupling problem is solvable by *generalized static state feedback*

$$u = F(e^{-s})x + G(s, e^{-s})v,$$

where

$$F(e^{-s}) = F_0 + F_1e^{-s} + \cdots,$$
$$G(s, e^{-s}) = G_0 + G_1(s)e^{-s} + \cdots,$$

with possible polynomial matrices $G_i(s), i \geq 1, G_0 = D_0^{-1}$ and constant matrices $F_i, i \in \mathbb{N}$.

Let us denote by $\Delta(s, e^{-s})$ the matrix

$$\Delta(s, e^{-s}) = \mathrm{diag}\,\{s^{n_1}e^{k_1s}, \ldots, s^{n_m}e^{k_ms}\}.$$

Then the transfer function matrix of the delay system may be factorized as

$$\begin{aligned} &T(s, e^{-s}) = \\ &\quad \Delta(s, e^{-s})^{-1}\left(D(e^{-s}) + W(s, e^{-s})\right), \\ &\quad W(s, e^{-s}) = W_1(s, e^{-s}) + W_2(s, e^{-s}), \end{aligned} \tag{6}$$

where $D(e^{-s}) = D_0 + D_1e^{-s} + \ldots$, $W_1(s, e^{-s})$ and $W_2(s, e^{-s})$ being respectively the strictly strong proper and the strictly weak proper parts in this unique decomposition (see (Rabah and Malabre, 1999)).

Theorem 2.5. If the delay system (1) is decouplable by static state feedback then D_0 is invertible and $D(e^{-s})$ is strong biproper.

Proof. If the system (1) is decouplable by static state feedback then it is decouplable by generalized state feedback and this implies that D_0 is invertible. This means that $D(e^{-s})$ is biproper in the strong sense. ∎

Theorem 2.6. If D_0 is invertible and $W_2(s, e^{-s}) = 0$, then (1) is decouplable by static state feedback.

Proof. If the condition of the theorem are satisfied then $K(s, e^{-s}) = (D(e^{-s}) + W_1(s, e^{-s})^{-1})$ is a strong biproper decoupling compensator. According to (Rabah and Malabre, 1999), it may be realizable by static state feedback. ∎

Note that the converse of this theorem is not true. For example, consider the decoupled system

$$T(s, e^{-s}) = \begin{bmatrix} s^{-2} + s^{-1}e^{-s} & 0 \\ 0 & s^{-1} \end{bmatrix}.$$

For this system, $D_0 = I$, $W_2(s, e^{-s}) \neq 0$ and $W_1(s, e^{-s}) = 0$.

3. DECOUPLING BY OUTPUT FEEDBACK

In order to simplify the notation, we are omitting sometimes the arguments of the corresponding function. The arguments are precised when it stands necessary.
We consider first the problem of decoupling by static output feedback of the form

$$u(s) = H(e^{-s})y(s) + G(e^{-s})v(s), \qquad (7)$$

with $H(e^{-s})$ and $G(e^{-s})$ proper in e^s. If the feedback is given by (7), then the corresponding input-output relation may be written as

$$y = T\left[I - HT\right]^{-1} Gv = \left[T^{-1} - H\right]^{-1} Gv.$$

Theorem 3.1. The system (1) is decouplable by static output feedback if and only if
i) It is decouplable by static state feedback
ii) The off-diagonal elements of the matrix

$$D(e^{-s})T(s, e^{-s})^{-1}$$

are proper in e^s and do not depend explicitly on s.

The condition ii) may be precised as follows. If we denote by $\alpha_{ij}(s, e^{-s})$ the off-diagonal elements, then ii) is equivalent to: $\alpha_{ij}(s, e^{-s}) = a_0 + a_1 e^{-s} + a_2 e^{-2s} + \ldots,\ i \neq j$, where a_k are constants depending on i and j.
Proof. The proof is given in the Annex. ∎

4. DECOUPLING BY GENERALIZED OUTPUT FEEDBACK

In many cases, static state feedback is not sufficient to decouple linear system with delay even when there exists a decoupling precompensator. Let us consider the system

$$T(s, e^{-s}) = \begin{bmatrix} s^{-3} & (s^{-4} + s^{-2})e^{-s} \\ 0 & s^{-1} \end{bmatrix}. \qquad (8)$$

It can be easily checked that there is no static state feedback law which decouples this system. However, $K(s, e^{-s}) = \begin{bmatrix} 1 & -(s^{-1} + s)e^{-s} \\ 0 & 1 \end{bmatrix}$ is a decoupling precompensator. This precompensator is weak but not strong biproper, which means that it cannot be realizable by static state feedback. For the output feedback decoupling the same situation occurs. Then we need, as for the state feedback law (see (Rabah and Malabre, 1999)), to extend the condition of realization. This is the aim of this section.
Let us give a preliminary result.

Lemma 1. Let $T(s, e^{-s})$ be decomposed as in (6). Then let us denote

$$\begin{aligned} T_1 &\stackrel{\text{def}}{=} \Delta^{-1}(D + W_1) = \\ &T\left[I + (D + W_1)^{-1}W_2\right]^{-1}. \end{aligned} \qquad (9)$$

Proof. This may be obtained by a simple calculation. ∎

Lemma 2. The system $T(s, e^{-s})$ is decouplable by generalized output feedback if and only if the system $T_1(s, e^{-s})$ is decouplable by static output feedback.

Proof. The proof is given in the Annex. ∎

Theorem 4.1. The system (1) is decouplable by generalized output feedback if and only if
i) D_0 is invertible (it is decouplable by generalized state feedback).
ii) The off-diagonal elements of the matrix

$$D(e^{-s})T_1(s, e^{-s})^{-1}$$

are proper in e^s and do not depend explicitly on s.

Proof. Suppose that the condition of the theorem are satisfied. i) implies that $D(e^{-s})$ is strong biproper (Theorem 2.5) and then $T_1(s, e^{-s}) = \Delta(s, e^{-s})^{-1}(D(e^{-s}) + W_1(s, e^{-s}))$ is decouplable by static state feedback (Theorem 2.6). Then the condition ii) and Theorem 3.1 give that $T_1(s, e^{-s})$ is decouplable by output static output feedback. This and Lemma 2 give the result.
Conversely, if the system $T(s, e^{-s})$ is decouplable by generalized output feedback, then the condition i) is obviously satisfied and the condition ii) is the consequence of Lemma 2. ∎
It is easy to see that the system (8), given at the begining of this section, is decouplable by generalized output feedback.

5. CONCLUSION

Generalized output feedback law is used to decouple linear systems with delays. The conditions given here are easy to verify. The counterpart of the general framework is that we need the derivative of the delayed new control. This requires smoothness of the new control.

6. ANNEX

Recall that we are omitting arguments of some functions. They are precised when it stands necessary.

6.1 Proof of the Theorem 3.1

Suppose that the conditions of the theorem are satisfied. Then (see (Rabah and Malabre, 1999)) $D(e^{-s})$ is strongly biproper and so is $D(e^{-s})^{-1}$. Let

$$\delta(s,e^{-s}) = \{\delta_1(s,e^{-s}),\ldots,\delta_m(s,e^{-s})\}$$

be the diagonal of $D(e^{-s})T(s,e^{-s})^{-1}$. Then

$$D(e^{-s})T(s,e^{-s})^{-1} - \delta(s,e^{-s})$$

is proper in e^s and does not depend explicitly on s. Let us denote

$$H \stackrel{\text{def}}{=} D^{-1}\left[DT^{-1} - \delta\right].$$

This gives

$$DT^{-1} - DH = \delta,$$

or

$$\left[T^{-1} - H\right]^{-1} D^{-1} = \delta^{-1}$$

Moreover, the diagonal matrix $\delta(s,e^{-s})^{-1}$ is strictly strong proper, because $\delta(s,e^{-s})$ is the diagonal of the matrix

$$(I + W(s,e^{-s})D(e^{-s})^{-1})\Delta(s,e^{-s}),$$

according to the formulae (6). Let now $\Lambda(e^{-s})$ be any diagonal matrix, proper in e^s. Then

$$\delta^{-1}\Lambda = \left[T^{-1} - H\right]^{-1} D^{-1}\Lambda$$

Putting $G(e^{-s}) = D(e^{-s})^{-1}\Lambda(e^{-s})$, the above relation means that the system is decouplable by the feedback

$$u(s) = H(e^{-s})y(s) + G(e^{-s})v(s),$$

according to the relation (7). The closed loop transfer function matrix is

$$T_d(s,e^{-s}) = \delta(s,e^{-s})^{-1}\Lambda(e^{-s}).$$

Conversely, suppose that the system is decouplable by the output feedback law (7). Then

$$\left[T(s,e^{-s})^{-1} - H(e^{-s})\right]^{-1} G(e^{-s}) = T_d(s,e^{-s})$$

with diagonal matrix $T_d(s,e^{-s})$. A simple calculation gives

$$DT^{-1} = DGT_d^{-1} + DH. \qquad (10)$$

In order to show the condition ii) of the theorem, it is sufficient to show that $D(e^{-s})G(e^{-s})$ is a diagonal matrix.

Note that (see (6)) $\Delta(s,e^{-s})T(s,e^{-s})$ is a weak biproper matrix. As $T(s,e^{-s})$ and $T_d(s,e^{-s})$ differ by a weak biproper matrix (they are equivalent at infinity and have the same structure at infinity), then the diagonal $\Delta(s,e^{-s})T_d(s,e^{-s})$ is also weak biproper. Then

$$B(s,e^{-s}) = T_d(s,e^{-s})^{-1}\Delta(s,e^{-s})^{-1} \qquad (11)$$

is a weak biproper diagonal matrix. From the decoupling condition we have

$$T(s,e^{-s})^{-1} = G(e^{-s})T_d(s,e^{-s})^{-1} + H(e^{-s})$$

and then

$$T^{-1}\Delta^{-1} = GT_d^{-1}\Delta^{-1} + H\Delta^{-1}.$$

From the expression (11) this equation may be written as

$$T^{-1}\Delta^{-1} = GB\Delta\Delta^{-1} + H\Delta^{-1}.$$

which gives

$$T^{-1}\Delta^{-1} = GB + H\Delta^{-1}. \qquad (12)$$

On the other hand, from (6) we get

$$T^{-1}\Delta^{-1} = (D+W)^{-1} \qquad (13)$$

From (12) and (13) we obtain

$$(D+W)^{-1} = GB + H\Delta^{-1}.$$

By identification, we can obtain

$$G(e^{-s})B_0(e^{-s}) = D(e^{-s})^{-1},$$

where $B_0(e^{-s})$ is the "constant" (according to the explicit dependence of s) part of the strong biproper matrix $B(s,e^{-s})$. Let us precise that $B_0(e^{-s})$ is diagonal because $B(s,e^{-s})$ is. Putting $\Lambda(e^{-s}) = B_0^{-1}(e^{-s})$ we obtain:

$$D(e^{-s})G(e^{-s}) = \Lambda(e^{-s}),$$

which is a diagonal matrix, proper in e^s. Consider now the expression of $D(e^{-s})T(s,e^{-s})^{-1}$ given by (10). The fact that $D(e^{-s})G(e^{-s}) = \Lambda(e^{-s})$ is diagonal proves that the off-diagonal elements of $D(e^{-s})T(s,e^{-s})^{-1}$ are the off diagonal elements of $D(e^{-s})H(e^{-s})$. This gives ii). ∎

6.2 Proof of the Lemma 2

As in both cases the system is decouplable by state feedback, $D(e^{-s})$ is biproper in the strong sense (and then in the weak sense), which means that $(D+W_1)^{-1}$ is well defined for $s > s_0$. Let $H(e^{-s})$ and $G_1(e^{-s})$ be proper in e^s matrices and let us denote $\Omega \stackrel{\text{def}}{=} (D+W_1)^{-1}W_2$. From Lemma 2 we have

$$T_1(I - HT_1)^{-1}G_1 =$$
$$= T(I + \Omega)^{-1}\left[I - HT(I+\Omega)^{-1}\right]^{-1})G_1$$
$$= T\left[(I - HT(I+\Omega)^{-1})(I+\Omega)\right]^{-1}G_1$$
$$= T\left[I - HT + \Omega\right]^{-1}G_1$$
$$= T(I - HT)^{-1}\left[I + \Omega(I-HT)^{-1}\right]^{-1}G_1.$$

Let us now put $G = \left[I + \Omega(I-HT)^{-1}\right]^{-1}G_1$ and $G_2 = G - G_1$. Then

$$T_1(I - HT_1)^{-1}G_1 = T(I-HT)^{-1}G,$$

with $G = \left[I + \Omega(I-HT)^{-1}\right]^{-1}G_1 = G_1 + G_2$. It is not difficult to see that G_2 is a function of two arguments (s, e^{-s}), weak proper in s if $G_1(e^{-s})$ is strong biproper in s. Suppose now that $u(s) = H(e^{-s})y(s) + G_1(e^{-s})v(s)$ decouples the system $T_1(s, e^{-s})$. This gives $T_1(I - HT_1)^{-1}G_1 = T_d$, where a non singular $T_d(s, e^{-s})$ is a diagonal matrix. Taking G as indicated, we get

$$T(I - HT)^{-1}G = T_d.$$

Conversely, suppose that $T(s, e^{-s})$ is decouplable by the generalized output feedback

$$u(s) = H(e^{-s})y(s) + G(s, e^{-s})v(s),$$

then $T(I - HT)^{-1}G = T_d$, with weak biproper $G(s, e^{-s})$. The transfer function matrix $T_d(s, e^{-s})$ may also be decomposed as $T(s, e^{-s})$ in (6):

$$T_d = \Delta^{-1}(D_d + W_{1d} + W_{2d})$$

with the same matrix Δ because of the equivalence at infinity, $W_{1d}(s, e^{-s})$ being strongly strictly proper and $W_{2d}(s, e^{-s})$ weakly strictly. This gives

$$T(I - HT)^{-1}G - \Delta^{-1}W_{2d} =$$
$$\Delta^{-1}(D_d + W_{1d}) \stackrel{\text{def}}{=} T_{1d},$$

with diagonal matrix T_{1d}. This gives

$$T_{1d} =$$
$$T(I - HT)^{-1} \times$$
$$\left[G - (I - HT)(D + W_1 + W_2)^{-1}W_{2d}\right],$$

where we used one more time the decomposition (6) for T. Let us denote

$$\Gamma \stackrel{\text{def}}{=} G - (I - HT)(D + W_1 + W_2)^{-1}W_{2d}$$

and $G_1 \stackrel{\text{def}}{=} [I + \Omega(I - HT)^{-1}]\Gamma$. This gives

$$T(I - HT)^{-1}\Gamma = T_{1d}$$

and

$$T(I - HT)^{-1}[I + \Omega(I - HT)^{-1}]^{-1}G_1 = T_{1d}.$$

Then using the same calculation as above we get

$$T_1(I - HT_1)^{-1}G_1 = T_{1d}.$$

which means that T_1 is decouplable by output feedback.

Let us now precise the structure of G_1. Multiplying by $\Delta(s, e^{-s})$ both parts, we obtain

$$(D + W_1)(I - HT_1)^{-1}G_1 = D_d + W_{1d},$$

and by identification $DG_1 = D_d$ and then

$$G_1(e^{-s}) = D^{-1}(e^{-s})D_d(e^{-s}).$$

Note that the fact that G_1 depends explicitly only on e^{-s} may be obtained directly from the construction of G_1. ∎

REFERENCES

Datta K. B., Hautus M. L. J. (1984). Decoupling of multivariable control systems over unique factorization domains. *SIAM J. Control and Optimization*, **22**, pp. 28–39.

Falb P. L., Wolovich W. A. (1967). Decoupling in the design and synthesis of multivariable control systems, *I.E.E.E. Trans. Autom. Contr.*, **AC-12**, No. 12, 651–659.

Rabah R., Malabre M. (1996). Structure at infinity for delay systems revisited. In: *IMACS and IEEE-SMC Multiconference CESA'96, Symposium on Modelling, Analysis and Simulation*, Lille, France, July 9-12, 87–90.

Rabah R., Malabre M. (1997). A note on decoupling for linear infinite dimensional systems.In: *Proc. 4-th IFAC Conf. on Syst. Structure and Contr.*, Bucharest, Oct. 23-25, 78–83.

Rabah R., Malabre M. (1999). The structure at infinity of linear delay systems and the row-by-row decoupling problem. In: Proceedings of the 7th IEEE Mediterranean Conference on Control and Automation, Haifa, Israel, June 28-30, 1999, pp. 1845–1854.(Submitted to System and Control Letters).

Sename O., Rabah R., Lafay J.-F. (1995), Decoupling without prediction of linear systems with delays: a structural approach. *Syst. Contr. Letters* **25**, 387–395.

Rekasius Z. V., Milzareck R. J. (1977), Decoupling without prediction of systems with delays. In *Proc. Joint. Automat. Control Conf.*, San Francisco.

Tzafestas S. G., Paraskevopoulos P. N. (1973). On the decoupling of multivariable control systems with time-delays, *Int. J. Control*, **17**, pp. 405-415.

A. C. Tsoi (1978). Recent advances in the algebraic system theory of delay differential equations. In: *Recent theoretical developments in control*, M. J. Gregson Ed., Academic Press, N. Y., pp. 67–127.

Wolovich W. A. (1975). Ouput feedback decoupling. *IEEE Transaction on Automatic Control*, **AC-20**, pp. 148–151.

www.elsevier.com/locate/ifac

STABILITY, CONTROL AND SMALL DELAYS

Jack K. Hale
School of Mathematics
Georgia Institute of Technology
Atlanta GA 30332
e-mail: hale@math.gatech.edu

Abstract. In the implementation of any stabilizing feedback control, it is very likely that time delays will occur. It is therefore important to understand the sensitivity of stability with respect to these delays. This paper is devoted to explaining the underlying mechanism for preserving stability for for such perturbations for neutral and retarded delay differential equations as well as difference equations. *Copyright © 2001 IFAC*

Keywords: Control (closed-loop), feedback stabilization, delay analysis.

A linear autonomous evolutionary equation in a Banach space Y is *strongly stabilizable* by a feedback control if the origin is exponentially stable when the control is applied instantaneously and this property is preserved under small variations in the time at which the feedback is implemented. There is a distinct difference between the situation when the feedback control acts on a finite dimensional subspace of Y and when it acts on an infinite dimensional subspace of Y. If the feedback control acts on an infinite dimensional subspace of Y, then it is possible that the system is not strongly stable; that is, there is a solution of the feedback system which does not approach zero exponentially for small deviations in the implementation time of the control. Using frequency domain techniques, Barman *et. al.* (1973) were one of the first to address this topic. A more recent and complete discussion has been given by Logemann and his colleagues (see Logemann (1998)). Hale *et. al.* (2001a, 2001b) have discussed this problem in a unified way by making use of the radius of the essential spectrum of the solution operator of the evolutionary equation. The feedback control can be chosen to act on a finite dimensional subspace of Y if this radius is less than one and it must act on an infinite dimensional subspace of Y otherwise. Results are presented on strong stabilization and nonstrong stabilization and give some applications to retarded and neutral functional differential equations and difference equations.

To explain more clearly, some definitions are required. If Y is a Banach space and $S : Y \to Y$ is a bounded linear operator, the *resolvent* set $\rho(S)$ of S is the set of elements λ of $\mathbb{C}$ such that $\lambda I - S$ has a bounded inverse. The *specturm* $\sigma(S)$ is $\mathbb{C} \setminus \rho(S)$. The *point spectrum* $P\sigma(S)$ is the set of $\lambda \in \sigma(S)$ with the property that the generalized eigenspace of λ,

$$\mathcal{M}_\lambda = \cup_{k\geq 1}\mathcal{N}((\lambda I - S)^k),$$

is finite dimensional, where $\mathcal{N}$ denotes null space. The *essential spectrum* $\sigma_e(S) = \sigma(S) \setminus P\sigma(S)$ of S is that part of the spectrum of S which cannot be removed by compact perturbations of S; that is,

$$\sigma_e(S) = \cap_{K\ compact}\, \sigma(S + K).$$

Recall that a compact operator S is one for which the closure of SB is compact for any bounded set $B \subset Y$. There are other definitions of the essential spectrum of S, but the important thing is that the radius $r(\sigma_e(S))$ of the essential spectrum of S is the same for all definitions.

If Y is finite dimensional, then any linear operator S on Y can be represented by a square matrix S. In this case, $\sigma(S) = P\sigma(S)$ is the set of eigenvalues of S.

If S is a compact operator on Y, then $\lambda \in \sigma(S)$, $\lambda \neq 0$, implies that $\lambda \in P\sigma(S)$. Therefore, $\{0\}$ is the only possible element of $\sigma_e(S)$. Of course, if Y is finite dimensional, then $\sigma_e(S) = \emptyset$. If Y is infinite dimensional, the $\sigma_\epsilon(S) = \{0\}$.

If $Sy = -y$ for all $y \in Y$ and Y is infinite dimensional, then $\sigma(S) = \sigma_e(S) = \{-1\}$ since every element of Y is an eigenvector for the eigenvalue -1, and, therefore, -1 is not an element of $P\sigma(S)$.

Let the evolutionary equation be defined on a Banach space Y and let $T(t) : Y \to Y$, $t \geq 0$, be the solution operator for $t \geq 0$; that is, for any $y \in X$, $T(t)y$, $t \geq 0$, is the solution which coincides with y at $t = 0$. Assume that $T(t)$ is a bounded linear operator for each t and is continuous in t. Since autonomy implies that $T(t + s) = T(t)T(s)$ for all $t \geq 0$, $s \geq 0$, the asymptotic behavior of solutions is determined by the asymptotic behavior of the iteratives of the map $S \equiv T(1) : Y \to Y$.

Several examples are now given to illustrate these concepts and, at the same time, to indictate the phenomenon that can occur with delays in a feedback control.

Example 1. For an ODE $\dot{x} = Ax$, $x \in I\!R^n$, the solution operator is e^{At}, $S = T(1) = e^A$, and $\sigma(S) = e^{\sigma(A)}$. The zero solution is exponentially stable if and only if $\lambda\sigma(A)$ implies that $Re\,\lambda < 0$.

Suppose that the ODE has been exponentially stabilized by the feedback control BFx; that is, the zero solution of the equation

$$\dot{x} = (A + BF)x \tag{1}$$

is exponentially stable. If there is a small delay in the feedback loop, then one must consider the delay differential equation

$$\dot{x}(t) = Ax(t) + BFx(t - \epsilon), \tag{2}$$

where $\epsilon > 0$ is a small constant. To obtain a solution of (2) for $t \geq 0$, an initial function must be specified on an interval which contains the interval $[-\epsilon, 0]$. Since ϵ varies and the initial space should be independent of ϵ, fix ϵ_0 and consider $\epsilon \in [0, \epsilon_0]$. The initial data is chosen from the space $C([-\epsilon_0, 0])$, $\epsilon_0 > 0$. Other spaces of initial data could be used (for example, $I\!R \times L^2(0, \epsilon_0)$) but the remarks below would still hold. For any initial data $y \in C([-\epsilon_0, 0])$, there is a unique continuous function $x(t, y)$, $t \geq -\epsilon_0$, which satisfies (2) for $t \geq 0$ and $x(\theta, y) = y(\theta)$, $\theta \in [-\epsilon_0, 0]$. The natural state space for (2) is $C([-\epsilon_0, o])$ and the solution operator of (2) should be a mapping on this space. This makes it very natural to let $(T_\epsilon(t)y)(\theta) = x(t + \theta, y)$, $\theta \in [-r, 0]$, $t \geq 0$, and refer to $T_\epsilon(t)$ as the solution operator on $C([-\epsilon_0, 0])$. For any $\epsilon \in [0, \epsilon_0]$, the solution $x(t, y)$ is continuously differentiable for $t \geq 0$ and the derivative is continuous in y. As a consequence of the Arzelá-Ascoli Theorem, the operator $T_\epsilon(t)$ is compact for $t \geq \epsilon_0$, $\epsilon \in [0, \epsilon_0]$. Therefore, if $S_\epsilon \equiv T_\epsilon(\epsilon_0)$, then S_ϵ is a compact operator with $\sigma_e(S_\epsilon) = \{0\}$. It is easy to see that $\mu \in P\sigma(S_\epsilon)$ if and only if $\mu = e^{\lambda \epsilon_0}$, where

$$det[\lambda I - A - BFe^{-\epsilon\lambda}] = 0.$$

Since $\sigma_e(S\epsilon) = \{0\}$, each element of $P\sigma(S_\epsilon)$ is continuous in ϵ and $\sigma(S_0) = \{0\} \cup \sigma(A + BF)$, we see that the system is strongly stabilizable by negative feedback.

Example 2. For the difference equation

$$x(t) - ax(t - 1) = 0$$

with $|a| \geq 1$, the origin is not exponentially stable. With a feedback $fx(t - 1)$, the origin of the feedback system

$$x(t) - (a + f)x(t - 1) = 0$$

is exponentially stable if $|a + f| < 1$. Notice that $|a| + |f| > 1$. If the feedback occurs at a time $1 - \epsilon$, then the difference equation

$$x(t) - ax(t - 1) - fx(t - 1 - \epsilon) = 0 \tag{3}$$

contains two delays. Now exponential stability is not obvious. One can show that there is a dense set U of ϵ such that, for any $\epsilon \in U$, there is a solution which is exponentially unbounded; that is, the system is not strongly stabilizable. An indication of the proof is now given. If $e^{\lambda t}$ is a solution of (3), $\lambda = \rho + i\tau$, then ρ, τ satisfy the characteristic equation

$$\begin{aligned} h(\rho, \tau, \epsilon) &\equiv ae^{-i\tau}e^{-\rho} + fe^{-i(1+\epsilon)\tau}e^{-(1+\epsilon)\rho} \\ &= 1. \end{aligned} \tag{4}$$

Therefore,

$$|a|e^{-\rho} + |f|e^{-(1+\epsilon)\rho} \geq 1. \tag{5}$$

The larger ρ, the smaller the left hand side of (5). Thus, to find solutions of (4) with the largest real part ρ, it is natural to try to choose τ so that $a^{-i\tau} = |a|$, $fe^{-i(1+\epsilon)\tau} = |f|$. This cannot be done exactly, but, if $1 + \epsilon$ is irrational, then, for any $\delta > 0$ and any $\theta_1, \theta_2 \in [0, 2\pi)$, Kronecker's approximation theorem implies that we can find τ such that

$$|\tau - \theta_1| < \delta, \quad |\tau(1 + \epsilon) - \theta_2| < \delta.$$

This shows that $sup\ Re\,\lambda = \rho_\epsilon > 0$, where

$$|a|e^{-\rho_\epsilon} + |f|e^{-(1+\epsilon)\rho_\epsilon} = 1.$$

Let U be the set of irrational numbers on $I\!R$. Since U is dense in $I\!R$, for every $\epsilon \in U$, there is an exponentially unstable solution of (3), proving the assertion.

The requirement that $1 + \epsilon$ be irrational is imposed to ensure that the function $h(\rho, \tau, \epsilon)$ in (4) is quasiperiodic in τ. This is not a restriction physically since one can prove that, if $\bar{\epsilon}$ is irrational, then $sup\,Re\lambda$ is continuous at $\bar{\epsilon}$.

Example 3. Consider the scalar neutral equation

$$\begin{aligned} &\frac{d}{dt}[x(t) + cx(t - 1) + dx(t - \alpha)] \\ &\qquad = ax(t - 1) + bx(t - \alpha), \end{aligned} \tag{6}$$

where a, b, c, d are constants. A function $x(t, y)$, $t \geq -4$, is a solution of (6) with initial data in $C([-r, 0])$, $r \geq max\{1, \alpha\}$, if $x(t, y)$ is defined and continuous for all $t \geq -r$ and the function

$$x(t) + cx(t - 1) + dx(t - \alpha)$$

is differentiable for $t > 0$ with a right hand derivative at $t = 0$ and (6) is satisfied for $t \geq 0$. Notice that the function $x(t)$ is not required to be differentiable. Define the solution operator $T_{a,b,c,d}(t)$ as in Example 1 and let $S_{a,b,c,d} = T_{a,b,c,d}(1)$.

Along with (6), consider the difference equation

$$x(t) - cx(t - 1) - dx(t - \alpha) = 0 \tag{7}$$

in the space $C_0 = \{y \in C : y(0) - cy(-1) - dy(-\alpha) = 0\}$. There is a unique solution of (7) with initial data $y \in C_0$. The solution operator $T_{c,d}(t)$, $t \geq 0$, is defined as in Example 1. Let $S_{c,d} = T_{c,d}(1)$. It is possible to show that

$$r(\sigma_e(S_{a,b,c,d})) = r(\sigma_e(S_{c,d})) \tag{8}$$

To see intuitively why this should be true, let

$$D_{c,d}x_t = x(t) - cx(t-1) - dx(t-\alpha)$$

and write (6) as

$$D_{c,d}x_t = D_{c,d}y + \int_0^t [ax(s-1)+bx(t-\sigma)]ds, \; t \geq 0.$$

This indicates that the integral term should correspond to a compact perturbation and therefore the essential spectrum should be determined by the nonhomogeneous difference equation

$$D_{c,d}x_t = D_{c,d}y.$$

There exists a $y_0 \in C([-r,0])$ such that $D_{c,d}y_0 = 1$. If $x_t = z_t + y_0 D_{c,d}y$, then $D_{c,d}z_t = 0$. Since the operator $y_0 D_{c,d}$ on $C([-r,0])$ is compact, the result is proved.

For (6) to be exponentially stable with variations in the delays, it is necessary that $|\mu| < 1$ for $\mu \in \sigma_e(S_{a,b,c,d})$. If this is true, then the largest growth rate of solutions is determined by elements of $P\sigma(S_{a,b,c,d})$ and these are continuous in the delays. It is possible to show that $\sigma_e(S_{c,d}) = \sigma(S_{c,d})$. From (8), if the neutral equation is exponentially stable for given delays, then it is exponentially stable for small variations in the delay if and only if the difference equation (7) enjoys the same property.

Using an argument similar to the one in Example 2, the origin of the difference equation is exponentially stable under variations in the delays if and only if $|c| + |d| < 1$. If $\tau = 1 + \epsilon$ and $dx(t-1-\epsilon)$, $d \neq 0$, $bx(t-1-\epsilon)$, $b \neq 0$, represent the feedback control terms, then the neutral equation is strongly stabilizable if and only if $|c| + |d| < 1$. Therefore, $|c| < 1$ and the uncontrolled system has the corresponding difference equation exponentially stable.

Example 4. Consider the linear wave equation

$$w_{tt} - w_{xx} = 0, \quad 0 < x < 1, \quad t > 0, \tag{9}$$

with the boundary conditions

$$w(0,t) = 0, \quad w_x(1,t) = -kw_t(1,t-r), \tag{10}$$

where $k > 0$, $r \geq 0$, are constants. System (9), (10) represents a boundary control problem with the feedback control function $kw_t(1,t-r)$.

Let $Y_0 = H^1_{bc}(0,1) \times L^2(0,1)$, where the subscript 'bc' denotes the homogeneous Dirichlet boundary condition at $x = 0$ and the Neumann boundary condition at $x = 1$. For $r = 0$ and initial data $(w_0, w_1) \in Y_0$, there is a unique solution $(w, w_t) \in Y_0$, $t \geq 0$, with $w(x,0) = w_0(x)$, $w_t(x,0) = w_1(x)$. Let $T_k(t)$ denote the solution operator.

For $k = 0$ and each $t > 0$, the radius of the essential spectrum of $T_0(t)$ is equal to 1. It also is not difficult to see that the origin is exponentially stable if $k > 0$; that is, the system is exponentially stabilizable with negative feedback if the control is implemented without a delay. The feedback function corresponds to a bounded perturbation (not a compact one) and it is to be expected that the stability properties may be sensitive to the implementation of the feedback with a small delay. This is actually the case.

Fix $h > 0$, let $r \in [0,h]$ and define

$$Y_h = C([-h,0], H^1_{bc}(0,1)) \times C([-h,0], L^2(0,1))$$

For any $(y_0, y_1) \in Y_h$, there is a unique solution w, w_t, $t \geq -h$, with initial data $(y_0, y_1) \in Y_h$. If $T_h(t)(y_0,y_1)(\theta)(w(t+\theta), w_t(t+\theta))$, $\theta \in [-h,0]$, then $T_h(t)$ is the solution operator for (9), (10). A function $w(x,t) = e^{\lambda t}v(x)$ is a solution of (9), (10) if and only if λ satisfies the equation

$$e^{\lambda} + e^{-\lambda} = -k(e^{\lambda} - e^{\lambda})e^{-\lambda r}$$

or, equivalently,

$$1 + ke^{-\lambda r} - ke^{-\lambda(2+r)} + e^{-2\lambda} = 0. \tag{11}$$

If there is a λ satisfying (11) with positive real part, then the introduction of the delay destabilizes the system and it is not strongly stabilizable by negative feedback. The situation is actually very bad as we state in the following

Proposition 1. *For system (9), (10), the following statements hold:*

(i) If $r = 0, k > 0$, the origin is exponentially stable.

(ii) If $r > 0, 0 < k < 1$, the origin is exponentially stable.

(iii) If $r > 0$, $k > 1$, there is a dense set of r for which the origin is exponentially unstable. Also, for any $\rho > 0$, there is an $r(\rho) \to 0$ as $\rho \to \infty$ such that there is a solution which becomes unbounded at the rate $e^{\rho t}$ as $t \to \infty$.

Notice that the characteristic equation (11) is the same one that would have been obtained for the difference equation

$$x(t) + kx(t-r) - kx(t-2-r) + x(t-2r) = 0.$$

It is not surprising that the essential spectrum of the original problem is determined by a difference equation. In fact, it is always possible to transform the problem of the wave equation (even with nonlinear boundary conditions) to a neutral differential equation (see Hale *et. al.* (1993) for references).

Proposition 1 indicates that using boundary control on the wave equation is not the proper physical model. The control function is required to stabilize an infinite number of modes. More effort should be devoted to develop models which put less emphasis on high frequency modes. It has been noted by Morgül (1995, 1998) that such an approach is possible as shown in the next example.

Example 5. Consider the strongly damped wave equation

$$w_{tt} - w_{xx} - cw_{xxt} = 0, \quad 0 < x < 1, \quad t > 0, \quad (12)$$

with the boundary conditions

$$\begin{aligned} w(0,t) &= 0, \\ w_x(1,t) + cw_{xt}(1,t) &= -kw_t(1,t-r), \end{aligned} \quad (13)$$

where $c > 0$, $k > 0$, $r \geq 0$, are constants. This equation takes into account the rate of change of strain and the control at $x = 1$ is applied to the total strain.

If $C_h = C([-h,0], H^1_{bc}(0,1))$ and $Y_h = C_h \times C_h$, then it is possible to follow the same type of argument as in Massatt (1983) to show that the system (12), (13) has a unique solution $\tilde{T}_{r,k}(t)(w_0, w_1)$, $t \geq 0$, for every $(w_0, w_1) \in Y_h$. In this setting, the boundary control function $-kw_t(1, t-r)$ is a compact perturbation. Therefore,

$$r(\sigma_e(\tilde{T}_{r,k}(t))) = r(\sigma_e(T_0(t))),$$

where $T_0(t)$ is the solution operator on $H^1_{bc}(0,1) \times H^1_{bc}(0,1)$ of the equation

$$w_{tt} - w_{xx} - cw_{xxt} = 0, \quad 0 < x < 1, \quad t > 0, \quad (14)$$

with the boundary conditions

$$w(0,t) = 0, \quad w_x(1,t) + cw_{xt} = 0. \quad (15)$$

The eigenvalues of (14), (15) can be computed directly. In fact, if one observes that $\lambda_k = (2k+1)^2\pi^2$ are the eigenvalues of

$$v_{xx} = -\lambda x, \quad v(0) = 0, \ v_x(1) = 0,$$

then the eigenvalues μ of (14), (15), are the solutions of the equation

$$\mu^2 + c\lambda_k\mu + \lambda_k = 0. \quad (16)$$

One can show that the solutions of (16) have negative real parts and that the only accumulation point is $-1/c$. The points not accumulating at $-1/c$ approach $-\infty$ as $k \to \infty$. Therefore,

$$r(\sigma_e(T_0(1))) = e^{-1/c} < 1.$$

Therefore, if one chooses the control $-kw_t(1,t)$ in such a way that the origin is exponentially stable, then it will remain so for the implementiation of the control at time $t - r$ for r small.

Similar difficulties occur with the beam equation (see Desch *et. al.* (1989)). If one introduces an additional term involving internal damping (see Sakawa (1984)), the system becomes stronly stabilizable.

Some general results are now presented on strong stabilization of NDDE (neutral delay differential equations) and RDDE (retarded delay differential equations). The complete proofs for even more general situations are contained in Hale *et. al.* (1993, 2001a, 2001b).

For any $h \geq 0$, let $Y_h = C([-h,0], I\!R^n)$ be the space of continuous functions from $[-h,0]$ to $I\!R^n$. For any continuous $x : [-h,0] \to I\!R^n$ and any $t \in [0,\infty)$, let $x_t \in Y_h$ be defined by $x_t(\theta) = x(t+\theta)$, $\theta \in [-h,0]$. Let $r = \{r_j, 1 \leq j \leq N\}$, $0 < r_j \leq h$, $A = \{A_j, 0 \leq j \leq N\}$, $B = \{B_j, 1 \leq j \leq N\}$, where each A_j, B_j is an $n \times n$ real matrix. Also, define the continuous linear operators $D_{r,B}, L_{r,A} : Y_h \to I\!R^n$ by

$$\begin{aligned} D_{r,B}y &= y(0) - \Sigma_{j=1}^N B_j y(t - r_j), \\ L_{r,B}y &= A_0 y(0) + \Sigma_{j=1}^N A_j y(t - r_j), \end{aligned} \quad (17)$$

for each $y \in Y_h$.

A NDDE is

$$\frac{d}{dt} D_{r,B}x_t = L_{r,A}x_t. \quad (18)$$

Given $y \in Y_h$, a solution of (18) is a function defined and continuous for $t \geq -h$, $x_0(\cdot, y) = y$, $Dx_t(\cdot, y)$ is continuously differentiable for $t > 0$ with a continuous right hand derivative at $t = 0$ and satisfies (18) for $t \geq 0$. Notice that it not required that the function $x(t,y)$ be differentiable. It is not difficult to prove the existence and uniqueness of a solution $x(t,y)$. The operator $T_{r,A,B}(t) : Y_h \to Y_h$ defined by $T_{r,A,B}(t)y = x_t(\cdot, y)$ is called the solution operator of (18).

For $B = 0$, we have a RDDE,

$$\dot{x}(t) = L_{r,A}x_t. \quad (19)$$

For linear equations, the assumption that all solutions approach zero as $t \to \infty$ has strong implications, in constrast to nonlinear equations.

Theorem 1. *If each solution of the linear NDDE approaches zero as $t \to \infty$, then*

(i) the origin is stable,

(ii) for any compact set $K \subset Y_h$ and any $\epsilon > 0$, there is a t_0 such that $|x_t(\cdot, y)| < \epsilon$ for $t \geq t_0$ and all $y \in K$; that is, the origin uniformly attracts compact sets of Y_h.

(iii) The set $\{x_t(\cdot, y), t \geq 0, y \in U\}$ is bounded if U is bounded.

For a RDDE and some NDDE, one can say even more.

Theorem 2. *For a RDDE, if each solution approaches zero as $t \to \infty$, then the origin is exponentially stable; that is, there are positive constants K, α such that*

$$|x_t(\cdot, y)| \leq Ke^{-\alpha t}|y|, \quad t \geq 0, \ y \in Y_h. \qquad (20)$$

The same conclusion is true for NDDE provided that $r(\sigma_e(T_{r,A,B}(1))) < 1$.

The conclusions in Theorem 2 are rather easy to prove since the hypotheses imply that the radius of the spectrum of the solution operator is less than 1.

These two results make it natural to see if there is a NDDE for which every solution approaches zero as $t \to \infty$ and the origin is not exponentially stable. If $c > 0$, it is an interesting exercise to show that the equation

$$\frac{d}{dt}[x(t) + x(t-1)] = -cx(t)$$

has this property if $c > 0$.

Along with a NDDE, consider the difference equation

$$D_{r,B}x_t = 0 \qquad (21)$$

with initial data in the space $Y_{h,D} = \{y \in Y_h : D_{r,B}y = 0\}$. Let $\tilde{T}_{r,B}(t)$ be the solution operator for (21) on $Y_{h,D}$. Following the spirit outlined in Example 3, one can prove the following important result relating the essential spectral radii of the solution operator of (18) to the one for (21).

Theorem 3.

$$r(\sigma_e(T_{r,A,B}(1))) = r(\sigma_e(\tilde{T}_{r,B}(1))). \qquad (22)$$

Consider the control problem

$$\begin{aligned}\frac{d}{dt}[D_{r,B}x_t - \Sigma_{j=1}^N C_j u_1(t - r_j)] = \\ Lx_t + G_0 x(t) + \Sigma_{j=1}^N G_j u_2(t - r_j)\end{aligned} \qquad (23)$$

and assume that u_1, u_2 are feedback controls given by

$$u_1(t) = Fx(t), \quad u_2(t) = Gx(t). \qquad (24)$$

Suppose that the controlled system has been exponentially stabilized. From Theorem 3, one obtains

Theorem 4. *The feedback control system (23), (24) is strongly stabilizable if and only the feedback controlled difference equation*

$$x(t) - \Sigma_{j=1}^N (B_j - C_j F)x(t - r_j) = 0 \qquad (25)$$

is strongly stabilizable.

If the difference equation (21) is not exponentially stable, then the following result shows it is of no advantage to use the feedback control u_1 in (23) (compare with Logeman *et. al.* (1996)).

Theorem 5. *Suppose that (21) is not exponentially stable (resp., exponentially unstable). If the control system (25) is exponentially stabilized by a feedback control $u(t) = Fx(t)$, then there exists a dense set E of $I\!R$ such that for $\epsilon_j \in E$, $j = 1, 2, \ldots, N$, the closed loop control system*

$$\begin{aligned}x(t) - \Sigma_{j=1}^N B_j x(t - r_j) \\ -\Sigma_{j=1}^n C_j Fx(t - r_j - \epsilon_j) = 0\end{aligned} \qquad (26)$$

is not exponentially stable (resp., exponentially unstable).

It is clear from the previous discussion that it is very important to determine the radius of the spectrum of the solution operator for difference equations. This is not an easy task. The following result of Hale *et. al.* (2001b) has proved to very effective and is a slight generalization of previous results (see Hale *et. al.* (1993) for references and Kaashoek *et. al.* (1994) for additional results).

Theorem 6. *For the difference equation (21) and $D_{r,B}$ given in (17), define*

$$\gamma_0 = \max_{\theta_j \in [0, 2\pi], 1 \leq j \leq N} r(\sigma(\Sigma_{j=1}^N e^{i\theta_j} B_j) \qquad (27)$$

If the components of $r = \{r_j, 1 \leq j \leq N\}$ are rationally independent, then (21), (17) is exponentially stable if and only if $\gamma_0 < 1$. Furthermore, if $\gamma_0 > 1$, then (21), (17) is exponentially unstable.

Notice that, for a scalar difference equation,

$$x(t) - \Sigma_{j=1}^N b_j x(t - r_j) = 0, \qquad (28)$$

we have

$$\gamma_0 = \Sigma_{j=1}^N |b_j|$$

Therefore, (28) is exponential stable for variations in the delays if and only if the sum of the absolute values of the coefficients is < 1.

Corollary 1. *Exponential stability of (18), (17) is preserved under small variations in the delays if and only if* $\gamma_0 < 1$, *where* γ_0 *is defined in (27).*

Theorem 6 was used to verify the statements in the previous examples regarding difference equations.

References

Barman, J.F., F.M. Callier and C. Desoer (1973) L^2-stability and L^2-instability of linear time invariant distributed feedback systems perturbed by a small delay in the loop. *IEEE Trans. Automat. Control*, **18**, 479-484.

Desch, W. and R.L. Wheeler, (1989). Destabilization due to delay in one dimensional feedback. *Int. Series Num. Math.* **91**, 61-83.

Hale, J.K. and S.M. Verduyn-Lunel (1993). *An Introduction to Functional Differential Equations.* Springer-Verlag, New York.

Hale, J.K. and S.M. Verduyn-Lunel (2001a). Effects of small delays on stability and control. *Operator Theory; Advances and Applications*, **122**, 275-301.

Hale, J.K. and S. Verduyn-Lunel (2001b). Strong stabilization of neutral functional differential equations. *IMA J. Math. Control Inform.*.

Kaashoek, M.A. and S.M. Verduyn-Lunel (1994). An integrability condition on the resolvent for hyperbolicity of the semigroup. *J. Differential Equations*, **112**, 374-406.

Logemann, H. (1998). Destabilization effects of small time delays on feedback controlled descriptor systems. *Linear Algebra Appl.*, **272**, 131-152.

Logemann, H. and S Townley (1996). The effect of small delays in the feedback loop on the stability of neutral equations. *Systems and Control Letters*, **27**, 267-274.

Massatt, P. (1983). Limiting behavior for strongly damped wave equations. *J. Differential Equations* **48**, 334-349.

Morgül, Ö. (1995). On stabilization and stability robustness against small delays of some damped wave equations. *IEEE Trans. Automat. Control*, **40**, 1626-1630.

Morgül, Ö. (1998). Stabilization and disturbance rejection for the wave equation. *IEEE Trans. Automat. Control*, **43**, 89-95.

Sakawa, Y. (1984). Feedback control of second order evolution equations with damping. *SIAM J. Cont. Opt.* **22**, 343-361.

ON CLOSED-LOOP STABILITY FOR MECHANICAL SYSTEMS WITH INPUT DELAYS

Dan Ivănescu * **Rogelio Lozano** **
Silviu-Iulian Niculescu **

* *Eurocontrol Experimental Centre, BP 15, 91222, Bretigny sur Orge, France, email:* Dan.Ivanescu@eurocontrol.int
** *HEUDYASIC, Université de Technologie de Compiègne, BP 529, 60205, Compiègne, FRANCE, Fax: (33) 3.44.23.44.77*
email: rlozano,silviu@hds.utc.fr

Abstract: This paper focuses on some closed-loop *delay-dependent* asymptotic stability problems for a large class of nonlinear underactuated mechanical systems subject to *delayed inputs*. Upper bounds on the delays size such that the closed-loop system stability is guaranteed are obtained in terms of some matrix inequalities. Using a time-domain approach, we show how to compute *quadratic* Liapunov functionals and functions for the corresponding analysis. *Copyright © 2001 IFAC*

Keywords: Nonlinear systems, delay systems, Liapunov functional.

1. INTRODUCTION

Delay systems represent a class of infinite-dimensional systems largely used to describe propagation and transport phenomena as well as heredity or population dynamics (see for instance Kolmanovskii and Myshkis 1992 and the references therein). Futhermore, it is well known that the existence of a delay in a physical model may induce instabilities or bad performances.

One of the easiest ways to control a system subject to delay inputs is to find the maximal bound on the delay size such that a stabilizing control law for the system free of delay can stabilize the delayed one. If this method was largely used for linear delay systems (see, for instance Niculescu *et al.* 1997, Ivanescu 2001 and the references therein), the nonlinear mechanical case is almost not considered in the literature.

In the sequel, we will show that for underactuated systems a delayed control can be used to stabilize the angular speed or, under some assumptions, the angular speed and also the angular position.

The stabilization algorithms applied to the delayed closed-loop systems are original, but related to the works of Ailon and Gil 2000 (who considered a linearized delay system) or Fantoni 2000 (for nonlinear mechanical models free of delays). The control inputs as well the convergence analysis are based on Liapunov theory and Barbalat's lemma.

The paper is organized as follows: some preliminary results are presented in section 2. The angular velocity control, and angular position and angular velocity control are given in Section 3. Some final remarks end the paper.

Throughout the paper we denote $\dot{q}_t(\theta) = \dot{q}(t+\theta)$. In the sequel, when there is no confusion, we will drop the explicit time dependence of variables on t for brevity.

2. PRELIMINARY

As we shall see, the stability results will be obtained using stability arguments based on the Barbalat's lemma. Note that this result is well known in control literature in the passivity theory (see Lozano *et al.* 2000 and the references therein).

Theorem 1. (Barbalat's lemma) Let f be a real function defined, and uniformly continuous on $[0, \infty)$ such that

$$\lim_{t\to\infty} | \int_0^t f(\theta)d\theta| < +\infty$$

Then: $\lim_{t\to\infty} f(t) = 0$

A natural consequence is the following:

Corollary 1. Let f be a nonnegative function defined on $[0, \infty)$ such that f is integrable on $[0, \infty)$, and uniformly continuous on the same interval. Then: $\lim_{t\to\infty} f(t) = 0$.

3. MAIN RESULTS

We present here two approaches based on Liapunov techniques for nonlinear systems combined with delay systems theory. In the case of non-delayed systems LaSalle's invariance theorem is used to prove the asymptotic stability. For delayed non-autonomous systems we make use of Barbalat's lemma and the Krasovskii theorem to prove that the system's trajectories asymptotically converge to a homoclinic orbit.

In the first case, for a mechanical nonlinear system model, we use a very simple control law such that the system's angular velocity converges to zero but does not necessarily bring the system to a desired angular position. The second approach, under some assumptions is such that the system's angular position converges also to zero.

3.1 *Angular velocity control*

We start with the first approach which concerns a general model of an n-degree of freedom rigid robot:

$$D(q)\ddot{q} + C(q, \dot{q})\dot{q} = u(t-\tau) \qquad (1)$$

where $q \in R^n$ represents the link angles, $D(q)$ is the inertia matrix (supposed to be symmetric with $D(q) > 0, \forall q$), $C(q, \dot{q})$ represents the Coriolis and centrifugal forces and $u \in R^n$ is the vector of the applied torques. An important property for such systems is that $\dot{D} - 2C$ is a skew symmetric matrix, i.e. $x^T(\dot{D} - 2C)x = 0 \forall x$.

This class of systems is used in modeling teleoperated manipulators Lee and Li, 2001, or mechanical systems moving in a horizontal plane having no potential energy (planar flexible-joint robot), see Fantoni, 2000 and the references therein. No gravitational forces are applied on the system.

Problem Statement

Consider the system (1), in closed-loop with the following controller:

$$u(t-\tau) = -K\dot{q}(t-\tau) \qquad (2)$$

where K is the (stabilizing) control gain matrix to be computed. We are interested in establishing conditions that allow the asymptotic stability of the angular velocity $\dot{q}$ but for delayed input (2) and find a bound τ^* on the delay τ such that this asymptotic stability is preserved for any delay $0 < \tau < \tau^*$. Here, the delay τ is supposed to be constant. Note that one can use it also as a *design parameter.* Then we have:

Theorem 2. The angular velocity for the system (1) in the presence of the delay τ can be controlled by the following feedback law : $u(t-\tau) = -K\dot{q}(t-\tau)$ for any delay τ and for any $K > 0$ satisfying the inequality:

$$K - \tau^*(\frac{1}{2}C^T D^{-1} C + \frac{3}{2} K D^{-1} K^T) > 0 \quad (3)$$

Proof: Let us define the energy of the system as its kinetic energy, namely $E = \frac{1}{2}\dot{q}^T(t)D(q)\dot{q}(t)$. Starting from the energy Liapunov function, we add more general quadratic forms which correspond to the delayed terms in the system. Finally consider $u(t) = -\dot{q}(t-\tau)$, and consider the following functional: $V(t, \dot{q}_t) = V_1 + V_2 + V_3$ where

$$V_1 = E = \frac{1}{2}\dot{q}^T(t)D(q)\dot{q}(t)$$

$$V_2(q, \dot{q}_t, t) = \frac{1}{2}\int_{-\tau}^{0}\left(\int_{t+\theta}^{t} \dot{q}(\xi)^T S_1 \dot{q}(\xi)d\xi\right)d\theta$$

$$V_3(q, \dot{q}_t, t) = \frac{1}{2}\int_{-2\tau}^{-\tau}\left(\int_{t+\theta}^{t} \dot{q}(\xi)^T S_2 \dot{q}(\xi)d\xi\right)d\theta$$

It is easy to verify that

$$\beta_1 ||\dot{q}||^2 \le V(\dot{q}_t, t), \qquad (4)$$

where $\beta_1 = \inf_q(\lambda_{min}(D)) > 0$.

Denote $\dot{V}(t, \dot{q}_t)$ the derivative of V. The derivative of V_1 :

$$\begin{aligned}\dot{V}_1(\dot{q}_t, t) &= \dot{q}^T D\ddot{q} + \frac{1}{2}\dot{q}^T \dot{D}\dot{q} \\ &= \dot{q}^T(u(t-\tau) - C\dot{q}) + \frac{1}{2}\dot{q}^T\dot{D}\dot{q} \\ &= -\dot{q}(t)^T K\dot{q}(t-\tau) \\ &= -\dot{q}^T K(\dot{q} - \dot{q} + \dot{q}(t-\tau))\end{aligned}$$

$$= -\dot q(t)^T K\dot q(t) + \dot q^T(t)K\int_{-\tau}^{0}\ddot q(t+\theta)d\theta$$

The derivatives of $V_2(\dot q_t, t)$ and $V_3(\dot q_t, t)$ are:

$$\dot V_2 = \tau\frac{1}{2}\dot q^T(t)S_1\dot q(t) - \frac{1}{2}\int_{-\tau}^{0}\dot q^T(t+\theta)S_1\dot q(t+\theta)d\theta$$

$$\dot V_3 = \tau\frac{1}{2}\dot q^T(t)S_2\dot q(t)$$
$$-\frac{1}{2}\int_{-\tau}^{0}\dot q^T(t+\theta-\tau)S_2 x(t+\theta-\tau)d\theta \quad (5)$$

Since $D(q(0)) > 0$ it follows:

$$\ddot q(\theta) = -D(q(\theta))^{-1}C(q(\theta))\dot q$$
$$-D(q(\theta))^{-1}K\dot q(\theta-\tau)). \quad (6)$$

With (6), the derivative of V_1 becomes:

$$\dot V_1(q, q_t, t) = -\dot q(t)^T K\dot q(t)$$
$$+\dot q^T(t)K\int_{-\tau}^{0}(-D(q(t+\theta))^{-1}C(q(t+\theta))\dot q(t+\theta))d\theta$$
$$+\dot q^T(t)\int_{-\tau}^{0}(-D(q(t+\theta))^{-1}K\dot q(t+\theta-\tau))d\theta. \quad (7)$$

To bound the integral terms we use the well-known inequality:

$$-2a^T b \le inf\{a^T X a + b^T X^{-1} b\},\ X = X^T > 0. \ (8)$$

We applied (8) and we obtain that, for any positive matrices R_1 and R_2:

$$-\dot q^T(t)K\int_{-\tau}^{0}D(q(t+\theta))^{-1}C(q(t+\theta))\dot q(t+\theta)d\theta$$
$$< \frac{1}{2}\int_{-\tau}^{0}[\dot q(t+\theta)^T C^T D^{-T}R_1 D^{-1}C\dot q(t+\theta)$$
$$+\dot q^T(t)KR_1^{-1}K^T\dot q(t)]d\theta = \tau\frac{1}{2}\dot q^T(t)KR_1^{-1}K^T\dot q(t)$$
$$+\frac{1}{2}\int_{-\tau}^{0}\dot q(t+\theta)^T C^T D^{-T}R_1 D^{-1}C\dot q(t+\theta)d\theta. \quad (9)$$

$$-\dot q^T(t)K\int_{-\tau}^{0}D^{-1}K(q(t+\theta)\dot q(t+\theta-\tau))d\theta$$
$$< \frac{1}{2}\int_{-\tau}^{0}[\dot q(t+\theta-\tau)^T K^T D^{-T}R_2 D^{-1}K\dot q(t+\theta-\tau)$$
$$+\dot q^T(t)KR_2^{-1}K^T\dot q(t)]d\theta = \tau\frac{1}{2}\dot q^T(t)R_2^{-1}\dot q(t)$$
$$+\frac{1}{2}\int_{-\tau}^{0}[\dot q(t+\theta-\tau)^T D^{-T}R_2 D^{-1}\dot q(t+\theta-\tau)d\theta. \quad (10)$$

If one chooses: $S_2(q) = K^T D^{-T}(q)R_2 D^{-1}(q)K$ and $S_1(q) = C^T D^{-T}R_1 D^{-1}C$ we obtain that the integral terms in (9) and (10) are identically with the integral terms of $\dot V_2$ and $\dot V_3$. With this choice, the derivative of V becomes:

$$\dot V = \dot V_1 + \dot V_2 + \dot V_3 < -\dot q^T(t)Kq(t)$$
$$+\tau\frac{1}{2}\dot q^T(t)KR_1^{-1}K^T\dot q(t) + \tau\frac{1}{2}\dot q^T(t)C^T D^{-T}R_1 D^{-T}C\dot q(t)$$
$$+\tau\frac{1}{2}\dot q^T(t)KR_2^{-1}K^T\dot q(t)$$
$$+\tau\frac{1}{2}\dot q^T(t)K^T D^{-T}(q)R_2 D^{-1}K(q)\dot q(t)$$

For simplicty we consider $R_1 = R_2 = D$:

$$\dot V < -\dot q^T(t)Kq(t) + \tau\frac{1}{2}\dot q^T(t)(C^T D^{-1}C + 3KD^{-1}K^T)\dot q(t)$$
$$= -\dot q^T(t)(K - \tau M)\dot q(t), \quad (11)$$

with $M = (\frac{1}{2}C^T D^{-1}C + \frac{3}{2}KD^{-1}K^T) > 0$. We prove now the asymptotic stability using Barbalat's lemma. Indeed (11) leads to $\dot V(t) \le 0$ if $K - \tau M > 0$ and to the following inequality:

$$\dot q^T(t)D\dot q(t) \le 2V(t) \le V(0) \quad (12)$$

since V is decreasing. In conclusion, any arbitrary non-trivial solution is uniformly bounded on $[0, +\infty)$ if $K - \tau M > 0$, i.e,

$$K - \tau M(q, \dot q) > \zeta I, \forall q, \dot q$$

It follows also that $\ddot q$ is also uniformly bounded on the same interval, implying that $\dot q$ is uniformly continuous on $[0, +\infty)$. Now, by integrating in 'θ' variable (11) we obtain:

$$V(t) - V(0) \le -\int_0^t \dot q^T(\theta)(K - \tau M)\dot q(\theta)d\theta$$

$$V(t) + \zeta\int_0^t \dot q^T(\theta)\dot q(\theta)d\theta$$

$$\le V(t) + \int_0^t \dot q^T(\theta)(K - \tau M)\dot q(\theta)d\theta \le V(0)$$

and $||\dot q||^2 \in L_1[0, \infty)$. By applying Barbalat lemma we obtain:

$$\lim_{t\to\infty}||\dot q(t)||^2 = 0, \quad (13)$$

which implies

$$\lim_{t\to\infty}||\dot q(t)||^2 = 0, \quad (14)$$

that is the asymptotic stability for the system's angular speed.

3.2 *Control of the angular position and angular speed*

In this section we will propose another approach based on Liapunov time domain analysis, using a similar strategy to the one developed in the previous section. This controller allows also, under some assumptions, the asymptotic stability for the angular position q of the system.

Consider the nonlinear system with delay input:

$$D(q)\ddot{q}(t) + C(q,\dot{q})\dot{q}(t) + G(q) = u(t-\tau) \quad (15)$$

This time, the controller will depend by q and $\dot{q}$. We consider the following feedback law:

$$u(t) = -S\dot{q}(t) - Sq(t) \quad (16)$$

where S is the (stabilizing) control gain matrix to be computed.

Then we have the following result:

Theorem 3. Consider the system (15) and consider the controller input given by (16).

Then, in the closed-loop, the angular position q and the angular velocity $\dot{q}$ converge to zero for all the delays τ with $0 < \tau < \tau^*$, if there exists S, Y and R positive matrices such that the following inequlities are verified:

$$\begin{bmatrix} -S + \tau^*\mathcal{E} + M^T R^{-1} M & \tau^*\mathcal{E} \\ \tau^*\mathcal{E}^T & -S - G + R + \tau^*\mathcal{E} \end{bmatrix} < 0 \quad (17)$$

$$\begin{bmatrix} \frac{1}{2}D(q) & \frac{1}{4}D(q) \\ \frac{1}{4}D(q) & S \end{bmatrix} > 0 \quad (18)$$

where $\mathcal{E} = SY^{-1}S + Y$ and $M = (\frac{1}{2}\dot{D} - C)$.

Remark 1. Since $D(q)$ is a positive matrix for all q, it follows that there always exists a symmetric and positive-definite matrix S such that the second matrix inequality above holds.

Remark 2. It is obviously that, when the system is free of delay there exist always two positive matrices $S > 0$ and $R > 0$ such that

$$\begin{bmatrix} -S & M^T \\ M & -R \end{bmatrix} < 0$$
$$-S - G + R < 0$$

Proof Let us propose the following Liapunov functional:

$$V(q,\dot{q},t) = V_1 + V_2 \quad (19)$$

where

$$V_1(q,\dot{q}) = \begin{bmatrix} \dot{q}^T & q^T \end{bmatrix} \begin{bmatrix} \frac{1}{2}D(q) & L^T(D(q)) \\ L(D(q)) & S \end{bmatrix} \begin{bmatrix} \dot{q} \\ q \end{bmatrix} + P(q),$$

$$V_2(q,\dot{q}_t,t) = \frac{1}{2}\int_{-\tau}^{0}\left(\int_{t+\theta}^{t} \dot{r}(\xi)^T X \dot{r}(\xi) d\xi\right) d\theta \quad (20)$$

with L such that $\begin{bmatrix} \frac{1}{2}D(q) & L^T(D(q)) \\ L(D(q)) & S \end{bmatrix} > 0.$

Here $D(q) = D^T(q) > 0$, S, and X are constant and positive matrices, with $r(t) = q(t) + \dot{q}(t)$. We consider the special case when $L(D(q)) = \frac{1}{4}D(q)$, which leads to the second matrix inequality in the statement of the theorem. P(q) is the potential energy. Note that P is related to $G(q)$ as follows:

$$G(q) = \frac{\delta P}{\delta q}. \quad (21)$$

In the sequel, we assume a linear form of $G(q)$:

$$G(q) = Gq$$

with G a positive matrix.

Taking the derivative of V_1 we obtain:

$$\begin{aligned} \dot{V}_1 &= \dot{q}^T D\ddot{q} + \frac{1}{2}\dot{q}^T(t)\dot{D}\dot{q} + \dot{q}^T Sq \\ &+ q^T S\dot{q} + q^T D\ddot{q} + \frac{1}{2}q^T\dot{D}\dot{q} + \dot{q}^T G(q) \\ &= \dot{q}^T(u - G(q) - C\dot{q}) + \frac{1}{2}\dot{q}^T\dot{D}\dot{q} \\ &+ \dot{q}^T Sq + q^T S\dot{q} + q^T(u - G - C\dot{q}) + \frac{1}{2}q^T\dot{D}\dot{q} + \dot{q}^T G(q) \end{aligned}$$

Since $\frac{1}{2}\dot{D} - C$ is a skew symmetric matrix we obtain that

$$\begin{aligned} \dot{V}(q,\dot{q},t) = &\dot{q}^T u + q^T u - q^T G(q) \\ &+ \dot{q}^T Sq + q^T S\dot{q} + q^T M\dot{q} \end{aligned} \quad (22)$$

with $M = (\frac{1}{2}\dot{D} - C)$.

We replace in equation (22) the input u by its form (16):

$$\begin{aligned} \dot{V}_1(q,\dot{q}) &= -\dot{q}^T S\dot{q}(t-\tau) + \dot{q}^T Sq - \dot{q}^T Sq(t-\tau) \\ &-q^T S\dot{q}(t-\tau) - q^T Sq(t-\tau) + q^T S\dot{q} \\ &+q^T M\dot{q} - q^T G(q). \\ &= -\dot{q}^T S\dot{q}(t-\tau) + \dot{q}^T S(q - q(t-\tau)) \\ &-q^T Sq(t-\tau) \\ &+q^T S(\dot{q} - \dot{q}(t-\tau)) + q^T M\dot{q} - q^T G(q). \\ &= -\dot{q}^T S\dot{q} + \dot{q}^T S\int_{-\tau}^{0}\ddot{q}(t+\theta)d\theta \\ &+\dot{q}^T S\int_{-\tau}^{0}\dot{q}(t+\theta)d\theta - q^T Sq \end{aligned}$$

$$+q^T S \int_{-\tau}^{0} \ddot{q}(t+\theta)d\theta + \dot{q}^T S \int_{-\tau}^{0} \ddot{q}(t+\theta)d\theta$$
$$+q^T M\dot{q} - q^T G(q) \quad (23)$$
$$= -\dot{q}^T S\dot{q} - q^T Sq + r^T S \int_{-\tau}^{0} \dot{r}(t+\theta)d\theta$$
$$+q^T M\dot{q} - q^T G(q)$$

Derivative of V_2 is:

$$\dot{V}_2 = \frac{1}{2}\tau r^T(t)Xr(t)$$
$$-\frac{1}{2}\int_{-\tau}^{0} \dot{r}^T(t+\theta)X\dot{r}(t+\theta)d\theta \quad (24)$$

From the inequalities (23) and (24) and for $r = q + \lambda\dot{q}$, the derivative of V becomes:

$$\dot{V} = \dot{V}_1 + \dot{V}_2 = -\dot{q}^T S\dot{q} - q^T Sq$$
$$+ r^T S \int_{-\tau}^{0} \dot{r}(t+\theta)d\theta + \frac{1}{2}\tau r^T(t)Xr(t)$$
$$-\frac{1}{2}\int_{-\tau}^{0} \dot{r}^T(t+\theta)X\dot{r}(t+\theta)d\theta \leq -\dot{q}^T S\dot{q} - q^T Sq$$
$$+\frac{1}{2}\tau r^T SY^{-1}Sr + \frac{1}{2}\int_{-\tau}^{0} \dot{r}^T(t+\theta)Y\dot{r}(t+\theta)d\theta$$
$$+\frac{1}{2}\tau r^T(t)Xr(t) - \frac{1}{2}\int_{-\tau}^{0} \dot{r}^T(t+\theta)X\dot{r}(t+\theta)d\theta$$
$$+ q^T Rq + \dot{q}^T M^T R^{-1} Mq^T.$$

for any positive matrices Y and R. Taking $Y \equiv X$ we obtain the inequality (17) of the Theorem after some numerical computations.

4. NUMERICAL EXAMPLE

We consider the inverted pendulum, who is one of the most popular laboratory experiments used for illustrated nonlinear control techniques. This system is motivated by applications as the control of rockets and the anti-seismic control of buildings.

We propose an stabilization algorithm for a 'cart pendulum system' but, in contrast to other authors, we use a delayed-controller and this controller is designed directly without partial feedback linearization.

The system can be written (see Fantoni 2000) in standard form:

$$D(q)\ddot{q} + C(q,\dot{q})\dot{q} + G(q) = u \quad (25)$$

where

$$q = \begin{bmatrix} x \\ \theta \end{bmatrix} \quad D(q) = \begin{bmatrix} M+m & ml\cos\theta \\ ml\cos\theta & ml^2 \end{bmatrix}$$
$$C = \begin{bmatrix} 0 & -ml\dot{\theta}\sin\theta \\ 0 & 0 \end{bmatrix} \quad G(q) = \begin{bmatrix} 0 \\ -mgl\sin\theta \end{bmatrix}$$

Here, M is the mass of the cart, m is mass of the pendulum, l is the distance from the pivot point to the center of the gravity, g is the gravity acceleration. The state vector q consists in x which is the distance of the cart center from its initial position and θ which is the pendulum angle.

In the sequel, we assume that $M = m = l = 1$ without any loss of generality. We consider that the system's velocity is bounded i.e. $||\dot{\theta}|| \leq \epsilon$ with $\epsilon < \infty$. Note that $M = (\frac{1}{2}\dot{D} - C) = \frac{1}{2}\begin{bmatrix} 0 & \dot{\theta}\sin\theta \\ -\dot{\theta}\sin\theta & 0 \end{bmatrix}$

<u>Example 1, $g = 0$, control of the angular speed</u>

From Theorem 1 the worst case is encountered when, in the norm, the matrices $C^T D^{-1} C$ and $KD^{-1}D$ take maximal values $\forall\theta$, i.e. $C^T D^{-1} C = \frac{\epsilon^2}{2} I$ and $D = \min_\theta \begin{bmatrix} 2 & cos\theta \\ cos\theta & 1 \end{bmatrix} = \begin{bmatrix} 2 & 0 \\ 0 & 1 \end{bmatrix}$.

The condition 3 from Theorem 1 becomes a LMI in (K, ϵ) for a fixed τ: Find ($K > 0$, $\epsilon > 0$) such that:

$$\begin{bmatrix} K - \tau\frac{\epsilon^2}{2} I & \tau K \\ \tau K & \tau\frac{2}{3} D \end{bmatrix} > 0 \quad (26)$$

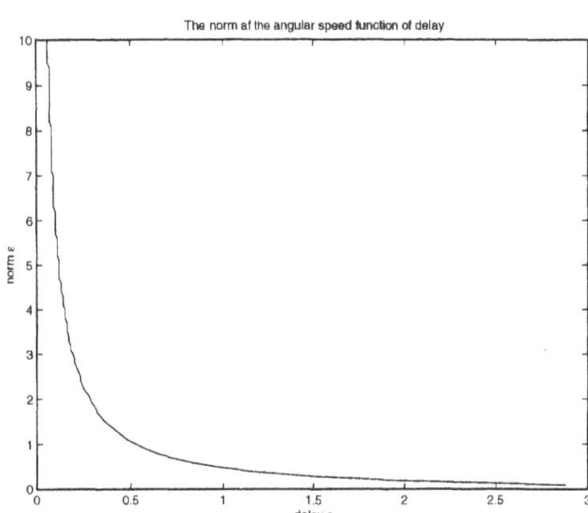

Fig. 1. Norm bound of the angular speed function of delay

The figure 1 represents the variation of $\epsilon = ||\theta||$ function of delay.

The results are sufficient ones. For a fixed $\epsilon = 2$ the best delay τ obtained using Theorem 1 is $\tau^* = 0.4$ sec. Simulations under Matlab Simulink environment shows that the maximal value for the delay is $\tau_m ax = 1.22$ sec.

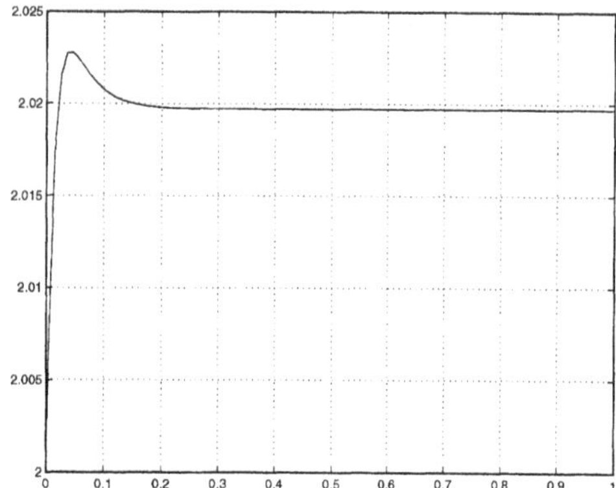

Fig. 2. Angular velocity for a maximal delay $\tau = 1.22$

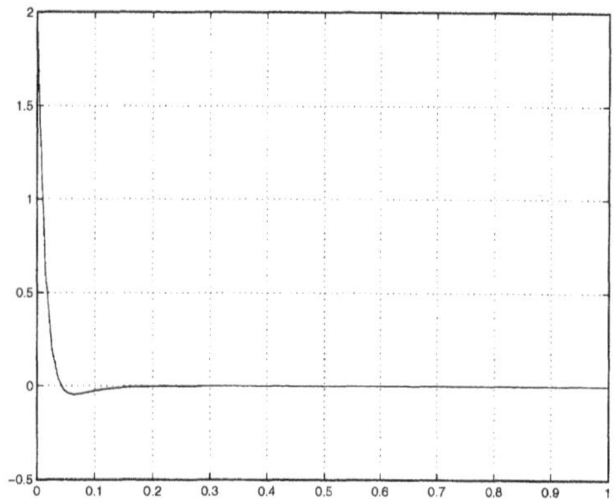

Fig. 3. Angular position for a maximal delay $\tau = 1.22$

Example 2, $g \neq 0$
control of the angular speed and position

We take (see Theorem 2) $Y = S$ and $R = I$. The condition (18) becomes

$$\begin{bmatrix} -S + \tau^* \lambda^2 S + M^T M & \tau^* S \\ \tau^* S & -S - G + I + \tau^* S \end{bmatrix} < 0 \quad (27)$$

and

$$M^T M = \frac{1}{4} \begin{bmatrix} \dot{\theta}^2 \sin^2 \theta & 0 \\ 0 & \dot{\theta}^2 \sin^2 \theta \end{bmatrix} \quad (28)$$

Looking to equation (27) the worst case is encountered when, in the norm, the matrix $M^T M$ takes its maximal value $\forall \theta$, i.e. $M^T M = \frac{\epsilon^2}{4} I$. According to this remark, the inequality (27) is feasible if the following one is:

$$\begin{bmatrix} -S + \tau^* \lambda^2 S + \frac{\epsilon^2}{4} I & \lambda \tau^* S \\ \lambda \tau^* S & -S - G + I + \tau^* S \end{bmatrix} < 0 \quad (29)$$

which is a LMI with the variables $S > 0, \epsilon$ for a fixed τ.

5. CONCLUSIONS

We have proved in this paper that the classical P and PD control laws for nonlinear underactuated mechanical systems stabilize the system even if some delays are present in the control input. We have established sufficient closed-loop stability criteria in terms of some matrix inequalities. Thus, some upper bounds for the allowed delays can be obtained. We have studied control schemes to stabilize the system angular speed as well as control strategies for both angular speed and position. Given that we have considered the nonlinearities of the system, the convergence results are achieved in the global sense.

REFERENCES

Ailon, A. and M. Gil (2000). "Stability analysis of a rigid robot with output-based controller and time delay", Systems & Control and Letters , **40** , pp 31-35.

Boyd, S., L. El Ghaoui, E. Feron and V. Balakrishnan (1994). *Linear Matrix Inequalities in System and Control Theory*, SIAM Studies in Applied Mathematics: Philadelphia, **15**.

Fantoni, I. (2000). *Non-linear Control for Underactued Mechanical Systems.* PhD Thesis, Heudiasyc, UTCompiegne.

Hale, J.K., and S.M. Verduyn Lunel (1993). *Introduction to Functional Differential Equations, Applied Mathematical Sciences*, vol. **99**, Springer-Verlag: New York.

Ivanescu, D., (2000). *Sur la stabilisation des systèmes avec retards. Théorie et applications.* Ph.D Thesis , Laboratoire d'Automatique de Grenoble , (in French).

Kolmanovskii V. B. and A. Myshkis (1992). *Applied Theory of Functional Differential Equations.* Kluwer, Dordrecht, the Netherlands.

Lee D. and P. Y. Li (2001). "Passive Control Approaches to Bilateral Teleoperated Manipulators: Theory and Experiments," *Proc. 2001 American Control Conference*, Anchorage, VA.

Lozano, R., B. Brogliato, O. Egeland and B. Maschke (2000). *Dissipative Systems Analysis and Control : Theory and Applications*, Springer Verlaag, Communications and Control Engineering: London.

Niculescu, S.-I. M. Fu and H. Li (1997). "Delay-dependent closed-loop stability of linear systems with input delays: An LMI approach," *36th Proc. IEEE Conf. Dec. Contr.*, San Diego, CL, 1997.

Niculescu, S.-I. and B. Brogliato (1995). "On force measurements time-delays in control of constrained manipulators," *Proc. IFAC Syst. Struct. & Contr.*, Nantes, France, 266-271.

Stépán, G. (1989). *Retarded dynamical systems: stability and characteristic function*, John Wiley & Sons.

www.elsevier.com/locate/ifac

CONTROL OF MICROWAVE-PROPELLED SAILS USING DELAYED MEASUREMENTS

E. Schamiloglu, C.T. Abdallah, G.L. Heileman,* D. Georgiev, J. Benford,*** G. Benford******

* *Department of Electrical and Computer Engineering, University of New Mexico, Albuquerque, NM 87131*
** *Department of Mechanical Engineering, University of New Mexico, Albuquerque, NM 87131*
*** *Microwave Sciences, Inc., 1041 Los Arabis Lane, Lafayette, CA 94549*
**** *Department of Physics and Astronomy, University of California, Irvine, 4129 Frederick Reines Hall, Irvine, CA 92697-4575*

Abstract: This paper is concerned with the control of carbon fiber sail structures that are being studied in a series of experiments at the Jet Propulsion Laboratory (JPL). The passive dynamic stability in the one-dimensional (1-D) case is studied in terms of the fixed points of the trajectories for the governing equations of motion. The simple 1-D model introduces the possibility of controlling a microwave-propelled sail using various nonlinear control strategies. In the case where the velocity is not available, we use a novel feedback that employs delays of the position measurements to stabilize the sail about an equilibrium position.

Keywords: microwave-propelled sails, delayed feedback.

1. INTRODUCTION

Microwave-propelled sails belong to a class of spacecrafts that promises to revolutionize future space travel. As an example, NASA's Gossamer Spacecraft Initiative focuses on developing spacecraft architectures for very large, ultra-lightweight apertures and structures. A goal of this initiative is to achieve breakthrough enhancements in mission capability and reductions in mission cost, primarily through revolutionary advances in structures, materials, optics, and adaptive and multifunctional systems. Solar and other types of sails will provide low-cost propulsion, station-keeping in unstable orbits, and precursor interstellar exploration missions. For a general introduction to solar sails and similar structures the reader is referred to McInnes (1999). For an introduction to the notion of beamed microwave power and its application to space propulsion the reader is referred to Benford and Dickinson (1995). This paper is concerned specifically with the stability and control of carbon fiber sails propelled using microwave radiation.

The notion of *beam-riding*, *i.e.*, the stable flight of a sail propelled by Poynting flux, places considerable demands upon a sail. Even if the beam is steady, a sail can wander off the beam if its shape becomes deformed, or if it does not have enough spin to keep its angular momentum aligned with the beam direction, in the face of perturbations. Generally, sails without structural elements cannot be flown if they are convex toward the beam, as the beam pressure would cause them to collapse. On the other hand, the beam pressure keeps concave shapes in tension, so concave shapes arise naturally while beam riding. They will resist sidewise motions if the beam moves off center, since a net sideways force restores the sail to its position. Therefore, we concentrate on a conical shape for the sail and study its dynamics in 1-D. We have previously shown using the Poincare-Bendixon theorem that such conical shapes will oscillate when a constant microwave power beam is used Abdallah et al. (2001). We present in this paper feedback controllers to show how to stabilize such shapes about a given height, using only position measurements.

This paper is organized as follows: Section 2 presents the 1-D system dynamics, while section 3 discusses the controller design using position measurements. Section 4 summarizes our conclusions and presents directions for future research.

2. 1-D PROBLEM STATEMENT

The problem we are concerned with is that of controlling a sail that is only allowed to move in the vertical direction, under the influence of a microwave power source placed either on earth or in orbit. The 1-D dynamics of the sail are given by Benford (2000):

$$\frac{1}{g}\frac{d^2z}{dt^2} = -1 + \frac{P}{P_0}\frac{\cos^2\theta}{(1+\frac{z}{R}\tan\Phi)^2}, \tag{1}$$

where g is the acceleration of gravity (constant in the appropriate units), $P_0$is the power necessary to overcome gravity, P is the power control input, z is the elevation of the sail's center of mass referenced to $z=0$, R is the beam radius, Φ is half of the total beam opening angle, and θ is the angle at which the microwave photons strike the sail. In the case of high altitudes, θ is close to zero and $\cos\theta = 1$. In addition, $\tan\Phi = 1$ and therefore, equation (1) is reduced to,

$$\frac{1}{g}\frac{d^2z}{dt^2} = -1 + \frac{P}{P_0}\frac{1}{(1+\frac{z}{R})^2}. \tag{2}$$

The equations of the sails are then written in state-space form as

$$\begin{bmatrix}\dot{x}_1\\ \dot{x}_2\end{bmatrix} = \begin{bmatrix}x_2\\ -g\end{bmatrix} + \begin{bmatrix}0\\ \frac{g}{P_0}\frac{1}{(1+x_1/R)^2}\end{bmatrix}u \tag{3}$$

where $x_1 = z$ and $x_2 = \dot{z}$, and $u = P$.

In the case where $\cos\theta$ is not close to 1, and letting

$$\cos\theta = \frac{z^2}{z^2+r^2}$$

and averaging over the disk radius r we obtain the following dynamics,

$$\frac{1}{g}\frac{d^2z}{dt^2} = -1 + \frac{P}{P_0}\ln\left(1+\frac{R^2}{z^2}\right). \tag{4}$$

A more typical fall-off in beam power is as $P = P^*\cos^m\theta$ where $m = 1, 2, \cdots$ is used to describe different fall-off rates. In a similar fashion to the previous development, and averaging over r we obtain the following dynamics,

$$\frac{1}{g}\frac{d^2z}{dt^2} = -1 + \frac{P}{P_0}\ln\left(1+\frac{R^2}{z^2}\right) \tag{5}$$

leading to the state-space representation

$$\begin{bmatrix}\dot{x}_1\\ \dot{x}_2\end{bmatrix} = \begin{bmatrix}x_2\\ -g\end{bmatrix} + \begin{bmatrix}0\\ \frac{g}{P_0}\ln(1+\frac{R^2}{x_1^2})\end{bmatrix}u \tag{6}$$

where $x_1 = z$ and $x_2 = \dot{z}$, and $u = P$. In either case (3) or (6), we obtain the following general system:

$$\begin{bmatrix}\dot{x}_1\\ \dot{x}_2\end{bmatrix} = \begin{bmatrix}x_2\\ -g\end{bmatrix} + \begin{bmatrix}0\\ h(x)\end{bmatrix}u,$$

where $x_1 = z$, $x_2 = \dot{z}$, and $u = P$. The term $h(x)$ denotes either $\frac{g}{P_0(1+x_1/R)^2}$ or $\frac{g}{P_0}\ln\left(1+\frac{R^2}{x_1^2}\right)$. Note that in either case $h(x)$ can never be zero as long as $x_1 \geq 0$. Note also, that $h(x)$ is only a function of x_1, and that it is invertible, two facts that will be important later.

3. CONTROL DESIGN USING DELAY FEEDBACK

Consider again the system (3) but with $u = P$ where P is a constant such that $P > P_0$. Note first that the only equilibrium point of the system (3) is at

$$x_e = \begin{bmatrix}x_{e1}\\ x_{e2}\end{bmatrix} = \begin{bmatrix}R\left[\sqrt{\frac{P}{P_0}}-1\right]\\ 0\end{bmatrix}. \tag{7}$$

On the other hand, and for system (6), we have the unique equilibrium point

$$x_e = \begin{bmatrix}x_{e1}\\ x_{e2}\end{bmatrix} = \begin{bmatrix}\frac{R}{\sqrt{e^{P_0/P}-1}}\\ 0\end{bmatrix}. \tag{8}$$

As discussed earlier, we showed in Abdallah et al. (2001); Schamiloglu et al. (2001) that using a fixed power P will result in oscillations about the equilibrium point x_e. On the other hand, and due to various practical factors, we do not have access to the measurements of the velocity $\dot{z}$.

Let us define the error as $e_1 = x_1 - x_{e1}$, $e_2 = \dot{e}_1$, and express the error system in state-space as

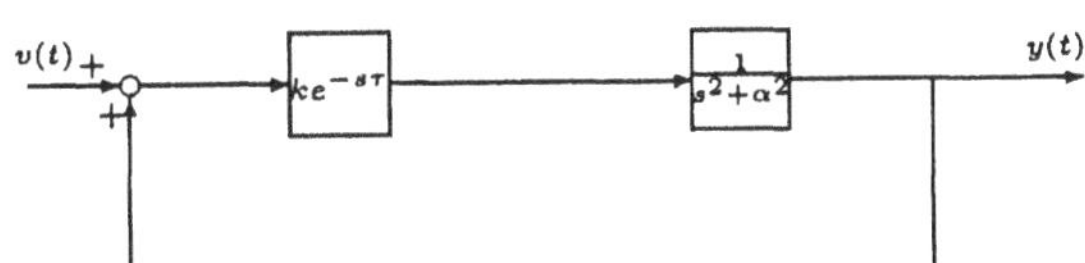

Fig. 1. Feedback stabilization of oscillatory systems.

$$\begin{bmatrix} \dot{e}_1 \\ \dot{e}_2 \end{bmatrix} = \begin{bmatrix} e_2 \\ -g \end{bmatrix} + \begin{bmatrix} 0 \\ h(e_1) \end{bmatrix} u,$$

For the system (9), and since $h(e_1) \neq 0$ for all e_1, we can use a feedback-linearization based controller Isidori (1995) as follows:

$$u(e) = \frac{1}{h(e_1)}[g - k_1 e_1 - k_2 e_2], \qquad (9)$$

where k_1 and k_2 are positive numbers that then make the closed-loop system stable with a characteristic polynomial $s^2 + k_2 s + k_1$. We can use the freedom in the choice of k_1 and k_2 to specify the closed-loop behavior of the system.

However, in our case, we do not have access to the velocity measurements needed to compute e_2. The question then becomes, is it possible to control the sail using only position measurements?

Calling on results from an earlier paper, we were able to show that one can stabilize various systems using time-delayed feedback Abdallah et al. (1993). In order to describe those results, assume we are given a linear time-invariant system

$$G(s) = \frac{1}{s^2 + \alpha^2} \qquad (10)$$

and the positive-feedback, time-delay compensator

$$C(s) = ke^{-s\tau} \qquad (11)$$

where $k < 0$ in a simple unity-feedback loop as shown in Figure 1, and such that the closed-loop system is given by:

$$\begin{aligned} T(s) &= \frac{G(s)C(s)}{1 - G(s)C(s)} \\ &= \frac{ke^{-s\tau}}{s^2 + \alpha^2 - ke^{-s\tau}} \end{aligned} \qquad (12)$$

In reference Abdallah et al. (1993), we showed that stability is guaranteed for

$$\begin{aligned} 0 < k < \frac{1 + 4n}{1 + 4n + 8n^2}\alpha^2 \\ \frac{2n\pi}{\sqrt{\alpha^2 - k}} < \tau < \frac{(2n+1)\pi}{\sqrt{\alpha^2 + k}} \end{aligned} \qquad (13)$$

for $n = 0, 1, \cdots$. let us then go back to our sail dynamics and consider the feedback-linearizable system

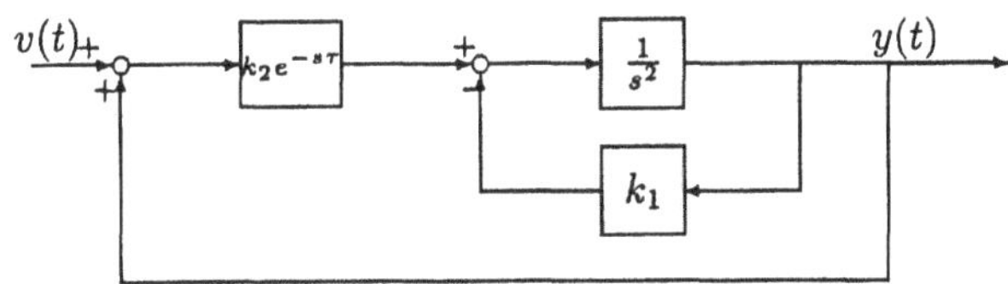

Fig. 2. Feedback stabilization of oscillatory systems.

$$\begin{bmatrix} \dot{e}_1 \\ \dot{e}_2 \end{bmatrix} = \begin{bmatrix} e_2 \\ -g \end{bmatrix} + \begin{bmatrix} 0 \\ h(e_1) \end{bmatrix} u,$$

and let us choose $u = \frac{1}{h(e_1)}[g + v]$. Let us also assume that only position measurements are available so that the measured output is: $y = e_1$. Thus, the open-loop transfer function from the new input v to the output e_1 is given by $G_0(s) = 1/s^2$. Next, consider the output feedback

$$v(t) = -k_1 e_1(t) + k_2 e_1(t - \tau) \qquad (14)$$

leading to the closed-loop transfer function

$$G_{cl}(s) = \frac{k_2 e^{-s\tau}}{s^2 + k_1 - k_2 e^{-s\tau}} \qquad (15)$$

so that by letting $k_1 = \alpha^2$, $k_2 = k$ we recover our earlier results.

The controller for the nonlinear system is then given by:

$$u(t) = \frac{1}{h(e_1)}[g - k_1 e_1(t) + k_2 e_1(t - \tau)] \qquad (16)$$

which depends only on position measurements.

4. CONCLUSIONS

In this paper we have studied the control problem of a conical-shaped microwave-propelled sails in the 1-D case, when only position measurements are available. The example is interesting in its own right but also suggests that many feedback-linearizable systems nay be controlled using output (and delayed output) measurements only rather than the more standard approaches of building nonlinear observers or using state-feedback controllers. The class of systems to which our technique applies is that for which the linearizing transformation depends only on the output.

In an ongoing extension of the results of this paper, we show that we may be able to use an all-delay feedback (as opposed to one that includes delayed and current output measurements) to achieve stability, thus more realistically modeling the case where output measurements incur a delay as they are transferred to the actuators.

REFERENCES

C.T. Abdallah, P. Dorato, J. Benitez-Read, and R. Byrne. Delayed positive feedback can stabilize oscillatory systems. In *Proceedings of the American Control Conference*, pages 3106–3107, 1993. San Francisco CA.

C.T. Abdallah, E. Schamiloglu, K.A. Miller, D. Georgiev, J. Benford, and G. Benford. Stability and control of microwave-propelled sails in 1-d. In *Proceedings 2001 Space Exploration and Transportation: Journey into the Future*, 2001. Albuquerque, NM.

J. Benford. Wireless power transmission for science applications. Technical report, Microwave Sciences, Interim Final Report Contract number NAS8-99135, Microwave Sciences, 1041 Los Arabis Lane, Lafayette, CA 94549, May 2000.

J. Benford and R. Dickinson. Space propulsion and power beaming using millimeter systems. In *Intense Microwave Pulses III, Proceedings SPIE*, volume 2557, page 179, 1995.

A. Isidori. *Nonlinear Control Systems.* Springer-Verlag, London, 3rd edition edition, 1995.

C. R. McInnes. *Solar Sailing: Technology, Dynamics, and Mission Applications.* Springer-Verlag, New York, 1999.

E. Schamiloglu, C.T. Abdallah, K.A. Miller, D. Georgiev, J. Benford, G. Benford, and G. Singh. 3-d simulations of rigid microwave-propelled sails including spin. In *Proceedings 2001 Space Exploration and Transportation: Journey into the Future*, 2001. Albuquerque, NM.

www.elsevier.com/locate/ifac

COMPARISON OF OBSERVER-BASED H_∞ CONTROL APPROACHES FOR TIME-DELAY SYSTEMS: APPLICATION TO A WIND TUNNEL MODEL

O. Sename * **E. Haro** *

* *Laboratoire d'Automatique de Grenoble*
INPG-CNRS (UMR5528)
ENSIEG-BP 46
38402 Saint Martin d'Hères Cedex, FRANCE
Fax : (33) 4.76.82.63.88, email : Olivier.Sename@inpg.fr

Abstract: H_∞ observer-based feedback control law is designed for linear time-delay systems. We wish to find two non coupled Riccati equations to solve independently the H_∞ control and the H_∞ observer design problems. An application to a wind tunnel model is provided to emphasize the results, particularly in comparison with existing results on H_∞ observer-based control. *Copyright © 2001 IFAC*

Keywords: Time-delay, Memoryless, H_∞-Control, Observers.

1. INTRODUCTION

The control of time-delay systems has been thoroughly studied, assuming that all the system state variables are known. However, in most practical situations, this condition is rarely satisfied and an observer has to be implemented.
The aim of this paper is to propose an H_∞ dynamic output feedback using the H_∞ observer developed in [Fattouh et al., 1999] and the H_∞ control approach proposed by Lee et al. [1994]. The proposed results will be delay independent. This will be compared with the observer-based H_∞ control approach proposed by Choi and Chung [1996] in the case of a physical process: a wind tunnel model, studied by Manitius [1984].
The outline of this paper is as follows. Section 2 states the problem. Section 3 is devoted to the results of Choi and Chung [1996] concerning H_∞ observer-based controller. In section 4 our proposed methodology to obtain H_∞ dynamic output feedback is described. The illustrative example, i.e., the wind tunnel model, is tackled in section 5. Finally, some concluding remarks end the paper.

2. PROBLEM STATEMENT

For sake of simplicity, we consider single delayed-state systems, even if some results are directly extended to the case of multiple time-delay systems. The system under consideration is then :

$$\begin{cases} \dot{x}(t) &= A_0x(t) + A_1x(t-h) + Bu(t) + Ew(t) \\ y(t) &= Cx(t) + Fw(t) \\ z(t) &= Dx(t) \\ \\ x(\theta) &= \phi(\theta); \qquad \theta \in [-h, 0] \end{cases} \tag{1}$$

where $x(t) \in \mathbb{R}^n$ is the state vector, $u(t) \in \mathbb{R}^r$ is the control input, $y(t) \in \mathbb{R}^p$ is the output measurement vector, $z(t) \in \mathbb{R}^m$ is the controlled output, $w(t) \in R^q$ is the square-integrable disturbance vector, $\phi(\theta) \in \mathcal{C}[-h, 0]$ is the functional initial condition and $h \in \mathbb{R}^+$ is the delay.
The main contributions to H_∞ control approaches are based on stabilization method by memoryless state feedback. The first one is based on the Lyapunov-Krasovskii theorem and leads to delay-independent result (see Lee et al. [1994]) as the

second one uses the Lyapunov-Razumikhin theorem and leads to delay-dependent result (see Su [1994]).

In this paper H_∞ control design is considered through an observer and a delay-independent framework, as in [Lee et al., 1994]. The control implementation will then rely on an H_∞ observer of the form given in [Fattouh et al., 1999]. The provided methodology will be compared to the observer-based controller proposed by Choi and Chung [1996].

3. H_∞ OBSERVER-BASED CONTROLLER

This part is devoted to the results of Choi and Chung [1996] where the observer-based controller is of the form:

$$\begin{cases} \dot{\hat{x}}(t) &= A_0\hat{x}(t) + A_1\hat{x}(t-h) + Bu(t) \\ & \quad -L(C\hat{x}(t) - y(t)) + EG\hat{x}(t) \\ u(t) &= K\hat{x}(t) \end{cases} \tag{2}$$

Note that the observer form includes a specific term $EG\hat{x}(t)$ that represents the coupling within the observer and the control. In an H_∞ framework it represents an estimation of the worst possible disturbance (see [Trentelman et al., 2001] for the linear non delay case).

The observer and the control are then obtained through two coupled Riccati equations including 7 parameters (2 matrices and 5 constants). Nevertheless the estimation error stability is guaranteed for closed-loop systems only and no performance (in terms of H_∞ gain) is ensured for the observer. On the other hand, such a performance is obtained for the closed-loop system, as stated below.

Theorem 1. Consider the time-delay system (1) (with $F = I_p$) and the observer-based controller (2). If both following algebraic Riccati equations

$$\begin{aligned} & A_0^T P_c + P_c A_0 \\ & -\frac{1}{\epsilon_c} P_c (BB^T - \frac{1}{\delta_c} A_1 A_1^T - \frac{1}{\gamma^2} EE^T) P_c \\ & +\epsilon_c(\delta_c I_n + D^T D + Q_c) \quad = 0 \\ & (A_0 + EG)P_o + P_o(A_0 + EG)^T \\ & -\frac{1}{\epsilon_o} P_o (C^T C - \frac{1}{\gamma^2} K^T K - \frac{\delta_o}{\gamma^2} I_n) P_o \\ & +\epsilon_o(\frac{\gamma^2}{\delta_o} A_1 A_1^T + EE^T + Q_o) \quad = 0 \end{aligned} \tag{3}$$

have positive definite solutions P_c and P_o for some positive constants γ, ϵ_c, δ_c, ϵ_o and δ_o, some positive definite matrices Q_c and Q_o, and with the control parameters :

$$K = -\frac{1}{\epsilon_c} B^T P_c, \quad G = \frac{1}{\gamma^2 \epsilon_c} E^T P_c$$

then, for all h:

1- Observer (2) is asymptotically stable and the observer gain is :

$$L = \frac{1}{\epsilon_o} P_o C^T$$

2- the closed-loop system is asymptotically stable and such that : $\|z\|_2 \leq \gamma \ \|w\|_2$.

4. H_∞ DYNAMIC OUTPUT FEEDBACK DESIGN

The proposed method is to design an observer independently of the control. The observer will be designed using the results of Fattouh et al. [1999] while the control will be obtained through Lee et al. [1994]'s results.

In this section, we consider H_∞ dynamic output feedback using Luenberger-type observer given by the following dynamical system:

$$\begin{aligned} \dot{\hat{x}}(t) &= A_0\hat{x}(t) + A_1\hat{x}(t-h) + Bu(t) \\ & \quad -L(C\hat{x}(t) - y(t)) \\ u(t) &= K\hat{x}(t) \end{aligned} \tag{4}$$

where $\hat{x}(t) \in \mathbb{R}^n$ is the estimated state of $x(t)$ and L (resp. K) is the $n \times p$ constant observer (resp. controller) gain matrix to be designed.

4.1 H_∞ *observer design*

The estimated error, defined as $e(t) := x(t) - \hat{x}(t)$, satisfies the following dynamical system:

$$\begin{aligned} \dot{e}(t) &= (A_0 - LC)e(t) + A_1 e(t-h) \\ & \quad +(E - LF)w(t) \end{aligned} \tag{5}$$

The transfer function $T_{ew}(s)$ between the disturbance and the error is then given by:

$$T_{ew}(s) = \left[sI_n - A_0 + LC - A_1 e^{-sh}\right]^{-1}(E - LF)$$

The following result ensures some H_∞ disturbance attenuation level for the observer.

Theorem 2. [Fattouh et al., 1999] Consider the time-delay system (1) and the observer (4). If the following algebraic Riccati equation

$$\begin{aligned} & A_0^T P_o + P_o A_0 + 2P_o(\gamma_o^2 A_1 A_1^T + EE^T)P_o \\ & -\frac{2}{\epsilon_o} C^T (I_p - \frac{1}{\epsilon_o} FF^T) C + \frac{2}{\gamma_o^2} I_n = 0 \end{aligned} \tag{6}$$

has positive definite solution P_o for some positive constants γ_o and ϵ_o then, for all h, system (4) is an H_∞ observer i.e. the solution of the functional differential equation (5) with $w(t) \equiv 0$ converges to zero asymptotically and, under zero initial condition, the H_∞-norm of the transfer function between the disturbance and the estimated error is bounded, that is

$$\lim_{t \to \infty} e(t) \to 0 \quad \text{for} \quad w(t) \equiv 0$$

$$\|T_{ew}(s)\|_\infty \leq \gamma_o$$

where $\|.\|_\infty$ is the H_∞-norm and $\gamma_o > 0$ is the disturbance attenuation level. In this case the observer gain is :

$$L = \frac{1}{\epsilon_o} P_o^{-1} C^T$$

This method only concerns H_∞ observer design. First the dynamical system (4) has a more usual form than the previous one in (2). Then only 2 parameters are here necessary to obtain an observer for which stability as well as H_∞ performance are guaranteed.

4.2 H_∞ *control design*

Now some solutions of the H_∞ memoryless control problem by state feedback have been obtained by Lee et al. [1994], Niculescu [1998]. In [Lee et al., 1994] single delayed-state systems are considered and a frequential approach is used to derive H_∞ delay independent stability criterion. In [Niculescu, 1998] the control is designed with special requirements on the pole placement for the closed-loop system. A Krasovskii approach is used and an LMI formulation is proposed to get the maximal bound on the time-delay such that the prescribed disturbance attenuation level and the α-stability requirements are obtained.
Here, another criterion for H_∞ asymptotic stability, given for multiple time-delay systems, will be considered. This result is an extension of the main result of Lee et al. [1994] concerning the H_∞ stabilization of single delayed-state systems.

Proposition 3. Consider the following linear multiple delayed-state system:

$$\begin{cases} \dot{x}(t) = A_0 x(t) + \sum_{i=1}^{N} A_i x(t-ih) + Bu(t) \\ \qquad +Ew(t) \\ z(t) = Dx(t) \\ x(t) = \phi(t); \quad t \in [-Nh, 0] \end{cases} \tag{7}$$

where $x(t) \in \mathbb{R}^n$ is the state vector, $w(t) \in \mathbb{R}^q$ is the $\mathcal{L}_2$ disturbance vector, $z(t) \in \mathbb{R}^m$ is the controlled output, $\phi(t) \in \mathcal{C}[-Nh, 0]$ is the initial functional condition vector, $h \in \mathbb{R}^+$ is the fixed delay duration (could be unknown) ($N \in \mathbb{N}$) and A, A_i, B, D and E, $i = 1, ..., N$, are real matrices with appropriate dimensions.
Given a positive scalar γ_c, if there exists two positive definite matrices P_c and Q such that

$$\begin{aligned} &(A_0 + BK)^T P_c + P_c(A_0 + BK) \\ &+P_c(\sum_{i=1}^{N} A_i Q^{-1} A_i^T) P_c + NQ \\ &+\frac{1}{\gamma_c{}^2} D^T D + P_c E E^T P_c < 0 \end{aligned} \tag{8}$$

with the controller K given by

$$K = -\frac{1}{2\epsilon_c} R^{-1} B^T P_c \tag{9}$$

for any positive definite matrix R and the constant ϵ_c, then the closed-loop multiple delayed-state system (7)-(9) is asymptotically stable for any value of the delay, and the following inequality holds:

$$\|D(sI_n - A - BK - \sum_{i=1}^{N} A_i e^{-sih})^{-1} E\|_\infty \leq \gamma_c \tag{10}$$

PROOF. This is only outlined here as it is directly obtained using [Fattouh et al., 1999, Lee et al., 1994]. It is first proved the asymptotic stability of the closed-loop system (7)-(9), and then the inequality (10) is shown.
Suppose $\gamma_c > 0$, $P_c = P_c{}^T > 0$ and $Q = Q^T > 0$ such that ARE (8) holds. Now let the Lyapunov functional $V(t)$ be:

$$V(t) := x^T(t) P_c x(t) + \sum_{i=1}^{N} \int_{t-ih}^{t} x^T(\theta) Q x(\theta) d\theta$$

One can show that there exist two positive scalars β_1 and β_2 such that

$$\beta_1 \|x(t)\|^2 \leq V(t) \leq \beta_2 \|x(t)\|^2 \tag{11}$$

and that the Lyapunov derivative of $V(t)$ along the trajectory of $x(t)$ is given by

$$\frac{dV(t)}{dt} = \bar{x}^T(t) \mathcal{A} \bar{x}(t)$$

where $\mathcal{A}$ is negative definite (see Fattouh et al. [1999]). Then

$$\frac{dV(t)}{dt} \leq -\lambda_{min}(\mathcal{A}) \|\bar{x}(t)\|^2 \tag{12}$$

Combining (11) and (12) one can prove the asymptotic stability of (7)-(9) using the Krasovskii theorem.
Now, let us define $A := A_0 + BK$, and the positive definite matrix S:

$$\begin{aligned} S := &-(A^T P_c + P_c A + NQ + \frac{1}{\gamma_c^2} D^T D + P_c E E^T P_c \\ &+P_c [A_1, A_2, .., A_N] \bar{Q} [A_1, A_2, .., A_N]^T P_c) \end{aligned} \tag{13}$$

and the matrices $W(jw)$ and $X(jw)$:

$$\begin{aligned} W(jw) :=\ & P_c [A_1, A_2, ..., A_N] \bar{Q} [A_1, A_2, ..., A_N]^T P_c \\ & +NQ - \sum_{i=1}^{N} A_i^T P_c e^{jwih} - \sum_{i=1}^{N} P_c A_i e^{-jwih} \\ X(jw) :=\ & (jwI_n - A - \sum_{i=1}^{N} A_i e^{-jwih})^{-1} \end{aligned}$$

where $\bar{Q} = diag\{Q^{-1}\}_N$, and $W(jw)$ is non-negative definite for all $w \in \mathbb{R}$.
Using the above definitions, equation (13) can be transformed into:

$$\begin{aligned} &-(I_n - E^T P_c X(-jw) E)^T (I_n - E^T P_c X(jw) E) \\ &= -I_n + E^T X^*(jw) \{W(jw) + S\} X(jw) E \\ &+\frac{1}{\gamma_c^2} E^T X^*(jw) D^T D X(jw) E \end{aligned} \tag{14}$$

The left hand side is non-positive definite for all $w \in \mathbb{R}$. With $T_{zw}(jw) := DX(jw)E$ we have then

$$-In + E^T X^*(jw)\{W(jw) + S\}X(jw)E + \frac{1}{\gamma_c^2} T_{zw}^*(jw)T_{zw}(jw) \leq 0 \quad (15)$$

and

$$\begin{aligned} &T_{zw}^*(jw)T_{zw}(jw) \\ &\leq \gamma_c^2 I_n - \gamma_c^2 E^T X^*(jw)\{W(jw) + S\}X(jw)E \quad (16) \\ &\leq \gamma_c^2 I_n \end{aligned}$$

for all $w \in \mathbb{R}$, ie., $\|T_{zw}\|_\infty \leq \gamma_c$. □

The H_∞ control design proposed in this subsection is independent from the observer, and 2 parameters are also necessary in order to find the matrix gain assuring the H_∞ specifications.

4.3 *H_∞ dynamic measurement feedback*

In this part both previous approaches are mixed to design an H_∞ dynamic output feedback controller. First, let us prove that the separation principle holds in this case. With $e(t) := x(t) - \hat{x}(t)$, let us consider the extended state:

$$x_e(t) = \begin{bmatrix} x(t) & e(t) \end{bmatrix}^T$$

The closed-loop system with observer and control (4) is:

$$\dot{x}_e(t) = \begin{bmatrix} A_0 + BK & -BK \\ 0 & A_0 - LC \end{bmatrix} x_e(t) + \begin{bmatrix} A_1 & 0 \\ 0 & A_1 \end{bmatrix} x_e(t-h) + \begin{bmatrix} E \\ E - LF \end{bmatrix} w(t) \quad (17)$$

This guarantees that the separation principle holds, as the characteristic polynomial of the extended system is:

$det(sI_n - A - BK - A_1 e^{-sh}) \times det(sI_n - A + LC - A_1 e^{-sh})$

If the observer and the control are designed separately then the closed-loop system with the dynamic measurement feedback is stable, given that the control and observer systems are stable and the eigenvalues of (17) can be obtained directly from them. We also have that the H_∞-norm of the transfer function between the disturbance and the controlled vector of (17), $\tilde{T}_{zw}$, is bounded as follows

$$\|\tilde{T}_{zw}\|_\infty \leq \|T_{zw}\|_\infty + \|H\|_\infty \|T_{ew}\|_\infty \quad (18)$$

where T_{zw} is the H_∞ control design result, T_{ew} is the H_∞ observer design result and H is defined by

$$H(s) = D[sI_n - A_0 - BK - A_1 e^{-sh}]^{-1} BK$$

Let us illustrate this result on the following example which is an application to a physical process.

5. APPLICATION TO A WIND TUNNEL MODEL

This example shows the efficiency of the proposed method to a feedback control problem arising in the regulation of a wind tunnel, as described in Manitius [1984]. It is a simplified mathematical model of the Mach number dynamic response to guide vane changes. The delay in one state variable represents the transportation time between the guide vanes of the fan and the test section of the tunnel. In steady-state operating conditions (fan speed, liquid nitrogen injection rate and gaseous-nitrogen vent rate) the dynamic response of the Mach number is given by the following system:

$$\begin{aligned} \dot{x}(t) &= \begin{bmatrix} -0.5091 & 0 & 0 \\ 0 & 0 & 1 \\ 0 & -36 & -9.6 \end{bmatrix} x(t) + \begin{bmatrix} 0 & -0.005956 & 0 \\ 0 & 0 & 0 \\ 0 & 0 & 0 \end{bmatrix} x(t-h) + \begin{bmatrix} 0 \\ 0 \\ 36 \end{bmatrix} u(t) + \begin{bmatrix} 0 \\ 0 \\ 1 \end{bmatrix} w(t) \\ y(t) &= \begin{bmatrix} 1 & 0 & 0 \end{bmatrix} x(t) + w(t) \\ z(t) &= \begin{bmatrix} 1 & 1 & 1 \end{bmatrix} x(t) \end{aligned}$$

$$x(t) = \phi(t); \qquad t \in [-h, 0]$$

where $h = 0.33sec.$, x_1 is the Mach number, x_2 is the guide vane angle and $x_3 = \dot{x}_2$.
The observability matrix is given by :

$$< \frac{C(\nabla)}{A(\nabla)} >= \begin{bmatrix} 1000 & 0 & 0 \\ -509.1 & -5.956\nabla & 0 \\ 259.2 & 3.032\nabla & -5.956\nabla \end{bmatrix} \times 10^{-3}$$

The above system is weakly observable as $rank < \frac{C(\nabla)}{A(\nabla)} >= 3$. However, it is not strongly observable as $< \frac{C(\nabla)}{A(\nabla)} >$ is not invertible over $\mathbb{R}[\nabla]$. Now $rank \begin{bmatrix} sI_2 - A(e^{-sh}) \\ C(e^{-sh}) \end{bmatrix} =$ is of rank 3 for all complex s, i.e. the system is spectrally observable.
On the other hand we know that the controllability matrix is

$$< A(\nabla)|B(\nabla) >= \begin{bmatrix} 0 & 0 & -0.2144\nabla \\ 0 & 36 & -345.6 \\ 36 & -345.6 & 2021.76 \end{bmatrix}$$

Thus, the system is not strongly controllable as the controllability matrix $< A(\nabla)|B(\nabla) >$ is not invertible over $\mathbb{R}[\nabla]$; but it is spectrally controllable given that $rank\,[sI_2 - A(e^{-sh})|B(e^{-sh})] =$ is also of rank 3 for all complex s, and therefore weakly controllable.

5.1 *Choi-Chung method*

Using the method of Choi and Chung [1996] we can obtain the following result. Let us choose the following parameters : $\epsilon_c = 0.0125$, $\epsilon_o = 0.0125$, $\delta_c = 2$, $\delta_o = 2.5$ and $Q_o = Q_c = I_2$. For $\gamma = 1.5$ (for the closed-loop system) , the ARE (3) has the following solution and the corresponding observer gain :

$$P_c = \begin{bmatrix} 0.0491 & 0 & 0 \\ 0 & 0.0409 & 0.0003 \\ 0 & 0.0003 & 0.0005 \end{bmatrix}, \quad K = \begin{bmatrix} 0 \\ -1.0001 \\ -1.5018 \end{bmatrix}^T$$

$$G = \begin{bmatrix} 0 & 0.0123 & 0.0185 \end{bmatrix}$$

$$P_o = \begin{bmatrix} 0.1006 & 0 & 0 \\ 0 & 0.0009 & -0.006 \\ 0 & -0.006 & 0.0977 \end{bmatrix}, L = \begin{bmatrix} -8.0451 \\ 0 \\ 0 \end{bmatrix}$$

The estimated error response $e(t)$ and the controlled output $z(t)$ are shown in figure 1 for $h = 0.33sec.$ For simulation purpose an initial value at time $t = 0sec.$ is used to generate a functional initial condition on $t \in [0, 0.5]$. The observer acts at $t = 0.5sec.$ A unit step disturbance is applied at $t = 10sec.$

On figure 2, it is shown that $\|T_{zw}(j\omega)\|_\infty \leq \gamma$ for all $\omega \in \mathbb{R}$, which corresponds to the disturbance attenuation of $z(t)$.

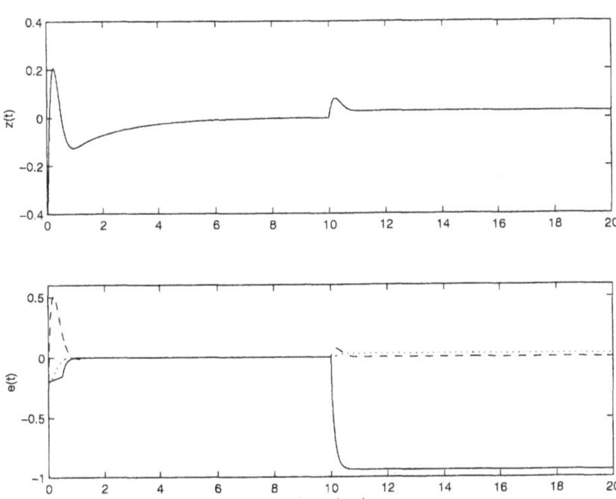

Fig. 1. Controlled output and estimated errors - Choi and Chung

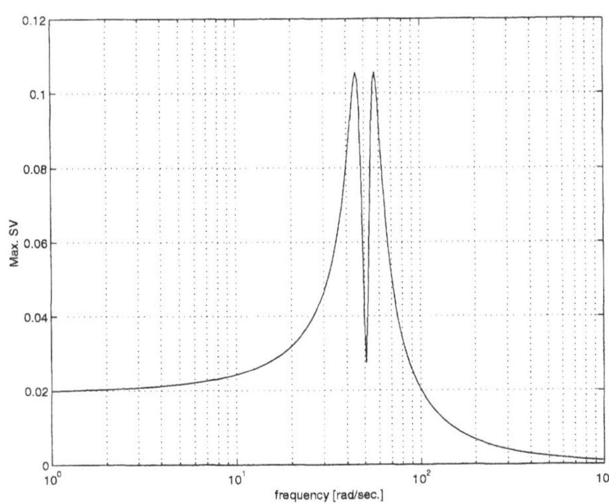

Fig. 2. Maximum SV of $T_{zw}(jw)$ - Choi and Chung

We can note in vector $e(t)$ of figure 1 that, as no robust property is guaranteed for the observer in this case, the disturbance attenuation for the state estimation error is not good. This can be appreciated in the maximum singular value frequency plot of $T_{ew}(j\omega)$ (Fig. 3). Finally the gain L in the observer is large and the convergence of the observer is guaranteed only for state feedback controlled-loop systems.

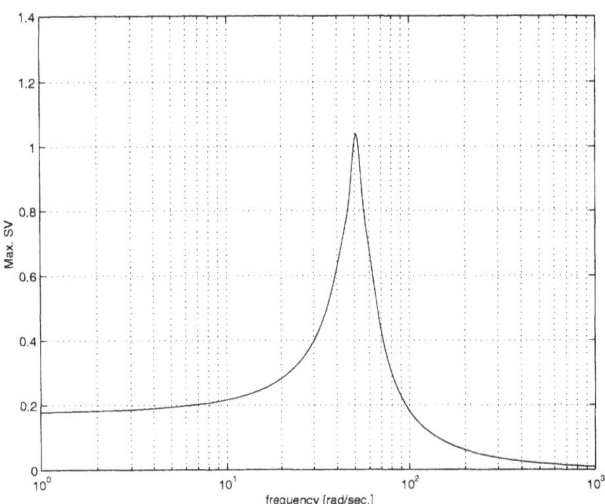

Fig. 3. Maximum SV of $T_{ew}(jw)$ - Choi and Chung

5.2 *Proposed method*

Now our methodology is applied to construct an observer independently of the control. The observer is of the form (4) and is obtained by solving one ARE (6). For $\epsilon_o = 3$ and $\gamma_o = 0.5$, the ARE (6) has the following solution and the corresponding observer gain :

$$P_o = \begin{bmatrix} 5.7398 & 0 & 0 \\ 0 & 0.0328 & 0 \\ 0 & 0 & 0.0009 \end{bmatrix} \times 10^4, \quad L = \begin{bmatrix} 0.5807 \\ 0 \\ 0 \end{bmatrix} \times 10^{-5}$$

Similarly, the controller for system (7) is designed by solving (8)-(9) with $\epsilon_c = 2.5$, $\gamma_c = 0.5$ and $R = I_3$. We obtain the following matrices:

$$P_c = \begin{bmatrix} 5.6742 & -0.3 & 0.0577 \\ -0.3 & 4.1756 & 0.0695 \\ -0.3 & 4.1756 & 0.0695 \end{bmatrix}, \quad K = \begin{bmatrix} -0.4156 \\ -0.5002 \\ -0.7514 \end{bmatrix}^T$$

The estimated error and controlled output responses are shown in figure 4 for the same $h = 0.33sec.$ and unit step disturbance at $t = 10sec.$

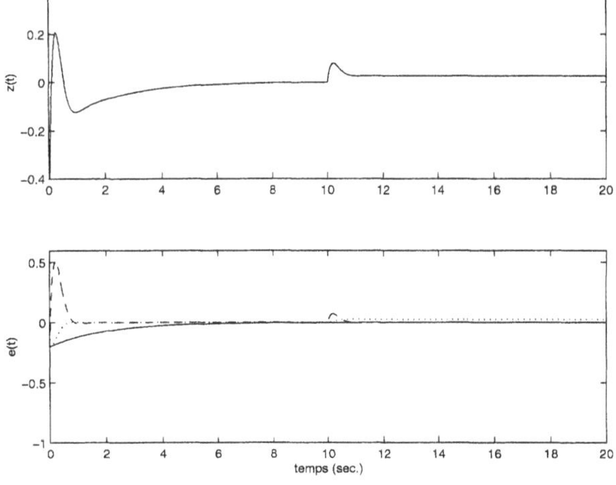

Fig. 4. Extended system estimated responses

In this case we note that the gain matrices L and K are reduced and that some fixed output disturbance attenuation level is guaranteed (here

$\|\widetilde{T}_{zw}\|_\infty = 0.105$) as we can see in figure 5. We can prove that $\|\widetilde{T}_{zw}\|_\infty$ satisfies:

$$\|\widetilde{T}_{zw}\|_\infty \leq 0.02 + (1.2)(0.105) = 0.146$$

The convergence of the observer is quite different to the previous case (see $e(t)$ in figures 4 and 1). Moreover, in Fig. 6 we can see that the frequency representation of $T_{ew}(jw)$ is better than in figure 3 (0.105 instead of 1.05). Finally one state estimation error has less important steady-state error due to the disturbance than in Choi and Chung's case.

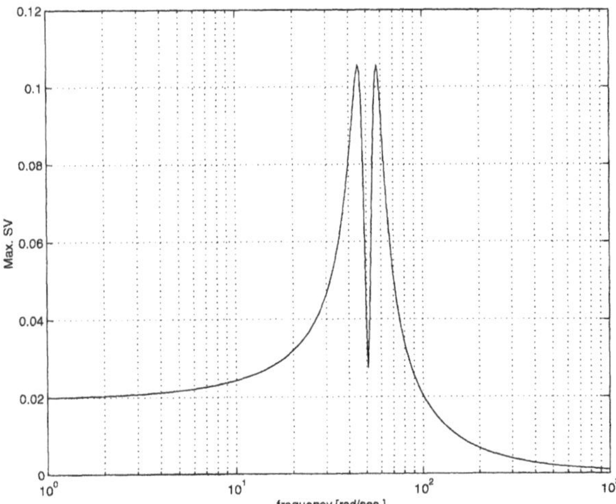

Fig. 5. Maximum SV of $\widetilde{T}_{zw}(jw)$

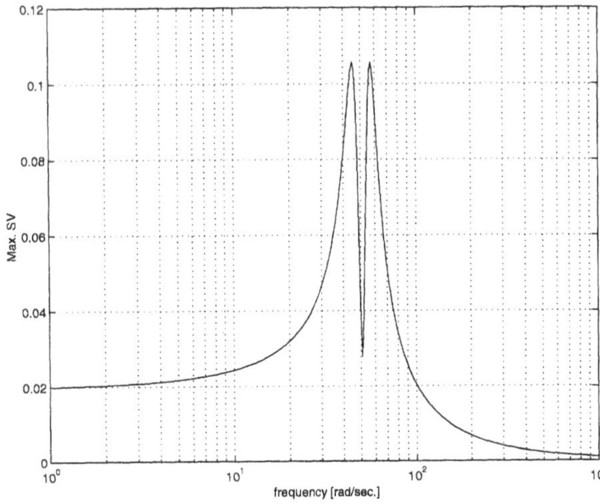

Fig. 6. Maximum SV of $T_{ew}(jw)$

6. CONCLUDING REMARKS

Using the results of Fattouh et al. [1999] and Lee et al. [1994] we have developed a delay independent dynamic feedback controlled system with good H_∞ control as well as H_∞ observer performance, by solving two independent Riccati equations with only two parameters for each one.

In Choi and Chung [1996] two coupled Riccati equations have to be solved, including 7 parameters, and H_∞ performance is ensured only for the control, not for the observer. Furthermore the observer designed in Fattouh et al. [1999] can also be used to perform other advanced H_∞ control laws.

Note also that sufficient conditions may be given in terms of some observability (resp. controllability) property for the resolution of the ARE (6) (resp. ARE (8)), as specified in Fattouh et al. [1999] (resp. in Lee et al. [1994]). Finally such an approach has been used to design H_∞ controllers and observers for linear systems with *point and distributed time-delays* using the LMI technique [Fattouh et al., 2000].

References

H. H. Choi and M. J. Chung. Observer-based h_∞ controller design for state delayed linear systems. *Automatica*, 32(7):1073–1075, 1996.

A. Fattouh, O. Sename, and J.-M. Dion. Robust observer design for time-delay systems: A riccati equation approach. *Kybernetika*, 35(6):753–764, 1999.

A. Fattouh, O. Sename, and J.-M. Dion. h_∞ controller and observer design for linear systems with point and distributed time-delays: An lmi approach. In *2nd IFAC Workshop on Linear Time Delay Systems*, Ancône, Italy, 2000.

J. H. Lee, S. W. Kim, and W. H. Kwon. Memoryless h^∞ controllers for state delayed systems. *IEEE Transactions on Automatic Control*, 39 (1):159–162, 1994.

A. Z. Manitius. Feedback controllers for a wind tunnel model involving a delay: Analytical design and numerical simulation. *IEEE Trans. on Automatic Control*, 29(12):1058–1068, 1984.

S.-I. Niculescu. h_∞ memoryless control with α-stability constraint for time-delay systems: An lmi approach. *IEEE Trans. on Automatic Control*, 43(5):739–743, 1998.

J.-H. Su. Further results on the robust stability of linear systems with a single time delay. *Systems & Control Letters*, 23:375–379, 1994.

H.L. Trentelman, A.A. Stoorvogel, and M. Hautus. *Control Theory for Linear Systems.* Springer, 2001.

www.elsevier.com/locate/ifac

STABILITY ANALYSIS OF ROLL GRINDING DELAY SYSTEM

LiHong Yuan, Veli-Matti Järvenpää, Erno Keskinen

Tampere University of Technology, Laboratory of Machine Dynamics
PO Box 589, FIN-33101 Tampere, Finland, Fax +358 3 3652307,
Email lihong.yuan@tut.fi

Abstract: In this paper the stability of a roll grinding delay system is studied. This delay system is characterised with a negative feedback and large imaginary parts in its poles. The authors analysed its stability by using the Nyquist's criterion. Time domain simulations were used for the verification of the stability analyses. *Copyright © 2001 IFAC*

Keywords: time delay, stability analysis, Nyquist's criterion, grinding system

NOMENCLATURE

c	damping coefficient
k	stiffness coefficient of the roll
k_n	stiffness coefficient of the contact
k_w	the wear coefficient
m	mass of the roll
Δr	surface shape error amplitude of the roll
s	the Laplace transform operator
r, R	radius of grindstone and roll
T	rotation time of the roll (the delay time)
w	width of grindstone
x	normal displacement of the roll
α, β	control factors of the delay effect
$\varepsilon(t)$	the total penetration function of the grindstone
ω_{stone}	angle frequency of grindstone
ω_n	surface shape error frequency

1. INTRODUCTION

The grinding process of paper machine rolls are applied either during the manufacturing phase of the rolls or in paper mills when the polymer covers of the rolls require maintenance to ensure the manufacturing quality of paper. In the grinding process, the cover of the roll is ground in a grinding machine and due to the removing polymer material the surface of the cover has the memory effect from previous rotations because the grinding path on the roll surface overlaps itself. The transversal profile of the roll surface has an undesired curved error shape after grinding. This delay effect will provide an important additional effect on the vibrations of the system and to describe the dynamic of this system, the delay factor must be involved in the differential equations.

In manufacturing engineering lots of theoretical analyses about chatter oscillations during roll grinding process has been carried out (see. e.g. Thomson 1992). Basically these analyses use a single degree of freedom delay equation and they are focused on tool's (grindstone's) oscillations with shape error sources on the surfaces of the tool and the workpiece. Now in this paper, instead of the tool the vibrating part studied is the workpiece i.e. the paper machine roll itself, because in this kind of grinding operation it is the most flexible component.

The general stability of delay systems is discussed in many modern control books and publications (e.g. Saaty, 1981). In previous contexts the authors have studied the stability of a grinding delay system with time domain analyses by conducting inverse Laplace transform or by using the method of steps (Zwillinger, 1989). However, it is more beneficial to know more about the stability of the grinding process by applying the methods used in control engineering. In this paper the stability of this grinding system is analysed by applying the Nyquist's criterion based on the Laplace transform function (without conducting the inverse transform as earlier).

2. ROLL GRINDING SYSTEM MODEL

The roll grinding system is composed of four main elements (figure 1):

- A roll, which is driven by its motor.
- A roll motor, which operates with the angular velocity about 0.2 Hz.
- A grindstone, which is driven by another motor. The penetration of the grindstone into the roll surface is about 0.2 mm.
- A grindstone motor, which drives the stone with angular velocity about 10 Hz. The rotation speed of the grindstone is much higher than the speed of the roll.

This system is a high-grinding system, which is used only in surface finishing operations. The external excitation source of this study is a roll, which has already a surface error shape expressed by a number of sinusoidal waves. The penetration of the grindstone into the cover layer is described by this shape error function. The roll is modelled as a simply supported flexural beam and the contact stiffness of the system is calculated according to the wear theory. The delay time is the rotation time of the roll (about 6 seconds) and it is constant.

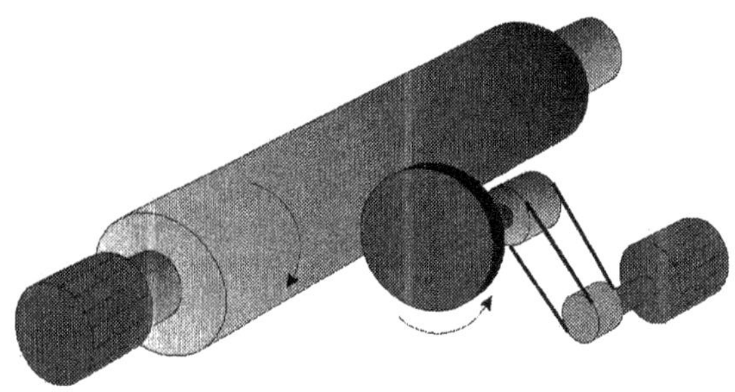

Fig. 1. The roll grinding procedure.

During the grinding process, the roll dynamical movement consists of the roll surface deformations and the vibration of the roll itself. The primary motions at the grinding point happen in the grinding contact's normal, tangential and torsional directions. In this paper, only the normal force is considered and the grindstone only gives a transversal feed without Z-axial movement. So the equation of motion describes the normal contact problem of the roll and the grindstone as a single degree of freedom system. The equation of motion of the system is

$$m\ddot{x} + c\dot{x} + kx = k_n\left[\varepsilon - \alpha\beta\varepsilon\,(t - T)\right] \quad (1)$$

with the initial conditions

$$x(0) = 0\,,\ \dot{x}(0) = 0\,, \quad (2)$$

and

$$\sin(\omega_n\, t) = 0 \ \text{ for } -T \le t < 0 \quad (3)$$

The delay time is the rotation time of the roll T. The penetration function is

$$\varepsilon(t) = -x(t) + \Delta r \sin(\omega_n\, t) \quad (4)$$

According to the classical wear theory, the contact stiffness coefficient k_n is also a function of the rotation time T

$$k_n = \frac{w}{k_w} \cdot \frac{2\pi R / T}{r\omega_{\text{stone}} - 2\pi R / T} \quad (5)$$

when $\omega_{stone} >> 2\pi R / T$

$$k_n \approx \frac{w}{k_w} \cdot \frac{R}{r\omega_{\text{stone}}} \cdot \frac{2\pi}{T} = \frac{k_{n0}}{T} \quad (6)$$

and now this can be rewritten

$$k_n(T) = \frac{k_{n0}}{T}\,,\ k_{n0} = \frac{wR}{k_w r} \cdot \frac{2\pi}{\omega_{\text{stone}}}$$

then the equation (1) becomes

$$\begin{aligned} &m\ddot{x} + c\dot{x} + (k + \frac{k_{n0}}{T})x - \alpha\beta\frac{k_{n0}}{T}x(t - T) = \\ &\frac{k_{n0}}{T}\Delta r \sin(\omega_n\, t) - \alpha\beta\frac{k_{n0}}{T}\Delta r \sin\{\omega_n\,(t - T)\} \end{aligned} \quad (7)$$

By using Laplace transform (Spiegel, 1965), the corresponding transfer function of the system (1) can be obtained

$$X(s) = \frac{k_n(1 - \alpha\beta e^{-sT})\Delta r \cdot \frac{\omega_n}{s^2 + \omega_n^2}}{ms^2 + cs + k + k_n - \alpha\beta\, k_n e^{-sT}} \quad (8)$$

and the transfer function (8) includes exponent components due to the time delay. The control block (Sinha, 1986, Dutton, 1997) diagram (SIMULINK) for this is illustrated in figure 7.

3. STABILITY ANALYSIS

Without solving the inverse Laplace transform of the system, the stability of the grinding system could be analysed by using stability criterions.

Let

$$G_1(s) = k_n /\left(ms^2 + cs + k + k_n\right) \qquad (9)$$

$$R(s) = \Delta r\omega_n /(s^2 + \omega_n^2) \qquad (10)$$

The closed-loop transfer function is

$$W_{c\text{-}lo}(s) = \frac{X(s)}{R(s)} = \frac{(1-\alpha\beta e^{-sT})G_1(s)}{1-\alpha\beta e^{-sT}G_1(s)} \qquad (11)$$

The close loop characteristic equation for s is transcendent and has the form of

$$F(s) = 1 - \alpha\beta e^{-sT} G_1(s) = 0 \qquad (12)$$

or

$$1 + (-\alpha\beta e^{-sT} G_1(s)) = 0 \qquad (13)$$

The condition of stability for a control system with delay is that the characteristic equation has no roots in the right-hand half-plane of the variable s. However, there is a transcendental function and infinite roots in the characteristic equation (12) or (13) for our grinding system with delay. This prevents using an algebraic stability criterion. The use of the argument principle with $s = i\omega$ leads to a conventional formulation of the Mikhailov's criterion for $F(i\omega)$ whose locus for a stable system with $0 \leq \omega < \infty$. For this delay system the formulation of the Nyquist criterion remains true with respect to the frequency locus or open-loop transfer function

$$W_{o-lo}(s) = -\alpha\beta e^{-sT} G_1(s) \qquad (14)$$

The Nyquist's stability criterion (Netushil, 1978) is more convenient way to study delay systems. According to the Nyquist's criterion, the closed-loop stability is determined from the system's open-loop transfer function. If the closed-loop system is stable, the open-loop gain and phase margin measures indicate the degree of stability and the adjustments required for improvement. So the open-loop transfer function is analysed. To guarantee stability, the Nyquist plot of open loop transfer function may not encircle the point (-1,j0) in the complex plane.

By using parameter values from an industrial case, the following Nyquist and Bode plots are illustrated in figures 2 to 4. The three cases are studied:

1) A case with delay. The open-loop transfer function is (14).

For the open-loop transfer function (14), there are some characteristics needed to notice here. Firstly, the time delay in a system will create a phase lag without altering the magnitude. This will reduce the system's stability margins and can make the system difficult to control. Secondly, in the open-loop transfer function (14), there is negative feedback, which is not a normal case and will increase phase values. Thirdly, the imaginary parts of the poles of in the open-loop transfer function (14) are much larger than the real parts with the industrial parameter values. This also makes the stability study more difficult. In this context the delay case will include all the above characteristics when it will be mentioned in later text.

2) A case when delay T is set to zero. The open-loop transfer function is

$$W_{o-lo}(s) = -\alpha\beta G_1(s) \qquad (15)$$

3) A case without delay. The open-loop transfer function is

$$W_{o-lo}(s) = G_1(s) \qquad (16)$$

This is the transfer function of the basic no-delay system.

In figure 2, in the no delay cases (transfer function (15) and (16)), the Nyquist curves do not encircle the point (-1, j0) and the system is stable with the parameters used.

In figures 2b and 3a (transfer function (14)) with the same parameters when the delay is T = 6 seconds system is unstable. It can also be seen that the damping gives influence on the stability. When the relative damping is 0.1 % (figure 2b), the curve encircles the point (-1, j0) and the system is unstable. When the relative damping is 0.2 % (figure 3a), the curve is almost pass over the point (-1, j0) and the system is close stable. When the relative damping is 0.3 %, the curve does not encircle the point (-1, j0) and the system is stable (figure 3b).

From the bode plots of figure 4, it can be seen that the magnitudes have same values in delay case and no delay cases. It means that the delay does not give effect on the magnitude (Dorf, 1995).But the phase lag is decreased a lot at large enough frequencies in the delay case compare to the both no delay cases. It means the delay has a dramatic main effect on the phase. Additionally the negative feedback and large imaginary parts of poles seems also contribute to the large phase in this system.

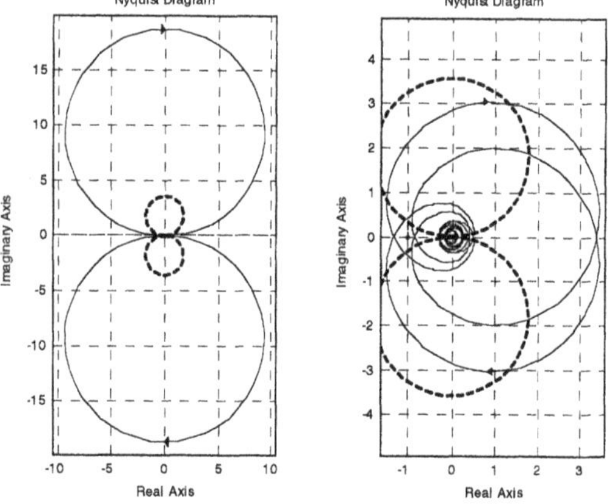

Fig. 2. Nyquist plots of the system.
a) no delay (—) and delay T set to zero (▪▪),
b) time delay set to zero (▪▪) and delay $T = 6$ sec and relative damping $\zeta = 0.1\%$ (—).

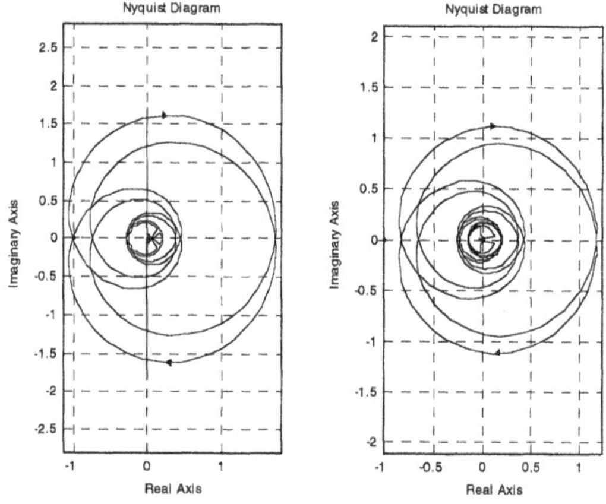

Fig. 3. Nyquist plots of the system. Delay $T = 6$ sec
a) ζ=0.3 %. b) ζ=0.2 %.

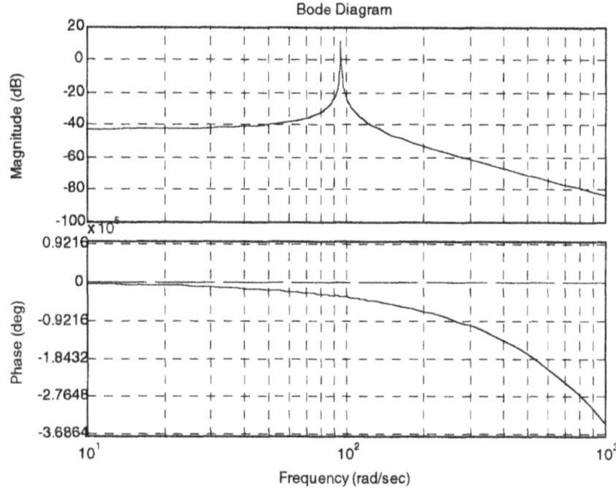

Fig. 4. Bode plots of the system. Relative damping ζ=0.1 %. No delay (--) and delay $T = 6$ sec (—)

The next interesting thing to know is to get some information of the general stability of the system with different delay time values i.e. the with different rotation times of the roll ground. It should be noted that the contact stiffness k_n is a function of delay T (equation 6). In figure 5 the stable and the unstable regions are presented as a bar plot. The solid black bar marks a stable region and a space between the black bars marks an unstable region (a "1" is set when the system is stable and a "0" is set when it is unstable). The time delay T changes from 5.5 to 6.5 seconds. Two different cases are shown with relative damping $\zeta = 0.2$ % and $\zeta = 0.3$ %. Each stable region seems to increase when the damping is getting bigger.

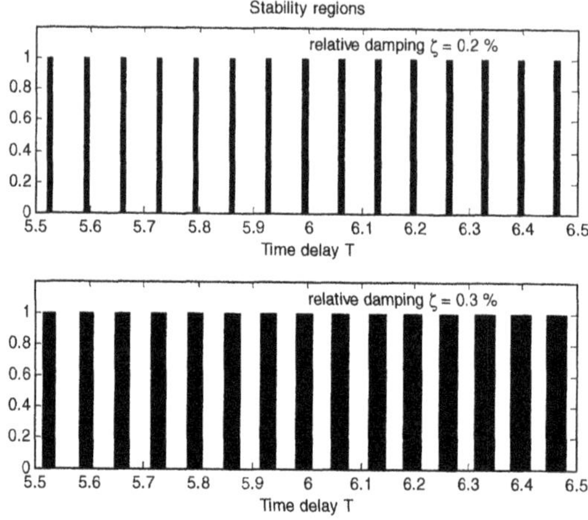

Fig. 5. Stable and unstable regions changing with time delay a) ζ= 0.2 % b) ζ= 0.3 %

For more information concerning the delay system stability, the phase margin and delay margin can measure the degree of stability of the system (Meirovitch, 1990) and can be computed by means of Nyquist plot and Bode plot in a Matlab program. The phase margin and the delay margin are shown as functions of relative damping when T = 6 seconds in figure 6. In this case the delay margin seems to be valid only around this delay time.

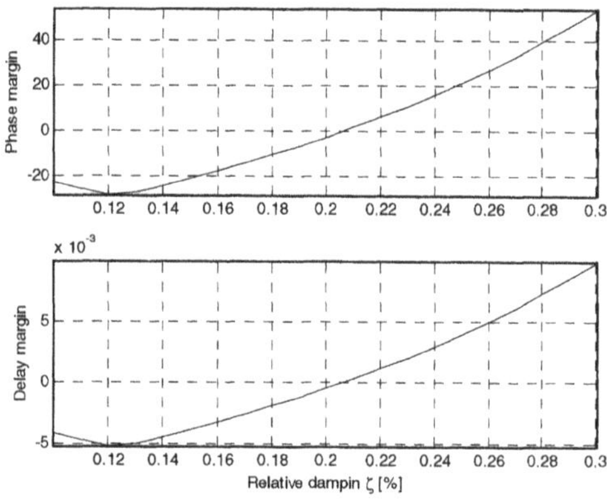

Fig. 6. Phase margin and delay margin changing with relative damping

4. SIMULINK MODEL SIMULATION RESULTS

By using the transfer function (8) and a sine signal as an input, the control block diagram presented in figure 7 is obtained. Two simulation results of the delay case are shown for verification purposes.

The stable and the unstable cases in figure 9 have the same time delay value $T = 6$ sec and relative damping values same as in the corresponding Nyquist plots in figures 2b and 3b.

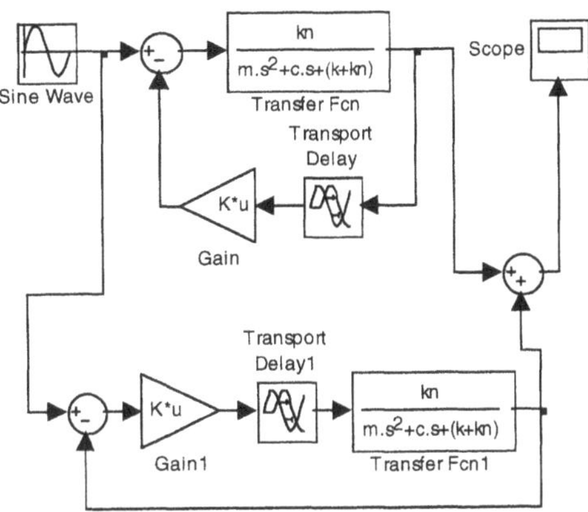

Fig. 7. SIMULINK diagram of the system.

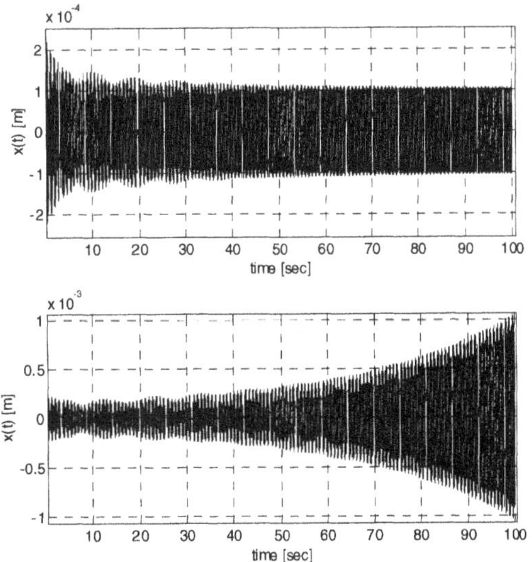

Fig. 8. Time domain plots of the system. Time delay $T = 6$ sec. a) Stable case ζ=0.3 %, b) Unstable case ζ=0.1 %

5.CONCLUSIONS

The stability of the roll grinding system with delay is observed by applying the Nyquist's criterion. The criterion seems work in the above special case well. The phase and delay margins are changing with damping. The method also seems to be accurate enough for industrial cases. Proper controllers could be added to improve the delay system stability in a future work.

ACKNOWLEDGEMENTS

The authors would like to express their gratitude to The Finnish Academy, which has provided funding for this very interesting project.

REFERENCES

Dorf R. C., Bishop R. H., *Modern Control Systems*, 7th Ed., Addison-Wesley, 1995

Dutton K., Thompson S., Barraclough B., *The Art of Control Engineering*, Addison-Wesley, 1997

Meirovitch L., *Dynamics and Control of Structures*, John Wiley & Sons, Inc., 1990

Netushil A., *Theory of Automatic Control*, Mir Publisher, 1978

Saaty T.L., *Modern Nonlinear Equations*, Dover, New York, 1981

Sinha N. K., *Control Systems*, CBS Publishing Japan Ltd., 1986

Spiegel M. R.., *Theory and Problems of Laplace Transforms*, Schaum Publishing Com.,1965. Ltd., 1986

Thompson R. A., *On the doubly regenerative Stability of a grinder. The effect of contact stiffness and wave filtering.* Trans. ASME 114:53-60,1992.

Zwillinger D., *Handbook of Differential Equations*, Academic Press,Inc.,1989

www.elsevier.com/locate/ifac

NUMERICAL SIMULATION OF DYNAMIC ROLLING CONTACT INCLUDING DELAY EFFECT FROM COVER LAYER

Veli-Matti Järvenpää, LiHong Yuan

Tampere University of Technology, Laboratory of Machine Dynamics
PO Box 589, FIN-33101 Tampere, Finland, Fax +358 3 3652307,
Email vmj@ruuvi.me.tut.fi

Abstract: The numerical solution of a rolling contact dynamics of two paper machine rolls is discussed. The delay effect of a polymer roll cover layer is included in the system and the vibrations of the contact are studied. The stability of a test delay case is determined by time domain analyses. *Copyright © 2001 IFAC*

Keywords: delay analysis, time domain analysis, rotors, dynamics, paper industry

1. INTRODUCTION

In this paper a numerical analysis of a classical rolling contact delay case is discussed. The technical systems of interest are the paper finishing units of paper mills. These units are used for processing the surface of the paper at the end of the paper manufacturing process. Typically a unit consists of a stack of two horizontal rolls in a pre-compressed contact (this is called as a nip contact) operated by a hydraulic closing mechanism and two electric motors. The paper web goes through this nip contact with the speed of 1000-1500 m/min, which corresponds to rotation speeds of 5-10 Hz for the rolls. The rolls are made from steel or cast iron, but polymer cover layers are usually used on them to smoothen the contact [Joukkio 1999]. The delay effect is introduced to the system when the polymer layer has a viscoelastic recovery time longer than the rotation period of the rolls. Some deformation on the layer surface still remains when the same layer point rolls through the nip contact again. To analyze this kind of system the following steps have been carried out. Firstly, the equations of motions of the rolls are expressed in rotating coordinate frames. Secondly, the rolling contact with damping is modeled by the contact dynamics formulation. Thirdly, an external excitation source is introduced to the contact and finally this excitation has been altered to carry the delay effect of the polymer cover layer. This system of equations is then solved in time domain by an appropriate numerical integration procedure.

2. ROTOR DYNAMICS MODEL OF ROLL

The rolls are modeled as simply supported beams with the lateral degrees of freedom u and v (figure 1). The dynamic behavior is described by the set of the analytical vibration modes (eigenvectors)

$$u = \sum u_i \, n_i(z) \qquad v = \sum v_i \, n_i(z) \tag{1}$$

where the modes are

$$n_i(z) = \sin\frac{i\pi z}{L} \tag{2}$$

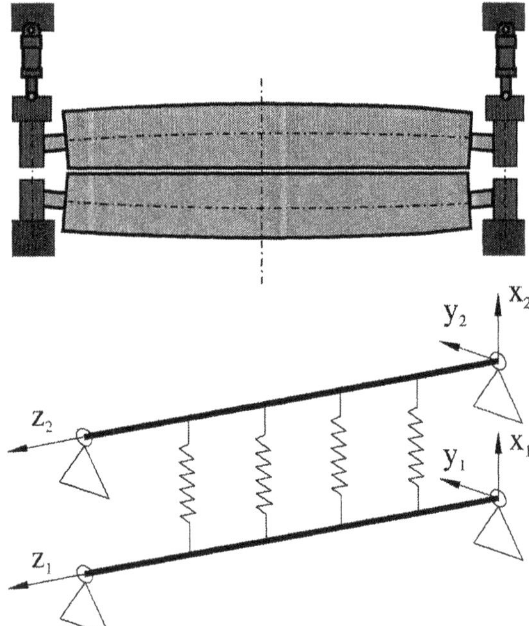

Fig. 1. The beam model of the rolls

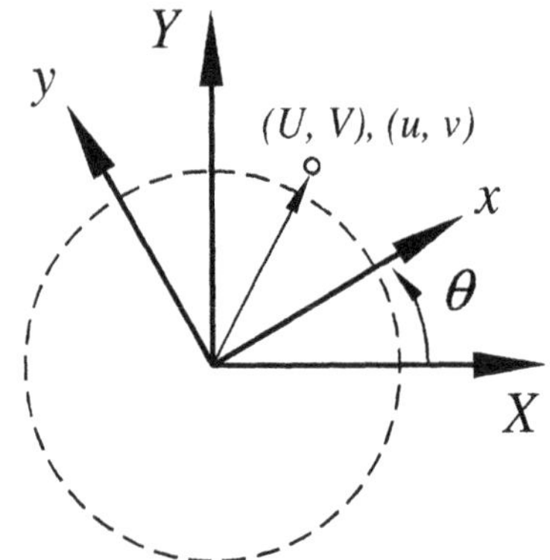

Fig. 2. The local and global coordinate systems.

Classically, in a rotor dynamics model it is more convenient to express the equations of motion in a rotating coordinate frame. The rotation about z-axis is defined as θ and the transformation between the local and global coordinate system (figure 2) is described as

$$\mathbf{S} = \mathbf{Rs} \tag{3}$$

where

$$\mathbf{S} = \begin{bmatrix} \cos\theta & -\sin\theta \\ \sin\theta & \cos\theta \end{bmatrix} \tag{4}$$

and the vectors

$$\mathbf{S} = \begin{bmatrix} U \\ V \end{bmatrix}, \quad \mathbf{s} = \begin{bmatrix} u \\ v \end{bmatrix} \tag{5}$$

are the global and the local displacement vector of a roll, respectively. In the case of two rolls in a nip contact the both rolls have their own local coordinate systems (figure 2).

The global velocities can be derived from (3)

$$\dot{\mathbf{S}} = \mathbf{A}\dot{\mathbf{s}} + \dot{\theta}\frac{\partial \mathbf{A}}{\partial \theta}\mathbf{s} \tag{6}$$

The kinetic energy of one roll is

$$K = ½\int_0^L \rho A \dot{\mathbf{S}} \cdot \dot{\mathbf{S}}\, dz + ½ J_z \dot{\theta}^2 \tag{7}$$

and the strain energy

$$E = ½\int_0^L EI\,(\frac{\partial^2 u}{\partial z^2} + \frac{\partial^2 v}{\partial^2 z})\,dz \tag{8}$$

By substituting (7) and (8) into the Lagrange's equation

$$\frac{d}{dt}(\frac{\partial K}{\partial \dot{\mathbf{y}}}) - \frac{\partial K}{\partial \mathbf{y}} + \frac{\partial E}{\partial \mathbf{y}} = \mathbf{f} \tag{9}$$

the equations of motion of one roll are put in the form of

$$\begin{gathered} \begin{bmatrix} m_i & \\ & m_i \end{bmatrix}\begin{bmatrix} \ddot{u}_i \\ \ddot{v}_i \end{bmatrix} + \begin{bmatrix} k_i & \\ & k_i \end{bmatrix}\begin{bmatrix} u_i \\ v_i \end{bmatrix} = \\ \begin{bmatrix} f_{xi} \\ f_{yi} \end{bmatrix} - 2\dot{\theta}\begin{bmatrix} & -m_i \\ m_i & \end{bmatrix}\begin{bmatrix} \dot{u}_i \\ \dot{v}_i \end{bmatrix} \\ + \dot{\theta}^2\begin{bmatrix} m_i & \\ & m_i \end{bmatrix}\begin{bmatrix} u_i \\ v_i \end{bmatrix} - \ddot{\theta}\begin{bmatrix} & -m_i \\ m_i & \end{bmatrix}\begin{bmatrix} u_i \\ v_i \end{bmatrix} \\ J_z \ddot{\theta} = f_\theta \\ -\sum m_i(\ddot{v}_i u_i - \ddot{u}_i v_i + 2\dot{\theta}\dot{u}_i u_i \\ + 2\dot{\theta}\dot{v}_i v_i + \ddot{\theta} u_i^2 + \ddot{\theta} v_i^2) \end{gathered} \tag{10}$$

where

$$\begin{aligned} m_i &= \frac{\rho A}{L}\int_0^L n_i \cdot n_i\, dz \\ k_i &= EI\int_0^L \frac{\partial^2 n_i}{\partial z^2} \cdot \frac{\partial^2 n_i}{\partial z^2} dz \end{aligned} \tag{11}$$

If a constant rotation speed is assumed with $\ddot{\theta} = 0$ and $\dot{\theta} = 0$, (10) is simplified as

$$\begin{bmatrix} m_i & \\ & m_i \end{bmatrix}\begin{bmatrix} \ddot{u}_i \\ \ddot{v}_i \end{bmatrix}+\begin{bmatrix} k_i & \\ & k_i \end{bmatrix}\begin{bmatrix} u_i \\ v_i \end{bmatrix}=\begin{bmatrix} f_{xi} \\ f_{yi} \end{bmatrix} \\ -2\omega\begin{bmatrix} & -m_i \\ m_i & \end{bmatrix}\begin{bmatrix} \dot{u}_i \\ \dot{v}_i \end{bmatrix}+\omega^2\begin{bmatrix} m_i & \\ & m_i \end{bmatrix}\begin{bmatrix} u_i \\ v_i \end{bmatrix} \quad (12)$$

The use of (12) instead of (10) is of course more beneficial because the delay time is now constant (the rotation period of rolls and the delay time is $2\pi/\omega$). This is also a physically reasonable selection due to the fast acceleration of the nip unit to its operational speed. The oscillations during the rapid start-up have little influence on the long-time stability of the system. But if for example a speed control of θ is studied naturally (10) must be implemented.

3. MODEL OF ROLLING CONTACT

In the case of two rolls in contact each roll has its equations of motion in the form of (12) or (10). Now, the rolling nip contact must be described between the rolls. This is more complicated because the roll equations are expressed in the rotating coordinate frame and the nip contact has always the same position in the global reference coordinate frame (figure 3). In this paper the contact dynamics formulation [Johnson 1985] [Keskinen 1994] is used for describing the contact. The idea is to introduce a contact force between the rolls as

$$N_k = k_n(U_{1x} - U_{2x}) \quad (13)$$

where U_{1x} and U_{2x} are the displacements of the first and the second roll in the contact direction, respectively and k_n is the stiffness of the polymer layer during the nip contact in the case of beams used. Similarly, a damping force can be introduced as

$$N_c = c_n(\dot{U}_{1x} - \dot{U}_{2x}) \quad (14)$$

So finally the system equations of two rolls in contact include firstly the rotor dynamic equations (12) for each rolls and secondly the equations (13) and (14) for the contact with the coordinate transformations (3) and (6). The solution of this system naturally requires a non-linear (iterative) time integration procedure to be used.

4. DELAY EQUATION

Ideally this kind of nip unit system should not have too much vibration problems if the (constant) rotation speed is selected properly inside appropriate driving

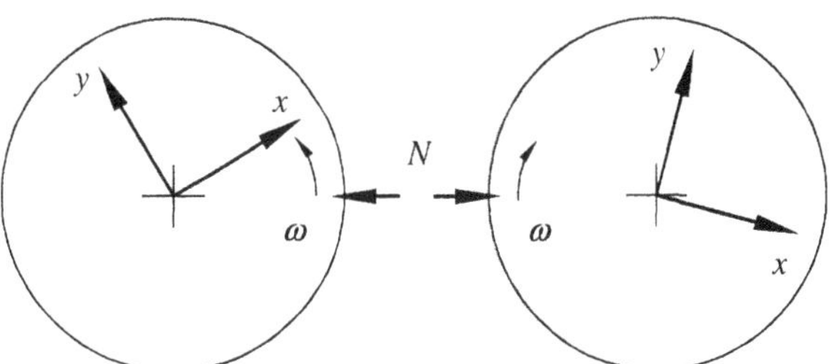

Fig. 3. The nip contact.

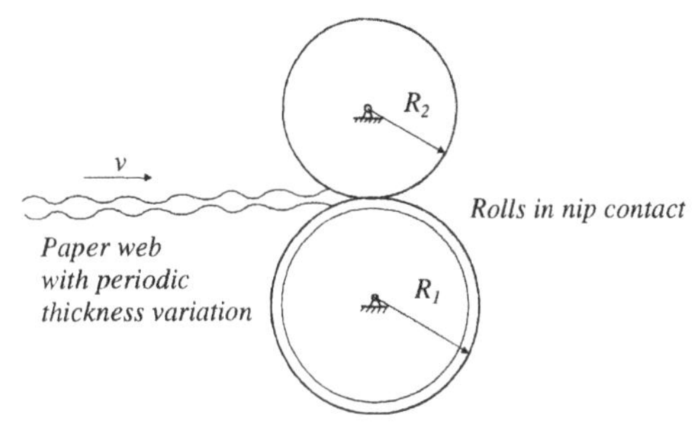

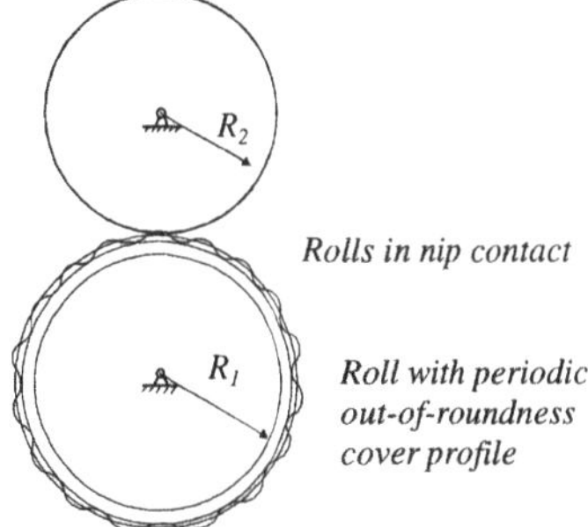

Fig. 4. Typical external excitation sources.

window i.e. 20 % away from closest eigenfrequency. But serious vibrations can still exist if the polymer cover layer carries an undesirable delay effect. To study numerically the stability of this rotor dynamics system with a delay, equation (10) should be modified accordingly. Typical normal vibration excitation sources in nip units are a periodic thickness variation of the paper web or a periodic surface error shape profile on roll covers (figure 4). Both have a sinusoidal characteristic and these excitation sources can be described as an external force

$$f(t) = k_n\, \Delta Z \sin \omega_w t \quad (15)$$

where in ΔZ is the variation amplitude and ω_w the corresponding excitation frequency. Now when the delay is included the external force system in the nip can be expressed as

$$\begin{aligned} N_f &= f(t) \\ &+ \lambda k_n\{U_{x1}(t-\tau) - U_{x2}(t-\tau)\} \\ &- f(t-\tau) \end{aligned} \quad (16)$$

where τ is the time delay and λ is the recovering constant of a polymer cover. So finally the total force system in the nip is the combination of equations (13), (14) and (16)

5. NUMERICAL RESULTS

A test example with appropriate parameter values was implemented for verification purposes. A constant rotation speed was selected (equation (12)) and only the first vibration mode was used (i.e. $i = 1$ in equation 2), so the equations of motion can be expressed as

$$\begin{bmatrix} m_1 & & & \\ & m_1 & & \\ & & m_2 & \\ & & & m_2 \end{bmatrix}\begin{bmatrix} \ddot{u}_1 \\ \ddot{v}_1 \\ \ddot{u}_2 \\ \ddot{v}_2 \end{bmatrix} + \begin{bmatrix} k_1 & & & \\ & k_1 & & \\ & & k_2 & \\ & & & k_2 \end{bmatrix}\begin{bmatrix} u_1 \\ v_1 \\ u_2 \\ v_2 \end{bmatrix}$$

$$= \begin{bmatrix} c_1 & -s_1 & & \\ s_1 & c_1 & & \\ & & c_2 & -s_2 \\ & & s_2 & c_2 \end{bmatrix}^T \left(\begin{bmatrix} -N_k \\ 0 \\ N_k \\ 0 \end{bmatrix} + \begin{bmatrix} -N_c \\ 0 \\ N_c \\ 0 \end{bmatrix} + \begin{bmatrix} -N_f \\ 0 \\ N_f \\ 0 \end{bmatrix} \right)$$

$$+ 2\omega \begin{bmatrix} & -m_1 & & \\ m_1 & & & \\ & & & -m_2 \\ & & m_2 & \end{bmatrix}\begin{bmatrix} u_1 \\ v_1 \\ u_1 \\ v_2 \end{bmatrix}$$

$$+ \omega^2 \begin{bmatrix} m_1 & & & \\ & m_1 & & \\ & & m_2 & \\ & & & m_2 \end{bmatrix}\begin{bmatrix} u_1 \\ v_1 \\ u_2 \\ v_2 \end{bmatrix} \qquad (18)$$

where

$$\begin{aligned} c_1 &= \cos\theta_1, \quad s_1 = \sin\theta_1 \\ c_2 &= \cos\theta_1, \quad s_2 = \sin\theta_2 \\ \theta_1 &= \theta, \quad \theta_2 = -\theta, \quad \theta = \omega t \end{aligned} \qquad (19)$$

The Newmark numerical time integration method was selected (see e.g. [Bathe 1996]). The solution of (18) requires a non-linear version of this method and the procedure described in [Gerandin 1997] was used. The delay equation was solved by the method of steps procedure by adjusting the time step according to the delay size. At the beginning of the simulation a short non-delay simulation was executed as a start-up procedure for the delay part. The nip stiffness k_n = $2.5 \cdot 10^8$ N/m and the delay time T is 0.1667 sec.

The case without delay is presented in figures 5 and 6. In this case only the external excitation of equation (16) is included and as can be seen the system is stable with the parameters used. When the delay is introduced to the system it becomes totally unstable as can be seen from figure 7. This is of course an

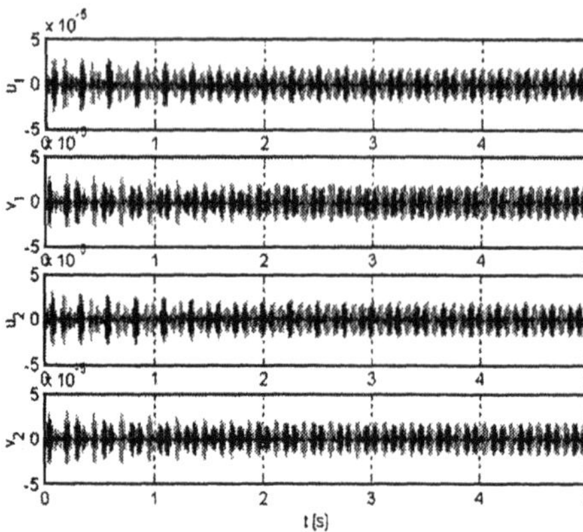

Fig. 5. The non-delay case. Rotating frame.

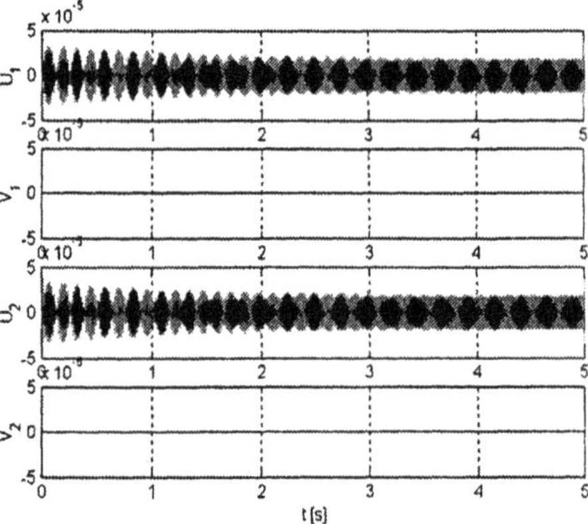

Fig. 6. The non-delay case. Global frame.

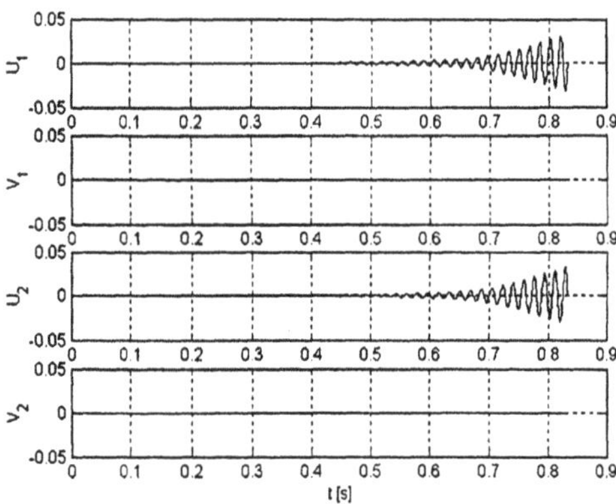

Fig. 7. The delay case. Damping 0.1 %.

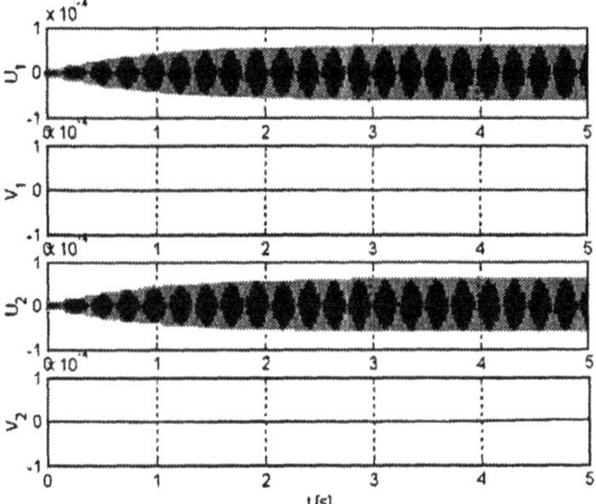

Fig. 8. The delay case. Damping 15 %.

expected result because the size of the time delay is huge and probably out of any reasonable region of stability. Unfortunately, the nature of this rolling contact system is that the stability cannot be reached by adjusting the rotation speed of the system, because this would require very unpractical rotation velocities for the rolls. More practical way is to change the system dynamics for example by increasing the damping. In figure 8 the relative damping of the nip contact is set to 15 % and the system becomes stable. Naturally, the damping value of 15 % is out of the rage of roll cover polymer material damping values and the structural damping of the system cannot be expected to be this high in reality. One possibility to solve this problem is to introduce an active damping mechanism construction to the system.

6. CONCLUSIONS

The numerical solution of the delay equation as a part of a non-linear system of equations described in this paper's rolling contact case seems to work well. The first main purpose of this solution procedure is to provide a tool for analyzing the stability of complex delay systems. An alternative use could be for measurement data verification to detect the presence of delays in the system measured. Further analyses are required to verify the accuracy of the numerical method of this paper.

REFERENCES

Bathe K.-J. , *Finite Element Procedures*, Prentice-Hall, Inc. , 1996

Gerandin M. & Rixen D., *Mechanical Vibrations*, John Wiley & Sons Ltd., 1997.

Johnson, K.L., *Contact Mechanics*, Cambridge University Press, 1985

Jokkio M., *Papermaking Part 3, Finishing, Papermaking Science and Technology, Book 10,* TAPPI Press, Fapet Oy Publ., Helsinki, Finland, 1999.

Keskinen, E., *A Contact Dynamics Formulation for Distributed Simulation of Flexible Hydromechanical Systems*, Doctoral Thesis, Tampere University of Technology, 1994

www.elsevier.com/locate/ifac

STABILIZATION OF LINEAR TIME-DELAY SYSTEMS WITH ANTI-STABLE EIGENVALUES UNDER AMPLITUDE AND RATE SATURATION

S. Tarbouriech* G. Garcia*,[1] I. Queinnec*

* *LAAS-CNRS, 7 Avenue du Colonel Roche, 31077 Toulouse cedex 4, France. E-mail: (tarbour,garcia,queinnec)@laas.fr*

Abstract: This paper focuses on the delay-independent stabilization problem for delayed systems subject to control amplitude and rate saturation. Considering the actuator modeled as an integrator presenting input and output saturation, a condition for the stabilization of an a priori given set of admissible initial states is expressed from certain saturation nonlinearities representation and Lyapunov-Krasovskii results. An algorithm based on the iterative solution of LMI-problems combined with convex optimization is proposed in order to compute the control law. *Copyright © 2001 IFAC*

Keywords: Delay Continuous-Time Systems, Amplitude and Rate Saturation, Stability, Lyapunov-Krasovskii Theorem.

1. INTRODUCTION

In various engineering systems, both time-delay phenomena and saturating actuators are commonly encountered and are frequently a source of problems such as limit cycles, parasitic equilibrium points or even lost of stability for the closed-loop system. The stabilization problem of uncertain systems with state-delay has attracted much interest (Mahmound and Al-Muthairi, 1994). The problem of uncertain systems stabilization with saturating controls has concurrently motivated an important effort of research due to its practical importance and its inherent difficulty: see, for example, (Tarbouriech and Garcia, 1997) and references herein. Moreover, in most of practical problems, not only limited amplitude signals but also limited speed reactions of control actuators are imposed by physical and technological constraints. If the negligence of amplitude control bounds may be source of instability for the closed-loop system, the rate saturation induces a phase lag that may have a high destabilizing effect. is also subject to rate saturation.

[1] Also with INSA, Toulouse, France.

The aim of this paper is to study the problem of stabilization of uncertain linear time-delay systems subject to both input position and rate constraints. At the authors knowledge, one of the first papers dealing with this kind of problem is (Tarbouriech and Garcia, 1999). Our objective is to propose a method for computing state feedback control laws that ensure the asymptotic stability of the closed-loop system with respect to a given set of admissible initial conditions. Considering the actuator modeled as an integrator presenting input and output saturation, a condition of closed-loop stabilization is formulated from certain saturation nonlinearities representation and Lyapunov-Krasovskii results. It should be pointed out that, in plus of the presence of delay, our approach allows effective saturation and no open-loop stability assumptions are required, unlike what is done in (Lin, 1997) and (Shewchun and Feron, 1997). Our approach also differs from (Tarbouriech and Garcia, 1999) in the sense that the set of admissible initial states to be stabilized is given a priori and from (Tarbouriech *et al.*, 1999), (Shewchun and Feron, 1997) and (Tyan and Bernstein, 1997) due to the presence of delay.

It is necessary to underline the importance of characterizing or estimating the set of admissible initial conditions which can be stabilized despite the saturation occurrence when addressing the problem of stabilization for linear time-delay systems with actuator amplitude saturation (Tarbouriech and Gomes da Silva Jr., 2000). Indeed, in the current case of both position and rate saturations, the necessity of knowing which initial conditions can be stabilized is primordial (Gomes da Silva Jr. and Tarbouriech, 2000). This fact is not taken into account in (Niculescu *et al.*, 1996) or in (Chen *et al.*, 1988), in which the approach, based on a certain majoration of the saturation function, may lead to non-desired behaviors depending on the initial condition.

Notation. $\Re^+$ is the set of non-negative real numbers. For any vector $x \in \Re^n$, $x_{(i)}$ denotes the ith component of x. The elements of a matrix $A \in \Re^{m \times n}$ are denoted by $A_{(i,l)}$, $i = 1, \ldots, m$, $l = 1, \ldots, n$. $A_{(i)}$ denotes the ith row of matrix A. For two symmetric matrices, A and B, $A > B$ means that $A - B$ is positive definite. A' denotes the transpose of A. I_m and 1_m respectively denote the m-order identity matrix and the vector in $\Re^m$ defined as: $1_m = [1 \ldots 1]'$. $\lambda_{max}(P)$ and $\lambda_{min}(P)$ respectively denote the maximal and minimal eigenvalues of matrix P. $\mathcal{C}_{\tau,n} = \mathcal{C}([-\tau, 0], \Re^n)$ denotes the Banach space of continuous vector functions mapping the interval $[-\tau, 0]$ into $\Re^n$ with the topology of uniform convergence. $\| \phi \|_c = \sup_{-\tau \leq t \leq 0} \| \phi(t) \|$ stands for the norm of a function $\phi \in \mathcal{C}_{\tau,n}$. $\mathcal{C}^v_{\tau,n} = \{\phi \in \mathcal{C}_{\tau,n} \; ; \; \| \phi \|_c < v, v > 0\}$. The saturation functions are classical set-valued functions defined $\forall \; x \in \Re^m$ as $sat_{u_0}(x_{(i)}) = sign(x_{(i)}) min(u_{0(i)}, |x_{(i)}|)$, $u_{0(i)} > 0, \forall i = 1, \cdots, m$.

2. PROBLEM STATEMENT

We consider a class of nonlinear systems which are obtained by cascading linear systems with actuator containing memory-free input and output nonlinearities of saturation type (see Fig.1).

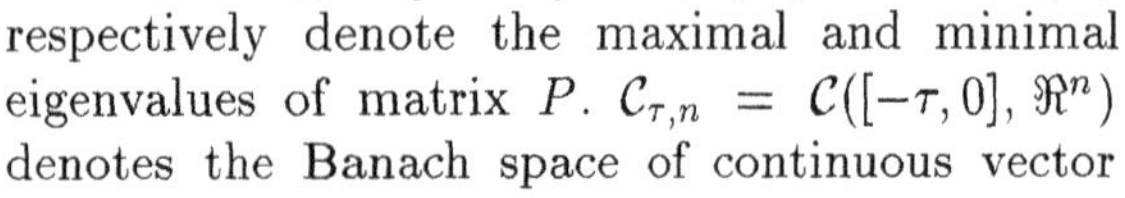

Fig. 1. Actuator modelling

The actuator under consideration is then modeled as an integrator with input ($u_a \in \Re^m$) and output ($y_a \in \Re^m$) saturations, described by:

$$\begin{cases} \dot{u}(t) = sat_{u_1}(u_a(t)) \\ y_a(t) = sat_{u_0}(u(t)) \end{cases} \tag{1}$$

The plant is a time-delay uncertain system defined as:

$$\begin{cases} \dot{x}(t) = (A + \Delta A)x(t) + (A_\tau + \Delta A_\tau)x(t - \tau) \\ \quad +(B_\theta + \Delta B_\theta)y_a(t - \theta) + (B + \Delta B)y_a(t) \\ y(t) = x(t) \end{cases} \tag{2}$$

with the initial conditions:

$$x(t_0 + \psi) = \phi_x(\psi), \forall \psi \in [-\hat{\tau}, 0], (t_0, \phi) \in \Re^+ \times \mathcal{C}^{v_x}_{\hat{\tau},n}$$
$$u(t_0 + \delta) = \phi_u(\delta), \forall \delta \in [-\hat{\tau}, 0], (t_0, \phi) \in \Re^+ \times \mathcal{C}^{v_u}_{\hat{\tau},m}$$
$$\hat{\tau} = \max(\tau, \theta), \; \tau > 0, \; \theta > 0$$

$x \in \Re^n$ and $u \in \Re^m$ are the states of the plant and of the actuator, respectively. $u_a \in \Re^m$ is the input of the actuator. $y_a \in \Re^m$ is both the control vector of the plant and the output of the actuator. Matrices A, B, A_τ and B_θ are real constant matrices of appropriate dimensions. Matrices ΔA, ΔB, ΔA_τ and ΔB_θ are unknown real norm-bounded matrices and represent time-varying parameter uncertainties (Dorato, 1987). They can be written as:

$$\left[\Delta A \; \Delta A_\tau \; \Delta B \; \Delta B_\theta \right] = DF(t) \left[E_1 \; E_2 \; E_3 \; E_4 \right]$$

where D, E_1, E_2, E_3 and E_4 are constant matrices defining how the uncertainty enters in the model and $F(t)$ belongs to the following set:

$$\mathcal{F} = \{F : \Re^+ \to \Re^{l \times r}; F(t)'F(t) \leq I_r, \forall t \geq 0\}.$$

Furthermore, pair (A, B) is supposed to be stabilizable. It is assumed that, in system (2), the dynamic matrix A has anti-stable eigenvalues, i.e., has some positive eigenvalues. In the cascaded system (1)-(2), the state of the plant (x) and the output of the actuator (y_a) are supposed available for control design purpose. Note that the positive vectors u_0 and u_1 may be viewed as bounds on the position and rate of the actuator state.

The objective in this paper is to derive a control $u_a(t)$ such that system (2) is asymptotically stable. Due to the presence of saturations, we have to characterize a set of admissible initial conditions $\phi_x(\psi)$, $\psi \in [-\hat{\tau}, 0]$, for which the closed-loop stability of system (2) is ensured. In this paper, a set of such initial conditions will be a priori fixed.

To solve this problem, we first define an augmented system associated with (2), with state:

$$z(t) = \begin{bmatrix} x(t) \\ u(t) \end{bmatrix} \in \Re^{n+m}$$

and consider, for the moment, $u_{0(i)} \to +\infty$ and $u_{1(i)} \to +\infty$, $\forall \; i = 1, \cdots, m$. Thus, in this case, one gets $u_a(t) = \dot{u}(t)$ and $y_a(t) = u(t)$. Equation (2) can be written as:

$$\begin{array}{l} \dot{z}(t) = (\mathbb{A} + \mathbb{D}F(t)\mathbb{E}_1)z(t) + \mathbb{B}\dot{u}(t) + \\ (\mathbb{A}_\tau + \mathbb{D}F(t)\mathbb{E}_2)z(t-\tau) + (\mathbb{A}_\theta + \mathbb{D}F(t)\mathbb{E}_3)z(t-\theta) \end{array} \tag{3}$$

where

$$\mathbb{A} = \begin{bmatrix} A & B \\ 0 & 0 \end{bmatrix}, \; \mathbb{A}_\tau = \begin{bmatrix} A_\tau & 0 \\ 0 & 0 \end{bmatrix}, \; \mathbb{A}_\theta = \begin{bmatrix} 0 & B_\theta \\ 0 & 0 \end{bmatrix}$$
$$\mathbb{B} = \begin{bmatrix} 0 \\ I_m \end{bmatrix}, \; \mathbb{D} = \begin{bmatrix} D \\ 0 \end{bmatrix}$$
$$\mathbb{E}_1 = \left[E_1 \; E_3 \right], \; \mathbb{E}_2 = \left[E_2 \; 0 \right], \; \mathbb{E}_3 = \left[0 \; E_4 \right]$$

The augmented system can be stabilized by state feedback. In this case we write:

$$u_a(t) = \dot{u}(t) = \mathbb{K}z(t) = \begin{bmatrix} K_1 & K_2 \end{bmatrix} z(t)$$

Lemma 1. If there exist matrices $\mathbb{K}$, $W = W' > 0$, $R_1 = R_1' > 0$, $R_2 = R_2' > 0$ and scalar $\varepsilon > 0$ such that:

$$\begin{bmatrix} W(\mathbb{A}+\mathbb{B}\mathbb{K})' + (\mathbb{A}+\mathbb{B}\mathbb{K})W + R_1 + R_2 + \varepsilon\mathbb{D}\mathbb{D}' & \mathbb{A}_\tau W & \mathbb{A}_\theta W & W\mathbb{E}_1' \\ W\mathbb{A}_\tau' & -R_1 & 0 & W\mathbb{E}_2' \\ W\mathbb{A}_\theta' & 0 & -R_2 & W\mathbb{E}_3' \\ \mathbb{E}_1 W & \mathbb{E}_2 W & \mathbb{E}_3 W & -\varepsilon I \end{bmatrix} < 0$$

then the closed-loop system (3) is asymptotically stable.

Proof. It readily follows from the study of the decrease of the Lyapunov function

$$V(z_t) = z(t)'W^{-1}z(t) + \int_{t-\tau}^{t} z(\xi)'W^{-1}R_1W^{-1}z(\xi)d\xi + \int_{t-\theta}^{t} z(\zeta)'W^{-1}R_2W^{-1}z(\zeta)d\zeta \quad (4)$$

along the trajectories of system (3), where z_t, $\forall\ t \geq t_0$ denotes the restriction of z to the interval $[t - \bar{\tau}, t]$ translated to $[-\bar{\tau}, 0]$, that is, $z_t(\psi) = z(t + \psi)\ \forall\ \psi \in [-\bar{\tau}, 0]$. See also (Esfahani and Petersen, 2000). □

We move now to the case where the control can saturate, that is, $u_{0(i)} < +\infty$ and $u_{1(i)} < +\infty$, $\forall\ i = 1, \cdots, m$. The control can be written as:

$$\dot{u}(t) = sat_{u_1}(K_1x(t) + K_2sat_{u_0}(u(t))) \quad (5)$$

Problem 1. Given a set $\mathcal{Z}_0 \subseteq \Re^{n+m}$, find matrices K_1 and K_2 such that:

i) System (1)-(2)-(5) is asymptotically stable $\forall \begin{bmatrix} \phi_x(\psi) \\ \phi_u(\psi) \end{bmatrix} \in \mathcal{Z}_0, \forall\ \psi \in [-\bar{\tau}, 0]$.

ii) When the system operates inside the region of linearity, i.e., in the region where no saturation occurs, a certain time-domain performance specification is fullfilled.

Remark 1. In absence of control constraints, i.e., u_0 and $u_1 \rightarrow +\infty$, the set of initial states that can be stabilized could tend towards $\Re^{n+m}$ if the delay-independent stability condition is satisfied (Esfahani and Petersen, 2000). Otherwise, if the control is limited (in position and rate), the set of stabilizable initial states is not the entire space $\Re^{n+m}$, except in the case where the open-loop system is stable and therefore global stabilization can be obtained (Niculescu *et al.*, 1996). Throughout the paper, since the open-loop system is assumed to be unstable, it is clear that we consider a set of stabilizable initial states different from $\Re^{n+m}$ and therefore strictly included in $\Re^{n+m}$.

3. MATHEMATICAL FRAMEWORK

Consider now system (1)-(2)-(5). Let us define:

- a vector $\alpha(z(t)) \in \Re^m$ with m components:

$$\alpha_{(i)}(z(t)) = \min\left(1, \frac{u_{0(i)}}{|[0\ I_m]_{(i)}z(t)|}\right)$$

- a vector $\beta(z(t)) \in \Re^m$ with m components:

$$\beta_{(i)}(z(t)) = \min\left(1, \frac{u_{1(i)}}{|\dot{u}_{(i)}(t)|}\right)$$

- the matrices:

$$\begin{aligned} &\mathbb{A}(\alpha(z(t))) = \begin{bmatrix} A & BD(\alpha(z(t))) \\ 0 & 0 \end{bmatrix} \\ &\mathbb{A}_\theta(\alpha(z(t-\theta))) = \begin{bmatrix} 0 & B_\theta D(\alpha(z(t-\theta))) \\ 0 & 0 \end{bmatrix} \\ &\mathbb{B}(\beta(z(t))) = \begin{bmatrix} 0 \\ D(\beta(z(t))) \end{bmatrix} \\ &\mathbb{K}(\alpha(z(t))) = \begin{bmatrix} K_1 & K_2D(\alpha(z(t))) \end{bmatrix} \\ &\mathbb{E}_1(\alpha(z(t))) = \begin{bmatrix} E_1 & E_3D(\alpha(z(t))) \end{bmatrix} \\ &\mathbb{E}_3(\alpha(z(t-\theta))) = \begin{bmatrix} 0 & E_4D(\alpha(z(t-\theta))) \end{bmatrix} \\ &\mathbb{A}_\tau \text{ and } \mathbb{E}_2 \text{ as previously defined} \end{aligned} \quad (6)$$

where $D(\alpha(z(t)))$ and $D(\beta(z(t)))$ are diagonal matrices whose diagonal elements are $\alpha_{(i)}(z(t))$, and $\beta_{(i)}(z(t))$, $i = 1, \cdots, m$, respectively. From the above definition, system (1)-(2)-(5) is equivalent to the following one:

$$\begin{aligned} \dot{z}(t) = &(\mathbb{A}(\alpha(z(t))) + \mathbb{D}F(t)\mathbb{E}_1(\alpha(z(t))) \\ &+\mathbb{B}(\beta(z(t)))\mathbb{K}(\alpha(z(t)))z(t) \\ &+ (\mathbb{A}_\tau + \mathbb{D}F(t)\mathbb{E}_2)\, z(t-\tau) \\ &+ (\mathbb{A}_\theta(\alpha(z(t-\theta))) + \mathbb{D}F(t)\mathbb{E}_3(\alpha(z(t-\theta))))\, z(t-\theta) \end{aligned} \quad (7)$$

Consider now lower bounds $\alpha_\ell = \begin{bmatrix} \alpha_{\ell(1)} & \cdots & \alpha_{\ell(m)} \end{bmatrix}'$ and $\beta_\ell = \begin{bmatrix} \beta_{\ell(1)} & \cdots & \beta_{\ell(m)} \end{bmatrix}'$ for $\alpha(z(t))$ and $\beta(z(t))$ defined above. By considering all the possible m-order vectors such that the ith entry takes the value 1 or $\alpha_{\ell(i)}$ (resp. 1 or $\beta_{\ell(i)}$), we can conclude that there exist 2^m different vectors associated to α_ℓ (resp. β_ℓ). By denoting each of these vectors $\gamma_j(\alpha_\ell)$ and $\gamma_j(\beta_\ell)$, $j = 1, \cdots, 2^m$, respectively, we can define the following matrices:

$$D_j(\alpha_\ell) = diag(\gamma_j(\alpha_\ell));\ D_j(\beta_\ell) = diag(\gamma_j(\beta_\ell)) \quad (8)$$

Furthermore, associated to α_ℓ and β_ℓ, we can define the following regions in $\Re^{n+m}$:

$$S(u_0, \alpha_\ell) = \left\{ z \in \Re^{n+m}; \left| \begin{bmatrix} 0 & I_m \end{bmatrix}_{(i)} z \right| \leq \frac{u_{0(i)}}{\alpha_{\ell(i)}} \right\} \quad (9)$$

$$S(u_1, \beta_\ell) = \bigcap_{j=1}^{2^m} S_j(u_1, \beta_\ell) \quad (10)$$

where

$$S_j(u_1, \beta_\ell) = \left\{ z \in \Re^{n+m}; \left| [K_1\ \ K_2D_j(\alpha_\ell)]z \right| \leq \frac{u_{1(i)}}{\beta_{\ell(i)}} \right\}$$

For all $z(t) \in S(u_0, \alpha_\ell) \cap S(u_1, \beta_\ell)$, one has:

$$\begin{aligned} &0 < \alpha_{\ell(i)} \leq \alpha_{(i)}(z(t)) \leq 1\ ,\quad i = 1, \cdots, m \\ &0 < \beta_{\ell(i)} \leq \beta_{(i)}(z(t)) \leq 1\ ,\quad i = 1, \cdots, m \end{aligned} \quad (11)$$

Hence, if $z(t) \in S(u_0, \alpha_\ell) \cap S(u_1, \beta_\ell)$, there exist nonnegative scalars $\lambda_j(z(t))$, $j = 1, \cdots, m$, $\delta_k(z(t-\theta))$, $k = 1, \cdots, 2^m$ and $\mu_q(z(t))$, $q = 1, \cdots, 2^m$ such that $\dot{z}(t)$ can be computed as:

$$\begin{aligned}\dot{z}(t) = \sum_{j=1}^{2^m} \lambda_j(z(t)) \left(\mathbb{A}_j(\alpha_\ell) + \mathbb{D}F(t)\mathbb{E}_{1j}(\alpha_\ell)\right) z(t) \\ + \left(\mathbb{A}_\tau + \mathbb{D}F(t)\mathbb{E}_2\right) z(t-\tau) \\ + \sum_{k=1}^{2^m} \delta_k(z(t-\theta)) \left(\mathbb{A}_{\theta k}(\alpha_\ell) + \mathbb{D}F(t)\mathbb{E}_{3k}(\alpha_\ell)\right) z(t-\theta) \\ + \sum_{q=1}^{2^m} \mu_q(z(t)) \mathbb{B}_q(\beta_\ell) \sum_{j=1}^{2^m} \lambda_j(z(t)) \mathbb{K}_j(\alpha_\ell) z(t)\end{aligned} \tag{12}$$

where

• $\sum_{j=1}^{2^m} \lambda_j(z(t)) = \sum_{k=1}^{2^m} \delta_k(z(t-\theta)) = \sum_{q=1}^{2^m} \mu_q(z(t)) = 1$,

• $\mathbb{A}_j(\alpha_\ell)$, $\mathbb{E}_{1j}(\alpha_\ell)$ and $\mathbb{K}_j(\alpha_\ell)$ are computed from (6) by replacing $D(\alpha(z(t))$ by $D_j(\alpha_\ell)$, given in (8),

• $\mathbb{A}_{\theta k}(\alpha_\ell)$ and $\mathbb{E}_{3k}(\alpha_\ell)$ are computed from (6) by replacing $D(\alpha(z(t-\theta))$ by $D_k(\alpha_\ell)$, which is similarly defined from (8),

• $\mathbb{B}_q(\beta_\ell)$ is computed from (6) by replacing $D(\beta(z(t))$ by $D_q(\beta_\ell)$, defined as in (8).

4. MAIN RESULTS

Aiming to solve Problem 1, we consider that the following data are a priori given.

• A set of initial conditions $\mathcal{Z}_0$ defined by a polyhedral set in $\Re^{n+m}$, described by its vertices:

$$\mathcal{Z}_0 = \mathrm{Co}\{v_1, \cdots, v_N\}, v_s \in \Re^{n+m}, \forall\ s = 1, \cdots, N$$

• A positive scalar η that represents the desired decay rate in the zone of linear behavior $S(u_0, 1_m) \cap S(u_1, 1_m)$. In other words, according to the definition of η-stability in (Mori *et al.*, 1982), the closed-loop system (3) is said η-stable in $S(u_0, 1_m) \cap S(u_1, 1_m)$.

Let us define for $j = 1, ..., 2^m$, $q = 1, ..., 2^m$:

$$\begin{aligned}\mathcal{M}(j,q) = R_1 + R_2 + \epsilon\mathbb{D}\mathbb{D}' + W \begin{bmatrix} I_n & 0 \\ 0 & D_j(\alpha_\ell) \end{bmatrix} \mathbb{A}' \\ + \mathbb{A} \begin{bmatrix} I_n & 0 \\ 0 & D_j(\alpha_\ell) \end{bmatrix} W + \mathbb{B}D_q(\beta_\ell)\mathbb{K} \begin{bmatrix} I_n & 0 \\ 0 & D_j(\alpha_\ell) \end{bmatrix} W \\ + W \begin{bmatrix} I_n & 0 \\ 0 & D_j(\alpha_\ell) \end{bmatrix} \mathbb{K}' D_q(\beta_\ell)\mathbb{B}'\end{aligned} \tag{13}$$

Proposition 1. Let ρ_1, ρ_2, η and c_0 positive given scalars. If there exist $W = W' > 0$, $R_1 = R_1' > 0$, $R_2 = R_2' > 0$, $\mathbb{K}$, $\varepsilon > 0$, $\lambda_0 > 0$ and vectors α_ℓ, β_ℓ satisfying the following matrix inequalities:

$$\begin{bmatrix} W(\mathbb{A}+\mathbb{B}\mathbb{K})' + (\mathbb{A}+\mathbb{B}\mathbb{K})W \\ +R_1 + R_2 + \varepsilon\mathbb{D}\mathbb{D}' + 2\eta W & \star & \star & \star \\ e^{\eta\tau} W\mathbb{A}_\tau' & -R_1 & 0 & \star \\ e^{\eta\theta} W\mathbb{A}_\theta' & 0 & -R_2 & \star \\ \mathbb{E}_1 W & e^{\eta\tau}\mathbb{E}_2 W & e^{\eta\theta}\mathbb{E}_3 W & -\varepsilon I \end{bmatrix} < 0 \tag{14}$$

$$\begin{bmatrix} \mathcal{M}(j,q) & \star & \star & \star \\ W\mathcal{A}_\tau' & -R_1 & 0 & \star \\ W\begin{bmatrix} I_n & 0 \\ 0 & D_k(\alpha_\ell) \end{bmatrix}\mathbb{A}_\theta' & 0 & -R_2 & \star \\ \mathbb{E}_1\begin{bmatrix} I_n & 0 \\ 0 & D_j(\alpha_\ell) \end{bmatrix}W & \mathbb{E}_2 W & \mathbb{E}_3\begin{bmatrix} I_n & 0 \\ 0 & D_k(\alpha_\ell) \end{bmatrix}W & -\varepsilon I \end{bmatrix} < 0 \tag{15}$$

$$\begin{bmatrix} W & \star \\ \alpha_{\ell(i)} \begin{bmatrix} 0 & I_m \end{bmatrix}_{(i)} W & u_{0(i)}^2 \end{bmatrix} \geq 0 \tag{16}$$

$$\begin{bmatrix} W & \star \\ \beta_{\ell(i)}\mathbb{K}_{(i)} \begin{bmatrix} I_n & 0 \\ 0 & D_j(\alpha_\ell) \end{bmatrix} W & u_{1(i)}^2 \end{bmatrix} \geq 0, \tag{17}$$

$$R_1 \leq \rho_1 W, \quad R_2 \leq \rho_2 W \tag{18}$$

$$\lambda_0 \leq W \leq c_0\lambda_0 \tag{19}$$

$$\lambda_0 - (1 + \tau\rho_1 + \theta\rho_2)v_s'v_s \geq 0,\ \forall s = 1, \cdots, N \tag{20}$$

$$0 < \alpha_{\ell(i)} \leq 1 \tag{21}$$

$$0 < \beta_{\ell(i)} \leq 1 \tag{22}$$

$\forall i = 1, \cdots, m$, $j = 1, \cdots, 2^m$, $k = 1, \cdots, 2^m$, $q = 1, \cdots, 2^m$, then the state feedback gain matrix $\mathbb{K}$ solves Problem 1.

Proof. Consider the ellipsoidal set defined as: $\mathfrak{E} = \{z \in \Re^{n+m}\ ;\ z'W^{-1}z \leq 1\}$. Consider the gain $\mathbb{K} = YW^{-1}$ and the region $S(u_0, \alpha_\ell) \cap S(u_1, \beta_\ell)$ defined in (9), (10) from vectors α_ℓ and β_ℓ satisfying (21), (22). The satisfaction of relations (14) to (20) has the following meaning.

• Relation (14) ensures that linear system (3) is η-stable (Tarbouriech and Gomes da Silva Jr., 2000).

• Relation (15) implies, by using convexity properties and Schur complement, that the time-derivative of the Lyapunov-Krasovskii function (4) satisfies $\dot{V}(z) \leq -\pi_3||z||^2 < 0$, $\pi_3 > 0$, along the trajectories of system (12) (Tarbouriech and Gomes da Silva Jr., 2000).

• Relations (16) and (17) ensure that $\mathfrak{E} \subseteq S(u_0, \alpha_\ell)$ and $\mathfrak{E} \subseteq S(u_1, \beta_\ell)$, respectively, and therefore that $\mathfrak{E} \subseteq S(u_0, \alpha_\ell) \cap S(u_1, \beta_\ell)$.

• Associated to relations (18) and (19), relation (20) guarantees that $\mathcal{Z}_0$ is included in the ball $\mathcal{B}(\delta) = \{\phi \in \Re^{n+m}\ ;\ ||\phi||_c^2 \leq \delta\}$ with $\delta = \frac{1}{\lambda_{max}(W^{-1}) + \tau\lambda_{max}(W^{-1}R_1W^{-1}) + \theta\lambda_{max}(W^{-1}R_2W^{-1})}$. In fact, from relations (18) to (19), we can prove that $\delta \geq \frac{\lambda_0}{1+\tau\rho_1+\theta\rho_2}$ and the condition of inclusion of $\mathcal{Z}_0$ in the ball $\mathcal{B}(\delta)$ follows.

Hence, for any initial condition taken in $\mathcal{Z}_0$, we can verify that we have $\dot{V}(z) \leq -\pi_3||z||^2$ or equivalently $z(t)'W^{-1}z(t) \leq V(z(t)) \leq V(z(t_0)) \leq 1$, that is, $z(t) \in \mathfrak{E}$ and therefore $z(t) \in S(u_0, \alpha_\ell) \cap S(u_1, \beta_\ell)$ from (16) - (17). Thus for any initial condition in $\mathcal{Z}_0$, the stability of the saturated system (1)-(2)-(5) is guaranteed. □

Remark 2. The Lyapunov functional (4) satisfies

$$\pi_1\|z(t)\|^2 \leq V(z_t) \leq \pi_2\|z_t\|_c^2$$

with $\pi_1 = \lambda_{\min}(W^{-1})$ and $\pi_2 = \lambda_{max}(W^{-1}) + \tau\lambda_{max}(W^{-1}R_1W^{-1})+\theta\lambda_{max}(W^{-1}R_2W^{-1})$. Thus, the following inclusion holds:

$$\mathcal{Z}_0 \subset \mathcal{B}(\delta) \subset \mathfrak{C} \subseteq S(u_0, \alpha_\ell) \cap S(u_1, \beta_\ell)$$

5. CONTROL LAW COMPUTATION

Proposition 1 gives a sufficient condition of existence of some feedback gain matrix $\mathbb{K}$ solution to Problem 1. But the decision variables appear in products in the matrix inequalities, which means that the computation of a feasible gain $\mathbb{K}$ solution of constraints (14)-(22) is a NP-hard problem. To overcome this difficulty, matrix inequalities may be recasted into LMIs by transformations, over-estimation and/or relaxation techniques. Associated to a priori given components of vectors α_ℓ and β_ℓ, the following lemmas propose some over-estimation of the nonlinear terms to obtain LMIs in the variables W, $Y = \mathbb{K}W^{-1}$, R_1, R_2, λ_0, ε.

Lemma 2. The term $\mathcal{M}(j, q)$ defined in (13) satisfies:

$$\mathcal{M}(j,q) \leq W\mathbb{A}'+\mathbb{A}W+\mathbb{B}D_q(\beta_\ell)Y+Y'D_q(\beta_\ell)\mathbb{B}'+R_1+R_2$$
$$+\left(\mathbb{A}W+\mathbb{B}D_q(\beta_\ell)Y\right)W^{-1}\left(W\mathbb{A}'+Y'D_q(\beta_\ell)\mathbb{B}'\right)+\varepsilon\mathbb{D}\mathbb{D}'$$
$$+\left(W\begin{bmatrix} I_n & 0 \\ 0 & D_j(\alpha_\ell)\end{bmatrix}-W\right)W^{-1}\left(\begin{bmatrix} I_n & 0 \\ 0 & D_j(\alpha_\ell)\end{bmatrix}W-W\right)$$

Proof. It readily follows from the use of the classical property, $2u'v \leq u'Xu+v'X^{-1}v$, $X = X' > 0$, where $X = W$, $u' = \left(\mathbb{A}W + \mathbb{B}D_q(\beta_\ell)Y\right)W^{-1}$ and $v = \begin{bmatrix} I_n & 0 \\ 0 & D_j(\alpha_\ell)\end{bmatrix}W - W$. □

Lemma 3. The inclusion of $\mathfrak{C}$ in $S(u_1, \beta_\ell)$ given by equation (17) may be overestimated by using the matrix inequality

$$\begin{bmatrix} W & \beta_{\ell(i)}Y'_{(i)} \\ \beta_{\ell(i)}Y_{(i)} & \dfrac{u_{1(i)}^2}{c_0}\end{bmatrix} \geq 0, \qquad (23)$$

Proof. It readily follows from noting that $D_j(\alpha_\ell) \leq I_m$ and by using relation (19). □

Even if the over-estimation allows to propose a set of LMIs to be satisfied (for given a priori α_ℓ, β_ℓ) to solve the problem, the conservatism of such a solution must not be hidden. Moreover, note that when no saturations are allowed ($\alpha_\ell = \beta_\ell = 1_m$), constraints (14)-(20) become LMIs in the decision variable W, Y, R_1, R_2, λ_0, ε. When it exists, a feasible solution of this linear problem may serve as the initialisation of an algorithm formed by a sequence of the following relaxation schemes:

Rs1 For given α_ℓ, W, $\mathbb{K}$, search for a feasible β_ℓ solution to constraints (14)-(15), (17)-(20), (22).

Rs2 For given β_ℓ, W, $\mathbb{K}$, search for a feasible α_ℓ solution to constraints (14)-(16), (18)-(21).

Rs3 For given α_ℓ, β_ℓ, $\mathbb{K}$, search for a feasible matrix W solution to constraints (14)-(20).

Rs4 For given α_ℓ, β_ℓ, W, search for the gain matrix $\mathbb{K}$ solution to constraints (14)-(20).

However, there are no a priori rules which could help to determine an optimal sequence to use these different relaxation schemes.

The second part of this section concerns the definition of some optimization criterion which allows to orient the solution to some desired direction. From given set of initial condition $\mathcal{Z}_0$, scalar η and bounds on the actuator u_0, u_1, Proposition 1 gives a sufficient condition to test the feasibility of the problem, i.e. of existence of a gain $\mathbb{K}$. However, when vectors α_ℓ and β_ℓ are given a priori, it may be impossible to find a feasible solution. A possibility then consists to consider a scaling factor μ, $\mu > 0$, and therefore to search for the maximum scaled set $\mu^*\mathcal{Z}_0$ that can be stabilized using the proposed approach and the fixed data. For this, inequality (20) may be replaced by:

$$\lambda_0 - (1 + \tau\rho_1 + \theta\rho_2)\nu v_s' v_s \geq 0,\ \forall s = 1,\cdots,N \qquad (24)$$

where $\nu = \mu^2$, and one can solve:

$$\begin{array}{c} \max \nu \\ \nu, \mathbb{K}, W, R_1, R_2, \lambda_0, \varepsilon, \alpha_\ell, \beta_\ell \\ \text{subject to LMIs } (14)-(19), (24), (21), (22) \end{array} \qquad (25)$$

The interest of such an optimization problem resides in the significance of $\mu^* = \sqrt{\nu^*}$.

- If $\mu^* < 1$, it means that the original problem given in terms of $\mathcal{Z}_0$, η, u_0, u_1 has no atteignable solution by using the optimization procedure derived from Proposition 1. The solution is then given by decreasing the size of $\mathcal{Z}_0$ by a scaling factor μ^*. But η, u_0 or u_1 could also be modified in order to increase the optimal value μ^* rather than to decrease the set of initial conditions.
- If $\mu^* \geq 1$, it means that the original problem has a solution. If μ^* is much larger than 1, it signifies that η may be increased, i.e., the performance requirements may be hardened, or that the actuators are overdimensioned with respect to the operating conditions.

6. NUMERICAL EXAMPLE

Consider the following numerical example borrowed and completed from (Towsen, 1982):

$$A = \begin{bmatrix} 1 & 1.5 \\ 0.3 & -2\end{bmatrix},\ A_\tau = \begin{bmatrix} 0 & -1 \\ 0 & 0\end{bmatrix},\ B = \begin{bmatrix} 10 \\ 1\end{bmatrix}$$
$$B_\theta = \begin{bmatrix} 1 \\ 0\end{bmatrix},\ D = \begin{bmatrix} 1 \\ 0\end{bmatrix},\ E_1 = \begin{bmatrix} 1 & 0\end{bmatrix}$$
$$E_2 = \begin{bmatrix} 0 & 1\end{bmatrix},\ E_3 = 1,\ E_4 = 0,\ \tau = 1,\ \theta = 0.5$$

with $\rho_1 = \rho_2 = 1$ and operating conditions:

- Decay rate: $\eta = 1$
- Actuator limitations: $u_0 = 15$, $u_1 = 50$
- 6 vertices of $\mathcal{Z}_0$: $v_i = \begin{bmatrix} \pm 10 \\ 0 \\ 0 \end{bmatrix}, \begin{bmatrix} 0 \\ \pm 10 \\ 0 \end{bmatrix}, \begin{bmatrix} 0 \\ 0 \\ \pm 10 \end{bmatrix}$

In this case, for the linear problem given by $\alpha_\ell = 1$, $\beta_\ell = 1$, one obtains $\mu^* = 0.4904$. It means that one cannot determine a gain $\mathbb{K}$ such that all the set of initial states $\mathcal{Z}_0$ is stabilized.

Let us now consider the sequence of relaxation schemes Rs1-2-3-4. Although the optimization criterion may be max ν for all relaxation schemes, one can also take min β_ℓ and min α_ℓ in relaxations 1 and 2, respectively. Since α_ℓ and β_ℓ may be associated to the saturation tolerance index, one intuitively concludes that the more they are decreased, the more saturation is allowed and the set of admissible initial conditions is increased. In this numerical example, the rate of convergence of the algorithm is largely increased by using such criteria. After 13 iterations of the sequence, one obtains: $\mu^* = 0.6829$, $\alpha_\ell = 0.8677$, $\beta_\ell = 0.4606$, $\mathbb{K} = \begin{bmatrix} -3.4201 & -1.7164 & -9.2484 \end{bmatrix}$. Saturation allowance allows to increase the size of the set of admissible initial conditions $\mu^* \mathcal{Z}_0$ with respect to the linear solution, but doesn't allow to stabilize all $\mathcal{Z}_0$. To stabilize all $\mathcal{Z}_0$, it is then necessary to decrease η or/and to increase u_0 and u_1. Different solutions such that μ^* is just over 1 are summarized in Table 1.

u_0	u_1	η	μ^*	α_ℓ	β_ℓ	iteration
20	75	1	1.0240	0.7934	0.3317	16
18	60	0.7	1.0120	0.8063	0.5767	26

Table 1. Cases allowing to stabilize $\mathcal{Z}_0$

7. CONCLUSION

The problem of stabilization by state feedback of multiple bounded inputs state space models with delays (both in the state and in the input vectors) and uncertainties was addressed. Conditions for the existence of a control law that guarantees the asymptotic stability of the closed-loop system for a given region of initial states in presence of both control amplitude and rate saturations was proposed. The occurence of position and rate saturations was effectively considered, which generally allows the stabilization of larger sets of inital conditions, although by increasing the complexity of the solvability conditions. Computation difficulties may then be overcome via the use of some relaxation schemes with LMI-based framework. The main sources of conservatism were induced by the use of differential inclusions to represent the saturated closed-loop system, by the actuator model considered in the approach or still by considering delay-independent stability.

8. REFERENCES

Chen, B.S., S.S. Wang and H.C. Lu. (1988). Stabilization of time-delay systems containing saturating actuators. *Int. J. Contr.* **47**, 867–881.

Dorato, P. (1987). *Robust Control.* IEEE Press Book.

Esfahani, S.H. and I.R. Petersen (2000). An lmi approach to output feedback guaranteed cost control for uncertain time-delay systems. *Int. J. Robust and Nonlinear Contr.* **10**, 157–174.

Gomes da Silva Jr., J.M. and S. Tarbouriech (2000). Local stabilization of linear systems under amplitude and rate saturating actuators. In: *39th IEEE Conference on Decision and Control (CDC).* Sydney, Australia.

Lin, Z. (1997). Semi-global stabiization of linear systems with position and rate-limited actuators. *Sys. & Contr. Letters* **30**, 1–11.

Mahmound, M.S. and N.F. Al-Muthairi (1994). Quadratic stabilization of continuous-time systems with state-delay and norm-bounded time-varying uncertainties. *IEEE Trans. Aut. Contr.* **39**(10), 2135–2139.

Mori, T., N. Fukuma and M. Kuwahara (1982). On an estimate of the decay rate for stable linear delay systems. *Int. J. Contr.* **36**, 95–97.

Niculescu, S.I., J.M. Dion and L. Dugard (1996). Robust stabilization for uncertain time-delay systems containing saturating actuators. *IEEE Trans. Aut. Contr.* **41**, 742–747.

Shewchun, J.M. and E. Feron (1997). High performance bounded control. In: *American Control Conference (ACC).* Vol. 5. Albuquerque (USA). pp. 3250–3254.

Tarbouriech, S. and G. Garcia (1997). *Control of uncertain systems with bounded inputs.* Vol. 227 of *LNCIS.* Springer-Verlag. Editors.

Tarbouriech, S. and G. Garcia (1999). Stabilization of uncertain time-delay systems with position and rate bounded actuators: an are approach. In: *European Control Conference (ECC).* Karlsruhe (Germany).

Tarbouriech, S. and J.M. Gomes da Silva Jr. (2000). Synthesis of controllers for continuous-time delay systems with saturating controls via lmis. *IEEE Trans. Aut. Contr.* **45**(1), 105–110.

Tarbouriech, S., G. Garcia and D. Henrion (1999). Local stabilization of linear systems with position and rate bounded actuators. In: *World IFAC Congress.* Vol. F. Beijing (China). pp. 459–464.

Towsen, A. (1982). Stable sample-data feedback control of dynamic systems with time-delays. *Int. J. Sys. & Sci.* **12**(12), 1379–1384.

Tyan, F. and D.S. Bernstein (1997). Dynamic output feedback compensation for linear systems with independent amplitude and rate saturation. *Int. J. Contr.* **67**(1), 89–116.

www.elsevier.com/locate/ifac

DELAY-INDEPENDENT CIRCLE CRITERION AND POPOV CRITERION.

Pierre-Alexandre Bliman *,[1]

* *INRIA, Rocquencourt BP 105, 78153 Le Chesnay cedex, France.*

Abstract: In the present paper are studied delay-independent versions of circle criterion and Popov criterion, for some nonlinear delay systems with sector-bounded nonlinearities. Necessary and sufficient conditions are obtained in terms of solvability of linear matrix inequalities. This constitutes an analogue to the famous results, valid for nonlinear delay-free systems. *Copyright © 2001 IFAC*

Keywords: Absolute stability, delay analysis, Lyapunov function, Lyapunov equation.

1. INTRODUCTION

Delays in control loops may lead to bad performance or instabilities, and the study of time-delay systems has attracted large interest. One of the problems one may have to deal with, is the possible uncertainties on the value of these delays. In order to tackle this robustness issue, but also for reasons of computational simplicity, stability criteria independent of the values of the delays have been proposed, mainly for linear delay systems. We are here concerned with the issue of delay-independent stability of delay systems with sector-bounded nonlinearities; in other words delay-independent absolute stability.

In the classical absolute stability theory for nonlinear, finite-dimensional (i.e. delay-free), systems, the usual results (circle criterion, Popov criterion) are equivalently expressed by conditions involving a linear matrix inequality (appearing when evaluating the evolution of a candidate Lyapunov function), or under frequency domain form. The linear matrix inequalities have the advantage to be solvable by sound numerical methods (Boyd *et al.*, 1994).

Since at least the early sixties, the circle criterion (Sandberg, 1964; Zames, 1966) and the Popov criterion (Popov *et al.*, 1962; Li, 1963) have been shown to be valid for delay systems. On the other hand, Krasovskii has proposed an extension of the Lyapunov method to these delay systems, leading in particular to some sufficient conditions for stability of linear and nonlinear delay systems independent of the value of the delay(s) (Krasovskii, 1963, Section 34.), and also (Hale, 1977; Boyd *et al.*, 1994; Verriest *et al.*, 1996; Verriest *et al.*, 1998). These conditions are expressed with linear matrix inequalities.

In (Bliman, 2001), we were able to provide a *necessary and sufficient* condition for delay-independent asymptotic stability of the *linear* delay systems, expressed as LMI, see Proposition 4 below. The latter are obtained by use of generalized Lyapunov-Krasovskii functionals parametrized by some positive definite matrices, which are precisely the unknowns of the LMIs.

Our goal in the present paper is to derive from the choice of appropriate classes of Lyapunov-Krasovskii functionals, LMI conditions *necessary and sufficient* for delay-independent absolute stability. The point of view which is adopted here is linked to the studies of systems over a ring (Kamen *et al.*, 1984a; Kamen *et al.*, 1984b), more

[1] Phone: (33) 1 39 63 55 68, Fax: (33) 1 39 63 57 86, Email: pierre-alexandre.bliman@inria.fr

precisely over the ring of polynomials in the variable $z = e^{-sh}$, which, due to the hypothesis of delay independence, may be considered as independent from the (usual) Laplace variable s. Indeed, the results stated below look like generalization of Kalman-Yakubovich-Popov Lemma to 2-D systems, with transfer involving not only a continuous-time frequency domain variable s, but also a discrete-time frequency domain variable z. These results may also be interpreted in terms of properties of systems with uncertain parameter z.

The problem is presented in Section 2 and some preparatory work is made. Definitions of delay-independent absolute stability are introduced in Section 2.1, and some LMIs are presented in Section 2.2 and motivated. The results are given in Section 3.

The matrices I_n, 0_n, $0_{n\times p}$ are resp. the $n \times n$ identity matrix and the $n \times n$ and $n \times p$ zero matrices. The symbol $\otimes$ denotes Kronecker product. The conjugate and transconjugate of M, are denoted M^T and M^*. By $\overline{\mathbb{D}}$ is denoted the closed unit ball in $\mathbb{C}$. By $\overline{\mathbb{C}^+}$ is meant the closed set of complex numbers with *nonnegative* real part. For brievity, we write SPR instead of "strictly positive real".

2. PRESENTATION OF THE PROBLEM AND PREPARATION

The goal of this paper is to study the absolute stability of the delay systems of the class

$$\begin{cases} \dot{x} = A_0 x(t) + A_1 x(t-h) + B_0 u(t) + B_1 u(t-h), \\ y(t) = C_0 x(t) + C_1 x(t-h),\ u(t) = \psi(t, y(t)) , \end{cases} \tag{1}$$

where, for $n, p, q \in \mathbb{N}$, $S \stackrel{\text{def}}{=} (A_0, A_1, B_0, B_1, C_0, C_1) \in \mathbb{R}^{n\times n} \times \mathbb{R}^{n\times n} \times \mathbb{R}^{n\times p} \times \mathbb{R}^{n\times p} \times \mathbb{R}^{q\times n} \times \mathbb{R}^{q\times n}$, and the nonlinearity $\psi : [0, +\infty) \times \mathbb{R}^q \to \mathbb{R}^p$ fulfills the following sector condition: there exists $K \in \mathbb{R}^{p\times q}$ such that

$$\forall t \in \mathbb{R}^+, \forall y \in \mathbb{R}^q, \quad \psi(t, y)^T (Ky - \psi(t, y)) \geq 0 . \tag{2}$$

Define the transfer matrix

$$H(s, z) \stackrel{\text{def}}{=} (C_0 + zC_1)(sI_n - A_0 - zA_1)^{-1}(B_0 + zB_1) .$$

2.1 *Frequency domain notions of delay-independent absolute stability*

By analogy with the notion of *strong* delay-independent stability (Niculescu *et al.*, 1996) or *pointwise* stability (Kamen, 1982; Kamen, 1983), one sets the two following properties of internal stability:

Definition 1. System (1) is said to be *internally* (or *weakly*) *delay-independently stable* (for short, $S \in \mathcal{S}$) if, for any $h \geq 0$,

$$\forall s \in \overline{\mathbb{C}^+}, \det(sI_n - A_0 - e^{-sh}A_1) \neq 0 .$$

System (1) is said to be *internally strongly delay-independently stable* ($S \in \mathcal{S}'$) if, for any $z \in \overline{\mathbb{D}}$,

$$\forall s \in \overline{\mathbb{C}^+}, \det(sI_n - A_0 - zA_1) \neq 0 .$$

■

One now defines the four properties of system (1) which will be studied.

Definition 2. System (1) is said to verify *(weak) delay-independent circle criterion* ($S \in \mathcal{S}_C$) if it is internally delay-independently stable and, for any $h \geq 0$,

$$I_p - KH(s, e^{-hs}) \text{ is SPR} .$$

System (1) is said to verify *strong delay-independent circle criterion* ($S \in \mathcal{S}'_C$) if it is internally strongly delay-independently stable and, for any $z \in \overline{\mathbb{D}}$,

$$I_p - KH(s, z) \text{ is SPR} .$$

■

Absolute stability of system (1) is ensured for any value of the delay $h \geq 0$, if the system verifies delay-independent circle criterion. Similarly, define

Definition 3. System (1) is said to verify *uniform delay-independent Popov criterion* ($S \in \mathcal{S}_P$) if it is internally delay-independently stable and if there exists a diagonal matrix $\eta \in \mathbb{R}^{p\times p}$ such that, for any $h \geq 0$,

$$I_p - (I + \eta s)KH(s, e^{-hs}) \text{ is SPR} .$$

System (1) is said to verify *strong uniform delay-independent Popov criterion* ($S \in \mathcal{S}'_P$) if it is internally strongly delay-independently stable and if there exists a diagonal matrix $\eta \in \mathbb{R}^{p\times p}$ such that, for any $z \in \overline{\mathbb{D}}$,

$$I_p - (I + \eta s)KH(s, z) \text{ is SPR} .$$

■

Obviously, when system (1) is *square* ($p = q$) with *time-invariant, decentralized* (Khalil, 1992) nonlinearity, its absolute stability is guaranteed for any value of the delay $h \geq 0$, if it verifies uniform delay-independent Popov criterion. The two previous notions are called *uniform*, because the choice of the Popov slope η has to be made uniformly with respect to $h \geq 0$. The method chosen here, of study of the strong properties, does not permit to treat the non-uniform case.

The sets $\mathcal{S}, \mathcal{S}', \mathcal{S}_C, \mathcal{S}'_C, \mathcal{S}_P, \mathcal{S}'_P$ of systems $S = (A_0, A_1, B_0, B_1, C_0, C_1)$ defined previously are subsets of $\mathbb{R}^{n\times n} \times \mathbb{R}^{n\times n} \times \mathbb{R}^{n\times p} \times \mathbb{R}^{n\times p} \times \mathbb{R}^{q\times n} \times \mathbb{R}^{q\times n}$, considered as a Banach space endowed with the product topology induced by the largest singular value. The interior (for this topology) is denoted by int.

Straighforward arguments show that the following relations hold: $\begin{array}{ccccc} \mathcal{S} & \subset & \mathcal{S}_C & \subset & \mathcal{S}_P \\ \cap & & \cap & & \cap \\ \mathcal{S}' & \subset & \mathcal{S}'_C & \subset & \mathcal{S}'_P \end{array}$.

2.2 *LMI problems associated to system* (1) *and motivation*

We now introduce some LMIs related to the class of systems (1). First, define, for $k \in \mathbb{N}$, the matrices $J^u_{n,k}, J^l_{n,k} \in \mathbb{R}^{kn\times(k+1)n}$ (*u,l as upper, lower*), $J_{n,k} \in \mathbb{R}^{2kn\times(k+1)n}$ by identities (3a). and similarly $J^u_{p,k}, J^l_{p,k}, J_{p,k}$. Then, define $J^u_k, J^l_k \in \mathbb{R}^{k(n+p)\times(k+1)(n+p)}$, $J_k \in \mathbb{R}^{2k(n+p)\times(k+1)(n+p)}$ as in (3b). Define at last $\tilde{J}_{n,k} \in \mathbb{R}^{3kn\times(k+2)n}$, $\tilde{J}_k \in \mathbb{R}^{3k(n+p)\times(k+2)(n+p)}$, by (3c).

Then, consider the three families of linear matrix inequalities $\mathbf{LMI}(k)$, $\mathbf{LMI}_C(k)$, $\mathbf{LMI}_P(k)$, indexed by the positive integer k (see next page).

Let us motivate why these LMIs appear naturally, and how they constitute prolongation of the results obtained with the usual Lyapunov-Krasovskii functionals. The key of the interpretation lies in the use of the augmented variables $\mathcal{X}_k(t) \in \mathbb{R}^{kn}$, $\mathcal{U}_k(t) \in \mathbb{R}^{kp}$, $\mathcal{Y}_k(t) \in \mathbb{R}^{kq}$, given by

$$\mathcal{X}_k(t) \stackrel{\text{def}}{=} \begin{pmatrix} x(t) \\ x(t-h) \\ \vdots \\ x(t-(k-1)h) \end{pmatrix}, \quad \mathcal{U}_k(t) \stackrel{\text{def}}{=} \begin{pmatrix} u(t) \\ u(t-h) \\ \vdots \\ u(t-(k-1)h) \end{pmatrix}, \quad \mathcal{Y}_k(t) \stackrel{\text{def}}{=} \begin{pmatrix} y(t) \\ y(t-h) \\ \vdots \\ y(t-(k-1)h) \end{pmatrix}.$$

$\mathcal{X}_k, \mathcal{U}_k, \mathcal{Y}_k$ are linked by the identities: $\dot{\mathcal{X}}_k = (I_k \otimes A_0)\mathcal{X}_k(t) + (I_k \otimes A_1)\mathcal{X}_k(t-h) + (I_k \otimes B_0)\mathcal{U}_k(t) + (I_k \otimes B_1)\mathcal{U}_k(t-h)$, $\mathcal{Y}_k(t) = (I_k \otimes C_0)\mathcal{X}_k(t) + (I_k \otimes C_1)\mathcal{X}_k(t-h)$. Moreover, their nested structure permits to write the identities (4).

Consider first the case where $K = 0$ (so $\psi \equiv 0$ and system (1) is linear). For matrices $P_k, Q_k \in \mathbb{R}^{kn\times kn}$ such that (P_k, Q_k) solves $\mathbf{LMI}(k)$, consider functional (5). Then, along the trajectories of (1), (8) holds, due to (4a). The negativity of R_k ensures that V_k is decreasing along the trajectories of S.

The Lyapunov-Krasovskii functional (5) is not adapted to the nonlinear case ($K \neq 0$). Rather, consider for $P_k \in \mathbb{R}^{kn\times kn}$, $Q_k \in \mathbb{R}^{k(n+p)\times k(n+p)}$ such that (P_k, Q_k) solves $\mathbf{LMI}_C(k)$, the functional (6). Using the identities in (4b) yields (9). Due to the sector inequality, the last quantity is nonpositive, and the functional $V_{C,k}$ is decreasing along the trajectories of (1).

When (1) is square with time-invariant decentralized nonlinearity, define (7). Then, (10) is true, by use of (4c). The functional (7) is hence decreasing along the trajectories of (1).

Remark that the functionals given by (5), (6), (7) are positive definite only if Q_k is itself positive definite. However, the results established below do not include this assumption, although we conjecture that there is no loss of generality in supposing Q_k is positive.

3. MAIN RESULTS

For linear systems, the following result has been established.

Proposition 4. (Bliman, 2001) int$\mathcal{S} = \mathcal{S}'$, and $S \in \mathcal{S}'$ *if and only if* there exists $k \in \mathbb{N}$ such that $\mathbf{LMI}(k)$ is solvable. ■

The results in this paper constitute an attempt to generalize the previous properties to absolute stability.

Proposition 5. The following identities hold: int$\mathcal{S}_C = \mathcal{S}'_C$, int$\mathcal{S}_P = \mathcal{S}'_P$. ■

Proposition 5 means that the systems fulfilling a strong property fulfill also the corresponding weak property. Reciprocally, for any system S fulfilling a weak property but not the strong one, there exist infinitesimal perturbations of the matrices $(A_0, A_1, B_0, B_1, C_0, C_1)$ under which, for some nonnegative value of the delay h, the system does not fulfill circle (resp. Popov) criterion. In other words, the strong notions are the counterpartXC of the weak ones, "robustified" with respect to parametric perturbations of the model.

The next result claims that the LMI conditions $\mathbf{LMI}_C(k)$, resp. $\mathbf{LMI}_P(k)$, not only imply that $S \in \mathcal{S}_C$, resp. $S \in \mathcal{S}_P$ (as suggested in Section 2.2), but also that $S \in \mathcal{S}'_C$, resp. $S \in \mathcal{S}'_P$. A crucial point here is that the converse is also true. We indicate rapidly in Section 3 how this fact is linked to a result on approximation of the solution of LMIs with coefficients analytical in $\overline{\mathbb{D}}$.

$$J^u_{n,k} \overset{\text{def}}{=} \begin{pmatrix} I_{kn} & 0_{kn\times n}\end{pmatrix},\ J^l_{n,k} \overset{\text{def}}{=} \begin{pmatrix} 0_{kn\times n} & I_{kn}\end{pmatrix},\ J_{n,k} \overset{\text{def}}{=} \begin{pmatrix} J^u_{n,k} \\ J^l_{n,k}\end{pmatrix} = \begin{pmatrix} I_{kn} & 0_{kn\times n} \\ 0_{kn\times n} & I_{kn}\end{pmatrix}, \tag{3a}$$

$$J^u_k \overset{\text{def}}{=} \begin{pmatrix} J^u_{n,k} & 0_{kn\times(k+1)p} \\ 0_{kp\times(k+1)n} & J^u_{p,k}\end{pmatrix}, J^l_k \overset{\text{def}}{=} \begin{pmatrix} J^l_{n,k} & 0_{kn\times(k+1)p} \\ 0_{kp\times(k+1)n} & J^l_{p,k}\end{pmatrix},\ J_k \overset{\text{def}}{=} \begin{pmatrix} J_{n,k} & 0_{2kn\times(k+1)p} \\ 0_{2kp\times(k+1)n} & J_{p,k}\end{pmatrix}, \tag{3b}$$

$$\tilde{J}_{n,k} \overset{\text{def}}{=} \begin{pmatrix} I_{kn} & & 0_{kn\times 2n} \\ 0_{kn\times n} & I_{kn} & 0_{kn\times n} \\ 0_{kn\times 2n} & & I_{kn}\end{pmatrix},\ \tilde{J}_k \overset{\text{def}}{=} \begin{pmatrix} \tilde{J}_{n,k} & 0_{3kn\times(k+2)p} \\ 0_{3kp\times(k+2)n} & \tilde{J}_{p,k}\end{pmatrix}. \tag{3c}$$

- **LMI**(k): $P_k, Q_k \in \mathbb{R}^{kn\times kn}$, $P_k = P_k^T > 0$, $Q_k = Q_k^T$, $R_k < 0$, where $R_k \in \mathbb{R}^{(k+1)n\times(k+1)n}$ is given by

$$R_k \overset{\text{def}}{=} J^T_{n,k} \begin{pmatrix} P_k(I_k\otimes A_0) + (I_k\otimes A_0)^T P_k + Q_k & P_k(I_k\otimes A_1) \\ (I_k\otimes A_1)^T P_k & -Q_k \end{pmatrix} J_{n,k}\,.$$

- **LMI**$_C(k)$: $P_k \in \mathbb{R}^{kn\times kn}$, $Q_k \in \mathbb{R}^{k(n+p)\times k(n+p)}$, $P_k = P_k^T > 0$, $Q_k = Q_k^T$, $R_{C,k} < 0$, where $R_{C,k} \in \mathbb{R}^{(k+1)(n+p)\times(k+1)(n+p)}$ is given by

$$\begin{aligned} R_{C,k} \overset{\text{def}}{=} J_k^T & \left[\left(\begin{array}{c|c} P_k(I_k\otimes A_0) + (I_k\otimes A_0)^T P_k & P_k(I_k\otimes A_1)\ P_k(I_k\otimes B_0)\ P_k(I_k\otimes B_1) \\ \hline (I_k\otimes A_1)^T P_k & \\ (I_k\otimes B_0)^T P_k & 0_{k(n+2p)} \\ (I_k\otimes B_1)^T P_k & \end{array}\right)\right. \\ & + \begin{pmatrix} 0_{2kn\times kp} \\ I_{kp} \\ 0_{kp}\end{pmatrix} \begin{pmatrix} (I_k\otimes KC_0) & (I_k\otimes KC_1) & 0_{kp\times 2kp}\end{pmatrix} + \begin{pmatrix} (I_k\otimes KC_0)^T \\ (I_k\otimes KC_1)^T \\ 0_{2kp\times kp}\end{pmatrix} \begin{pmatrix} 0_{kp\times 2kn} & I_{kp} & 0_{kp}\end{pmatrix} \\ & \left. -2 \begin{pmatrix} 0_{2kn\times kp} \\ I_{kp} \\ 0_{kp}\end{pmatrix} \begin{pmatrix} 0_{kp\times 2kn} & I_{kp} & 0_{kp}\end{pmatrix}\right] J_k + J_k^{uT} Q_k J_k^u - J_k^{lT} Q_k J_k^l\,. \end{aligned}$$

- **LMI**$_P(k)$: $\eta \in \mathbb{R}^{p\times p}$, $P_k \in \mathbb{R}^{kn\times kn}$, $Q_k \in \mathbb{R}^{k(n+p)\times k(n+p)}$, η diagonal, $P_k = P_k^T > 0$, $Q_k = Q_k^T$, $R_{P,k} < 0$, where $R_{P,k} \in \mathbb{R}^{(k+1)(n+p)\times(k+1)(n+p)}$,

$$\begin{aligned} R_{P,k} \overset{\text{def}}{=} R_{C,k} + \tilde{J}^T_{k-1} & \left[\begin{pmatrix} (I_{k-1}\otimes \eta KC_0A_0)^T \\ (I_{k-1}\otimes \eta KC_0A_1)^T \\ 0_{(k-1)n\times(k-1)p} \\ (I_{k-1}\otimes \eta KC_0B_0)^T \\ (I_{k-1}\otimes \eta KC_0B_1)^T \\ 0_{(k-1)p}\end{pmatrix} \begin{pmatrix} 0_{(k-1)p\times 3(k-1)n} & I_{(k-1)p} & 0_{(k-1)p\times 2(k-1)p}\end{pmatrix}\right. \\ & + \begin{pmatrix} 0_{3(k-1)n\times(k-1)p} \\ I_{(k-1)p} \\ 0_{2(k-1)p\times(k-1)p}\end{pmatrix} \begin{pmatrix} (I_{k-1}\otimes \eta KC_0A_0) & (I_{k-1}\otimes \eta KC_0A_1) & 0_{(k-1)p\times(k-1)n} & (I_{k-1}\otimes \eta KC_0B_0) & (I_{k-1}\otimes \eta KC_0B_1) & 0_{(k-1)p}\end{pmatrix} \\ & + \begin{pmatrix} 0_{3(k-1)n\times(k-1)p} \\ I_{(k-1)p} \\ 0_{2(k-1)p\times(k-1)p}\end{pmatrix} \begin{pmatrix} 0_{(k-1)p\times(k-1)n} & (I_{k-1}\otimes \eta KC_1A_0) & (I_{k-1}\otimes \eta KC_1A_1) & 0_{(k-1)p} & (I_{k-1}\otimes \eta KC_1B_0) & (I_{k-1}\otimes \eta KC_1B_1)\end{pmatrix} \\ & \left. + \begin{pmatrix} 0_{(k-1)n\times(k-1)p} \\ (I_{k-1}\otimes \eta KC_1A_0)^T \\ (I_{k-1}\otimes \eta KC_1A_1)^T \\ 0_{(k-1)p} \\ -(I_{k-1}\otimes \eta KC_1B_0)^T \\ -(I_{k-1}\otimes \eta KC_1B_1)^T\end{pmatrix} \begin{pmatrix} 0_{(k-1)p\times 3(k-1)n} & I_{(k-1)p} & 0_{(k-1)p\times 2(k-1)p}\end{pmatrix}\right] \tilde{J}_{k-1}\,. \end{aligned}$$

Proposition 6. There exists $k \in \mathbb{N}$ such that **LMI**$_C(k)$ (resp. **LMI**$_P(k)$) is solvable *if and only if* $S \in \mathcal{S}'_C$ (resp. $\mathcal{S}'_P$). ∎

From Propositions 5 and 6, it is hence possible to *characterize* the sets $\mathcal{S}_C$, $\mathcal{S}_P$ as follows.

Theorem 7. There exists a positive integer k such that **LMI**$_C(k)$ (resp. **LMI**$_P(k)$) is feasible, *if and only if* all the systems in a (sufficiently small) neighborhood of $S = (A_0, A_1, B_0, B_1, C_0, C_1)$ verify delay-independent circle criterion (resp. uniform delay-independent Popov criterion). ∎

In their present state, the results provided above do not provide decidible algorithms, as there is no indication on how big k has to be tested. On the other hand, checking that a system belongs to $\mathcal{S}'_C$ or $\mathcal{S}'_P$ amounts to check scaled small gain condition involving a structured singular value with one repeated scalar block and one full block: the only numerical alternative necessitates frequency

$$J_{n,k}\mathcal{X}_{k+1}(t) = \begin{pmatrix} \mathcal{X}_k(t) \\ \mathcal{X}_k(t-h) \end{pmatrix}, \tag{4a}$$

$$J_k^u \begin{pmatrix} \mathcal{X}_{k+1}(t) \\ \mathcal{U}_{k+1}(t) \end{pmatrix} = \begin{pmatrix} \mathcal{X}_k(t) \\ \mathcal{U}_k(t) \end{pmatrix}, \; J_k^l \begin{pmatrix} \mathcal{X}_{k+1}(t) \\ \mathcal{U}_{k+1}(t) \end{pmatrix} = \begin{pmatrix} \mathcal{X}_k(t-h) \\ \mathcal{U}_k(t-h) \end{pmatrix}, \; J_k \begin{pmatrix} \mathcal{X}_{k+1}(t) \\ \mathcal{U}_{k+1}(t) \end{pmatrix} = \begin{pmatrix} \mathcal{X}_k(t) \\ \mathcal{X}_k(t-h) \\ \mathcal{U}_k(t) \\ \mathcal{U}_k(t-h) \end{pmatrix}, \tag{4b}$$

$$\tilde{J}_{n,k}\mathcal{X}_{k+2}(t) = \begin{pmatrix} \mathcal{X}_k(t) \\ \mathcal{X}_k(t-h) \\ \mathcal{X}_k(t-2h) \end{pmatrix}, \; \tilde{J}_k \begin{pmatrix} \mathcal{X}_{k+2}(t) \\ \mathcal{U}_{k+2}(t) \end{pmatrix} = \begin{pmatrix} \mathcal{X}_k(t) \\ \mathcal{X}_k(t-h) \\ \mathcal{X}_k(t-2h) \\ \mathcal{U}_k(t) \\ \mathcal{U}_k(t-h) \\ \mathcal{U}_k(t-2h) \end{pmatrix}. \tag{4c}$$

$$V_k(\mathcal{X}_k|_{[t-h;t]}) \stackrel{\text{def}}{=} \mathcal{X}_k^T(t)P_k\mathcal{X}_k(t) + \int_{t-h}^{t} \mathcal{X}_k^T(\tau)Q_k\mathcal{X}_k(\tau)\, d\tau . \tag{5}$$

$$V_{C,k}(\mathcal{X}_k|_{[t-h;t]}) \stackrel{\text{def}}{=} \mathcal{X}_k^T(t)P_k\mathcal{X}_k(t) + \int_{t-h}^{t} \begin{pmatrix} \mathcal{X}_k(\tau) \\ \mathcal{U}_k(\tau) \end{pmatrix}^T Q_k \begin{pmatrix} \mathcal{X}_k(\tau) \\ \mathcal{U}_k(\tau) \end{pmatrix} d\tau . \tag{6}$$

$$V_{P,k}(\mathcal{X}_k|_{[t-h;t]}) \stackrel{\text{def}}{=} V_{C,k}(\mathcal{X}_k|_{[t-h;t]}) + +2\sum_{k'=0}^{k-2}\sum_{i=1}^{p} \eta_i K_i \int_0^{y_i(t-k'h)} \psi_i(v)\, dv . \tag{7}$$

$$\begin{aligned} \frac{d[V_k(\mathcal{X}_k|_{[t-h;t]})]}{dt} &= \begin{pmatrix} \mathcal{X}_k(t) \\ \mathcal{X}_k(t-h) \end{pmatrix}^T \begin{pmatrix} P_k(I_k\otimes A_0)+(I_k\otimes A_0)^TP_k+Q_k & P_k(I_k\otimes A_1) \\ (I_k\otimes A_1)^TP_k & -Q_k \end{pmatrix} \begin{pmatrix} \mathcal{X}_k(t) \\ \mathcal{X}_k(t-h) \end{pmatrix} \\ &= \mathcal{X}_{k+1}^T(t)R_k\mathcal{X}_{k+1}(t) , \end{aligned} \tag{8}$$

$$\begin{aligned} \frac{d}{dt}[V_{C,k}(\mathcal{X}_k|_{[t-h;t]})] &= \dot{\mathcal{X}}_k^T(t)P_k\mathcal{X}_k(t) + \mathcal{X}_k^T(t)P_k\dot{\mathcal{X}}_k(t) + \begin{pmatrix} \mathcal{X}_k(t) \\ \mathcal{U}_k(t) \end{pmatrix}^T Q_k \begin{pmatrix} \mathcal{X}_k(t) \\ \mathcal{U}_k(t) \end{pmatrix} - \begin{pmatrix} \mathcal{X}_k(t-h) \\ \mathcal{U}_k(t-h) \end{pmatrix}^T Q_k \begin{pmatrix} \mathcal{X}_k(t-h) \\ \mathcal{U}_k(t-h) \end{pmatrix} \\ &= -\mathcal{U}_k^T(t)((I_k\otimes K)\mathcal{Y}_k(t)-\mathcal{U}_k(t)) + \begin{pmatrix} \mathcal{X}_{k+1}(t) \\ \mathcal{U}_{k+1}(t) \end{pmatrix}^T R_{C,k} \begin{pmatrix} \mathcal{X}_{k+1}(t) \\ \mathcal{U}_{k+1}(t) \end{pmatrix} . \end{aligned} \tag{9}$$

$$\begin{aligned} \frac{d}{dt}[V_{P,k}(\mathcal{X}_k|_{[t-h;t]})] &= \frac{d}{dt}[V_{C,k}(\mathcal{X}_k|_{[t-h;t]})] + 2\mathcal{U}_{k-1}^T(t)(I_{k-1}\otimes \eta K)\dot{\mathcal{Y}}_{k-1}(t) \\ &= -\mathcal{U}_k^T(t)((I_k\otimes K)\mathcal{Y}_k(t)-\mathcal{U}_k(t)) + \begin{pmatrix} \mathcal{X}_{k+1}(t) \\ \mathcal{U}_{k+1}(t) \end{pmatrix}^T R_{P,k} \begin{pmatrix} \mathcal{X}_{k+1}(t) \\ \mathcal{U}_{k+1}(t) \end{pmatrix} , \end{aligned} \tag{10}$$

---

gridding. As a matter of fact, the problem under study is a NP-hard problem.

To conclude, we give a general idea of the principle of the proof. As an example, the proof of delay-independent circle criterion contains mainly the three following steps.

1. One first shows that $S \in \mathcal{S}'_C$ if and only if, for all $z \in \overline{\mathbb{D}}$, there exists $P(z) = P(z)^* > 0$, such that (11) holds. This is deduced from the continuous time version of Kalman-Yakubovich-Popov lemma.

2. On the other hand, for $k \in \mathbb{N}$ and $P_k \in \mathbb{R}^{kn\times kn}$ a symmetric positive definite matrix, there exists Q_k symmetric such that (P_k, Q_k) is solution of $\mathbf{LMI}_C(k)$ if and only if, for any $z \in \overline{\mathbb{D}}$, a solution of (11) is given by formula (12). This is essentially a consequence of the discrete time version of Kalman-Yakubovich-Popov lemma.

3. When (11) admits for all z in $\overline{\mathbb{D}}$ a positive definite solution $P(z)$, the existence of a solution of the form (12) comes from the fact that the latter may be closely related to the existence of a solution *polynomial* in z, z^*, and *of degree* $k-1$. The conclusion then comes the from the fact that (11) admits solutions analytic wrt z, z^* in $\overline{\mathbb{D}}$, see (Delchamps, 1984) and (Bose Ed., 1985, see Lemma p. 134).

4. REFERENCES

P.-A. Bliman (2001). LMI characterization of the strong delay-independent stability of delay systems via quadratic Lyapunov-Krasovskii functionals, *Systems and Control Letters* **43** no 4, 263–274

N.K. Bose Ed. (1985). *Multidimensional systems theory. Progress, directions and open problems in multidimensional systems*, Mathe-

$$\begin{pmatrix} (A_0 + zA_1)^* P(z) + P(z)(A_0 + zA_1) & (C_0 + zC_1)^* K^T + P(z)(B_0 + zB_1) \\ K(C_0 + zC_1) + (B_0 + zB_1)^* P(z) & -2I_p \end{pmatrix} < 0 \,. \tag{11}$$

$$P(z) = \frac{1}{1 + |z|^2 + \cdots + |z|^{2(k-1)}} \begin{pmatrix} I_n \\ zI_n \\ \vdots \\ z^{k-1} I_n \end{pmatrix}^* P_k \begin{pmatrix} I_n \\ zI_n \\ \vdots \\ z^{k-1} I_n \end{pmatrix} \,. \tag{12}$$

matics and its Applications 16, D. Reidel Publishing Co, Dordrecht

S. Boyd, L. El Ghaoui, E. Feron, V. Balakrishnan (1994). *Linear matrix inequalities in system and control theory*, SIAM Studies in Applied Mathematics vol. 15

D.F. Delchamps (1984). Analytic feedback control and the algebraic Riccati equation, *IEEE Trans. Automat. Control* **AC-29** no 11 1031–1033

J.K. Hale (1977). *Theory of functional differential equations*, Applied Mathematical Sciences 3, Springer Verlag, New York

E.W. Kamen (1982). Linear systems with commensurate time delays: stability and stabilization independent of delay. *IEEE Trans. Automat. Control* **27** no 2 367–375

E.W. Kamen (1983). Correction to "Linear systems with commensurate time delays: stability and stabilization independent of delay", *IEEE Trans. Automat. Control* **28** no 2 248–249

E.W. Kamen, P.P. Khargonekar (1984a). On the control of linear systems whose coefficients are functions of parameters, *IEEE Trans. Automat. Control* **AC-29** no 1, 25–33

E.W. Kamen, P.P. Khargonekar, A. Tannenbaum (1984b). Pointwise stability and feedback control of linear systems with noncommensurate time delays, *Acta Applicandae Mathematicae* **2** (1984) 159–184

H.K. Khalil (1992). *Nonlinear systems*, Macmillan Publishing Company

N.N. Krasovskii (1963). *Stability of motion*, Stanford University Press

X.-J. Li (1963). On the absolute stability of systems with time lags, *Chinese Math.* **4**, 609-626

S.-I. Niculescu, J.-M. Dion, L. Dugard, H. Li (1996). Asymptotic stability sets for linear systems with commensurable delays: a matrix pencil approach, *IEEE/IMACS CESA'96*, Lille, France

V.M. Popov, A. Halanay (1962). On the stability of nonlinear automatic control systems with lagging argument, *Automat. Remote Control* **23**, 783-786

I.W. Sandberg (1964). A frequency domain condition for stability of feedback systems containing a single time-varying nonlinear element, *Bell Sys. Tech. J.* Part II **43**, 1601-1608

E.I. Verriest, M.K.H. Fan (1996). Robust stability of nonlinearly perturbed delay systems, *Proc. of 35th IEEE Conf. on Decision and Control*, Kobe, Japan, 2090–2091

E.I. Verriest, W. Aggoune (1998). Stability of nonlinear differential delay systems, *Mathematics and computers in Simulation* **45**, 257–267

G. Zames (1966). On the input-output stability of nonlinear time-varying feedback systems, *IEEE Trans. Automat. Contr.* Part I **11** no 2, 228-238; Part II **11** no 3, 465-477

www.elsevier.com/locate/ifac

ON DELAY DEPENDENT ROBUST STABILITY OF NEUTRAL SYSTEMS

S.A. Rodrìguez•, J.-M. Dion•, L. Dugard• and D. Ivănescu*

• Laboratoire d'Automatique de Grenoble (INPG CNRS UJF)
ENSIEG, BP 46, 38402, St. Martin d'Hères, FRANCE.
Author to be contacted: Jean Michel Dion
e-mail : Jean-Michel.Dion@inpg.fr
* Laboratoire HEUDIASYC, Université de Technologie de Compiègne, BP 529, 60205, Compiègne, FRANCE.

Abstract: This paper investigates the robust stability of linear neutral delay systems. The uncertain time delay neutral systems under consideration are described by functional differential equations of neutral type with norm bounded nonlinear uncertainties and unknown constant delays. The novelty here is to consider a delay dependent robust stability approach leading to less conservative results than when dealing with robust stability independent of the delay time. The analysis is performed via Lyapunov-Krasovskii functional approach. Sufficient conditions for delay-dependent robust stability are given in terms of the existence of positive definite solutions of linear matrix inequalities. The proposed stability analysis extends previous results obtained either for delay-dependent stability without robustness issues or for robust delay-independent stability.

Keywords : Time-Delay System, Neutral System, Robust Stability, Linear Matrix Inequalities.

1 INTRODUCTION

A great variety of systems can be modeled by retarded functional differential equations [Kolmanovskii and Myshkis 1999], i.e. the future states depend not only on the present, but also on the past history. Aftereffect is a natural component of the dynamic systems in many fields: mechanics, physics, biology, chemistry, economics, information theory, etc. Even if the system itself does not have internal delays, closed loop systems may involve delay phenomena, because of actuators, sensors and computation time [Richard 1998]. The theory and mathematical tools for such systems have been significantly developed in, for instance, [Bellman and Cooke 1963], [Hale and Verduyn Lunel 1993], [Kolmanovskii and Myshkis 1999].

Among time-delay systems, an interesting class is the class of neutral systems characterized by the fact that the delay argument occurs also in the derivative of the "state variables". Some examples of such neutral systems are given in [Brayton 1976], [Niculescu and Brogliato 1995], [Logemann and Townley 1996], [Mounier et al. 1997], [Bellen et al. 1999], [Hu and Davison 2000].

Several works have been concerned with the stability analysis of neutral systems either in the time domain approach, see for example: [Slemrod and Infante 1972], [Hale and Verduyn Lunel 1993], [Niculescu and Brogliato 1995], [Richard et al. 1997], [Verriest and Niculescu 1997], or in the frequency domain approach, see for example: [Chen 1995], [Verriest and Niculescu 1997].

In the above mentioned studies, the attention was mainly focused in giving conditions for delay independent stability. These conditions are conservative in the case where the delays are unknown. It is then of interest to consider delay-dependent stability analysis. Some results have been given in the frequency domain [Chen 1995] and in the time domain [Kolmanovskii 1996], [Ivănescu et al. 2000].

In practice, the model parameters are not precisely known and it is of interest to study the robustness of the stability with respect to parameter uncertainties, see [Li and de Souza 1997], [Kharitonov 1998]. Some neutral systems can be represented by uncertain models, see for instance [Kolmanovskii and Myshkis 1999], (lossless transmission line models may have uncertain parameters).

The objective of this paper is to study the stability analysis of linear neutral systems in a delay-dependent framework incorporating robustness issues. The delays are assumed to be unknown and constant and the uncertainties may be time varying and nonlinear. One obtains sufficient delay-dependent stability conditions via the Lyapunov-Krasovskii functional approach. The main result is expressed in terms of an easy to check Linear Matrix Inequality.

This paper is organized as follows: Section 2 gives some preliminaries and states the problem. Model transformation is discussed in section 3. The main stability result is given in section 4. Some final remarks end the paper.

NOTATION. Here I_m is the identity matrix of dimension m. $x \in R^n$, $\|\cdot\|$ denotes the Euclidean norm of x. For a real number $\tau > 0$, $\mathcal{C}_\tau = \mathcal{C}([-\tau,0],R^n)$ will be the Banach space of continuous vector functions $\varphi : [-\tau,0] \to R^n$ with the supremum norm $\|\varphi\|_\mathcal{C} = \sup_{-\tau\leq t\leq 0}\|\varphi(t)\|$. $\mathcal{C}_{\tau,\nu}$ denotes the open set $\varphi \in \mathcal{C}_\tau$ with $\|\varphi\|_\mathcal{C} < \nu$. "$\cdot$" on the variable and $\frac{d}{dt}(\cdot)$ denote the right-hand derivative at t. $\mathcal{C}_\tau^1 = \mathcal{C}^1([-\tau,0],R^n)$ denote the Banach space of continuous differentiable functions $\varphi : [-\tau,0] \to R^n$ with $\dot{\varphi} \in \mathcal{C}_\tau$ and the norm $\|\varphi\|_{\mathcal{C}^1} = \|\varphi\|_\mathcal{C} + \|\dot{\varphi}\|_\mathcal{C}$. The function x_t denote the restriction of x to interval $[t-\tau,t]$ so that x_t is an element of $\mathcal{C}_\tau$ defined by $x_t(\theta) = x(t+\theta)$ for $-\tau \leq \theta \leq 0$.

2 PROBLEM STATEMENT

Consider the following neutral type system, written in the form proposed by J.K. Hale and M.A. Cruz [Hale and Cruz 1970], [Hale and Verduyn Lunel 1993], [Kolmanovskii and Myshkis 1999]:

$$\dot{\mathcal{D}}x_t = f(t, x_t(-\tau_d)), \\ \mathcal{D}\varphi := [\varphi(0) - g(t, \varphi(-\tau))], \; t \geq t_0, \tag{1}$$

$$\text{with } x_{t_0} \equiv \phi, \; \phi \in \mathcal{C}_{\bar{\tau}}, \tag{2}$$

where the state (x, x_t) is an element of $R^n \times \mathcal{C}_{\bar{\tau}}$ with $\bar{\tau}$ the maximum of the delays $\{\tau, \tau_d\}$ and $f, g \; : \; R^+ \times \mathcal{C}_{\bar{\tau},\nu} \rightarrow R^n$ are supposed to be continuous, bounded

$$\|f(t,\phi)\| \leq \mu, \; \|g(t,\phi)\| \leq \mu, \; t \geq t_0, \; \phi \in \mathcal{C}_{\bar{\tau},\nu},$$

and locally Lipschitz with respect to the second argument with $f(t, 0) \equiv g(t, 0) \equiv 0$, so that $x = 0$ is a solution of the System (1)-(2) in order to have an equilibrium point at the origin, and the operator $\mathcal{D} : R^+ \times \mathcal{C}_{\bar{\tau}} \rightarrow R^n$, is supposed to be atomic at zero [Hale and Verduyn Lunel 1993].

We denote the solutions of the System (1)-(2) by $x(t_0, \phi)$ where $x_{t_0}(t_0, \phi) \equiv \phi$. The value of $x(t_0, \phi)$ at t is denoted by $x(t) = x(t, t_0, \phi)$.

Recall the definition of stability.

DEFINITION 1 *The zero solution of the System (1)-(2) is said to be stable if, for each $\varepsilon > 0$ and $t_0 \geq 0$, there exist $\delta = \delta(t_0, \varepsilon) > 0$ such that if $\|\phi\|_{\mathcal{C}} < \delta$, then $\|x(t, t_0, \phi)\| < \varepsilon$ for all $t \geq t_0$. The zero solution is called asymptotically stable if it is stable and the value δ can be chosen independently of ε such that, in addition, $\|x(t, t_0, \phi)\| \rightarrow 0$ as $t \rightarrow +\infty$ holds.*

Let $V : R^+ \times \mathcal{C}_{\bar{\tau}} \rightarrow R^+$ be a continuous functional; the upper right-hand derivative of V along the solution of the System (1)-(2) is defined by

$$\dot{V}(t,\phi) = \lim_{h \rightarrow 0+} \sup \frac{1}{h}[V(t+h, x_{t+h}(t,\phi)) - V(t,\phi)]$$

We will use the following Lyapunov-Krasovskii functional approach:

THEOREM 1 (Hale and Verduyn Lunel 1993) *Consider the neutral system defined by (1)-(2). Assume that $\mathcal{D}$ is stable and that there exist non decreasing continuous functions $v_i \; : \; R^+ \rightarrow R^+$, $i = 1, 2, 3$ such that $v_i(0) = 0$ and $v_i(r) > 0$, for all $r > 0$ and $i = 1, 2, 3$. Then, the zero solution of (1)-(2) is asymptotically stable if there exists a continuous functional $V : R^+ \times \mathcal{C}_{\bar{\tau}} \rightarrow R^+$ such that:*

1. *$v_1(\|\mathcal{D}\varphi\|) \leq V(t, \varphi) \leq v_2(\|\varphi\|_{\mathcal{C}})$ and*
2. *$\dot{V}(t, x_t) \leq -v_3(\|\mathcal{D}x_t\|)$, for all $t \geq t_0$.*

The present paper considers the following fairly general class of uncertain neutral systems, where the linear neutral differential equations include time-varying nonlinear uncertainties, the positive delays τ and τ_d are not assumed to be known, but constant,

$$\dot{\mathcal{D}}_1 x_t = Ax(t) + Bx(t-\tau_d) + \Delta A(t, x_t(0)) + \\ +\Delta B(t, x_t(-\tau_d)), \\ \mathcal{D}_1\varphi := [\varphi(0) - (C + \Delta C)\varphi(-\tau)], \; t \geq t_0, \tag{3}$$

$$\text{with } x_{t_0} \equiv \phi, \; \dot{x}_{t_0} \equiv \dot{\phi}, \; \{\phi, \varphi\} \in \mathcal{C}^1_{\bar{\tau}}, \tag{4}$$

where $x_t = \{x(t+\theta) : \theta \in [-\bar{\tau}, 0], \bar{\tau} := \max\{\tau, \tau_d\} > 0\}$, the known matrices A, B, and C are constant, ΔC is an uncertainty constant matrix and $\Delta A(\cdot,\cdot)$, $\Delta B(\cdot,\cdot)$ are some gain bounded smooth functions; $C + \Delta C$ constant implies that the operator $\mathcal{D}_1 : R^+ \times \mathcal{C}^1_{\bar{\tau}} \rightarrow R^n$ is continuous and atomic at zero [Hale and Verduyn Lunel 1993]. The unknown nonlinear mappings $\Delta A, \Delta B \; : \; R^+ \times \mathcal{C}_{\bar{\tau},\nu} \rightarrow R^n$ and the uncertainty matrix ΔC are gain bounded smooth functions described by

$$\begin{aligned} \Delta A(t, x_t(0)) &= E_a\delta_a(t, x_t(0)), \; \delta_a^T(t,y)\delta_a(t,y) \\ &\leq y^T W_a^T W_a y, \; \forall \; y \in R^n, \; \forall \; t \in R, \\ \Delta B(t, x_t(-\tau_d)) &= E_b\delta_b(t, x_t(-\tau_d)), \; \delta_b^T(t,y)\delta_b(t,y) \\ &\leq y^T W_b^T W_b y, \; \forall \; y \in R^n, \; \forall \; t \in R, \\ \Delta C x_t(-\tau) &= E_c\delta_c(x_t(-\tau)), \; \delta_c^T(y)\delta_c(y) \\ &\leq y^T W_c^T W_c y, \; \forall \; y \in R^n, \; \forall \; t \in R, \end{aligned} \tag{5}$$

with known matrices E_a, E_b and $E_c := \Delta C$. The matrices W_a, W_b and W_c are given weighting matrices. The unknown mappings δ_a and δ_b, and $\delta_c(x_t(-\tau)) := x(t-\tau)$ satisfy the conditions

$$\delta_a(t, 0) \equiv 0, \delta_b(t, 0) \equiv 0, \delta_c(0) \equiv 0, \tag{6}$$

so that $x = 0$ is a solution of the Neutral Differential Equation (3)-(4). If it is required to have smooth solutions of (3)-(4) then assume that the initial function ϕ satisfies the sewing condition

$$\dot{\phi}(0) = A\phi(0) + B\phi(-\tau_d) + C\dot{\phi}(-\tau) + \\ \Delta A(t_0, \phi(0)) + \Delta B(t_0, \phi(-\tau_d)) + \Delta C\dot{\phi}(-\tau)$$

This weak condition on initial conditions implies the continuity of $\dot{x}(t)$ for $t \geq t_0$ and then $x(t)$ is differentiable on $(t_0 - \bar{\tau}, \infty)$.

Two important stability problems are associated to the Neutral System (3)-(6):

Robust stability problem: [Niculescu 1998]

Find conditions, if they exist, to ensure the asymptotic stability (*independent of the delays*) of the Neutral System (3)-(6) for all nonlinear functions $\Delta A(\cdot,\cdot)$, $\Delta B(\cdot,\cdot)$.

Delay-dependent stability problem: [Ivănescu et al. 2000] Find a bound τ_d^*, if it exists, on the delay $\tau_d > 0$, such that the asymptotic stability of (3)-(6) when $\Delta A(\cdot,\cdot)$, $\Delta B(\cdot,\cdot)$ and ΔC are identically zero, is preserved for any $\tau > 0$ (delay independent) and for $\tau_d \leq \tau_d^*$ (delay dependent).

No bounds are considered in the delay τ because the Schur-Cohn stability of the operator $\mathcal{D}_1$ (ensured only by the Schur-Cohn stability of the matrix $(C + \Delta C)$, [Cruz and Hale 1969]) which is a necessary condition for stability, impies the stability for any value of τ (see Ivănescu et al. 2000 or Hale and Verduyn Lunel 1993).

In this paper, we are concerned with the following mixed problem:

Delay-dependent robust stability problem: Find a bound τ_d^*, if it exists, on the dela y $\tau_d > 0$ and conditions to ensure the asymptotic stabilit y of the Neutral System (3)-(6), for $\tau > 0$ and for $\tau_d \leq \tau_d^*$.

Different methods have been considered to study the stability of the solutions of such systems [Niculescu et al. 1997] and among them, the direct Lyapunov method (Razumikhin or Krasovskii approaches) [Hale and Verduyn Lunel 1993], [Niculescu et al. 1997], [Kolmanovskii and Myshkis 1999]. It reduces the stability problem to the construction of appropriate functionals V, defined along the systems solutions. We use this methodology to study robust stability and delay-dependent stability problems simultaneously. The delay-dependent robust stability conditions are given in terms of the existence positive definite solutions of some linear matrix inequality (LMI).

3 MODEL TRANSFORMATION

Let us consider now the Neutral System (3)-(6); it can rewritten as:

$$\dot{\mathcal{D}}_1 x_t = Ax(t) + Bx(t - \tau_d) + F(t, x_t), \tag{7}$$

with $F(t, x_t) := \Delta A(t, x_t(0)) + \Delta B(t, x_t(-\tau_d))$. Then, using the Leibnitz's rule $(x(t - \tau_d) = x(t) - \int_{-\tau_d}^{0} \dot{x}(t + \theta)d\theta)$, the equation (3) can be rewritten as

$$\dot{\mathcal{D}}_1 x_t = (A + B)x(t) - B \int_{-\tau_d}^{0} \dot{x}(t + \theta)d\theta + F(t, x_t)$$

Since $(C + \Delta C)$ is constant, using (3) and following [Kharitonov 1998] and [Ivănescu et al. 2000] one gets, in the new variable ξ

$$\begin{aligned} \dot{\mathcal{D}}_1 \xi_t = (A + B)\xi(t) - B \int_{-\tau_d}^{0} [A\xi(t + \theta) + \\ +B\xi(t + \theta - \tau_d) + (C + \Delta C)\dot{\xi}(t + \theta - \tau) + \\ +F(t + \theta, \xi_{t+\theta})]d\theta + F(t, \xi_t), \end{aligned} \tag{8}$$

where we used the fact that the initial condition $\phi \in \mathcal{C}_{\tau}^1$, implies $\dot{\mathcal{D}}_1 \varphi = \dot{\varphi}(0) - (C + \Delta C)\dot{\varphi}(-\tau)$. This leads to

$$\begin{aligned} \dot{\mathcal{D}}_1 \xi_t = (A + B)\xi(t) - \\ -B(C + \Delta C)\xi(t - \tau) + \\ +B(C + \Delta C)\xi(t - \tau - \tau_d) - \\ -B \int_{-\tau_d}^{0} [A\xi(t + \theta) + B\xi(t + \theta - \tau_d) + \\ +F(t + \theta, \xi_{t+\theta})]d\theta + F(t, \xi_t), \end{aligned} \tag{9}$$

where the Leibnitz's rule was used again. Finally replacing F and using Equations (5) in the last equation, we get the following equation:

$$\begin{aligned} \dot{\mathcal{D}}_1 \xi_t = (A + B)\xi(t) - \\ -B(C + \Delta C)\xi(t - \tau) \\ +B(C + \Delta C)\xi(t - \tau - \tau_d) + \\ +E_a \delta_a(t, \xi_t(0)) + \\ +E_b \delta_b(t, \xi_t(-\tau_d)) - B \int_{-\tau_d}^{0} [A\xi(t + \theta) + \\ +B\xi(t + \theta - \tau_d) + E_a \delta_a(t + \theta, \xi_t(\theta)) + \\ +E_b \delta_b(t + \theta, \xi_{t-\tau_d}(\theta))]d\theta. \end{aligned} \tag{10}$$

It is not very difficult to check that every solution of Neutral System (3) and the Equation (9) is also solution of the Equation (10), then the stability of (10) implies the stability of (3) [Kharitonov 1998], [Gu and Niculescu 1999].

In the next section, the stability analysis of the Transformed Model (10) is performed in terms of the variable x.

4 DELAY-DEPENDENT ROBUST STABILITY

The Lyapunov-Krasovskii functional approach, with the help of Theorem 1, is used to prove the following main result:

THEOREM 2 *The Neutral System (3)-(6) is delay-dependent robustly asymptotically stable for any $\tau > 0$ and $\tau_d \leq \tau_d^*$ if*

1. *$A_1 := A + A_d$ is a Hurwitz stable matrix;*
2. *$C_1 := C + \Delta C$ is a Schur-Cohn stable matrix;*
3. *The difference operator $\mathcal{D}_1 \varphi := [\varphi(0) - (C + \Delta C)\varphi(-\tau)]$ is atomic at zero;*
4. *there exist a real positive number τ_d^* and positive definite matrices P, $S_i > 0$, $i = \overline{1, 7}$ such that the following LMI holds:*

$$\Gamma := \begin{pmatrix} PA_1 + A_1^T P + K(\tau_d^*) & PE(\tau_d^*) & (PA + K(\tau_d^*))C_1 & 0 & PBC_1 & \sqrt{\tau_d^*} PBA & \sqrt{\tau_d^*} PB^2 \\ E^T(\tau_d^*)P & -I_{4n} & 0 & 0 & 0 & 0 & 0 \\ C_1^T(A^T P + K(\tau_d^*)) & 0 & C_1^T(K(\tau_d^*))C_1 - S_1 + S_7 & 0 & 0 & 0 & 0 \\ 0 & 0 & 0 & K'(\tau_d^*) & 0 & 0 & 0 \\ C_1^T B^T P & 0 & 0 & 0 & -S_7 & 0 & 0 \\ \sqrt{\tau_d^*} A^T B^T P & 0 & 0 & 0 & 0 & -S_3 & 0 \\ \sqrt{\tau_d^*} (B^2)^T P & 0 & 0 & 0 & 0 & 0 & -S_4 \end{pmatrix} < 0, \tag{11}$$

where

$$K(\tau_d^*) := S_1 + S_2 + \tau_d^*(S_3 + S_4 + S_5),$$
$$K'(\tau_d^*) := W_b^T W_b - S_2 + \tau_d^* S_6, \quad (12)$$
$$E(\tau_d^*) := \begin{pmatrix} E_a & E_b & \sqrt{\tau_d^*}BE_a & \sqrt{\tau_d^*}BE_b \end{pmatrix}.$$

Notice that condition 1 is necessary and directly follows from the satisfaction of condition 4, condition 2 consists in checking that the eigenvalues of C_1 are inside the unit disc. Condition 3 is verified, due to the fact that C_1 is constant and finally, condition 4 can be checked by using LMI Tools [Boyd et al. 1994].

PROOF
Consider the Lyapunov-Krasovkii functional

$$\mathrm{V}(\varphi) := V_1(\varphi) + V_2(\varphi) + V_3(\varphi),\ \varphi \in \mathcal{C}_{\bar{\tau}}, \quad (13)$$

where

$$\mathrm{V}_1(\varphi) := \mathcal{D}_1^T \varphi P \mathcal{D}_1 \varphi, \quad (14)$$

$$V_2(\varphi) := \sum_{i=1}^{2} \int_{-\tau_i}^{0} \varphi^T(\theta) S_i \varphi(\theta) d\theta + \int_{-\tau_d}^{0} [\int_{\theta}^{0} \varphi^T(\vartheta) S_3 \varphi(\vartheta) d\vartheta] d\theta,\ \tau_1 := \tau,\ \tau_2 := \tau_d, \quad (15)$$

$$V_3(\varphi) := \int_{-\tau_d}^{0} [\int_{\theta-\tau_d}^{0} \varphi^T(\vartheta) S_4 \varphi(\vartheta) d\vartheta] d\theta + \sum_{i=5}^{6} \int_{-\tau_d}^{0} [\int_{\theta}^{0} \psi_i^T(\vartheta) S_i \psi_i(\vartheta) d\vartheta] d\theta + \int_{-\tau-\tau_d}^{-\tau} \varphi^T(\theta) S_7 \varphi(\theta) d\theta, \quad (16)$$

and the functionals, ψ_i, are $\psi_5 := \varphi$, $\psi_6(\vartheta) := \varphi(\vartheta - \tau_d)$.

For the functional V, we can construct $\dot{V}$ along the trajectories of the Transformed Model (10) in terms of x, if φ is replaced by x_t in the right hand side of $V(\varphi)$, pass to x, differentiate in t (so that $\dot{x}$ appears only for the current (not delayed) value of t), substitute $\dot{x}(t)$ from the Transformed Model (10), pass from x to x_t, and replace x_t by φ. Having this procedure in mind, the expressions for V and $\dot{V}$ are often straightforwardly written in terms of x, keeping in mind the transition from x to φ, as proposed by [Kolmanovskii and Myshkis 1999]. Then we have:

$$\dot{V}_1(x_t) = \mathcal{D}_1^T x_t P \frac{d\mathcal{D}_1 x_t}{dt} + \frac{d\mathcal{D}_1^T x_t}{dt} P \mathcal{D}_1 x_t, \quad (17)$$

$$\dot{V}_2(x_t) = \sum_{i=1}^{2} [x^T(t) S_i x(t) - x^T(t-\tau_i) S_i x(t-\tau_i)] + \int_{-\tau_d}^{0} [x_t^T(0) S_3 x_t(0) - x_t^T(\theta) S_3 x_t(\theta)] d\theta, \quad (18)$$

$$\begin{aligned}\dot{V}_3(x_t) = &\int_{-\tau_d}^{0} [x_t^T(0) S_4 x_t(0) - x_{t-\tau_d}^T(\theta) S_4 x_{t-\tau_d}(\theta)] d\theta \\ &+ x^T(t-\tau) S_7 x(t-\tau) - x^T(t-\tau-\tau_d) S_7 x(t-\tau-\tau_d) \\ &+ \int_{-\tau_d}^{0} [x_t^T(0) S_5 x_t(0) - x_t^T(\theta) S_5 x_t(\theta) \\ &+ x_t^T(-\tau_d) S_6 x_t(-\tau_d) - x_{t-\tau_d}^T(\theta) S_6 x_{t-\tau_d}(\theta)] d\theta \end{aligned} \quad (19)$$

Substituting the value of $\frac{d\mathcal{D}_1 x_t}{dt}$ given by (10) in (17) and setting $A_1 := A + B$, $C_1 := C + \Delta C$ we get

$$\begin{aligned}\dot{V}_1(x_t) = &\mathcal{D}_1^T x_t P A_1 x(t) + x^T(t) A_1^T P \mathcal{D}_1 x_t \\ &+ 2\mathcal{D}_1^T x_t P [E_a \delta_a(t, x_t(0)) + E_b \delta_b(t, x_t(-\tau_d))] \\ &- \mathcal{D}_1^T x_t P B [C_1 x(t-\tau) - C_1 x(t-\tau-\tau_d)] \\ &- [x^T(t-\tau) C_1^T - x^T(t-\tau-\tau_d) C_1^T] B^T P \mathcal{D}_1 x_t \\ &- 2\mathcal{D}_1^T x_t P B \int_{-\tau_d}^{0} [A x_t(\theta) + B x_{t-\tau_d}(\theta) \\ &+ E_a \delta_a(t+\theta, x_t(\theta)) + E_b \delta_b(t+\theta, x_{t-\tau_d}(\theta))] d\theta. \end{aligned} \quad (20)$$

Using the following well known inequality

$$-2a^T b \leq \inf_{X>0} \{a^T X a + b^T X^{-1} b\ \forall\ a,\ b \in R^n\} \quad (21)$$

and (5), we have directly:

$$\begin{aligned} &2\mathcal{D}_1^T x_t P E_a \delta_a(t, x_t(0)) \leq \\ &\mathcal{D}_1^T x_t P E_a E_a^T P \mathcal{D}_1 x_t + x^T(t) W_a^T W_a x(t), \\ &2\mathcal{D}_1^T x_t P E_b \delta_b(t, x_t(-\tau_d)) \leq \\ &\mathcal{D}_1^T x_t P E_b E_b^T P \mathcal{D}_1 x_t + x^T(t-\tau_d) W_b^T W_b x(t-\tau_d) \end{aligned} \quad (22)$$

Now using again the Inequality (21), we get:

$$\begin{aligned} &-2\mathcal{D}_1^T x_t P B \int_{-\tau_d}^{0} A x_t(\theta) d\theta \leq \\ &\int_{-\tau_d}^{0} [\mathcal{D}_1^T x_t P B A R_1^{-1} A^T B^T P \mathcal{D}_1 x_t \\ &+ x_t^T(\theta) R_1 x_t(\theta)] d\theta, \\ &-2\mathcal{D}_1^T x_t P B \int_{-\tau_d}^{0} B x_{t-\tau_d}(\theta) d\theta \leq \\ &\int_{-\tau_d}^{0} [\mathcal{D}_1^T x_t P B^2 R_2^{-1} (B^2)^T P \mathcal{D}_1 x_t \\ &+ x_{t-\tau_d}^T(\theta) R_2 x_{t-\tau_d}(\theta)] d\theta, \end{aligned} \quad (23)$$

for all positive square matrices R_1, R_2, and

$$-2\mathcal{D}_1^T x_t PB \int_{-\tau_d}^{0} E_a\delta_a(t+\theta, x_t(\theta))d\theta \leq$$
$$\int_{-\tau_d}^{0} [\mathcal{D}_1^T x_t PBE_aE_a^T B^T P\mathcal{D}_1 x_t + x_t^T(\theta)W_a^T W_a x_t(\theta)]d\theta,$$
$$-2\mathcal{D}_1^T x_t PB \int_{-\tau_d}^{0} E_b\delta_b(t+\theta, x_{t-\tau_d}(\theta)d\theta \leq$$
$$\int_{-\tau_d}^{0} [\mathcal{D}_1^T x_t PBE_bE_b^T B^T P\mathcal{D}_1 x_t + x_{t-\tau_d}^T(\theta)W_b^T W_b x_{t-\tau_d}(\theta)]d\theta, \quad (24)$$

$$\Omega := \begin{pmatrix} Q(\tau_d^*) - S(\tau_d^*)R^{-1}S^T(\tau_d^*) & (PA+K(\tau_d^*))C_1 & 0 & PBC_1 \\ C_1^T(A^TP+K(\tau_d^*)) & C_1^T(K(\tau_d^*))C_1 - S_1 + S_7 & 0 & 0 \\ 0 & 0 & W_b^TW_b - S_2 + \tau_d^*S_6 & 0 \\ C_1^TB^TP & 0 & 0 & -S_7 \end{pmatrix} < 0, \quad (32)$$

(22)-(24) allows to get a bound for $\dot{V}_1$, and if we choose $R_1 := S_3$, $R_2 := S_4$,

$$\begin{cases} S_5 = W_a^TW_a, \text{ if } W_a^TW_a > 0, \\ S_5 > W_a^TW_a, \text{ in another case,} \end{cases} \quad (25)$$

$$\begin{cases} S_6 = W_b^TW_b, \text{ if } W_b^TW_b > 0, \\ S_6 > W_b^TW_b, \text{ in another case,} \end{cases} \quad (26)$$

then we have a bound for $\dot{V}$ with (18) and (19):

$$\begin{aligned} \dot{V}(x_t) \leq\ & \mathcal{D}_1^T x_t PA_1 x(t) + x^T(t-\tau)S_7 x(t-\tau) \\ & + x^T(t)A_1^T P\mathcal{D}_1 x_t - x^T(t-\tau-\tau_d)S_7 x(t-\tau-\tau_d) \\ & - \mathcal{D}_1^T x_t PB[C_1 x(t-\tau) - C_1 x(t-\tau-\tau_d)] \\ & - [x^T(t-\tau)C_1^T - x^T(t-\tau-\tau_d)C_1^T]B^T P\mathcal{D}_1 x_t \\ & + \mathcal{D}_1^T x_t PE_aE_a^T P\mathcal{D}_1 x_t + x^T(t)W_a^T W_a x(t) \\ & + \mathcal{D}_1^T x_t PE_bE_b^T P\mathcal{D}_1 x_t + x^T(t-\tau_d)W_b^T W_b x(t-\tau_d) \\ & + \int_{-\tau_d}^{0} \mathcal{D}_1^T x_t PBAS_3^{-1}A^TB^TP\mathcal{D}_1 x_t d\theta \\ & + \int_{-\tau_d}^{0} \mathcal{D}_1^T x_t PB^2S_4^{-1}(B^2)^TP\mathcal{D}_1 x_t d\theta \\ & + \int_{-\tau_d}^{0} \mathcal{D}_1^T x_t PBE_aE_a^TB^TP\mathcal{D}_1 x_t d\theta \\ & + \int_{-\tau_d}^{0} \mathcal{D}_1^T x_t PBE_bE_b^TB^TP\mathcal{D}_1 x_t d\theta \\ & + \sum_{i=1}^{2}[x^T(t)S_i x(t) - x^T(t-\tau_i)S_i x(t-\tau_i)] \\ & + \tau_d \sum_{i=3}^{5} x^T(t)S_i x(t) + \tau_d x^T(t-\tau_d)S_6 x_(t-\tau_d). \end{aligned} \quad (27)$$

Using the following identities, we can rewrite each expression containing the vector function $x(t)$ in (27) as an expression containing $\mathcal{D}_1 x_t$ and $x(t-\tau)$:

$$\begin{aligned} x^T(t)S_i x(t) = & \mathcal{D}_1^T x_t S_i \mathcal{D}_1 x_t + \mathcal{D}_1^T x_t S_i C_1 x(t-\tau) \\ & + x^T(t-\tau)C_1^T S_i \mathcal{D}_1 x_t + x^T(t-\tau)C_1^T S_i C_1 x(t-\tau) \end{aligned} \quad (28)$$

$$\begin{aligned} & \mathcal{D}_1^T x_t PA_1 x(t) + x^T(t)A_1^T P\mathcal{D}_1 x_t = \\ & \mathcal{D}_1^T x_t(A_1^TP + PA_1)\mathcal{D}_1 x_t + \mathcal{D}_1^T x_t PA_1C_1 x(t-\tau) \\ & + x^T(t-\tau)C_1^TA_1^TP\mathcal{D}_1 x_t, \end{aligned} \quad (29)$$

With all these inequalities and identities, there exists a real positive number $\tau_d^* \geq \tau_d$ such that:

$$\dot{V}(x_t) \leq \omega^T\Omega\omega, \quad (30)$$

where the vector ω and the matrix Ω are

$$\omega := \begin{pmatrix} \mathcal{D}_1 x_t \\ x(t-\tau) \\ x(t-\tau_d) \\ x(t-\tau-\tau_d) \end{pmatrix}, \quad (31)$$

with $S(\tau_d^*) := (S(\tau_d^*)_1, S(\tau_d^*)_2)$,

$$Q(\tau_d^*) := PA_1 + A_1^TP + K(\tau_d^*),$$
$$S(\tau_d^*)_1 := \begin{pmatrix} \sqrt{\tau_d^*}PBA & \sqrt{\tau_d^*}PB^2 & PE_a \end{pmatrix},$$
$$S(\tau_d^*)_2 := \begin{pmatrix} PE_b & \sqrt{\tau_d^*}PBE_a & \sqrt{\tau_d^*}PBE_b \end{pmatrix},$$
$$R^{-1} := \begin{pmatrix} -S_3^{-1} & 0 & 0 \\ 0 & -S_4^{-1} & 0 \\ 0 & 0 & I_{4n} \end{pmatrix},$$
$$K(\tau_d^*) := S_1 + S_2 + \tau_d^*(S_3 + S_4 + S_5).$$

Thus, if the Matrix Inequality (11) holds, it follows (via an appropriate Schur transformation [Boyd et al. 1994]) that: $\Omega < 0$.

$\mathcal{D}_1$ is stable by Assumption (2), see [Cruz and Hale 1969]. Then, the robust asymptotic stability of (3)-(6) is ensured by Theorem 1 for all delays $\tau > 0$, $\tau_d \leq \tau_d^*$, [Hale and Verduyn Lunel 1993].

REMARK 1 *When the uncertainties $\Delta A(\cdot,\cdot)$, $\Delta B(\cdot,\cdot)$, ΔC are zero, we can choose the Lyapunov-Krasovskii functional (13) $V(\varphi) := V_1(\varphi) + V_2(\varphi) + V_3(\varphi)$ with V_1 defined in (14) and*

$$V_2(\varphi) := \int_{-\tau}^{0} \varphi^T(\theta)S_1\varphi(\theta)d\theta + \int_{-\tau_d}^{0}[\int_{\theta}^{0} \varphi^T(\vartheta)S_3\varphi(\vartheta)d\vartheta]d\theta,$$

$$V_3(\varphi) := \int_{-\tau_d}^{0}[\int_{\theta-\tau_d}^{0} \varphi^T(\vartheta)S_4\varphi(\vartheta)d\vartheta]d\theta + \int_{-\tau-\tau_d}^{-\tau} \varphi^T(\theta)S_7\varphi(\theta)d\theta.$$

And we get, as a particular case, the result of [Ivănescu et al. 2000].

REMARK 2 *If we are interested only in the delay independent robust stability of Neutral System (3)-(6), the model transformation can be avoided, choosing $V(\varphi) := V_1(\varphi) + V_2(\varphi)$ with V_1 defined in (14) and*

$$V_2(\varphi) := \sum_{i=1}^{2}\int_{-\tau_i}^{0} \varphi^T(\theta)S_i\varphi(\theta)d\theta,\ \tau_1 := \tau,\ \tau_2 := \tau_d,$$

leading to similar stability results as proposed in [Niculescu 2001].

5 CONCLUDING REMARKS

In this paper, w econsidered the dela y dependent robust stabilit y of neutral systems. We got sufficien t conditions chec kablein the LMI framework. We proposed a fairly general Lyapunov-Krassovskii functional which includes, as particular cases, functionals developed in previous works. A more general class of neutral systems including nonlinearities in the operator $\mathcal{D}$ of (1) satisfying the atomicity condition could be considered. Along the same lines an interesting topic for further studies is the delay dependent robust stabilization of neutral systems.

6 BIBLIOGRAPHY

1. Bellen A., N. Guglielmi and A.E. Ruehli, "Methods for Linear Systems of Circuits Delays Differential Equations of Neutral Type", IEEE Trans. Circuits and Systems, vol. 46, 1, pp. 212-216, 1999.
2. Bellman R. and K.L. Cooke, "Differential Difference Equations", Academic Press, New York, 1963.
3. Boyd S., L.El Ghaoui, E. Feron and V. Balakrishnan, "Linear Matrix Inequalities in System and Control Theory", Society for Industrial and Applied Mathematics, 1994.
4. Brayton R., "Nonlinear Oscillations in a Distributed Network", Quart. Appl. Math., vol 24, 289-301, 1976.
5. Chen J., "On Computing the Maximal Delay Intervals for Stability of Linear Delay Systems" IEEE Trans. Automat. Contr., vol 40, pp. 1087-1093, 1995.
6. Cruz M.A. and J.K. Hale, "Asymptotic Behavior of Neutral Functional Differential Equations" Arch. Rat. Mech. Anal. Vol. 34, pp 331-353, 1969.
7. Gu K. and S.I. Niculescu, "Additional Eigenvalues in Transformed Time-Delay Systems", Proc. 38th IEEE Conf. Dec. Contr., Phoenix, AZ, Decembre 1999.
8. Hale J.K. and S.M. Verduyn Lunel, "Introduction to Functional Differential Equations", Springer-Verlag, 1993.
9. Hale J.K. and M.A. Cruz, "Existence, Uniqueness and Continuous Dependence for Hereditary Systems", Ann. Mat. Pura Appl. vol. 85, pp. 63-82, 1970.
10. Hu G. and E.J. Davison, "Real Stability Radii for Linear Neutral Delay Systems", 2nd IFAC Workshop on Linear Time Delay Systems, Ancona, Italy, pp. 106-110, septembre 11-13, 2000.
11. Ivănescu D., S. Niculescu, L. Dugard and J.M. Dion, "On Delay Dependent Stability for Linear Neutral Systems" Proc. IFAC Workshop on Linear Time Delay Systems, Ancona Italy, pp. 117-122, 2000.
12. Kharitonov V.L., "Robust Stability Analysis of Time-Delay Systems": a survey, Proc. IFAC Syst. Struct. Contr., Nantes, France, pp. 1-12, 1998.
13. Kolmanovskii V.B., "The Stability of Hereditary Systems of Neutral Type", J. Appl. Maths. Mechs., vol. 60, No 2, pp. 205-216, 1996.
14. Kolmanovskii V.B. and A.D. Myshkis, "Introduction to the Theory and Applications of Functional Differential Equations", Kluwer Academic Publishers, 1999.
15. Li X. and C.E. de Souza, "Delay Dependent Robust Stability and Stabilization of Uncertain Time Delay Systems: A Linear Matrix Inequality Approach", IEEE Trans. Aut. Contr. 42, pp. 144-148, 1997.
16. Logemann H. and S. Townley, "The Effect of Small Delays in the Feedback Loop on the Stability of Neutral Systems", Syst. Contr. Lett. 27 pp. 267-274, 1996.
17. Mounier H., P. Rouchon and J. Rudolph, "Some Examples of Linear Systems with Delays", RAIRO-APII-JESA (Journal Européen des Systè mes Automatisés), 31, no 6, pp. 911-925, 1997.
18. Niculescu S.I., "On the Stability of a Class of Uncertain Linear Neutral Systems: An LMI Approach with Applications". Proc. IFAC Syst. Struct. Contr. Nantes France, pp. 25-29, July 1998.
19. Niculescu S.I., "On Robust Stability of Neutral Systems, Special issue Time-Delay Systems", Kybernetica, vol. 37,3, pp. 253-264, 2001, to appear.
20. Niculescu S.I. and B. Brogliato , "On Force Measurements Time-delays in Control of Constrained Manipulators", Proc. IFAC Syst. Struct. Contr. Nantes France, pp. 266-271, 1995.
21. Niculescu S.I., E.I. Verriest, L. Dugard and J.M. Dion, "Stability and Robust Stability of Time-Delay Systems: A guided Tour", in Stability and Control of Time-Delay Systems, (L. Dugard and E.I. Verriest, Eds.), LNCIS, vol. 228, Springer-Verlag London, pp. 1-71, 1997.
22. Richard J.P., "Some Trends and Tools for the Study of Time Delay Systems", Proc. IEEE-IMACS Conference CESA'98, Comp. Engineering in Syst. Applications, pp. 27-43, Nabeul, Tunisia, April 1998.
23. Richard J.P., A. Goubet-Bartholoméüs, Ph. A. Tchangani and M. Dambrine, "Nonlinear Delays Systems: Tools for quantitative approach to stabilization" in Stability and Control of Time-Delay Systems, (L. Dugard and E.I. Verriest, Eds.), LNCIS, vol. 228, Springer-Verlag London, pp. 218-240, 1997.
24. Slemrod M. and E.F. Infante, "Asymptotic Stability Criteria for Linear Systems of Differential Difference Equations of Neutral Type and their Discrete Analogues", J.Math. Anal. Appl., 38, pp.399-415, 1972.
25. Verriest E.I. and S.I. Niculescu, "Delay-Independent Stability of Linear Neutral Systems: A. Riccati Equation Approach, in Stability and Control of Time-Delay Systems", (L. Dugard and E.I. Verriest, Eds.), LNCIS, vol. 228, Springer-Verlag London, pp. 92-100, 1997.

www.elsevier.com/locate/ifac

ROBUST $\mathcal{H}_\infty$ CONTROL FOR UNCERTAIN DISCRETE-TIME STATE-DELAYED LINEAR SYSTEMS WITH MARKOVIAN JUMPING PARAMETERS [1]

Reinaldo M. Palhares * **Carlos E. de Souza** ** **Pedro L. D. Peres** ***

* *Graduate Program in Electrical Engineering*
Pontifical Catholic University of Minas Gerais
Av. D. José Gaspar 500, 30535-610 Belo Horizonte, MG, Brazil
reinaldo@pucminas.br

** *Department of Systems and Control*
Laboratório Nacional de Computação Científica – LNCC
Av. Getúlio Vargas 333, 25651-070 Petrópolis, RJ, Brazil
csouza@lncc.br

*** *School of Electrical and Computer Engineering*
University of Campinas
CP 6101, 13081-970 Campinas, SP, Brazil
peres@dt.fee.unicamp.br

Abstract: This paper addresses the problem of robust $\mathcal{H}_\infty$ state-feedback control design for uncertain discrete-time linear systems with multiple time delays at the states and Markovian jumping parameters. The jumping parameters are assumed to be available and the uncertainties are supposed to belong to convex bounded domains (polytope type uncertainty). Delay-independent sufficient conditions assuring robust stochastic stability and a prescribed $\mathcal{H}_\infty$ disturbance attenuation for the closed-loop uncertain discrete-time linear system with multiple time delays and Markovian jumping parameters are established in terms of linear matrix inequalities, which have the advantage that can be implemented numerically very efficiently. *Copyright © 2001 IFAC*

Keywords: Robust $\mathcal{H}_\infty$ control; time delay systems; Markovian jumping parameters; linear matrix inequalities; discrete-time systems.

1. INTRODUCTION

Robust $\mathcal{H}_\infty$ control has been addressed by several approaches in many papers dealing with uncertain linear systems in both continuous and discrete-time. One may say that the two main existing strategies are the one based on Riccati equations, and the linear matrix inequality (LMI) approach.

Recently, $\mathcal{H}_\infty$ control techniques have been applied to uncertain linear systems subject to time-delays (among others, one may cite (Song *et al.*, 1999), (de Souza and Li, 1999), (Kim and Park, 1999)). As a fact, time-delay are frequently encountered in many engineering systems and can be source of undesirable behaviors including instability (see (Mahmoud, 2000) and references therein).

On the other hand, linear systems with Markovian jumping parameters have been deeply investigated in the last decades. Stochastic models provide interesting new representation for random component failures and abrupt changes in parameters that can appear in target tracking, robotics, power systems, large flexible structures, etc. In the context of control of discrete-time Markovian jump linear systems, it is worth to mention the approaches presented in (Costa *et al.*, 1999), (Costa and Marques, 1998).

A realistic control design must cope with all these features to assure closed-loop stability and performance. Only very recently a few papers have tackle the problem of robust control for uncertain state-delayed systems with Markovian jumping parameters, most of them in the context of continuous-time systems as in (Benjelloun and Boukas, 1998), (Benjelloun *et al.*, 1999) and (Palhares *et al.*, 2001). The problem of $\mathcal{H}_\infty$ control for discrete-time jump linear systems with time delay has been addressed by means of LMI based strategies and memoryless state feedback control in (Cao and Lam, 1999) (precisely known systems) and

[1] This work was supported in part by *Conselho Nacional de Desenvolvimento Científico e Tecnológico* - CNPq, Brazil, under grants 300596/98-7/PQ for the first author, 301653/96-8/PQ and PRONEX no. 15/98 - Control of Dynamical Systems, for the second author, and 304604/89-5/PQ for the third author.

(Shi *et al.*, 1999) (linear systems with norm-bounded uncertainties).

In this paper, sufficient LMI conditions for the robust stochastic stabilizability with an $\mathcal{H}_\infty$ performance bound are given for discrete-time uncertain linear systems with convex bounded uncertainties, multiple time delays and Markovian jumping parameters (which are assumed to be available). In this sense, the paper extend some previous results to the case of polytope type uncertainty, proposing a Markovian jump linear state feedback control law with both a memoryless and a "memory" term which is capable to accomplish with more stringent specifications. The control gains are obtained from LMI solvability problems, which can be numerically addressed by means of polynomial time procedures.

The notation used in this paper is quite standard. $\mathcal{E}(\cdot)$ stands for the mathematical expectation. $\mathcal{Z}^n$ denotes the space of all sequences mapping $\mathbb{Z}^+ \mapsto \mathbb{R}^n$, ℓ_2 denotes the subset of all sequences $\xi \in \mathcal{Z}^n$, which satisfy $\|\xi\|_{\ell_2} \triangleq \left(\sum_{t=0}^{\infty}\|\xi(t)\|^2\right)^{1/2} < \infty$, where $\|\cdot\|$ stands for the Euclidean vector norm. In a stochastic setting, $\ell_2[\Omega,\mathcal{F},\mathcal{P}]$ is the space over the probability space $(\Omega,\mathcal{F},\mathcal{P})$ with norm $\|\cdot\|_{\ell_2[\Omega,\mathcal{F},\mathcal{P}]} \triangleq \mathcal{E}\left\{\sum_{t=0}^{\infty}\|\cdot\|^2\right\}^{1/2}$. For a symmetric block matrix, the symbol $*$ denotes the submatrices that lie below the main block-diagonal, whereas diag$\{\cdots\}$ stands for a block-diagonal matrix.

2. PROBLEM FORMULATION

Consider the following class of uncertain discrete-time stochastic system on an appropriate probability space $(\Omega,\mathcal{F},\mathcal{P})$ with multiple time delays given by:

$$x(k+1) = A_0(\theta_k)x(t) + \sum_{j=1}^{q} A_k(\theta_k)x(t-\tau_j) + B_1(\theta_k)w(t) + B_2(\theta_k)u(t) \quad (1)$$

$$z(k) = C_0(\theta_k)x(k) + \sum_{j=1}^{q} C_j(\theta_k)x(k-\tau_j) + D_1(\theta_k)w(t) + D_2(\theta_k)u(t) \quad (2)$$

$$\theta_0 = i_0; \quad x(k) = \phi(k) \quad (3)$$

with $k = -\bar{\tau}, -\bar{\tau}+1, \ldots, 0$, $\bar{\tau} = \max_j(\tau_j)$ where $x(k) \in \mathbb{Z}^n$ is the state vector, $y(k) \in \mathbb{Z}^r$ is the measurements vector, $w(k) \in \mathbb{Z}^m$ is the noise signal vector, $z(k) \in \mathbb{Z}^p$ is the controlled output, $\{\phi(k)\}$ is a given known initial vector sequence, and $\tau_j \geq 0$, $j = 1,\ldots,q$, are constant time delays. The noise signal w is assumed to be an arbitrary signal in ℓ_2. $\{\theta_k;\ k = 0,1,\ldots\}$ is a discrete-time Markov chain with finite state-space $\mathcal{S} = \{1,\ldots,N\}$ and transition probability matrix $\Lambda = [\lambda_{ij}], i,j = 1,\ldots,N$, namely $\lambda_{ij} \triangleq \text{Prob}\{\theta_{k+1} = j \,|\, \theta_k = i\}$ with $\lambda_{ij} \geq 0$, $\forall i,j \in \mathcal{S}$ and $\sum_{j=1}^{N}\lambda_{ij} = 1$.

The set $\mathcal{S}$ comprises the various operating modes of system (1)-(3) and for each possible valor of $\theta_k = i$, $i \in \mathcal{S}$, one can denote the matrices associated with the "i-th mode" by

$$A_{0i} \triangleq A_0(\theta_k = i); \quad A_{ji} \triangleq A_j(\theta_k = i),\ j = 1,\ldots,q;$$

$$B_{1i} \triangleq B_1(\theta_k = i); \quad B_{2i} \triangleq B_2(\theta_k = i);$$

$$C_{0i} \triangleq C_0(\theta_k = i); \quad C_{ji} \triangleq C_j(\theta_k = i),\ j = 1,\ldots,q$$

$$D_{1i} \triangleq D_1(\theta_k = i); \quad D_{2i} \triangleq D_2(\theta_k = i)$$

For any $i \in \mathcal{S}$, the system matrices are assumed to be unknown (uncertain) but belonging to a known convex bounded polyhedral, namely

$$(A_{0i},\ldots,A_{qi},B_{1i},B_{2i},C_{0i},\ldots,C_{qi},D_{1i},D_{2i}) \in \mathcal{D}_i$$

with $i \in \mathcal{S}$. Moreover, $\mathcal{D}_i$ is a polytope described by ν vertices[2] as follows:

$$\mathcal{D}_i \triangleq \Big\{(A_{0i},\ldots,D_{2i}),\ \big|\ (A_{0i},\ldots,D_{2i}) = \sum_{h=1}^{\nu}\alpha_h\left(A_{0ih},A_{jih},B_{1ih},B_{2ih},C_{0ih},C_{jih},D_{1ih},D_{2ih}\right); \quad \alpha_h \geq 0;\ \sum_{h=1}^{\nu}\alpha_h = 1\Big\}$$

Throughout this paper it is assumed that the jumping process $\{\theta_k\}$ is accessible, i.e. the operating mode of the Markovian linear system (1)-(3) is known for every $k \geq 0$ (this is a standard assumption; see, for example, (de Souza and Fragoso, 1996)).

The robust control problem addressed in this paper consists on the design of a Markovian jump linear state feedback control law such that the closed-loop uncertain state delayed system described in (1)-(3) presents robust stochastic stability and a prespecified $\mathcal{H}_\infty$ guaranteed performance irrespective of the uncertainty and the time delays.

Since the jumping process $\{\theta_k\}$, $k \geq 0$ as well as the delayed states at each operating mode $\theta_k = i, i \in \mathcal{S}$ are supposed to be available, the following control law is proposed:

$$u(k) = K_0(\theta_k)x(k) + \sum_{j=1}^{q} K_j(\theta_k)x(k-\tau_j) \quad (4)$$

Note that the control law (4) differs from most of the existing state-feedback control strategies which are memoryless (de Souza and Li, 1999), (Kim and Park, 1999), (Cao and Lam, 1999), (Shi *et al.*, 1999).

Considering (4), the closed-loop system is given by

$$x(k+1) = \tilde{A}_0(\theta_k)x(k) + \sum_{j=1}^{q} \tilde{A}_j(\theta_k)x(t-\tau_j) + B_1(\theta_k)w(k) \quad (5)$$

[2] To avoid a cumbersome notation, the uncertainty domains $\mathcal{D}_i$ are supposed to have the same number of vertices ν.

$$z(k) = \tilde{C}_0(\theta_k)x(k) + \sum_{j=1}^{q} \tilde{C}_j(\theta_k)x(t-\tau_j) + D_1(\theta_k)w(t) \quad (6)$$

where

$$\tilde{A}_0(\theta_k) = A_0(\theta_k) + B_2(\theta_k)K_0(\theta_k) \quad (7)$$

$$\tilde{C}_0(\theta_k) = C_0(\theta_k) + D_2(\theta_k)K_0(\theta_k) \quad (8)$$

$$\tilde{A}_k(\theta_k) = A_k(\theta_k) + B_2(\theta_k)K_k(\theta_k),\ k = 1,\ldots,q \quad (9)$$

$$\tilde{C}_k(\theta_k) = C_k(\theta_k) + D_2(\theta_k)K_k(\theta_k),\ k = 1,\ldots,q \quad (10)$$

Throughout the paper, $\tilde{A}_{0ih}$, $\tilde{A}_{jih}$, $\tilde{C}_{0ih}$, $\tilde{C}_{jih}$, B_{1ih} D_{1ih}, $h = 1,\ldots,\nu$ denote the matrices $\tilde{A}_{0i}$, $\tilde{A}_{ji}$, $\tilde{C}_{0i}$, $\tilde{C}_{ji}$, B_{1i}, D_{1i} evaluated at each of the vertices of the polytope $\mathcal{D}_i$, $\theta_k = i$, $i \in \mathcal{S}$.

Before stating formally the $\mathcal{H}_\infty$ control problem to be addressed in this paper, consider the following definition of stochastic stability and robust stochastic stability.

Definition 1. The Markovian jump linear state-delayed system (5) is said to be stochastically stable if for $w \equiv \mathbf{0}$ its solution $x(k)$ is such that $\mathcal{E}\{\sum_{k=0}^{\infty} \|x(k)\|^2\} < \infty$, for all finite initial condition $(x(0),\theta_0)$. The system (5) is said to be robustly stochastically stable if it is stochastically stable for all admissible uncertainties. ■

Now, the $\mathcal{H}_\infty$ control problem addressed in this paper is presented:

Robust $\mathcal{H}_\infty$ control problem with Markovian jumping parameters and multiple time delays:

Given a scalar $\gamma > 0$, determine a Markovian jump linear control of the form (4) such that the closed-loop system (5)-(6) is robustly stochastically stable over the entire uncertainty domains $\mathcal{D}_i$ and ensures a prescribed level γ of $\mathcal{H}_\infty$ noise attenuation, i.e., under zero-initial conditions and for any non-zero $w \in \ell_2$

$$\|z\|_{\ell_2[\Omega,\mathcal{F},\mathcal{P}]} \le \gamma \|w\|_{\ell_2},\ \forall\ (A_{0i},\ldots,D_{2i}) \in \mathcal{D}_i$$

$\forall i \in \mathcal{S}$. In this situation, the closed-loop system (5)-(6) is said to have a guaranteed γ level of noise attenuation.

Notice that, in fact, the solvability of the above problem is not centered only on assuring the robust stochastic stability with an $\mathcal{H}_\infty$ performance for the closed-loop system, but also on establishing a delay-independent condition for the robust stochastic stability.

3. MAIN RESULTS

In the sequel, a sufficient condition for the closed-loop system (5)-(6) to have a guaranteed γ level of noise attenuation is established in terms of the solvability of a set of LMIs.

Lemma 1. Consider the closed-loop system (5)-(6) and let $\gamma > 0$ be a given scalar. If there exist symmetric matrices $P_i \in \mathbb{R}^{n\times n}$ $i = 1,\ldots,N$, and $Y_{ji} \in \mathbb{R}^{n\times n}$, $j = 1,\ldots,q$, $i = 1,\ldots,N$, such that

$$\begin{bmatrix} \Upsilon_{1i} & \Upsilon_{2ih} \\ * & \Upsilon_{3i} \end{bmatrix} > \mathbf{0},\ i = 1,\ldots,N;\ h = 1,\ldots,\nu \quad (11)$$

where

$$\Upsilon_{1i} = \operatorname{diag}\{P_1,\ldots,P_N,\mathbf{I}\} \quad (12)$$

$$\Upsilon_{2ih} = \begin{bmatrix} \varphi_{1i}\tilde{A}_{0ih} & \cdots & \varphi_{1i}\tilde{A}_{qih} & \varphi_{1i}\tilde{B}_{1ih} \\ \vdots & & \vdots & \vdots \\ \varphi_{1i}\tilde{A}_{0ih} & \cdots & \varphi_{1i}\tilde{A}_{qih} & \varphi_{1i}\tilde{B}_{1ih} \\ \tilde{C}_{0ih} & \cdots & \tilde{C}_{qih} & \tilde{D}_{1ih} \end{bmatrix} \quad (13)$$

$$\varphi_{pi} = \sqrt{\lambda_{pi}P_p},\quad p = 1,\ldots,N$$

$$\Upsilon_{3i} = \operatorname{diag}\left\{P_i - \sum_{j=1}^{q} Y_{ji},\ Y_{1i},\ \ldots,\ Y_{qi},\ \gamma^2\mathbf{I}\right\} \quad (14)$$

then the uncertain closed-loop system has a guaranteed γ level of noise attenuation. ■

Proof: First, it is shown that if the closed-loop system (5) with $w(t) \equiv \mathbf{0}$ satisfies the conditions of the lemma, then the robust stochastic stability of the state-delayed system of (5) is assured. For that, let the Lyapunov functional

$$V[x_{\bar{\tau}}(k),\theta_k] \triangleq x'(k)P(\theta_k)x(k) + \sum_{\beta=k-\tau_1}^{k-1} x'(\beta)Y_1(\theta_k)x(\beta) + \cdots + \sum_{\beta=k-\tau_q}^{k-1} x'(\beta)Y_q(\theta_k)x(\beta) \quad (15)$$

where $P(\theta_k) \in \mathbb{R}^{n\times n}$ and $Y_j(\theta_k) \in \mathbb{R}^{n\times n}$, $j = 1,\ldots,q$, are symmetric positive definite matrices to be determined, $x_{\bar{\tau}}(k)$ denotes the vector $[x'(k)\ \ x'(k-1)\ \cdots\ x'(k-\bar{\tau})]'$, $P(\theta_k) = P_i$ and $Y_j(\theta_k) = Y_{ji}$ when $\theta_k = i$, $i \in \mathcal{S}$. Moreover, note that there exist positive scalars $\mu(\theta_k)$ and $\nu(\theta_k)$ such that

$$\mu(\theta_k)\|x(k)\|^2 \le V[x_{\bar{\tau}}(k),\theta_k] \le \nu(\theta_k) \sup_{\varphi\in[-\bar{\tau},0]} \|x(k+\varphi)\|^2 \quad (16)$$

Particularly, one can select, $\mu(\theta_k) = \lambda_{min}(P(\theta_k))$, $\nu(\theta_k) = \lambda_{max}(P(\theta_k)) + \sum_{j=1}^{q}\lambda_{max}(Y_j(\theta_k))$, $\theta_k \in \mathcal{S}$. Where, $\lambda_{min}(M)$ and $\lambda_{max}(M)$ denote the minimum eigenvalue of matrix M and the maximum eigenvalue of matrix M, respectively.

Considering (5) with $w(t) \equiv 0$ and any arbitrary initial condition, one has

$$\mathcal{E}\left\{\Delta V[x_{\bar{\tau}}(k),\theta_k] \,\middle|\, x_{\bar{\tau}}(k),\theta_k\right\}$$

$$\triangleq \mathcal{E}\left\{V[x_{\bar{\tau}}(k+1),\theta_k] \,\middle|\, x_{\bar{\tau}}(k),\theta_k\right\} - V[x_{\bar{\tau}}(k),\theta_k]$$

$$= x'(k)\left(\tilde{A}_0'(\theta_k)\tilde{P}(\theta_k)\tilde{A}_0(\theta_k) - P(\theta_k) + \sum_{j=1}^{q} Y_j(\theta_k)\right)x(k)$$

$$+2\sum_{j=1}^{q} x'(k)\tilde{A}_0'(\theta_k)\widetilde{P}(\theta_k)\tilde{A}_j(\theta_k)x(k-\tau_j)$$
$$-\sum_{j=1}^{q} x'(k-\tau_j)Y_j(\theta_k)x(k-\tau_j)$$
$$+\sum_{j=1}^{q}\sum_{r=1}^{q} x'(k-\tau_j)\tilde{A}_j'(\theta_k)\widetilde{P}(\theta_k)\tilde{A}_r(\theta_k)x(k-\tau_r) \quad (17)$$

with $\widetilde{P}(\theta_k) \triangleq \sum_{p=1}^{N} \lambda_{ip}P(\theta_k)$, and $\widetilde{P}(\theta_k) = \widetilde{P}_i$, when $\theta_k = i,\ i \in \mathcal{S}$.

Defining $\xi(t) = [x'(t) \quad x'(t-\tau_1) \quad \cdots \quad x'(t-\tau_q)]'$, (17) can be rewritten in the following quadratic form

$$\mathcal{E}\left\{\Delta V[x_{\bar{\tau}}(k),\theta_k] \,\middle|\, x_{\bar{\tau}}(k),\theta_k\right\} = -\xi'(k)Q(\theta_k)\xi(k)$$

where

$$Q(\theta_k) \triangleq \begin{bmatrix} Q_1(\theta_k) & Q_2(\theta_k) \\ Q_2'(\theta_k) & Q_3(\theta_k) \end{bmatrix}$$

$$Q_1(\theta_k) = P(\theta_k) - \tilde{A}_0'(\theta_k)\widetilde{P}(\theta_k)\tilde{A}_0(\theta_k) - \sum_{j=1}^{q} Y_j(\theta_k)$$

$$Q_2(\theta_k) = [-\tilde{A}_0'(\theta_k)\widetilde{P}(\theta_k)\tilde{A}_1(\theta_k) \quad \cdots \quad -\tilde{A}_0'(\theta_k)\widetilde{P}(\theta_k)\tilde{A}_q(\theta_k)]$$

$$Q_3(\theta_k) = \begin{bmatrix} T_{(1,1)}(\theta_k) & \cdots & T_{(1,q)}(\theta_k) \\ \vdots & \ddots & \vdots \\ * & \cdots & T_{(q,q)}(\theta_k) \end{bmatrix}$$

$$T_{(j,j)}(\theta_k) = Y_j(\theta_k) - \tilde{A}_j'(\theta_k)\widetilde{P}(\theta_k)\tilde{A}_j(\theta_k), \quad j = 1,\ldots,q$$

$$T_{(j,t)}(\theta_k) = -\tilde{A}_j'(\theta_k)\widetilde{P}(\theta_k)\tilde{A}_t(\theta_k),$$

$$j = 1,\ldots,q-1; \quad t = j+1,\ldots,q; \ t \neq j$$

Considering that the LMIs of (11) are affine in the system matrices, it can be easily verified using Schur's complements that (11) ensure that $Q(\theta_k) > 0$ over the entire uncertainty domain $\mathcal{D}_i$. Note that the LMIs of (11) also imply that $P(\theta_k)$ and $Y_j(\theta_k)$, $j = 1,\ldots,q$ are positive definite matrices. Thus for $x(k) \neq \mathbf{0}$

$$\mathcal{E}\left\{\Delta V[x_{\bar{\tau}}(k),\theta_k] \,\middle|\, x_{\bar{\tau}}(k),\theta_k\right\} = -\xi'(k)Q(\theta_k)\xi(k)\frac{V[x_{\bar{\tau}}(k),\theta_k]}{V[x_{\bar{\tau}}(k),\theta_k]} < 0$$

$$\frac{\mathcal{E}\left\{V[x_{\bar{\tau}}(k+1),\theta_k] \,\middle|\, x_{\bar{\tau}}(k),\theta_k\right\} - V[x_{\bar{\tau}}(k),\theta_k]}{V[x_{\bar{\tau}}(k),\theta_k]} \leq \psi - 1 < 0 \quad (18)$$

where, using the bounds in (16), ψ is a scalar satisfying

$$\psi = 1 - \min_{\theta_k\in\mathcal{S}} \frac{\lambda_{min}(Q(\theta_k))\|\xi(k)\|^2}{\nu(\theta_k)\sup_{\varphi\in[-\bar{\tau},0]}\|x(k+\varphi)\|^2} < 1$$

thus from (18)

$$0 < \frac{\mathcal{E}\left\{V[x_{\bar{\tau}}(k+1),\theta_k] \,\middle|\, x_{\bar{\tau}}(k),\theta_k\right\}}{V[x_{\bar{\tau}}(k),\theta_k]} \leq \psi$$

and $0 < \psi < 1$. Now, the stability criterion follows from the proof of Theorem 1 in (Shi *et al.*, 1999), i.e.

$$\lim_{t\to\infty} \mathcal{E}\left\{\sum_{k=0}^{t} \|x_{\bar{\tau}}(k)\|^2 \,\middle|\, x(0),\theta_0\right\} \leq x'(0)Mx(0)$$

where M is a bounded matrix given by

$$M = \frac{P(\theta_0)}{\max_{\theta_k\in\mathcal{S}}\left(\|P(\theta_k)\| + \sum_{j=1}^{q}\|Y_j(\theta_k)\|\right)(1-\psi)}$$

hence assuring that the closed-loop system (5) with $w(t) \equiv \mathbf{0}$ is stochastically stable for all admissible uncertainties.

To establish the $\mathcal{H}_\infty$ performance for the system (5)-(6), first notice that since the closed-loop system is robustly stochastically stable over the entire uncertainty domain $\mathcal{D}_i$ and $w \in \ell_2$, then $z \in \ell_2[\Omega,\mathcal{F},\mathcal{P}]$. Next, assuming zero initial conditions for the closed-loop system (5)-(6), let the following performance index

$$\mathcal{J} = \mathcal{E}\left\{\sum_{t=0}^{\infty}\left(z'(k)z(k) - \gamma^2 w'(k)w(k)\right)\right\}$$
$$= \mathcal{E}\left\{\sum_{k=0}^{\infty}\left(z'(k)z(k) - \gamma^2 w'(k)w(k)\right)\right\}$$
$$+\mathcal{E}\left\{\sum_{k=0}^{\infty}\left(\Delta V[x_{\bar{\tau}}(k),\theta_k] \,\middle|\, x_{\bar{\tau}}(k),\theta_k\right)\right\}$$
$$-\lim_{k\to\infty}\mathcal{E}\{V[x(k),\theta_k]\} + V[x(0),\theta_0]$$

where $V[x_{\bar{\tau}}(k),\theta_k]$ is as in (15).

Assuming zero initial conditions and stability, it can be obtained that $V[x(0),\theta_0] = 0$ and $\lim_{k\to\infty} \mathcal{E}\{V(x(k))\} \to 0$ and, using (5) and (6), one gets

$$\mathcal{J} = \mathcal{E}\Big\{\sum_{k=0}^{\infty}\Big[x'(k)\Big(\mathcal{F}_{00}(\theta_k)) - P(\theta_k)$$
$$+\sum_{j=1}^{q} Y_j(\theta_k) + \mathcal{G}_{00}(\theta_k)\Big)x(k)$$
$$+2\sum_{j=1}^{q} x'(k)\left(\mathcal{F}_{0j}(\theta_k) + \mathcal{G}_{0j}(\theta_k)\right)x(k-\tau_j)$$
$$-\sum_{j=1}^{q} x'(k-\tau_j)Y_j(\theta_k)x(k-\tau_j)$$
$$+\sum_{j=1}^{q}\sum_{r=1}^{q} x'(k-\tau_j)\left(\mathcal{F}_{jr}(\theta_k) + \mathcal{G}_{jr}(\theta_k)\right)x(k-\tau_r)$$
$$+2x'(k)\left(\mathcal{H}_0(\theta_k) + \tilde{C}_0'(\theta_k)D_1(\theta_k)\right)w(k)$$

$$+2\sum_{j=1}^{q} x'(k-\tau_j)\left(\mathcal{H}_j(\theta_k)+\tilde{C}_j'(\theta_k)D_1(\theta_k)\right)w(k)$$
$$-w'(k)\left(\gamma^2\mathbf{I}-D_1'(\theta_k)D_1(\theta_k)\right.$$
$$\left.\left.-B_1'(\theta_k)\widetilde{P}(\theta_k)B_1(\theta_k)\right)w(k)\right]\Bigg\} \quad (19)$$

where

$$\mathcal{F}_{tv}(\theta_k) \triangleq \tilde{A}_t'(\theta_k)\widetilde{P}(\theta_k)\tilde{A}_v(\theta_k)$$
$$\mathcal{G}_{tv}(\theta_k) \triangleq \tilde{C}_t'(\theta_k)\tilde{C}_v(\theta_k)$$
$$\mathcal{H}_t(\theta_k) \triangleq \tilde{A}_j'(\theta_k)\widetilde{P}(\theta_k)B_1(\theta_k)$$

with appropriate indices. Defining

$$\eta(k) \triangleq [x'(k) \quad x'(k-\tau_1) \quad \cdots \quad x'(k-\tau_q) \quad w'(k)]'$$

one can rewrite (19) in the following form

$$\mathcal{J} = -\mathcal{E}\left\{\sum_{t=0}^{\infty}\eta'(t)W(\theta_k)\eta(t)\right\}$$

where

$$W(\theta_k) \triangleq \begin{bmatrix} W_1(\theta_k) & W_2(\theta_k) \\ W_2'(\theta_k) & W_3(\theta_k)\end{bmatrix}$$

$$W_1(\theta_k) = P(\theta_k) - \tilde{A}_0'(\theta_k)\widetilde{P}(\theta_k)\tilde{A}_0(\theta_k) - \sum_{j=1}^{q} Y_j(\theta_k) - \tilde{C}_0'(\theta_k)\tilde{C}_0(\theta_k)$$

$$W_2(\theta_k) = -\left[\tilde{A}_0'(\theta_k)\widetilde{P}(\theta_k)\tilde{A}_1(\theta_k)+\tilde{C}_0'(\theta_k)\tilde{C}_1(\theta_k) \quad \cdots \quad \tilde{A}_0'(\theta_k)\widetilde{P}(\theta_k)\tilde{A}_q(\theta_k)+\tilde{C}_0'(\theta_k)\tilde{C}_q(\theta_k) \quad \tilde{A}_0'(\theta_k)\widetilde{P}(\theta_k)B_1(\theta_k)+\tilde{C}_0'(\theta_k)D_1(\theta_k)\right]$$

$$W_3(\theta_k) = \begin{bmatrix} H_{(1,1)}(\theta_k) & \cdots & H_{(1,q)}(\theta_k) & -F_1(\theta_k) \\ \vdots & \ddots & \vdots & \vdots \\ * & \cdots & H_{(q,q)}(\theta_k) & -F_q(\theta_k) \\ * & \cdots & * & \gamma^2\mathbf{I}-G(\theta_k)\end{bmatrix}$$

$$G(\theta_k) = D_1'(\theta_k)D_1(\theta_k) + B_1'(\theta_k)\widetilde{P}(\theta_k)B_1(\theta_k)$$
$$F_j(\theta_k) = \tilde{A}_j'(\theta_k)\widetilde{P}(\theta_k)B_1(\theta_k) + \tilde{C}_j'(\theta_k)D_1(\theta_k)$$
$$H_{(j,j)}(\theta_k) = T_{j,j}(\theta_k) - \tilde{C}_j'(\theta_k)\tilde{C}_j(\theta_k), \quad j=1,\ldots,q$$
$$H_{(j,t)}(\theta_k) = -T_{j,t}(\theta_k) + \tilde{C}_j'(\theta_k)\tilde{C}_t(\theta_k),$$
$$j=1,\ldots,q-1; \quad t=j+1,\ldots,q; \; t\neq j$$

In the light of the above, it can be verified, using Schur's complement, that

$$W(\theta_k) = \Upsilon_3(\theta_k) - \Upsilon_2(\theta_k)'\Upsilon_1(\theta_k)^{-1}\Upsilon_2(\theta_k)$$

where $\Upsilon_1(\theta_k) = \Upsilon_{1i}$, $\Upsilon_2(\theta_k) = \Upsilon_{2i}$ and $\Upsilon_3(\theta_k) = \Upsilon_{3i}$, when $\theta_k = i$, $i \in \mathcal{S}$, are as defined in (12)-(14), respectively.

Hence, it can be readily verified that the LMIs of (11) ensure that $W(\theta_k) > 0$ over the entire uncertainty domain $\mathcal{D}_i$. This implies that $\mathcal{J} < 0$, for any non-zero $w \in \ell_2$ and over the entire uncertainty domain $\mathcal{D}_i$, i.e. the closed-loop system has a guaranteed γ level of noise attenuation, which concludes the proof. ■

Remark 1. *It should be noted that as the conditions of Lemma 1 do not use any information about the size of the time-delays τ_j, $j=1,\ldots,q$, this lemma ensures that the closed-loop system of (5)-(6) is stochastically stable with a prescribed γ level of noise attenuation over the entire uncertainty domain $\mathcal{D}_i$ and for any finite time-delays $\tau_j \geq 0$, $j=1,\ldots,q$.*

The next theorem provides a sufficient condition for the existence of a Markovian jump linear control law assuring a robust $\mathcal{H}_\infty$ performance for discrete-time linear systems with real polytope-type parameter uncertainty and multiple state delays.

Theorem 1. Consider the system (1)-(3) and let $\gamma > 0$ be a given scalar. If for all $i \in \mathcal{S}$ there exist symmetric matrices $R_i \in \mathbb{R}^{n\times n}$, and $M_{ki} \in \mathbb{R}^{n\times n}$, $k=1,\ldots,q$, with $R_i > \mathbf{0}$, and matrices, $Z_{0i} \in \mathbb{R}^{m\times n}$ and $Z_{ki} \in \mathbb{R}^{m\times n}$, satisfying the LMIs

$$\begin{bmatrix} \Phi_{1i} & \Phi_{2ih} \\ \Phi_{2ih}' & \Phi_{3i}\end{bmatrix} > \mathbf{0}, \quad \forall i \in \mathcal{S}, \forall h = 1,\ldots,\nu \quad (20)$$

where

$$\Phi_{1i} = \operatorname{diag}\{R_1,\ldots,R_N,\mathbf{I}\}$$

$$\Phi_{2ih} = \begin{bmatrix} \mathcal{A}_{01ih} & \cdots & \mathcal{A}_{q1ih} & \sqrt{\lambda_{1i}}B_{1ih} \\ \vdots & & \vdots & \vdots \\ \mathcal{A}_{0Nih} & \cdots & \mathcal{A}_{qNih} & \sqrt{\lambda_{Ni}}B_{1ih} \\ \mathcal{B}_{0ih} & \cdots & \mathcal{B}_{qih} & \tilde{D}_{1ih}\end{bmatrix}$$

$$\mathcal{A}_{tjih} = \sqrt{\lambda_{ji}}(A_{tih}R_i + B_{2ih}Z_{ti})$$
$$\mathcal{B}_{tih} = C_{tih}R_i + D_{2ih}Z_{ti},$$
$$j=1,\cdots,q; \quad t=0,\cdots,q$$

$$\Phi_{3i} = \operatorname{diag}\left\{R_i - \sum_{j=1}^{q} M_{ji},\, M_{1i},\,\ldots,\,M_{qi},\,\gamma^2\mathbf{I}\right\}$$

then the robust $\mathcal{H}_\infty$ control problem for system (1)-(3) is solvable. Moreover, under the above conditions, a suitable controller is given as in (4) for $\theta_k = i$, $i \in \mathcal{S}$, where

$$K_{0i} = Z_{0i}R_i^{-1}, \quad K_{ji} = Z_{ji}R_i^{-1}, \quad j=1,\ldots,q \quad (21)$$

■

Proof: Assume that for all $i \in \mathcal{S}$ there exist matrices $R_i > \mathbf{0}$, M_{ji}, $j=1,\ldots,q$, Z_{0i} and Z_{ji}, satisfying (20). From the (2,2)-block of (20) it follows that $M_{ji} > \mathbf{0}$, $j=1,\ldots,q$. Let the LMIs of (20) be pre- and post-multiplied by the transformation matrix given by

$$\operatorname{diag}\{R_1^{-1},\ldots,R_N^{-1},\mathbf{I},J,\mathbf{I}\}, \quad \forall i \in \mathcal{S}$$

where J is a block diagonal matrix of $(q+1)$ blocks, each one given by R_i^{-1}, yielding

$$\begin{bmatrix} \Lambda_{1i} & \Lambda_{2ih} \\ \Lambda'_{2ih} & \Lambda_{3i} \end{bmatrix} > \mathbf{0} \, \forall i \in \mathcal{S}, \forall h = 1, \ldots, \nu \quad (22)$$

where

$$\Lambda_{1i} = \text{diag}\{R_1^{-1}, \ldots, R_N^{-1}, \mathbf{I}\}$$

$$\Lambda_{2ih} = \begin{bmatrix} \mathcal{R}_{01ih} & \cdots & \mathcal{R}_{q1ih} & \sqrt{\lambda_{1i}} R_1^{-1} B_{1ih} \\ \vdots & & \vdots & \vdots \\ \mathcal{R}_{0Nih} & \cdots & \mathcal{R}_{qNih} & \sqrt{\lambda_{Ni}} R_N^{-1} B_{1ih} \\ \mathcal{S}_{0ih} & \cdots & \mathcal{S}_{qih} & \tilde{D}_{1ih} \end{bmatrix}$$

$$\mathcal{R}_{tjih} = \sqrt{\lambda_{ji}} (R_j^{-1} A_{tih} + R_j^{-1} B_{2ih} Z_{ti} R_i^{-1})$$

$$\mathcal{S}_{tih} = C_{tih} + D_{2ih} Z_{ti} R_i^{-1}$$

$$j = 1, \cdots, q; \quad t = 0, \cdots, q$$

$$\Lambda_{3i} = \text{diag}\{R_i^{-1} - \sum_{j=1}^{q} R_i^{-1} M_{ji} R_i^{-1}, \\ R_i^{-1} M_{1i} R_i^{-1}, \ldots, R_i^{-1} M_{qi} R_i^{-1}, \gamma^2 \mathbf{I}\}$$

Defining the matrices

$$P_i \triangleq R_i^{-1}; \; K_{0i} \triangleq Z_{0i} R_i^{-1}$$

$$K_{qi} \triangleq Z_{qi} R_i^{-1}; \; Y_{ji} \triangleq R_i^{-1} M_{ji} R_i^{-1} \quad (23)$$

it is straightforward to verify, using the matrices given in (7)-(10), that (22) is equivalent to the LMIs of (11). Hence, one concludes from Lemma 1 that the Markovian jumping linear control given as (4) where K_{0i} and K_{ji}, $j = 1, \ldots, q$, are as in (23), assures that the closed-loop system has a guaranteed γ level of noise attenuation. ∎

With the results of Theorem 1, the robust $\mathcal{H}_\infty$ control problem for system (1)-(3) can be solved by testing the feasibility of the LMIs (20). Note that any feasible solution to (20) yields a suitable Markovian jump linear robust control. The smallest γ attenuation level such that the conditions of Theorem 1 hold can be readily obtained from the optimal solution of the following LMI optimization problem:

$$\min_{R_i, M_{ji}, Z_{0i}, Z_{ji}, \delta} \delta$$

$$\text{subject to} \begin{cases} (20), & \forall i \in \mathcal{S}, \; h = 1, \ldots, \nu, \\ j = 1, \ldots, q, & \text{with } \gamma^2 \triangleq \delta \end{cases}$$

The minimum value of gamma is given by $\gamma^* = \sqrt{\delta^*}$, where δ^* is the optimal value of δ and the associated optimal control matrices are as in (21).

Note that Theorem 1 provides a robust $\mathcal{H}_\infty$ control which assures robust stochastic stability and a guaranteed $\mathcal{H}_\infty$ performance for the closed-loop system for all admissible uncertainties independently of the size of the time-delays.

4. CONCLUSIONS

LMI sufficient conditions have been proposed to solve the problem of robust $\mathcal{H}_\infty$ state feedback control for uncertain discrete-time state-delayed linear systems with Markovian jumping parameters.

5. REFERENCES

Benjelloun, K. and E. K. Boukas (1998). Mean square stochastic stability of linear time-delay systems with Markovian jumping parameters. *IEEE Transactions on Automatic Control* **43**(10), 1456–1460.

Benjelloun, K., E. K. Boukas and O. L. V. Costa (1999). $\mathcal{H}_\infty$ control for linear time-delay systems with Markovian jumping parameters. In: *Proceedings of the 38th IEEE Conference on Decision and Control.* Phoenix, AZ. pp. 1567–1572.

Cao, Y. Y. and J. Lam (1999). Stochastic stabilizability and $\mathcal{H}_\infty$ control for discrete-time jump linear systems with time-delay. *Journal of The Franklin Institute* **336**(8), 1263–1281.

Costa, O. L. V. and R. P. Marques (1998). Mixed $\mathcal{H}_2/\mathcal{H}_\infty$-control of discrete-time Markovian jump linear systems. *IEEE Transactions on Automatic Control* **43**(1), 95–100.

Costa, O. L. V., E. O. Assumpção Filho, E. K. Boukas and R. P. Marques (1999). Constrained quadratic state feedback control of discrete-time Markovian jump linear systems. *Automatica* **35**(4), 617–626.

de Souza, C. E. and M. D. Fragoso (1996). Robust $\mathcal{H}_\infty$ filtering for uncertain Markovian jump linear systems. In: *Proceedings of the 35th IEEE Conference on Decision and Control.* Kobe, Japan. pp. 4808–4813.

de Souza, C. E. and X. Li (1999). Delay-dependent robust $\mathcal{H}_\infty$ control of uncertain linear state-delayed systems. *Automatica* **35**(7), 1313–1321.

Kim, J. H. and H. B. Park (1999). $\mathcal{H}_\infty$ state feedback control for generalized continuous/discrete time-delay system. *Automatica* **35**, 1443–1451.

Mahmoud, M. S. (2000). *Robust Control and Filtering for Time-Delay Systems.* Control Engineering Series. Marcekl Dekker, Inc.. New York.

Palhares, R. M., C. E. de Souza and P. L. D. Peres (2001). Robust $\mathcal{H}_\infty$ control for uncertain state-delayed linear systems with Markovian jumping parameters. In: *European Control Conference – ECC 2001.* Porto, Pt.

Shi, P., E. K. Boukas and R. K. Agarwal (1999). Control of Markovian jump discrete-time systems with norm bounded uncertainty and unknown delay. *IEEE Transactions on Automatic Control* **44**(11), 2139–2144.

Song, S. H., J. K. Kim, C. H. Yim and H. C. Kim (1999). $\mathcal{H}_\infty$ control of discrete-time linear systems with time-varying delays in state. *Automatica* **35**, 1587–1591.

www.elsevier.com/locate/ifac

IMPROVED PERFORMANCE FOR OPEN-CHANNEL HYDRAULIC SYSTEMS USING INTERMEDIATE MEASUREMENTS

Xavier Litrico [*,1] **Vincent Fromion** [**,1]
Gérard Scorletti [***]

[*] *Cemagref, UR IRMO, B.P. 5095, 34033 Montpellier Cedex 1, France. E-mail:* xavier.litrico@cemagref.fr
[**] *INRA, LASB, 2 place Viala, 34060 Montpellier, France. E-mail:* fromion@ensam.inra.fr
[***] *LAP-ISMRA, 6 Bd. du Maréchal Juin, 14050 Caen, France. E-mail:* scorletti@greyc.ismra.fr

Abstract: The article investigates the performance of feedback controlled open-channel systems represented by a linear advection-diffusion equation. In a previous paper, we analyzed the maximal achievable performance of an open-channel system when only the regulated variable is measured and perturbations are assumed unmeasured. We developed an H_∞ scheme allowing to achieve this maximum performance. In this paper, we consider the more realistic situations, where other intermediate variables of the open-channel system are measured and some perturbations are known in the future. Introducing futur perturbation knowledge is traditionally handled in the predictive control approach. We point out that it is possible to design "predictive" controller using the standard H_∞ approach. The approach is developed for a generic SIMO system and applied to the river control.

Keywords: Water, Fluid flow control, H_∞ control, Rational approximation, Time Delay Systems, Predictive control, Robustness.

1. INTRODUCTION

Faced to the increasing demand in water savings, hydraulic engineers use automatic control techniques in order to obtain a better performance in real-time operation of open-channel systems. The open-channel system considered in this paper is a river controlled by a dam situated upstream where the action variable is the upstream discharge and the controlled variable is the downstream discharge. The river is used to deliver water from the upstream dam to various consumers pumping water along the reach (farmers irrigating their fields, industries, etc.). The objective of the controller is to keep the measured downstream discharge close to a target despite users' withdrawals. Since the numerous withdrawals cannot all be measured, the main control problem is here the rejection of unmeasured perturbations.

River flow control has already been treated in literature using different approaches: classical SISO controllers (Papageorgiou and Messmer, 1989), SISO predictive adaptive control (Foss et al., 1989), model predictive control (Sawadogo et al., 1991). However, to the best of our knowledge, no predictive method explicitly took into account the robustness of the control law.

[1] Partially supported by the joint research program INRA-Cemagref ASS AQUAE n° 02 on the control of time-delay hydraulic systems

In a previous paper, we analyzed the maximal achievable performance of an open-channel system when only the regulated variable is measured and perturbations are assumed unmeasured (Litrico and Fromion, 2001). We developed an H_∞ scheme allowing to achieve this maximum performance for the SISO case. The main goal of this paper is to investigate the effect of additional information on the performance: can additional information improve the performance in terms of perturbations rejection? Different piece of information will be considered: measurement of intermediate discharges in the river, measurement of some perturbations and use of prediction of future perturbations in order to obtain a predictive controller.

Designing control law taking into account the future behavior of perturbations when it is known in advance is traditionally achieved by the so-called Model Predictive control approach (Bitmead et al., 1990). Nevertheless, the robustness or even the stability analysis of the design controller is not *a priori* taken into account. Note that ensuring a prescribed robustness margin is an important constraint in our problem. The H_∞ approach allows to design robust control laws. Nevertheless, it does not take advantage to the futur behavior knowledge of the perturbations.

Our purpose is to combine the advantage of the Model Predictive approach and the H_∞ methods. We propose a simple extension of the H_∞ approach in order to explicitly take into account the futur behavior of the perturbation.

This paper is organized as follows. The problem under consideration is presented in section 2: the system is presented and the specifications are detailed. In section 3, we investigate the impact of the control law structure choice over the achieved performance. In the next section, we propose a simple extension of a standard H_∞ criterion in order to design a robust predictive control law. The design of such a controller is performed in the section 5: the proposed structure dramatically improves performance since in contrast with more traditional structures, the users' desired water supply is almost exactly satisfied.

2. PROBLEM STATEMENT

2.1 Considered problem

The paper will focus on the analysis of a controlled river of total length X with a finite number of discharge measurement points at given locations x_i, $0 < x_1 < \ldots < x_n = X$ (see figure 1). The discharge is supposed to be measured at different locations along the river reach and there is a finite number of intermediate pumping stations distributed along the reach which provide water to consumers (typically farmers who irrigate fields). The objective of the controller is to act on the upstream discharge $u(t) = q(0,t)$ in order to keep the measured downstream discharge $y_n(t) = q(X,t)$ close to a target despite unmeasured users' withdrawals $w_i(t)$, using intermediate discharge measurements $y_i(t) = q(x_i, t)$.

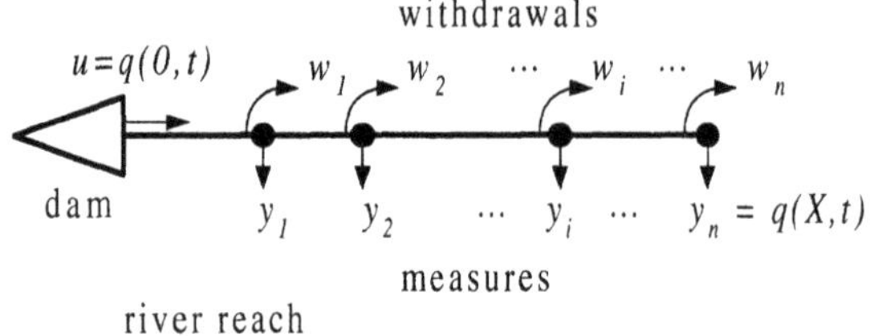

Fig. 1. Dam-river system considered

In other terms, the control objective is to use the upstream discharge u in order to keep the downstream discharge y_n as constant as possible, which means that the control should attenuate the perturbations w_i. This is a problem of regulation or desensitivity. This objective should be attained with a lower bound on gain and phase margins.

2.2 Incorporating information

We already showed in (Litrico and Fromion, 2001) that the performance of an open-channel system when the regulated variable is measured and perturbations are assumed unmeasured is strongly limited. We investigate in this paper different ways to bypass this limitation:

- use additional measurement points (discharge measurement in the river),
- use measurement of disturbances (measurement of withdrawals),
- use predictions of disturbances (predictive control).

The performance is here considered in terms of attenuation of perturbations acting on the system. This requirement can be formalized by direct constraints on the transfer function between the perturbation w_i and the output y_n $T_{w_i \to y_n}$, which will be characterized in the sequel, for each perturbation w_i by the maximum frequency ω_{s_i} such that

$$\omega_{s_i} = \max\{\omega_1 \,:\, |T_{w_i \to y_n}(j\omega)| < 1, \; \forall \omega < \omega_1\}$$

2.3 Hayami model of a single reach

The Hayami equation is a linear partial derivative equation representing the discharge transfer in a river reach around an equilibrium point:

$$\frac{\partial q}{\partial t} + C\frac{\partial q}{\partial x} - D\frac{\partial^2 q}{\partial x^2} = 0 \qquad (1)$$

where

- q is the discharge [m^3/s],
- C the celerity coefficient [m/s] and
- D the diffusion coefficient [m^2/s].

The relation between upstream and downstream discharge can also be expressed as a transfer function using Laplace transform. Thus, fixing a downstream limit condition of the type $\lim_{x\to\infty}\frac{\partial q}{\partial x}=0$, one gets the Hayami transfer function:

$$G(s,x)=e^{\left(\frac{C-\sqrt{C^2+4Ds}}{2D}\right)x} \qquad (2)$$

with x the length of the reach [m] and s the Laplace variable.

Remarks:
1. It can be shown that $G(s)$ is stable and belongs to the class of transfer functions defined by Callier and Desoer (1978).
2. A first order with delay is a good approximation of the Hayami transfer function for low frequencies. This approximation is usually considered in hydrology (Dooge et al., 1987).
3. This transfer function has the property $G(s,x_1+x_2)=G(s,x_1)G(s,x_2)$ which enables to consider the system as a series of subsystems if there are some intermediate measurement points.

3. PERFORMANCE VS INFORMATION

The question discussed in this section is the following: "Does additional information improve the performance of the controlled system considered in the paper?".

We first recall the result on the maximal achievable performance, illustrated in the case of a time-delay system; we then investigate the performance improvement obtained by each proposed point. The applications are given in the next section.

3.1 Maximal performance

As already shown in a previous paper (Litrico and Fromion, 2001), there is a structural limitation to the maximal achievable performance of open-channel hydraulic systems. Let us consider that the model of the river is a first order system with time-delay (the same argument can be used for the general case), controlled by a SISO controller as in figure 2.

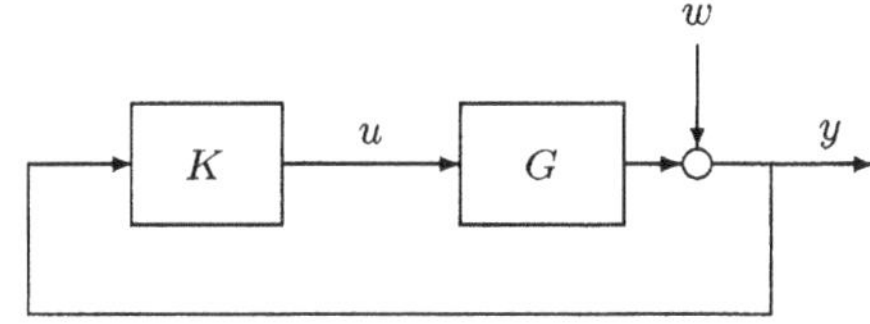

Fig. 2. SISO control and unmeasured perturbation

Following the result of Khargonekar and Poolla (1986), the complementary sensitivity function can be written as $T(s)=T_0(s)e^{-\tau s}$, where $T_0(s)$ is the rational part of the complementary sensitivity function. Since for our example and in the best case $T_0(s)=1$ (perfect tracking), the sensitivity function becomes $T_{w_n\to y_n}(s)=1-e^{-\tau s}$. We then deduce a limit on the maximal bandwidth

$$|T_{w_n\to y_n}(j\omega)|=1\Rightarrow\left|\sin\frac{\omega\tau}{2}\right|=0.5$$

which implies $\omega_{s_n}<\frac{\pi}{3\tau}\approx\frac{1}{\tau}$.

3.2 Incorporating measurement points

Incorporating additional measurement points in the river does not improve the global performance of the controlled system, but enables to improve the perturbations rejection at some points and the robustness of the controller. To show this, we just consider the case of an additional measurement point at the middle of a river (corresponding to the river in figure 1 with $n=2$). The general case can be treated in the same way. Let us consider a river represented by rational system with a time-delay $G_r(s)e^{-\tau s}$. The intermediate measurement separates the system into two subsystems G_1 and G_2 with the same time-delay, i.e. $\tau/2$ (see figure 3).

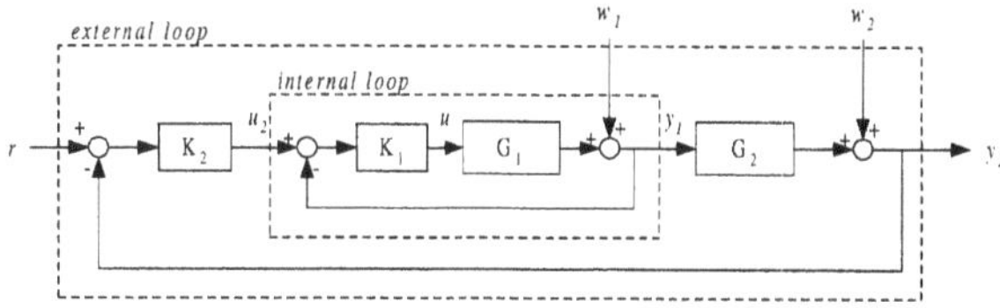

Fig. 3. An intermediate measurement point leads to a cascade controller

For this specific "cascade" system, the internal loop is given by:

$$y_1=T_1u_2+S_1w_1$$

with $T_1=G_1K_1(1+G_1K_1)^{-1}$ and $S_1=1-T_1$.

For the external loop:

$$y_2=T_{r\to y_2}r+S_{w_2\to y_2}(w_2+S_1G_2w_1)$$

with $T_{r\to y_2}=K_2G_2T_1(1+K_2G_2T_1)^{-1}$ and $T_{w_2\to y_2}=1-T_{r\to y_2}$.

Since $T_1(s)=T_{0,1}(s)e^{-\tau/2s}$ and $T_1(s)G_2(s)\propto e^{-\tau s}$, we have $T_{r\to y_2}(s)=T_{0,r\to y_2}(s)e^{-\tau s}$, which implies the same constraint on the maximal achievable performance for rejecting perturbation w_2 than for the classical SISO controller (i.e. $\omega_s\leq 1/\tau$). However, the internal loop can reject higher frequency perturbations w_1. In fact, the sensitivity to G_2w_1 is now $S_1T_{w_2\to y_2}$ (to be compared to $T_{w_2\to y_2}$ in the SISO case). Assuming a perfect tracking in both loops, one gets the following sensitivity:

$$|S_1(j\omega)T_{w_2\to y_2}(j\omega)| < 1 \Rightarrow 4\left|\sin\frac{\omega\tau}{4}\sin\frac{\omega\tau}{2}\right| < 1$$
$$\Rightarrow \sin^2\frac{\omega\tau}{4}\left|\cos\frac{\omega\tau}{2}\right| < \frac{1}{8}$$

which implies the inequality $x - x^3 < \frac{1}{8}$, with $x = \left|\cos\frac{\omega\tau}{2}\right|$. Solving for x gives $x > 0.93$, which leads to $\omega_s\tau < 1.5$. This corresponds to a 50 % increase of the maximum bandwidth (compare with the result obtained in section 3.1). In this case, only the performance on intermediate perturbations is improved.

It should be noted that intermediate measurements are also interesting when the system is not perfectly known: the robustness of the controlled system is improved when another loop is introduced (see Horowitz (1963)).

3.3 About measuring perturbations

For the sake of clearness, we first discuss the control design for rejecting output system perturbation. Let us focus on the case when the hydraulic system is modeled by equation:

$$y = e^{-\tau s}u + w \quad (3)$$

where w is the perturbation input and u the control input. Assume that we want to achieve the best rejection for any perturbation input w with a feedforward control law $u = Fy$. This problem can be formulated as minimizing the H_∞ norm of the transfer function between w and y, that is:

$$\inf_{F(s)\in\mathbf{H}_\infty}\left\|1 - F(s)e^{-\tau s}\right\|_\infty = \gamma$$

Following Tannenbaum (1992), $\gamma = 1$, which implies that perturbation rejection can be achieved only on a finite interval of frequencies. This first claim is a theoretical point of vue, and does not indicate clear bound on the size of the interval. As a matter of fact, the interval is directly limited by the model uncertainty. To illustrate this fact, let us consider that the delay of system (3) belongs to:

$$\tau \in [(1-\Delta_\tau)\tau_0, (1+\Delta_\tau)\tau_0]$$

and let us assume that the measured perturbation is a sinusoid *i.e.* $w(t) = a\sin(\omega_p t)$. Let us furthermore assume that the perturbation is known in the future, say τ_0 secondes in advance. Thus it is possible to perfectly reject it with the feedforward control $F(s)$ such that $F(j\omega_p) = e^{\tau_0 j\omega_p}$ which introduces with respect to the perturbation signal a lag effect given by $\phi = \tau_0\omega_p$.

Due to the fact that $\tau \in [(1-\Delta_\tau)\tau_0, (1+\Delta_\tau)\tau_0]$, the lag effect is given by

$$\Delta\phi \in [-\Delta_\tau\tau_0\omega_p,\ \Delta_\tau\tau_0\omega_p]$$

For the frequency ω_p^* such that $\Delta_\tau\tau_0\omega_p^* = \pi$, the feedforward law can increase by 2 the perturbation effect. Therefore, the uncertainty level on the delay constrains the maximal bandwidth where feedforward allows to improve performance.

The above point is also valid in closed-loop cases. In the one hand, following Khargonekar and Poolla (1986), the closed-loop system contains explicitly the time-delay, in the other hand, the frequency interval where the sensitivity function is less than one is also constrained by the delay. Thus, the frequency where the previous argument is used is in "open-loop".

When the sensitivity function is less than one, the feedback "shrinks" the uncertainty, which allows to improve the quality of feedforward loop. We thus consider in the sequel a controller structure with a feedback loop and a feedforward loop, in order to use the perturbation measurement for improving rejection.

4. CONTROLLER DESIGN

4.1 A rational approach

In the case when the future behavior of the perturbation is not known in the future, the considered control problem can be stated as a standard mixed sensitivity optimal control problem, using the H_∞ norm:

$$\gamma = \inf_K\left\|\begin{pmatrix} W_1 T_{w_n\to y_n} \\ W_2 K T_{w_n\to y_n}\end{pmatrix}\right\|_\infty$$

where W_1 and W_2 are weighting functions taking into account performance and robustness requirements. We select W_1 and W_2 in order to obtain the required robustness margins when $\gamma \leq 1$. Shaping $T_{w_n\to y_n}$ and $KT_{w_n\to y_n}$ allows (i) to constrain the rejection time and to limit the control energy (ii) to ensure a desired modulus margin and a desired time-delay margin.

In our case, and due to the irrational nature of the system, criteria (4.1) cannot be solved in an easy way. To bypass this, we solve problem (4.1) using a rational approximation of (2) (see Litrico and Fromion (2001)).

4.2 Predictive control

The vector of perturbations w is now assumed to be known in advance, with a prediction horizon equal to h. We propose to extend the previous H_∞ criterion in order to design a control law using the future behavior of the perturbation. Note that, to our best knowledge, this problem has not been considered in previous works. In order to simplify the problem, we consider the design of a discrete time control. In this case, the future value of the perturbation is given by a finite dimensional vector $[w(k), \ldots, w(k+h)]$: in the continuous time case, the corresponding vector is infinite dimensional. The control law has the corresponding structure:

$$u = K \begin{bmatrix} w(k) \\ \vdots \\ w(k+h) \\ y(k) \end{bmatrix} = \left[K_{ol} | K_{cl} \right] \begin{bmatrix} w(k) \\ \vdots \\ w(k+h) \\ \hline y(k) \end{bmatrix}$$

The corresponding closed loop system is repre-

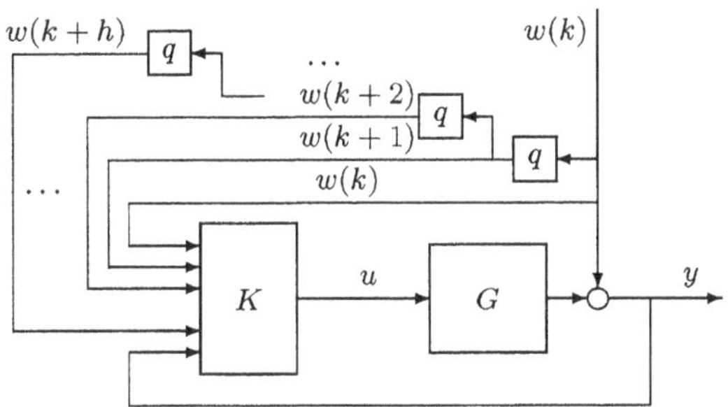

Fig. 4. Closed loop system with "predictive" control law

sented figure 4. Let us denote $T_{w \to y}$ the transfer function from w to y et let us denote $T_{w \to u}$ the transfer function from w to u. The corresponding H_∞ problem can be written as:

$$\min_K \left\| \begin{pmatrix} W_1 T_{w \to y} \\ W_2 T_{w \to u} \end{pmatrix} \right\|_\infty \tag{4}$$

The difficulty arises from the fact $T_{w \to y}$ and $T_{w \to y}$ are non causal transfer function since for instance:

$$T_{w \to y}(q) = (I - GK_{cl})^{-1} \left(I + GK_{ol} \left[1 \; q \; \cdots \; q^h \right]\right)$$

Nevertheless, note that:

$$\left\| \begin{pmatrix} W_1 T_{w \to y} \\ W_2 T_{w \to u} \end{pmatrix} \right\|_\infty = \left\| \begin{pmatrix} W_1 T_{w \to y} z^{-h} \\ W_2 T_{w \to u} z^{-h} \end{pmatrix} \right\|_\infty .$$

In this case, $T_{w \to y} z^{-h}$ and $T_{w \to y} z^{-h}$ are causal transfer functions. As a consequence solving (4) is equivalent to the following problem:

$$\min_K \left\| \begin{pmatrix} W_1 T_{w \to y} z^{-h} \\ W_2 T_{w \to u} z^{-h} \end{pmatrix} \right\|_\infty \tag{5}$$

which is a standard H_∞ problem (Doyle et al., 1989).

In fact, the previous criteria does not take into account any direct constraint on the sensitivity function *i.e.* $T_{w_n \to y_n}$, and therefore no constraint on the open-loop system bandwidth. In order to take into account desensitivity requirements, we add to the previous criteria a perturbation (see the augmented system depicted in figure 5), which models the unknown part of the perturbations which act on the system.

As specified in the above scheme, the controller can now use "measures" of future perturbations

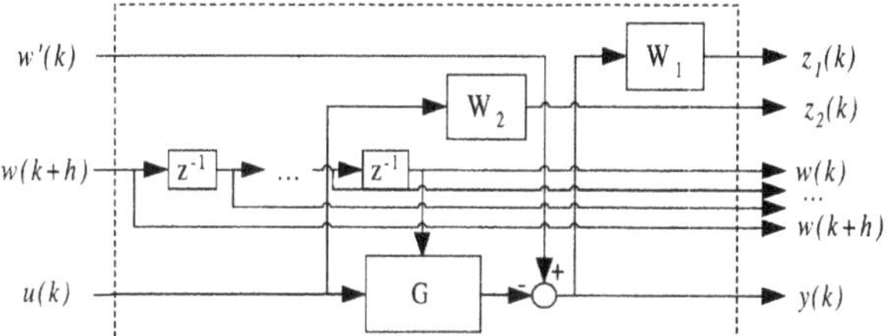

Fig. 5. Augmented system

(in fact, predictions) for real-time control. However, this problem cannot be solved with classical H_∞ algorithms using Riccati equations (Doyle et al., 1989) since the criteria is singular (the perturbations measurement is perfectly known). This technical problem can be bypassed by using the LMI based solver for the H_∞ optimization problem of the Matlab LMI control toolbox (Gahinet and Apkarian, 1994).

5. APPLICATION

The numerical applications of this paper are done for a river of 18 km, with 2 possible intermediate measurement points: y_1 at 6 km and y_2 at 12 km, the controlled point being the discharge y_3 at 18 km (*i.e.* $n = 3$ in figure 1). The dynamic parameters $C = 0.7$ m/s and $D = 170$ m^2/s correspond to average values for sloping rivers at low flows, and give for the whole river a delay close to 7 hours.

The figure 6 shows the Bode plot of the whole river (18 km), which is close to the one of a first order with delay.

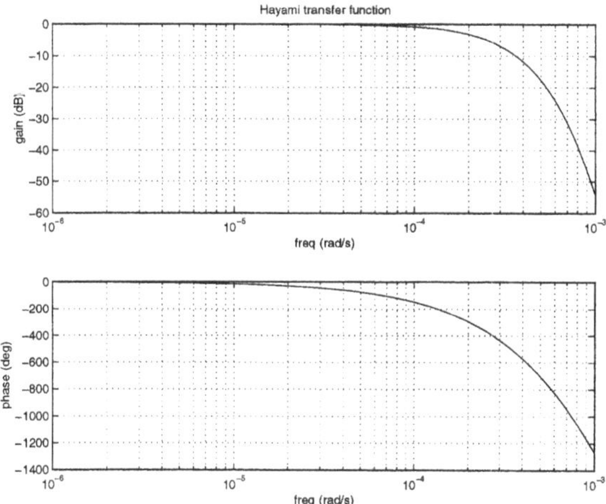

Fig. 6. Bode plot of $G(s)$

Five different controllers were designed and compared on two different scenarios on the system described above. The two scenarios consist in a disturbance step of 1 m^3/s applied at w_1 for the first scenario and w_3 for the second one. The results are summarized in table 1 where the chosen indicator is the volume of water not delivered at the downstream end of the river. As was shown above, except for the predictive controller, the use of additional measurement point in the river or of perturbation measurement does not improve the performance for a perturbation acting at the downstream end (w_3) of the river. It is interesting to note that the three controllers SISO, SISO$_w$ (using perturbation measurement), MISO (using intermediate measurement in the river) and MISO$_w$ (using both) behave in the same way. However, the performance in the first scenario is clearly improved when additional information is used. And finally, the knowledge of future perturbations enables to almost perfectly reject them (MISO predictive controller).

Table 1. Not delivered water (in 10^3 m^3) for the different controllers and scenarios

	scen. w_1	scen. w_3
SISO	44.3	44.4
SISO$_w$	23.8	44.4
MISO	23.9	44.4
MISO$_w$	23.8	44.4
MISO pred.	-2.9	2.1

Figure 7 shows the results in simulation for a withdrawal of 1 m^3/s at the first measurement point, for three different controllers, the MISO predictive controller using predicted perturbations, the MISO controller taking into account intermediate measurements (y_1 and y_2) and the SISO controller using only the downstream measurement point (y_3).

The effect of incorporating information in the controller is here clearly visible in terms of water savings.

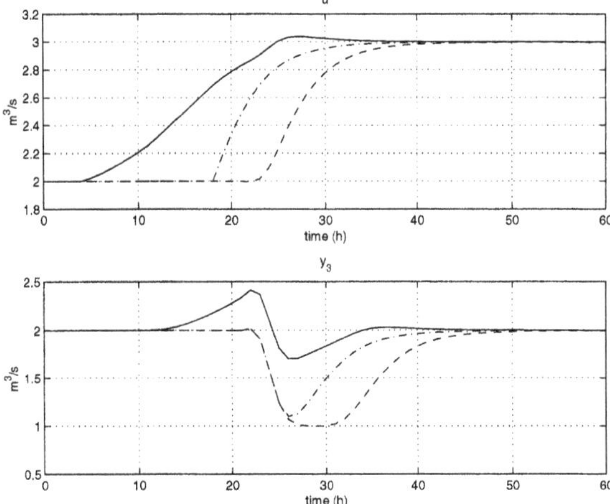

Fig. 7. Control u and controlled output y_3 for the three H_∞ controllers: MISO predictive controller (solid), MISO controller (dash-dotted) and SISO controller (dashed), scenario w_1

6. CONCLUSION

This paper shows that the H_∞ framework is suited to design controllers for open-channel systems represented by a linear advection-diffusion equation. The proposed method is flexible and enables to take into account more information when available. To this purpose, we extend standard H_∞ criterion in order to design "predictive" control law.

Our approach mixes the advantages of the H_∞ approach (a priori robustness and simplicity) with the advantages of the predictive approach (explicitly taking into account the future behavior or perturbations and/or references).

REFERENCES

R.R. Bitmead, M. Gevers, and V. Wertz. *Adaptative Optimal Control.* Prentice Hall International Series in Systems and Control Engineering. Prentice Hall, 1990.

F.M. Callier and C.A. Desoer. An algebra of transfer functions for distributed linear time-invariant systems. *IEEE Trans. on Circuits and Systems*, CAS-25(9):651–662, 1978.

J.C.I. Dooge, J.J. Napiórkowski, and W.G. Strupczewski. The linear downstream response of a generalized uniform channel. *Acta Geophysica Polonica*, XXXV(3): 279–293, 1987.

J.C. Doyle, K. Glover, P.P. Khargonekar, and B.A. Francis. State-space solutions to standard H_2 and H_∞ control problems. *IEEE Trans. on Automatic Control*, 34(8): 831 - 847, 1989.

B. Foss, J.E. Haug, J. Alne, and S. Aam. User experience with on-line predictive river flow regulation. *IEEE Trans. on Power Systems*, 4(3):1089–1094, 1989.

P. Gahinet and P. Apkarian. A linear matrix inequality approach to H_∞ control. *International Journal of Robust and Nonlinear Control*, 4:421–448, 1994.

I.M. Horowitz. *Synthesis of Feedback Systems.* Academic Press, New York, 1963.

P.P. Khargonekar and K. Poolla. Robust stabilization of distributed systems. *Automatica*, 22(1):77–84, 1986.

X. Litrico and V. Fromion. About optimal performance and approximation of open-channel hydraulic systems. In *40th IEEE Conf. on Decision and Control*, Orlando, 2001. accepted.

M. Papageorgiou and A. Messmer. Flow control of a long river stretch. *Automatica*, 25(2):177–183, 1989.

S. Sawadogo, A.K. Achaibou, and J. Aguilar-Martin. An application of adaptative predictive control to water distribution systems. In *IFAC, ITAC 91*, Singapour, 1991.

A. Tannenbaum. *Frequency domain methods for the H_∞-optimization of distributed systems*, volume 185 of *Lecture Notes in Control and Information Sciences*, pages 242–278. Springer-Verlag, 1992.

www.elsevier.com/locate/ifac

ROBUST STABILITY OF TIME DELAY SYSTEMS: THEORY

John Chiasson * **Chaouki Abdallah** **,[1]

* *ECE Dept, University of Tennessee, Knoxville TN 37996, USA, chiasson@utk.edu*
** *ECE Dept, University of New Mexico, Alburquerque NM 87131-1356, USA, chaouki@eece.unm.edu*

Abstract: Given that a time-delay system is stable for some delay $h_0 > 0$, a procedure is given to find the stability interval $[h_1^*, h_2^*]$ such that $h_0 \in [h_1^*, h_2^*]$ and for all h satisfying $h_1^* < h < h_2^*$ the system is stable. Further, the system is shown to be unstable if $h = h_1^*$ or $h = h_2^*$. It is then shown how this can be applied to test the robust stability (with respect to delay values) of a Smith-Predictor based controller. *Copyright © 2001 IFAC*

Keywords: Time Delay, Stability, Smith Predictor

1. INTRODUCTION

This paper considers the robust stability of time-delay systems of the form

$$\frac{d^n y}{dt^n} + \sum_{i=0}^{n-1}\sum_{j=0}^{m} a_{ij}\frac{d^i}{dt^i}y(t - jh) = 0. \qquad (1)$$

Following Kamen (Kamen, 1982), the characteristic equation for (1) with $z = e^{-sh}$ is

$$\begin{aligned} a(s,z) &= s^n + \sum_{i=0}^{n-1}\sum_{j=0}^{m} a_{ij}s^i z^j = 0 \qquad z = e^{-sh} \\ &= s^n + \sum_{i=0}^{n-1} a_i(z)s^i = 0 \qquad (2) \end{aligned}$$

$$a_i(z) \triangleq \sum_{j=0}^{m} a_{ij}z^j \text{ for } i = 0, 1, \ldots n-1.$$

This is the so-called commensurate delay case since all delays are integer multiples of $h > 0$. For a fixed delay $h > 0$, the polynomial $a(s, e^{-hs})$ is said to be *asymptotically stable* if and only if (iff)

$$a(s, e^{-hs}) \neq 0 \qquad \text{for } \mathrm{Re}(s) \geq 0. \qquad (3)$$

where $\mathrm{Re}(s)$ denotes the real part of s. Condition (3) implies asymptotic stability of (1) (see (Bellman and Cooke, 1963)(Hale and Lunel, 1993)(Diekmann and Walther, 1995)(Nicolescu, 2001)).

Kamen (Kamen, 1982) introduced the notion of a system being *asymptotically stable independent of delay*. In this formulation, the polynomial $a(s, z)$ is said to be asymptotically stable independent of delay (i.o.d.) iff (Kamen, 1982)

$$a(s, e^{-hs}) \neq 0 \qquad \text{for } \mathrm{Re}(s) \geq 0, h \geq 0. \qquad (4)$$

Condition (4) implies asymptotic stability for the solutions of (1). Algebraic tests exist to check for stability i.o.d. are given in (Chiasson and Lee, 1985)(Kamen, 1982). Given that the system is stable when $h = 0$, tests that determine the value h^* such that the system is stable for $0 \leq h < h^*$ and unstable when $h = h^*$ are given in (Hertz and Zeheb, 1984*a*)(Hertz and Zeheb, 1984*b*)(Chiasson, 1988).

In this paper, it is not assumed that the system is necessarily stable for $h = 0$. Specifically,

[1] The research of C.T. Abdallah is partially supported by NSF INT-9818312

given that a time-delay system is stable for some delay $h_0 > 0$, a procedure is given to find the stability interval (h_1^*, h_2^*) such that $h_0 \in (h_1^*, h_2^*)$ and for all $h \in (h_1^*, h_2^*)$ the system is stable. Further, it is shown that the system is unstable if $h = h_1^*$ or $h = h_2^*$. This is motivated by the fact that in some cases, the introduction of delay may have stabilizing effects (see for example (Petterson and Werner, 1991)(Niculescu and Abdallah, 1999)(C.T. Abdallah and Byrne, 1993)) or by the Smith predictor structure as described later in the paper.

The remainder of the paper is divided as follows: In section 2 we prove the main result and illustrate via an example taken from (Petterson and Werner, 1991). Section 3 discusses the robust stability of the Smith Predictor control structure using the results of section 2.

2. MAIN RESULTS

Following (Hertz and Zeheb, 1984*b*)(Chiasson and Lee, 1985)(Chiasson, 1988), the idea is to note that if a system is stable for some $h_0 > 0$, then as the system transitions from a stable to an unstable regime (as the delay h is either increased or decreased), it must have poles that cross over to the $j\omega$ axis. In other words, at the point of transition, there must be roots of the form $s = j\omega, z = e^{-j\omega h}$ ($\mathrm{Re}(s) = 0, |z| = 1$) of (2).

To exploit the above idea, an auxiliary polynomial $\tilde{a}(s, z)$ is defined as

$$\begin{aligned}\tilde{a}(s,z) &\triangleq z^m a(-s, 1/z)\\ &= z^m(-s)^n + z^m\left(\sum_{i=0}^{n-1}\sum_{j=0}^{m} a_{ij}(-s)^i(1/z)^j\right)\\ &= z^m(-1)^n s^n + \sum_{i=0}^{n-1}\sum_{j=0}^{m}(-1)^i a_{ij}s^i z^{m-j}\end{aligned}$$

Suppose at $h = h^*$, there is an $s^* = j\omega^*$ and a corresponding $z^* = e^{-j\omega^* h^*}$ such that

$$a(s^*, e^{-j\omega^* h^*}) = a(s^*, z^*) = 0$$

Then taking the complex conjugate of $a(s^*, e^{-j\omega^* h^*})$ and multiplying through by $(e^{-j\omega^* h^*})^m$ results in

$$\begin{aligned}(e^{-j\omega^* h^*})^m a(-s^*, e^{j\omega^* h^*}) &= (z^*)^m a(-s^*, 1/z^*)\\ &= \tilde{a}(s^*, z^*) = 0\end{aligned}$$

That is, for any value of delay h^* resulting in roots on the $j\omega$ axis, there is a $(s^*, z^*) \triangleq (j\omega^*, e^{-j\omega^* h^*})$ that must be a *common zero* of the two polynomials $a(s^*, z^*), \tilde{a}(s^*, z^*)$. In other words,

$$a(s^*, z^*) = \tilde{a}(s^*, z^*) = 0$$

for some (s^*, z^*) satisfying $\mathrm{Re}(s^*) = 0, |z^*| = 1$.

This observation will be combined with the following lemma to get the desired test.

Lemma (Datko) Let $\sigma(h) = \sup\{\mathrm{Re}(s) \mid a(s, e^{-hs}) = 0\}$. Then $\sigma(h)$ is a continuous function of h for $h \geq 0$.

Proof This is a result of the work of Datko (Datko, 1978). See also the more general results of Cooke and Ferreira (Cooke and Ferreira, 1983)

Given that the time-delay system (3) is stable for some delay value $h_0 > 0$, we now define h_1^*, h_2^* where $h_1^* < h_0 < h_2^*$, such that the systems remains stable for h satisfying $h_1^* < h < h_2^*$ and the system is unstable if $h = h_1^*$ or $h = h_2^*$. Note that (3) being stable for some $h = h_0 > 0$ implies $a(0, 1) \neq 0$ as seen by setting $s = 0$ in (3).

Definition Let $\{(s_i, z_i)$ for $i = 1, \cdots k\}$ be the common zeros of $\{a(s, z), \tilde{a}(s, z)\}$ for which $\mathrm{Re}(s_i) = 0, s_i \neq 0$, and $|z_i| = 1$. For each such pair (s_i, z_i), for a given h_0, let $h_{1i}^* = \max\{h \in \Re \mid h < h_0, z_i = e^{-s_i h}\}$ and $h_{2i}^* = \min\{h \in \Re \mid h > h_0, z_i = e^{-s_i h}\}$. Then define

$$\begin{aligned}h_1^* &= \min\{h_{1i}^*\}\\ h_2^* &= \min\{h_{2i}^*\}.\end{aligned}$$

With this definition, the main result is now stated and proven.

Theorem It is assumed that the time-delay system (1) is stable for some $h_0 > 0$, that is,

$$a(s, e^{-h_0 s}) \neq 0 \qquad \text{for } \mathrm{Re}(s) \geq 0. \qquad (5)$$

Then

$$a(s, e^{-hs}) \neq 0 \qquad \text{for } \mathrm{Re}(s) \geq 0 \text{ and } h_1^* < h < h_2^*$$

with h_1^*, h_2^* defined as above. Further, the system is not stable for $h = h_1^*, h_2^*$, that is,

$$\begin{aligned}a(s, e^{-h_1^* s}) &= 0 \qquad \text{for some } s \text{ with } \mathrm{Re}(s) = 0\\ a(s, e^{-h_2^* s}) &= 0 \qquad \text{for some } s \text{ with } \mathrm{Re}(s) = 0\end{aligned}$$

Given that the system is stable for $h = h_0$, it follows that $\sigma(h_0) < 0$. By the above lemma, increasing h above h_0, there is an h_2 which is the first value of the delay $h > h_0$ satisfying $\sigma(h_2) = 0$. For this value of h_2, there is an s_2 such that

$$a(s_2, e^{-s_2 h_2}) = 0, \ \mathrm{Re}(s_2) = 0.$$

That is, $s_2 = j\omega_2, z_2 = e^{-j\omega_2 h_2}$($|z_2| = 1$). Further, $s_2 \neq 0$ as $s_2 = 0$ would imply that $a(0, 1) = 0$ contradicting the assumption (5). Clearly, (s_2, z_2) also satisfies $\tilde{a}(s_2, z_2) = 0$ and thus $h_2 = h_2^*$. On the other hand, by definition, $\sigma(h_2^*) = 0$ so that the system is unstable when $h = h_2^*$. Similar arguments show there is an h_1^* such that the system is stable for $h_1^* < h \leq h_0$ and unstable when $h = h_1^*$.

Remark A simple variation of the above proof shows that if the time-delay system (3) is unstable for some delay value $h_0 > 0$, then the system remains unstable for h satisfying $h_1^* \leq h \leq h_2^*$ where h_1^*, h_2^* are defined with respect to h_0 as in Definition 2.

Example Consider the system given by

$$G(s, e^{-hs}) = \frac{b(s, e^{-hs})}{a(s, e^{-hs})} = \frac{1 - e^{-hs}}{s + 1 - e^{-hs}}.$$

Clearly $a(s, e^{-hs})\big|_{s=0} = a(0,1) = 0$ so that $a(s, e^{-hs})$ is *unstable independent of delay.* However, the numerator of $G(s, e^{-hs})$ also has a zero at $s = 0$ so that

$$G(0,1) = \lim_{s \to 0} \frac{b(s, e^{-hs})}{a(s, e^{-hs})} = \lim_{s \to 0} \frac{1 - e^{-hs}}{s + 1 - e^{-hs}} = 1.$$

Consequently, $s = 0$ is not a pole of $G(s)$ [2]. Next, it is shown that $c(s, e^{-hs}) \triangleq a(s, e^{-hs})/b(s, e^{-hs})$ has no zeros in the *open* right-half plane for $h \geq 0$. Proceeding, define

$$c(s, z) = \frac{a(s, z)}{b(s, z)} = \frac{s + 1 - z}{1 - z}$$

$$\tilde{c}(s, z) = zc(-s, 1/z) = \frac{z(-s+1) - 1}{z - 1}.$$

So if $G(s, e^{-hs})$ has a zero in the rhp for some $h > 0$ then there must be a pair (s^*, z^*) $c(s^*, z^*) = \tilde{c}(s^*, z^*) = 0$ with $\mathrm{Re}(s^*) = 0, |z^*| = 1$. Eliminating s from $c(s,z) = \tilde{c}(s,z) = 0$ shows that any such z^* must satisfy

$$r(z) = 1 - z = 0.$$

However, $z = 1$ requires $s = 0$ and the pair $(0, 1)$ is not a zero $G(s, e^{-hs})$. Consequently, $G(s, e^{-hs})$ is stable independent of delay.

2.1 Stabilization by Delayed Output Feedback

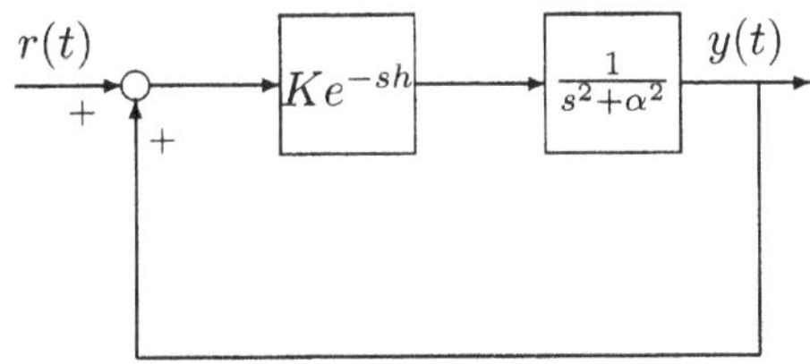

Fig. 1. Feedback stabilization of oscillatory systems.

Figure 1 shows a feedback system (from (Petterson and Werner, 1991)(Niculescu and Abdallah, 1999) (Nicolescu, 2001)) whose open-loop transfer function is $G_0(s) = 1/(s^2 + \alpha^2)$. To compare with the results in (Petterson and Werner, 1991)(Niculescu and Abdallah, 1999), the compensator is chosen to be the positive feedback $G_c(s) = Ke^{-hs}$, $K > 0$ resulting in the closed-loop transfer function

$$G_{CL}(s) = \frac{Ke^{-hs}}{s^2 + \alpha^2 - Ke^{-hs}}.$$

[2] Such pole-zero cancellations for time-delay systems were pointed out to the first author by Mark Spong.

The interest here is to study the stability of this system, that is, the roots of the closed-loop characteristic polynomial $a(s, e^{-hs}) \triangleq s^2 + \alpha^2 - Ke^{-hs}$.

Case 1. $0 < K < \alpha^2$.

This system is clearly unstable for $h = 0$ having poles on the $j\omega$ axis at $\pm j\sqrt{\alpha^2 - K}$. The polynomials $a(s,z), \tilde{a}(s,z)$ are given by

$$a(s,z) \triangleq s^2 + \alpha^2 - Kz$$
$$\tilde{a}(s,z) \triangleq z\left(s^2 + \alpha^2 - K/z\right) = zs^2 + z\alpha^2 - K.$$

Eliminating s^2 gives $z^2K - K = 0$ or $z = \pm 1$. The four common zeros are therefore

$$\{(s_k, z_k) \mid a(s_k, z_k) = \tilde{a}(s_k, z_k) = 0\} = \left\{(\pm j\sqrt{(\alpha^2 - K)}, 1), (\pm j\sqrt{(\alpha^2 + K)}, -1)\right\}.$$

Solving for the delay values corresponding to these common zeros results in

$$1 = e^{-hj\sqrt{(\alpha^2-K)}}$$
$$\Longrightarrow h = h_{1m} = \frac{m2\pi}{\sqrt{(\alpha^2 - K)}},\ m = 0, 1, 2, ...$$
$$-1 = e^{-hj\sqrt{(\alpha^2+K)}}$$
$$\Longrightarrow h = h_{2n} = \frac{(2n+1)\pi}{\sqrt{(\alpha^2 + K)}},\ n = 0, 1, 2, ...$$

The set of delays $\{h_{1m}, h_{2n} \mid m = 0, 1, 2, ...\ n = 0, 1, 2, ...\}$ can be arranged in increasing order. The above theorem implies that for *all* delay values between any two adjacent delays in this set, the system is either stable or unstable. Of course, for $h \in \{h_{1m}, h_{2n} \mid m = 0, 1, 2, ...\ n = 0, 1, 2, ...\}$, the system is unstable having poles on the $j\omega$ axis. According to Theorem 2, to determine whether a particular interval corresponds to the system being stable requires finding a specific h in the interval for which the system is stable for this delay value. To do so, consider the first two values of h, that is, $m = 0$ and $n = 0$ resulting in $h = 0$ and $h = \pi/\sqrt{(\alpha^2 + K)}$. For h small, $a(s, e^{-hs}) \triangleq s^2 + \alpha^2 - Ke^{-hs} \approx s^2 + \alpha^2 - K(1 - hs) = s^2 + Khs + \alpha^2 - K$. As $\alpha^2 < K$ in this case, it follows that $a(s, e^{-hs})$ is stable for $h > 0$ small enough. Consequently, by the above theorem, the system is stable for all h satisfying $0 < h < h_{20} = \pi/\sqrt{(\alpha^2 + K)}$. For $h < h_{20}$ the system is stable while for $h = h_{20}$ there are two poles $\left(s_0 = \pm j\sqrt{(\alpha^2 + K)}\right)$ on the $j\omega$ axis.

Case 2. $K > \alpha^2$

A similar analysis as in case 1 shows the system must be unstable for $0 \leq h < \pi/\sqrt{(\alpha^2 + K)}$.

Example The system $G_0(s) = 1/(s^2 + \alpha^2)$ given in (Petterson and Werner, 1991)(Niculescu and Abdallah, 1999) is now considered from a different point of view. This system has a statespace realization given by

$$\frac{dx}{dt} = \begin{bmatrix} 0 & 1 \\ \omega^2 & 0 \end{bmatrix} x + \begin{bmatrix} 0 \\ 1 \end{bmatrix} u$$
$$y = c \begin{bmatrix} 1 & 0 \end{bmatrix} x.$$

Approximate the derivative $\dot{y} \approx (y(t) - y(t-h))/h$. and choose the control as

$$u \triangleq - \begin{bmatrix} k_1 & k_0 \end{bmatrix} \begin{bmatrix} c \\ cA \end{bmatrix}^{-1} \begin{bmatrix} y(t) \\ \frac{y(t) - y(t-h)}{h} \end{bmatrix} + r$$

Note that as $\lim_{h \to 0} (y(t) - y(t-h))/h = \dot{y}(t)$ it is straightforward to show this system is stable for $h = 0$ as the closed loop poles can be arbitrarily placed through the choice of k_0, k_1. The closed-loop transfer function

$$\frac{Y(s, e^{-hs})}{R(s, e^{-hs})} = \frac{1}{s^2 + \omega^2 + k_0 + \frac{k_1}{h}(1 - e^{-sh})}$$
$$= \frac{1}{a(s, e^{-hs})}$$

With $k_0 = a_0 - \omega^2, k_1 = a_1$

$$a(s, 1) = \lim_{h \to 0} a(s, e^{-hs}) = s^2 + a_1 s + a_0$$

which is stable for $a_0 > 0, a_1 > 0$. Now rewrite $a(s, e^{-sh}) = s^2 + \alpha^2 - Ke^{-sh}$ where $\alpha^2 = \omega^2 + k_0 + k_1/h = a_0 + k_1/h, K = k_1/h$. (Note that $K < \alpha^2$). The polynomials $a(s, z), \tilde{a}(s, z)$ are given by

$$a(s, z) \triangleq s^2 + \alpha^2 - Kz$$
$$\tilde{a}(s, z) \triangleq z\left(s^2 + \alpha^2 - K/z\right) = zs^2 + z\alpha^2 - K.$$

The system is stable for $h = 0$ and, as h is increased, there is a first value h^* for which $a(s, e^{-h^* s})$ has roots on the imaginary axis. This means that when $h = h^*$ the polynomials $a(s, z)$ and $\tilde{a}(s, z)$ must have a common zero (s^*, z^*) satisfying $\text{Re}(s^*) = 0, |z^*| = 1$. To find the common zeros, s^2 is eliminated giving $z^2 K - K = 0$ or $z = \pm 1$. The four common zeros are therefore

$$\{(s_k, z_k) \mid a(s_k, z_k) = \tilde{a}(s_k, z_k) = 0\}$$
$$= \left\{(\pm j\sqrt{(\alpha^2 - K)}, 1), (\pm j\sqrt{(\alpha^2 + K)}, -1)\right\}$$
$$= \left\{(\pm j\sqrt{a_0}, 1), (\pm j\sqrt{a_0 + 2a_1/h}, -1)\right\}.$$

Solving for the delay values corresponding to these common zeros results in

$$1 = e^{-hj\sqrt{(\alpha^2 - K)}}$$
$$\Longrightarrow h = h_{1m} = \frac{m2\pi}{\sqrt{(\alpha^2 - K)}}, \quad m = 0, 1, 2, ...$$
$$-1 = e^{-hj\sqrt{(\alpha^2 + K)}}$$
$$\Longrightarrow h = h_{2n} = \frac{(2n+1)\pi}{\sqrt{(\alpha^2 + K)}}, \quad n = 0, 1, 2, ...$$

The delay $h_{10} = 0$ is eliminated since it has already been shown the system is stable for $h = 0$. Consequently, the first value of h correpond-ing to roots on the imaginary axis is $h_{20} = \pi/\sqrt{a_0 + 2k_1/h}$. The system is stable for

$$0 \leq h < \frac{\pi}{\sqrt{a_0 + 2a_1/h}}$$

or

$$0 \leq h < \frac{-a_1 + \sqrt{a_1^2 + \pi^2 a_0}}{a_0}$$

Setting $a_0 = r^2, a_1 = 2r$ places the closed-loop poles at $-r$ when $h = 0$. The above bounds on the delay become $0 \leq h < \left(-2 + \sqrt{4 + \pi^2}\right)/r = 1.724/r$. Consequently, this shows that the further these poles are placed in the left-half plane, the smaller the range of stability in the delay h.

3. ROBUST STABILITY OF THE SMITH PREDICTOR

The Smith Predictor control structure shown in Figure 2 provides a methodology to stabilize a time-delay system that has a pure process delay in the open-loop plant ((Power and Simpson, 1978), p.230).

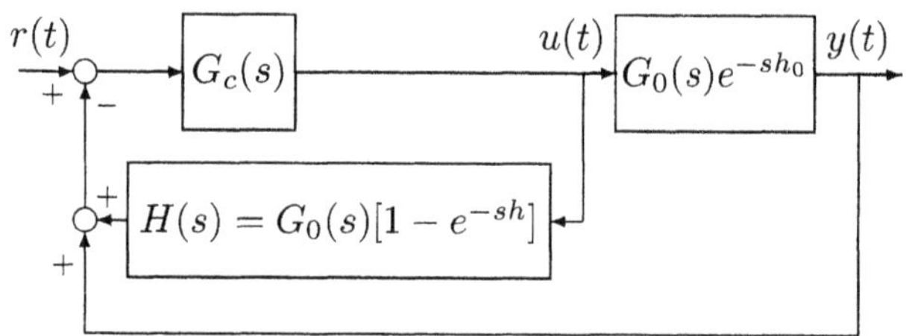

Fig. 2. The Smith-predictor feedback system.

Here the open-loop system is $G_0(s)e^{-sh_0}$ and application of the feedback $H(s) = G_0(s)[1 - e^{-hs}]$ *with* $h = h_0$ results in the closed-loop system transfer function

$$\frac{G_c(s)G_0(s)}{1 + G_c(s)G_0(s)} e^{-sh_0}. \tag{6}$$

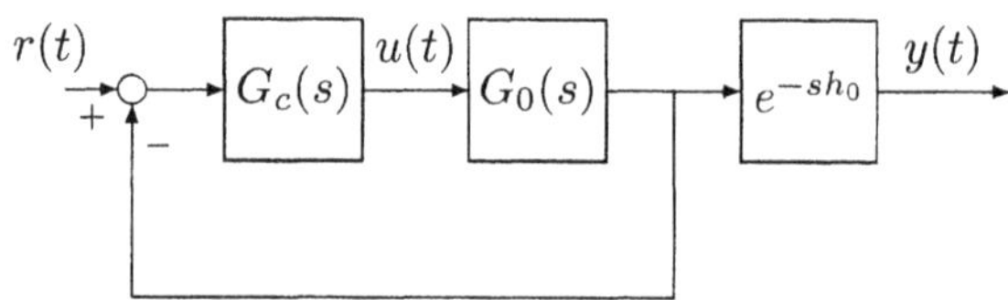

Fig. 3. The equivalent closed-loop system when $h = h_0$.

Thus, if the delay h_0 is known precisely, it is sufficient to stabilize the *delay-free* system $G_0(s)$ using the compensator $G_c(s)$. However, if the delay h is uncertain, then the perfect cancellation of the Smith predictor structure does not materialize. Instead, the closed-loop transfer function is given by

$$\frac{G_c(s)G_0(s)}{1 + G_c(s)G_0(s)[1 - (e^{-sh_0} - e^{-sh})]} e^{-sh_0}. \tag{7}$$

With $G_c(s)G_0(s) = n(s)/d(s)$, the characteristic equation of (7) may be written as

$$a(s, e^{-h_0 s}, e^{-hs}) \triangleq a_0(s) + a_1(s)\left[e^{-sh_0} - e^{-sh}\right] \tag{8}$$

where $a_0(s) = n(s)+d(s), a_1(s) = -n(s)$. Here the delays h_0, h are noncommensurate in general. By design, the compensator $G_c(s)$ is chosen so that (8) (or equivalently (6)) is stable for $h = h_0$, that is, $a_0(s)$ is stable. The interest here is to determine the range of values of h about h_0 for which (8) is stable. As h is increased (or decreased) in value from h_0, there is a first value h^* for which the roots of (8) are on the $j\omega$ axis, that is,

$$a(s, e^{-h_0 s}, e^{-h^* s}) \triangleq a_0(s) + a_1(s)\left(e^{-sh_0} - e^{-sh^*}\right) = 0 \tag{9}$$

for some $s^* = j\omega^*$. To exploit this idea set $z_0 = e^{-sh_0}, z = e^{-sh}$ and define

$$\begin{aligned} a(s, z_0, z) &= a_0(s) + a_1(s)(z_0 - z) \\ \tilde{a}(s, z_0, z) &= z_0 z a(-s, 1/z_0, 1/z) \\ &= z_0 z\left[a_0(-s) + a_1(-s)(1/z_0 - 1/z)\right] \\ &= z_0 z a_0(-s) + a_1(-s)(z - z_0). \end{aligned}$$

Equation (9) implies that for $h = h^*$, the polynomials $\{a(s, z_0, z), \tilde{a}(s, z_0, z)\}$ must have a common zero (s^*, z_0^*, z^*) satisfying $\mathrm{Re}(s^*) = 0, |z_0^*| = 1, |z^*| = 1$.

Solving $a(s, z_0, z) = 0$ for z_0 gives

$$z_0 = z - \frac{a_0(s)}{a_1(s)} \tag{10}$$

and substituting this into $\tilde{a}(s, z_0, z)$ results in

$$\begin{aligned} &\tilde{a}\left(s, z - \frac{a_0(s)}{a_1(s)}, z\right) \\ &= \frac{1}{a_1(s)}\left(z^2 a_0(-s)a_1(s) - a_0(s)a_0(-s)z + a_1(-s)a_0(s)\right) \\ &= \frac{a_0(s)a_0(-s)}{a_1(s)}\left(z^2\frac{a_1(s)}{a_0(s)} - z + \frac{a_1(-s)}{a_0(-s)}\right). \end{aligned}$$

As $a_0(s)$ is stable by design, $a_0(s)$ and $a_0(-s)$ are not zero on the $j\omega$ axis. Further, if $a_1(s)$ has a root on the $j\omega$ axis, then this root cannot correspond to a root of $a(s, z_0, z) = a_0(s) + a_1(s)(z_0 - z) = 0$ as this would obviously imply $a_0(s)$ has a root on $j\omega$ axis.

The problem now is to find the roots of

$$z^2\frac{a_1(s)}{a_0(s)} - z + \frac{a_1(-s)}{a_0(-s)} = 0. \tag{11}$$

for $\mathrm{Re}(s^*) = 0, |z^*| = 1$.

Solving (11) for z gives

$$z = \frac{1 \pm \sqrt{1 - 4\frac{a_1(s)}{a_0(s)}\frac{a_1(-s)}{a_0(-s)}}}{2\frac{a_1(s)}{a_0(s)}} \tag{12}$$

and substituting this into (10) gives

$$z_0 = \frac{-1 \pm \sqrt{1 - 4\frac{a_1(s)}{a_0(s)}\frac{a_1(-s)}{a_0(-s)}}}{2\frac{a_1(s)}{a_0(s)}} \tag{13}$$

Lemma For those values of ω for which $|a_1(j\omega)/a_0(j\omega)| < 1/2$, equation (11) with $z = e^{-hj\omega}$ cannot be zero for any h.

Proof With $z = e^{-hj\omega}$ and $\frac{a_1(j\omega)}{a_0(j\omega)} = \left|\frac{a_1(j\omega)}{a_0(j\omega)}\right| e^{j\angle\frac{a_1(j\omega)}{a_0(j\omega)}}$, equation (11) can be rewritten as

$$e^{-hj\omega}\frac{a_1(j\omega)}{a_0(j\omega)} - 1 + e^{hj\omega}\frac{a_1(-j\omega)}{a_0(-j\omega)} = 0$$

or

$$2\left|\frac{a_1(j\omega)}{a_0(j\omega)}\right|\cos\left(h\omega + \angle\frac{a_1(j\omega)}{a_0(j\omega)}\right) = 1.$$

Consequently, for those values of ω for which $|a_1(j\omega)/a_0(j\omega)| < 1/2$, there can be no roots of (11) on the $j\omega$ axis.

Corollary If for all ω, $|a_1(j\omega)/a_0(j\omega)| < 1/2$, then the system (8) is stable independent of the delay value h.

Proof Follows directly from lemma 3.

3.1 Stability test

This lemma shows that one need only consider (13)(12) for $|a_1(j\omega)/a_0(j\omega)| > 1/2$. Then, with h_0 the *fixed* nominal value of the delay and $z_0 = e^{-jh_0\omega}, s = j\omega$, one solves (13) for $\omega \in \{\omega : |a_1(j\omega)/a_0(j\omega)| > 1/2\}$. That is, solve

$$e^{-jh_0\omega} = \frac{-1 \pm j\sqrt{4\left|\frac{a_1(j\omega)}{a_0(j\omega)}\right|^2 - 1}}{2\frac{a_1(j\omega)}{a_0(j\omega)}}, \tag{14}$$

for $\omega_i \in \{\omega : |a_1(j\omega)/a_0(j\omega)| > 1/2\}$. With $z_1 = e^{-jh\omega}, s = j\omega$ in (12), each of the solutions ω_i of (14) are then substituted into

$$e^{-jh\omega} = \frac{1 \pm j\sqrt{4\left|\frac{a_1(j\omega)}{a_0(j\omega)}\right|^2 - 1}}{2\frac{a_1(j\omega)}{a_0(j\omega)}} \tag{15}$$

and solved for the corresponding $h_{ik}, k = 1, 2, \ldots$. Then for each i, one selects the h_{ik_1} that is closest and less than h_0 and the h_{ik_2} that is closest and greater than h_0. Finally, $h_1^* = \max_i\{h_{ik_1}\}, h_2^* = \min_i\{h_{ik_2}\}$.

Example Let the system model be $G_0(s) = 1/s$ and the controller $G_c(s) = K > 0$ in Figure 2. The closed-loop characteristic polynomial is $a(s, e^{-h_0 s}, e^{-hs}) = s + K + K\left(e^{-h_0 s} - e^{-hs}\right)$. Here $a_0(s) = s + K, a_1(s) = K$ resulting in $a_1(j\omega)/a_0(j\omega) = K/(j\omega + K)$. Substituting into equation (14) results

$$
\begin{aligned}
e^{-jh_0\omega} &= \frac{-1 \pm j\sqrt{4\frac{K^2}{\omega^2+K^2}-1}}{2K/(j\omega+K)} \\
&= -\frac{1}{2} \mp \frac{\omega}{2K}\sqrt{4\frac{K^2}{\omega^2+K^2}-1} \\
&\quad + j\left(-\frac{\omega}{2K} \pm \frac{1}{2}\sqrt{4\frac{K^2}{\omega^2+K^2}-1}\right) \\
&= e^{j\gamma(\omega)}
\end{aligned}
$$

With $h_0 = .25, K = .1$, this can be solved numerically by finding the roots of $r(\omega) \triangleq -h_0\omega - \gamma_i(\omega) = 0$ for $i = 1, 2$. It turns out that $r(\omega)$ has no roots so this system is stable for all $h \geq 0$.

Example Consider a variation of the previous example where $a(s, e^{-h_0 s}, e^{-hs}) = s + K - K\left(e^{-h_0 s} - e^{-hs}\right)$ with $K > 0$. Here $a_0(s) = s + K, a_1(s) = -K$ resulting in $a_1(j\omega)/a_0(j\omega) = -K/(j\omega + K)$. Substituting into equation (14) results in

$$
\begin{aligned}
e^{-jh_0\omega} &= \frac{-1 \pm j\sqrt{4\frac{K^2}{\omega^2+K^2}-1}}{-2K/(j\omega+K)} \\
&= \frac{1}{2} \pm \frac{\omega}{2K}\sqrt{4\frac{K^2}{\omega^2+K^2}-1} \\
&\quad + j\left(\frac{\omega}{2K} \mp \frac{1}{2}\sqrt{4\frac{K^2}{\omega^2+K^2}-1}\right) \\
&= e^{j\gamma_i(\omega)}, i = 1, 2
\end{aligned}
$$

With $h_0 = .25, K = .1$, this can be solved numerically by finding the roots of $r(\omega) \triangleq -h_0\omega - \gamma_i(\omega) = 0$ for $i = 1, 2$. Doing so results in the single root $\omega = .0976$ giving $z_0 = e^{-jh_0\omega} \mid_{\omega=.0976} = 0.9997 - 0.024398j$. Substituting this into (15) (or using equation (10)) gives $z = -0.00029767 - 1.0003975789j$. Finally, h is obtained from solving

$$
e^{-hj\omega} \mid_{\omega=.0976} = -0.00029767 - 1.0003975789j
$$

giving $h = h_2^* = 16.1$ seconds. That is, for $0 \leq h < 16.1$ it follows

$$
a(s, e^{-h_0 s}, e^{-hs}) \neq 0 \text{ for } \mathrm{Re}(s) \geqq 0
$$

and

$$
a(s, e^{-h_0 s}, e^{-h_2^* s}) = 0 \text{ for some } s = j\omega.
$$

Finally, a recent paper (Stojić and L.S.Draganović, 2001) also considers a complementary problem of the robustness of the Smith predictor due to perturbation $\Delta G_0(s)$ (the non delayed part of the plant) and using a finite power series expansion of the assumed known delay term $e^{-h_0 s}$.

REFERENCES

Bellman, R. and K.L. Cooke (1963). *Differential-Difference Equations.* New York: Academic.

Chiasson, J. (1988). A method for computing the interval of delay values for which a differential-delay system. *IEEE Transactions on Automatic Control* **33**(12), 1176–1178.

Chiasson, J.N., S.D. Brierley and E.B. Lee (1985). A simplified derivation of the zeheb-walach 2-d stability test with applications to time-delay systems. *IEEE Transactions on Automatic Control.*

Cooke, K.L. and J.M. Ferreira (1983). Stability conditions for linear retarded functional differential equations. *Journal of Mathematical Analysis and Applications.*

C.T. Abdallah, P. Dorato, J. Benitez-Read and R. Byrne (1993). Delayed positive feedback can stabilize oscillatory systems. In: *Proceedings of the American Control Conference.* pp. 3106–3107. San Francisco CA.

Datko, R. (1978). A procedure for determination of the exponential stability of certain differential-difference equations. *Quarterly Applied Mathematics.*

Diekmann, O., S. A. van Gils-S. M. Verduyn Lunel and H.O. Walther (1995). *Delay Equations.* Springer-Verlag.

Hale, J.K. and S.M. Verduyn Lunel (1993). *Introduction to Functional Differential Equations.* Springer-Verlag.

Hertz, D., E.I. Jury and E. Zeheb (1984*a*). Simplified analytic stability test for systems with commensurate time delays. *IEE Proceedings.* part D.

Hertz, D., E.I. Jury and E. Zeheb (1984*b*). Stability independent and dependent of delay for delay differential systems. *J. Franklin Institute.*

Kamen, E.W. (1982). Linear systems with commensurate time delays: Stability and stabilization independent of delay. *IEEE Transactions on Automatic Control* pp. 367–375.

Nicolescu, S.I. (2001). *Delay effects on Stability.* Springer-Verlag. Berlin.

Niculescu, S.I. and C. T. Abdallah (1999). Delay effects on static ouput feedback stabilization. In: *Proceedings of the IEEE CDC.* pp. 53–54. Sydney, Australia.

Petterson, B.J., R.D. Robinett and J.C. Werner (1991). Lag-stabilized force feedback damping. *SAMD91-0194, UC-406.*

Power, H.M. and R.J. Simpson (1978). *Introduction to Dynamics and Control.* McGraw-Hill.

Stojić, M.R., M.S. Matjević and L.S.Draganović (2001). A robust smith predictor modified by internal models for integrating process with dead time. *IEEE Transactions on Automatic Control* **46**, 1293–1298.

www.elsevier.com/locate/ifac

FREQUENCY DOMAIN STABILITY CRITERION FOR DISCRETE DELAY SYSTEMS

Erik I. Verriest

School of Electrical and Computer Engineering, Georgia Institute of Technology, Atlanta, Georgia, USA
erik.verriest@ece.gatech.edu

Abstract: This paper extends earlier work on stability of discrete time delay systems, and derives an analog to the known frequency sweep for continuous time delay systems. *Copyright © 2001 IFAC*

Keywords: Time Delay Systems, Stability, Lyapunov-Krasovskii, Frequency Sweep, Delay-Difference Systems, Lyapunov Theory.

1. INTRODUCTION

In this paper the general form of the linear delay difference equation

$$x_{k+1} = Ax_k + Bx_{k-n} \tag{1.1}$$

with arbitrary delay, n, is considered. Unlike the situation in continuous time where the Riccati method only supplied sufficient conditions for stability (Verriest and Ivanov, 1994), it was shown in (Ivanov and Verriest, 1994) and (Verriest and Ivanov, 1995) that the discrete time theory leads to sufficient conditions for *instability* as well. This is yet another approach to the problem of delay-independent stability for discrete delay systems. For other approaches, see Kaskurewicz and Bhaya (1993), Kolmanovski (1995), Kolmanovski *et al.* (1999), Kuruklis (1994), Lee *et al.* (1992), Mori *et al.* (1982), Rodionov (1996), and Xu *et al.*(2001). We first define our notion of robust (in)stability.

Definition: i) The delay-difference system (1.1) is *robustly asymptotically stable* if for any number of delay steps $n \geq 0$ the dynamical system is asymptotically stable.
ii) The delay-difference system (1.1) is *robustly unstable* if for all $n \geq 0$ the dynamical system is unstable.

The following sufficient conditions for robust asymptotic stability are established in Verriest and Ivanov (1995) and Ivanov and Verriest (1998).

Theorem 1.1. The system (1.1) is robustly asymptotically stable, if either of the following two conditions hold
i) there exists a triple of positive definite (symmetric) matrices P, R and W such that

$$\begin{aligned} A'PBW^{-1}B'PA + A'PA + W + \\ + B'PB + R = P, \end{aligned} \tag{1.2}$$

ii) there exists a triple of positive definite (symmetric) matrices Π, S and Z such that

$$\begin{aligned} B'\Pi AZ^{-1}A'\Pi B + B'\Pi B + Z + \\ + A'\Pi A + S = \Pi. \end{aligned} \tag{1.3}$$

Some other sufficient conditions can be derived from Theorem 1.1.

Corollary 1.1. The system (1.1) is asymptotically stable, if either of the following is satisfied:

i) there exist positive definite (symmetric) matrices P, R, Λ and Ω solving the pair

$$\begin{cases} A'\Lambda^{-1}A + \Omega + R = & P \\ B\Omega^{-1}B' + \Lambda & = P^{-1}, \end{cases}$$

ii) there exist positive definite (symmetric) matrices Π, S, Γ and Σ solving the pair

$$\begin{cases} B'\Sigma^{-1}B + \Gamma + S = & \Pi \\ A\Gamma^{-1}A' + \Sigma & = \Pi^{-1}. \end{cases}$$

Corollary 1.2. The system (1.1) is robustly asymptotically stable, if there exists positive scalars p and q such that

$$\frac{1}{p} + \frac{1}{q} = 1, \tag{1.4}$$

and positive definite symmetric matrices X and Q satisfying a generalized Lyapunov equation

$$pA'XA + qB'XB + Q = X.$$

Comment: This statement nicely generalizes Lyapunov's lemma for regular difference systems. The latter is however necessary and sufficient.

2. FREQUENCY SWEEP

In (Verriest *et al.*, 1993) a different type of criterion, involving a frequency sweep, was derived for the continuous time delay system. In this section an analogous formulation for the discrete time stability problem of delay systems is derived. First the discrete equivalent to the celebrated *bounded real lemma* is introduced.

Consider the minimal discrete time system with $A \in \mathbb{R}^{n\times n}, B \in \mathbb{R}^{n\times m}, C \in \mathbb{R}^{p\times n}$ and $D \in \mathbb{R}^{p\times m}$.

$$x_{k+1} = Ax_k + Bu_k \tag{2.1}$$

$$y_k = Cx_k + Du_k. \tag{2.2}$$

Its irreducible transfer matrix is

$$T(z) = C(zI - A)^{-1}B + D. \tag{2.3}$$

Define the norm $||T(z)||_\infty = \max_{|z|=1} ||T(z)||_2$.

Fact: Bounded Real Lemma (see Singh (1987) and references therein).
The following are equivalent:

(1) A is Schur-Cohn stable and $||T(z)||_\infty < 1$.

(2) There exist $P = P' > 0$ and L, W, such that

$$A'PA + C'C + L'L = P \tag{2.4}$$

$$A'PB + C'D + L'W = 0 \tag{2.5}$$

$$B'PB + D'D + W'W = I. \tag{2.6}$$

An interpretation of the second set of conditions of this lemma is as follows: First by Kalman's extension of Lyapunov's lemma, for $\mathcal{A} \in \mathbb{R}^{\overline{n}\times\overline{n}}$ and $\mathcal{C} \in \mathbb{R}^{\overline{p}\times\overline{n}}$ any two of the following statements imply the third:
i) There exists $\mathcal{P} = \mathcal{P}' > 0$ such that

$$\mathcal{A}'\mathcal{P}\mathcal{A} + \mathcal{C}'\mathcal{C} = \mathcal{P},$$

ii) The matrix $\mathcal{A}$ is Schur-Cohn stable,
iii) The pair $(\mathcal{A}, \mathcal{C})$ is observable.

Now apply the above to the matrices

$$\mathcal{A} = \begin{bmatrix} A & B \\ 0 & 0 \end{bmatrix}, \qquad \mathcal{C} = \begin{bmatrix} C & D \\ L & W \end{bmatrix}.$$

The observability of this augmented pair implies the equivalence of the Schur-Cohn stability of $\begin{bmatrix} A & B \\ 0 & 0 \end{bmatrix}$, hence of A, with the existence of a positive definite matrix $\mathcal{P} = \begin{bmatrix} P & R \\ R' & Q \end{bmatrix}$ satisfying the Lyapunov equation:

$$\begin{bmatrix} A' & 0 \\ B' & 0 \end{bmatrix} \begin{bmatrix} P_0 & R_0 \\ R_0' & Q_0 \end{bmatrix} \begin{bmatrix} A & B \\ 0 & 0 \end{bmatrix} + \begin{bmatrix} C' \\ D' \end{bmatrix} [C \; D] = \begin{bmatrix} P_0 & R_0 \\ R_0' & Q_0 \end{bmatrix}.$$

If $z = 0$ is not a zero of $T(z)$ (see Kailath, 1980), then the matrix $\begin{bmatrix} A & B \\ C & D \end{bmatrix}$ has full rank. Moreover, if in addition the original (A, C) is observable, then the matrix

$$\begin{bmatrix} zI_n - A & -B \\ 0 & zI_m \\ C & D \end{bmatrix},$$

has full rank for all z in $\mathbb{C}$. By the Popov-Belevitch-Hautus criterion this implies the observability of the augmented pair

$$\left(\begin{bmatrix} A & B \\ 0 & 0 \end{bmatrix}, [C \; D] \right),$$

and therefore also of the augmented pair

$$\left(\begin{bmatrix} A & B \\ 0 & 0 \end{bmatrix}, \begin{bmatrix} C & D \\ L & W \end{bmatrix} \right)$$

for all L and W of compatible dimension.

Thus if (A, C) is observable and $z = 0$ is not a zero of $T(z)$, the bounded real lemma states that there exist specific L and W such that the satisfaction of the augmented Lyapunov equation for $R = 0$ and $Q = I$ is equivalent to the Schur-Cohn stability of the matrix A.

A sufficient condition in the frequency domain follows:

Theorem 2.1. If A is Schur-Cohn stable, and $\|(e^{j\theta}I - A)^{-1}B\|_\infty < 1$, then the discrete delay system is robustly asymptotically stable.

Proof: By virtue of the bounded real lemma, for $C = I$ and $D = 0$, the stability of A, together with the infinity norm condition on $T(z) = (zI - A)^{-1}B$ imply the existence of $P = P' > 0$, L and W such that

$$\begin{bmatrix} A'PA - P + I & A'PB \\ B'PA & B'PB - I \end{bmatrix} = \begin{bmatrix} -L'L & -L'W \\ -W'L & -W'W \end{bmatrix} \leq 0.$$

Hence,

$$\begin{bmatrix} A'PA - P + (1-\epsilon)I & A'PB \\ B'PA & B'PB - (1-\epsilon)I \end{bmatrix} = \begin{bmatrix} -L'L - \epsilon I & -L'W \\ -W'L & -W'W + \epsilon I \end{bmatrix} < 0.$$

Choosing $Q = (1-\epsilon)I$ for $|\epsilon|$ sufficiently small yields the sufficient condition for robust asymptotic stability (2.5), expressed as an LMI, which is equivalent to the Riccati equation. □

Note that since the unit circle $|z| = 1$ is a compact set, this frequency sweep test for discrete delay system stability is simpler to execute (finite sweep) than its continuous time counterpart. The complexity issues are discussed by Toker and Özbayin (1996).

How far is the condition expressed in the above theorem from being necessary? For continuous time delay systems, the sufficient condition was improved to a NASC by Chen and Latchman (1995) and Chen *et al.* (1995), using frequency arguments.

In the theorem formulation below, $\rho(M) = \max\{|\lambda| : \lambda \in \mathrm{Spec}(M)\}$ is the spectral radius of the matrix M.

Theorem 2.2. The system (1.1) is robustly asymptotically stable if

i) A is asymptotically stable,
ii) $\forall\theta: \quad \rho((e^{j\theta}I - A)^{-1}B) < 1$.
If in addition there exists a point z_0 on the unit circle where $|(z_0 I - A)^{-1}B| < 1$, then the conditions are also necessary.

Proof: Sufficiency
Let $\mathbf{D}$ denote the domain $|z| \geq 1$, the closure of the outside of the unit disk. If A is Schur-Cohn stable, then $(zI - A)^{-1}B$ is analytic on $\mathbf{D}$. Thus also for all $n > 0$, $(zI - A)^{-1}Bz^{-n}$ is analytic on $\mathbf{D}$. Since the spectral radius is a subharmonic function of z,

$$\sup_{z\in\mathbf{D}} \rho\left[(zI - A)^{-1}Bz^{-n}\right] = \max_\theta \rho[(e^{j\theta}I - A)^{-1}B].$$

The condition ii) implies then that $\forall z \in \mathbf{D}$, and $\forall n \in Z_+$,

$$\det\left[(zI - A)^{-1}Bz^{-n} - I\right] \neq 0$$

or equivalently,

$$\frac{\det\left[zI - A - Bz^{-n}\right]}{\det(zI - A)} \neq 0, \quad \forall z \in \mathbf{D},\ \forall n \in Z_+,$$

thus implying robust asymptotic stability.
Necessity: is shown by contradiction. Assume that on the contrary, $\sup_\theta \rho[(e^{j\theta}I - A)^{-1}B] \geq 1$. First, consider the case $\sup \rho(\theta) = 1$. Then $\exists\theta_1$ such that $\rho[(e^{j\theta_1}I - A)^{-1}B] = 1$. In turn this implies that there exists a vector u and a scalar ϕ such that

$$(e^{j\theta_1}I - A)^{-1}Bu = e^{-j\phi}u,$$

and hence $\det\left[e^{j\theta_1}I - A - Be^{j\phi}\right] = 0$. There are two possibilities: If $\frac{\phi}{2\pi}$ is irrational, then by Jacobi's theorem (Devaney, 1986) the orbit of $e^{-j(\theta_1 + k\phi)}$, k integer, is dense in the unit circle $|z| = 1$. Consequently, for any $\epsilon' > 0$, there exist integers $n(\epsilon')$ and $k(\epsilon')$ such that

$$|n(\epsilon')\theta_1 - 2\pi k(\epsilon') - \phi| < \epsilon'$$

By continuity of the determinant function, for any ϵ there exists an ϵ' such that

$$\det\left[e^{j\theta_1}I - A - Be^{-n(\epsilon')\theta_1}\right] < \epsilon.$$

Thus the delay system $x_{k+1} = Ax_k + Bx_{k-n(\epsilon')}$ is ϵ-close to instability.
If $\frac{\phi}{2\pi}$ is rational, the upper semicontinuity of ρ implies the existence of an irrational $\tilde{\theta}_1$ such that $\rho[(e^{j\tilde{\theta}_1} - A)^{-1}B] = 1$, reducing the problem to the previous case.
The case where $\sup \rho[(e^{j\theta} - A)^{-1}B] = \rho_1 > 1$ proceeds similarly. Since $\exists\theta_0$ such that $\rho[(e^{j\theta_0} - A)^{-1}B] = \rho_0 < 1$, by continuity, there must exist

an irrational θ_2 such that $\rho[e^{j\theta_2} - A)^{-1}B] = 1$, then argue as in the previous case. □

3. CONCLUSIONS

A frequency domain interpretation of the stability condition was presented. Its usefulness stems from the fact that the frequency sweep, here over a compact set as opposed to the continuous time equivalent, is easily generated.

Acknowledgement

The support of the NSF-CNRS collaborative grant INT-9818312 is gratefully acknowledged.

4. REFERENCES

Chen, J. and H.A. Latchman, Frequency sweeping tests for stability independent of delay, *IEEE Transactions on Automatic Control*, **40**, No. 9, 1640-1645.

Chen, J., D. Xu and B. Shafai (1995). On sufficient conditions for stability independent of delay, *IEEE Transactions on Automatic Control*, **40**, No. 9, 1675-1680.

Devaney, R.L. (1986). *An Introduction to Chaotic Dynamical Systems*, Benjamin-Cummings.

Ivanov A.F. and E.I. Verriest (1994). Robust stability of delay-difference equations, in *Systems and Networks: Mathematical Theory and Applications*, (U. Helmke, R. Mennicken and J. Saurer, eds.), Univ. of Regensburg Press, 725-726.

Ivanov A.F. and E.I. Verriest (1998). Stability and Existence of Periodic Solutions in Discrete Delay Systems, Research Report 32/98 School of Information Technology and Mathematical Sciences, University of Ballarat, Victoria, Australia.

Kailath, T. (1980). *Linear Systems*, Prentice Hall.

Kaszkurewicz, E. and A. Bhaya (1993). A delay-independent robust stability condition for linear discrete time systems, *Proceedings of the 32th Conference on Decision and Control*, San Antonio, TX, 3459-3460.

Kolmanovskii, V.B. (1995). On the application of the Lyapunov second method to the Volterra difference equations, *Automation and Remote Control*, **56**, No. 11, 1545-1556.

Kolmanovskii, V.B., J.-F. Lafay and J.-P. Richard (1999). Riccati equations in stability theory of diference equations with memory, *Proc. 5-th European Control Conference*, Karlsrühe, Germany, DA-12,2.

Kuruklis, S. A. (1994). The asymptotic stability of $x_{n+1} - ax_n + bx_{n-k} = 0$, *J. Math Anal. Appl.*, **188** , 719-731.

Lee, C.-H., T.-H. S. Li and F.-C. Kung (1992). D-stability analysis for discrete systems with a time delay, *Systems & Control Letters*, **19**, 213-219.

Mori, T., N. Fukuma and M.M. Kuwahara, (1982). Delay-independent stability criteria for discrete-delay systems, *IEEE Transactions on Automatic Control*, **27**, No. 4, 964-966.

Rodionov, A.M. (1996). A method for investigating the stability of differential and discrete equations with delay, *Automation and Remote Control*, **57**, No. 12, 1720-1727.

Singh, V. (1987). A new proof of the discrete-time bounded-real lemma and lossless bounded-real lemma," *IEEE Transactions on Circuits and Systems*, **34**, No. 8, 960-962.

Toker O. and H. Özbay (1996). Complexity issues in robust stability of linear delay-differential systems, *Mathematics of Control, Signals, and Systems*, **9**, No. 4, 386-400.

Verriest, E.I., M.K. Fan and J. Kullstam (1993). Frequency domain robust stability criteria for linear delay systems, *Proceedings of the 32nd IEEE Conference on Decision and Control*, San Antonio, TX, 3473-3478.

Verriest, E.I. and A.F. Ivanov (1994). Robust stability of systems with delayed feedback, *Circuits, Systems, and Signal Processing*, **13** , No. 2-3, 213-222.

Verriest, E.I. and A.F. Ivanov (1995). Robust stability of delay-difference equations, *Proceedings of the 34th IEEE CDC*, New Orleans, LA, 386-391.

Xu, S., J. Lam, and C. Yang (2001). Quadratic stability and stabilization of uncertain linear discrte-time systems with state delay, *Systems and Control Letters*, **43**, 77-84.

www.elsevier.com/locate/ifac

AN IMPROVED STABILITY CRITERION FOR SYSTEMS WITH DISTRIBUTED DELAYS*

Keqin Gu

Department of Mechanical & Industrial Engineering
Southern Illinois University at Edwardsville
Edwardsville, Illinois 62026, USA, E-mail: kgu@siue.edu

Abstract: This article improves the formulation of discretized Lyapunov functional method for systems with distributed delays. The main idea is to apply Jensen's inequality and variable elimination in matrix inequalities. The resulting stability criterion is much simpler, and converges to the analytical result much faster. The new formulation is also applicable to a wider class of systems. *Copyright © 2001 IFAC*

Keywords: Stability, Time-delay, Linear Matrix Inequality

1. INTRODUCTION

Time-delay systems are frequently encountered in engineering, biology, economy, and other areas (Hale and Verduyn Lunel, 1993). In the wake of intensive research on the robust stability and control theory, the stability and control of time-delay systems received renewed interests. The development of efficient computational algorithm for non-smooth convex optimization problem (Nesterov and Nemirovskii, 1994), which made it possible to efficiently solve Linear Matrix Inequalities (LMI) (Boyd, et. al., 1994; Gahinet, et. al., 1995), inspired intensive activities to formulate such problems in an LMI form. For a comprehensive survey, see Niculescu, et. al. (1997), Kolmanovskii, *et. al.* (1999a) and Kharitonov (1998).

There are important practical applications for systems with distributed delays as documented in, e.g., Hale and Verduyn Lunel (1993). However, there are much fewer results available to check the stability of such systems. Among them in recent literature, Kolmanovskii and Richard (1997) proposed a simple stability criterion. Gu, *et. al.* (2001) used discretized Lyapunov functional method to check the stability of systems with distributed delay and piecewise constant coefficients.

In this article, the discretized Lyapunov functional method proposed in Gu, *et. al.* (2001) is improved using Jensen's inequality and variable elimination technique of matrix inequality. As the result, the convergence to the analytical stability limit is accelerated. The formulation is also applicable to a wider class of systems.

2. PROBLEM SETUP

Consider the system with distributed delays

$$\dot{x}(t) = A_0(t)x(t) + \int_{-r}^{0} A(t,\theta)x(t+\theta)d\theta \qquad (1)$$

where the coefficient matrix A is piecewise independent of θ

$$A(t,\theta) = A^i(t), \quad -r_i \leq \theta < -r_{i-1},\ i = 1,2,...,K \qquad (2)$$

with r_i's satisfy

$$0 = r_0 < r_1 < r_2 < ... < r_K = r \qquad (3)$$

In other words,

* This work is partially supported by National Science Foundation Grant INT-9818312

$$\dot{x}(t) = A_0(t)x(t) + \sum_{i=1}^{K} A^i(t) \int_{-r_i}^{-r_{i-1}} x(t+\theta)d\theta \tag{4}$$

The system matrices are uncertain, and are bounded by a known compact set Ω

$$(A_0(t), A^1(t), A^2(t), ..., A^K(t)) \in \Omega \text{ for all } t \geq 0 \tag{5}$$

For the sake of simplicity, we will often suppress the time dependence of the sysmtem matrices, and write A_0, A^i, and $A(\theta)$ instead of $A_0(t)$, $A^i(t)$ and $A(t,\theta)$. Compared to Gu, *et. al.* (2001), this setting allows the first term in the right hand side of (1) to be nonzero, and allows uncertainty.

A general quadratic Lyapunov-Krasovskii functional will be used

$$\begin{aligned} V(\phi) &= \phi^T(0)P\phi(0) + 2\phi^T(0)\int_{-r}^{0} Q(\xi)\phi(\xi)d\xi \\ &+ \int_{-r}^{0}[\int_{-r}^{0} \phi^T(\xi)R(\xi,\eta)\phi(\eta)d\eta]d\xi \\ &+ \int_{-r}^{0} \phi^T(\xi)S(\xi)\phi(\xi)d\xi \end{aligned} \tag{6}$$

where

$$P = P^T \in \mathrm{R}^{n\times n} \tag{7}$$

and for all $-r \leq \xi \leq 0$ and $-r \leq \eta \leq 0$,

$$Q(\xi) \in \mathrm{R}^{n\times n} \tag{8}$$

$$S(\xi) = S^T(\xi) \in \mathrm{R}^{n\times n} \tag{9}$$

$$R(\xi,\eta) = R^T(\eta,\xi) \in \mathrm{R}^{n\times n} \tag{10}$$

Take derivative of V with respect to time along the trajectory of the system (1), integration by parts when necessary, one can write $\dot{V}$ as a quadratic functional of ϕ as follows:

$$\begin{aligned} &\dot{V}(\phi) \triangleq \dot{V}(x_t)|_{x_t=\phi} \\ &= -\phi_{0r}^T \begin{pmatrix} \Delta_{00} & Q(-r) \\ Q^T(-r) & S(-r) \end{pmatrix} \phi_{0r} \\ &+2\phi_{0r}^T \int_{-r}^{0} \begin{pmatrix} \Gamma^0(\xi) \\ \Gamma^1(\xi) \end{pmatrix} \phi(\xi)d\xi \\ &- \int_{-r}^{0} \phi^T(\xi)\frac{dS(\xi)}{d\xi}\phi(\xi)d\xi \\ &- \int_{-r}^{0} d\xi \int_{-r}^{0} \phi^T(\xi)[\frac{\partial R(\xi,\eta)}{\partial \xi} + \frac{\partial R(\xi,\eta)}{\partial \eta} \\ &-A^T(\xi)Q(\eta) - Q^T(\xi)A(\eta)]\phi(\eta)d\eta \end{aligned} \tag{11}$$

where x_t is defined by $x_t(\theta) = x(t+\theta)$, $-r \leq \theta \leq 0$, and

$$\phi_{0r}^T = \left(\phi^T(0) \;\; \phi^T(-r) \right)$$

$$\Delta_{00} = -PA_0 - A_0^T P - Q(0) - Q^T(0) - S(0)$$

$$\Gamma^0(\xi) = A_0^T Q(\xi) + PA(\xi) - \frac{dQ(\xi)}{d\xi} + R(0,\xi)$$

$$\Gamma^1(\xi) = -R(-r,\xi)$$

Lemma 1. The system described by (4) and (5) is aymptotically stable if there exists a Lyapunov-Krasovskii functional (6) such that the Lyapunov-Krasovskii functional satisfies

$$V(\phi) \geq \varepsilon||\phi(0)||^2 \tag{12}$$

and its derivative along the trajectory of the system satisfies

$$\dot{V}(\phi) \leq -\varepsilon||\phi(0)||^2 \tag{13}$$

for some $\varepsilon > 0$.

Proof. This is a special case of the well known Lyapunov-Krasovskii stability Theorem, see, for example, Theorem 2.1, Chapter 5 of Hale and Verduyn Lunel (1993). The only thing needs to be shown is that there exists a sufficiently large $K > 0$ such that

$$V(\phi) \leq K \max_{-r\leq\theta\leq 0} ||\phi(\theta)||^2$$

But this is obvious since $V(\phi)$ is a bounded quadratic functional. ■

In this paper, the stability of the system will be tested by the satisfaction of condition (12), which will be referred to as the Lyapunov-Krasovskii functional condition, and condition (13), which will be referred to as the Lyapunov-Krasovskii derivative condition. To test these two conditions using a digital computer, we will try to write them in the form of linear matrix inequalities (LMI) (Boyd, et. al., 1994; Gahinet, et. al., 1995) such that efficient interior point algorithms are available (Nesterov and Nemirovskii, 1994).

To write the stability conditions in LMI, we discretize the interval $[-r, 0]$ into N nonuniform segments, such that $A(\xi)$ is independent of ξ in each segment. Specifically, let $0 = \theta_0 > \theta_1 > ... > \theta_N = -r$ be chosen such that

$$-r_i = \theta_{N_i}, \; i = 0, 1, 2, ..., K$$

thus the intervals $[-r, 0]$ is divided into N segments $[\theta_p, \theta_{p-1}]$ of length

$$h_p = \theta_{p-1} - \theta_p$$

$p = 1, 2, ..., N_i$. Adopt the convention

$$N_0 = 0$$

Then

$$0 = N_0 < N_1 < N_2 < ... < N_K = N$$

The parameters of Lyapunov-Krasovskii functional are again chosen to be piecewise linear, *i.e.*, for $0 \leq \alpha \leq 1$ and $p = 1, 2, ..., N$

$$Q(\theta_p + \alpha h_p) \triangleq Q^{(p)}(\alpha) = (1-\alpha)Q_p + \alpha Q_{p-1}$$
$$S(\theta_p + \alpha h_p) \triangleq S^{(p)}(\alpha) = (1-\alpha)S_p + \alpha S_{p-1}$$

and for $0 \le \alpha \le 1$, $0 \le \beta \le 1$, $p = 1, 2, ..., N$, $q = 1, 2, ..., N$,

$$R(\theta_p + \alpha h_p, \theta_q + \beta h_p) \triangleq R^{(pq)}(\alpha, \beta)$$
$$= \begin{cases} (1-\alpha)R_{pq} + \beta R_{p-1,q-1} + (\alpha-\beta)R_{p-1,q}, & \text{for } \alpha \ge \beta \\ (1-\beta)R_{pq} + \alpha R_{p-1,q-1} + (\beta-\alpha)R_{p,q-1}, & \text{for } \alpha < \beta \end{cases}$$

3. LYAPUNOV-KRASOVSKII FUNCTIONAL CONDITION

In this section, we will derive the LMI which guarantees the satisfaction of the Lyapunov-Krasovskii functional condition. Let

$$\phi^{(p)}(\alpha) = \phi(\theta_p + \alpha h_p) \tag{14}$$

We have

Proposition 2. The Lyapunov-Krasovskii functional $V(\phi)$ as expressed in (6) to (10), with Q, S and R piecewise linear, satisfies

$$V(\phi)$$
$$= \int_0^1 \phi_{0\alpha}^T \begin{pmatrix} P & \tilde{Q} \\ \tilde{Q}^T & \tilde{R} \end{pmatrix} \phi_{0\alpha} d\alpha$$
$$+ \sum_{p=1}^{N} \int_0^1 \phi^{(p)T}(\alpha) S^{(p)}(\alpha) \phi^{(p)}(\alpha) h_p d\alpha \tag{15}$$
$$\ge \int_0^1 \phi_{0\alpha}^T \begin{pmatrix} P & \tilde{Q} \\ \tilde{Q}^T & \tilde{R} + \tilde{S} \end{pmatrix} \phi_{0\alpha} d\alpha \tag{16}$$

if

$$S_p > 0, \qquad p = 0, 1, ..., N \tag{17}$$

where

$$\tilde{Q} = (\, Q_0 \; Q_2 \; ... \; Q_N \,) \tag{18}$$
$$\tilde{R} = \begin{pmatrix} R_{00} & R_{01} & ... & R_{0N} \\ R_{10} & R_{11} & ... & R_{1N} \\ \vdots & \vdots & \ddots & \vdots \\ R_{N0} & R_{N1} & ... & R_{NN} \end{pmatrix} \tag{19}$$
$$\tilde{S} = \text{diag}\left(\frac{1}{\tilde{h}_0} S_0 \; \frac{1}{\tilde{h}_1} S_1 \; ... \; \frac{1}{\tilde{h}_N} S_N \right) \tag{20}$$
$$\tilde{h}_p = \max\{h_p, h_{p+1}\},\ p = 1, 2, ..., N-1 \tag{21}$$
$$\tilde{h}_0 = h_1, \tag{22}$$
$$\tilde{h}_N = h_N \tag{23}$$

and

$$\phi_{0\alpha}^T = (\, \phi^T(0) \; \Psi^T(\alpha) \,)$$

$$\Psi(\alpha) = \begin{pmatrix} \psi_{(1)}(\alpha) \\ \psi_{(2)}(\alpha) + \psi^{(1)}(\alpha) \\ \psi_{(3)}(\alpha) + \psi^{(2)}(\alpha) \\ \vdots \\ \psi_{(N)}(\alpha) + \psi^{(N-1)}(\alpha) \\ \psi^{(N)}(\alpha) \end{pmatrix} \tag{24}$$

$$\psi^{(p)}(\alpha) = h_p \int_0^\alpha \phi^{(p)}(\beta) d\beta \tag{25}$$
$$\psi_{(p)}(\alpha) = h_p \int_\alpha^1 \phi^{(p)}(\beta) d\beta \tag{26}$$
$$p = 1, 2, ..., N$$

Therefore the Lyapunov-Krasovskii functional condition (12) is satisfied if (17) and

$$\begin{pmatrix} P & \tilde{Q} \\ \tilde{Q}^T & \tilde{R} + \tilde{S} \end{pmatrix} > 0 \tag{27}$$

are satisfied.

Proof. The first step (15) is a special case of Proposition 1 in Gu, *et. al.* (2001). To prove (16), we can bound the last term of (15) as follows:

$$\sum_{p=1}^{N} V_{S\text{P}} \triangleq \sum_{p=1}^{N} \int_0^1 \phi^{(p)T}(\alpha) S^{(p)}(\alpha) \phi^{(p)}(\alpha) h_p d\alpha$$
$$= \int_0^1 \alpha h_1 \phi^{(1)T}(\alpha) \frac{1}{h_1} S_0 h_1 \phi^{(1)}(\alpha) d\alpha$$
$$+ \int_0^1 (1-\alpha) h_N \phi^{(N)T}(\alpha) \frac{1}{h_N} S_N h_N \phi^{(N)}(\alpha) d\alpha$$
$$+ \sum_{p=1}^{N-1} \int_0^1 [\alpha h_{p+1} \phi^{(p+1)T}(\alpha) \frac{1}{h_{p+1}} S_p h_{p+1} \phi^{(p+1)}(\alpha)$$
$$+ (1-\alpha) h_p \phi^{(p)T}(\alpha) \frac{1}{h_p} S_p h_p \phi^{(p)}(\alpha)] d\alpha$$
$$\ge \int_0^1 \alpha h_1 \phi^{(1)T}(\alpha) \frac{1}{\tilde{h}_0} S_0 h_1 \phi^{(1)}(\alpha) d\alpha$$
$$+ \int_0^1 (1-\alpha) h_N \phi^{(N)T}(\alpha) \frac{1}{\tilde{h}_N} S_N h_N \phi^{(N)}(\alpha) d\alpha$$
$$+ \sum_{p=1}^{N-1} \int_0^1 [\alpha h_{p+1} \phi^{(p+1)T}(\alpha) \frac{1}{\tilde{h}_p} S_p h_{p+1} \phi^{(p+1)}(\alpha)$$
$$+ (1-\alpha) h_p \phi^{(p)T}(\alpha) \frac{1}{\tilde{h}_p} S_p h_p \phi^{(p)}(\alpha)] d\alpha \tag{28}$$

In view of (17), we can use Lemma 5 in Gu (2001) in each term above to obtain

$$\sum_{p=1}^{N} V_{S\text{P}} \ge \int_0^1 \Psi^T(\alpha) \tilde{S} \Psi(\alpha) d\alpha$$

from which (16) follows. Sufficiency of (17) and (27) is clearly implied by (16). ■

The above can be viewed as an extension to the corresponding result in Gu (2001) (Condition I in

Proposition 7) to the case of nonuniform mesh. A result similar to condition II in Gu (2001) is also possible, but will not be presented here due to space limitation.

4. LYAPUNOV-KRASOVSKII DERIVATIVE CONDITION

Define

$$M_p = \min\{i \mid N_i \geq p\} \tag{29}$$

We have

$A(\xi) = A^{M_p}$ independent of ξ for $\xi \in (\theta_p, \theta_{p-1})$.

We will write

$$A_p = A^{M_p} \tag{30}$$

for the sake of convenience. Then, after the discretization, dividing the integration interval $[-r, 0]$ into N segments $[\theta_p, \theta_{p-1}]$, $p = 1, 2, ..., N$, the derivative $\dot{V}$ of the Lyapunov-Krasovskii functional V along the trajectory of the system described by (1) to (5) may be written as

$$\begin{aligned}
&\dot{V}(\phi) \\
&= -\phi_{0r}^T \Delta \phi_{0r} + 2\phi_{0r}^T \int_0^1 [D^s + (1-2\alpha)D^a]\hat{\phi}(\alpha)d\alpha \\
&\quad - \int_0^1 \hat{\phi}^T(\alpha) S_d \hat{\phi}(\alpha) d\alpha \\
&\quad + 2(\int_0^1 \hat{\phi}(\alpha)d\alpha)^T \int_0^1 [E^s + (1-2\alpha)E^a]\hat{\phi}(\alpha)d\alpha \\
&\quad - (\int_0^1 \hat{\phi}(\alpha)d\alpha)^T R_{ds} (\int_0^1 \hat{\phi}(\alpha)d\alpha) \\
&\quad - \int_0^1 [\int_0^\alpha \begin{pmatrix} \hat{\phi}^T(\alpha) & \hat{\phi}^T(\beta) \end{pmatrix} \\
&\qquad \begin{pmatrix} 0 & R_{da} \\ R_{da}^T & 0 \end{pmatrix} \begin{pmatrix} \hat{\phi}(\alpha) \\ \hat{\phi}(\beta) \end{pmatrix} d\beta] d\alpha
\end{aligned} \tag{31}$$

where

$$\Delta = \begin{pmatrix} -PA_0 - A_0^T P - Q_0 - Q_0^T - S_0 & Q_N \\ Q_N^T & S_N \end{pmatrix} \tag{32}$$

$$D^s = \begin{pmatrix} D_{01}^s & D_{02}^s & ... & D_{0N}^s \\ D_{11}^s & D_{12}^s & ... & D_{1N}^s \end{pmatrix} \tag{33}$$

$$D^a = \begin{pmatrix} D_{01}^a & D_{02}^a & ... & D_{0N}^a \\ D_{11}^a & D_{12}^a & ... & D_{1N}^a \end{pmatrix} \tag{34}$$

$$D_{ip}^s = \frac{h_p}{2}[\Gamma^i(\theta_p + 0) + \Gamma^i(\theta_{p-1} - 0)]$$

$$D_{ip}^a = \frac{h_p}{2}[\Gamma^i(\theta_p + 0) - \Gamma^i(\theta_{p-1} - 0)]$$

or more explicitly

$$\begin{aligned}
D_{0p}^s &= \frac{h_p}{2}[A_0^T(Q_p + Q_{p-1}) + 2PA_p \\
&\quad + (R_{0p} + R_{0,p-1})] - (Q_{p-1} - Q_p)
\end{aligned} \tag{35}$$

$$D_{1p}^s = -\frac{h_p}{2}[R_{Np} + R_{N,p-1}]$$

$$D_{0p}^a = \frac{h_p}{2}[A_0^T(Q_p - Q_{p-1}) + (R_{0p} - R_{0,p-1})]$$

$$D_{1p}^a = \frac{h_p}{2}[R_{N,p-1} - R_{Np}] \tag{36}$$

$$S_d = \text{diag}\begin{pmatrix} S_{d1} & S_{d2} & ... & S_{dN} \end{pmatrix} \tag{37}$$

$$S_{dp} = S_{p-1} - S_p \tag{38}$$

$$E^s = \begin{pmatrix} E_{11}^s & E_{12}^s & ... & E_{1N}^s \\ E_{21}^s & E_{22}^s & ... & E_{2N}^s \\ \vdots & \vdots & \ddots & \vdots \\ E_{N1}^s & E_{N2}^s & ... & E_{NN}^s \end{pmatrix} \tag{39}$$

$$E^a = \begin{pmatrix} E_{11}^a & E_{12}^a & ... & E_{1N}^a \\ E_{21}^a & E_{22}^a & ... & E_{2N}^a \\ \vdots & \vdots & \ddots & \vdots \\ E_{N1}^a & E_{N2}^a & ... & E_{NN}^a \end{pmatrix} \tag{40}$$

$$E_{pq}^s = \frac{h_p h_q}{2} A_p^T (Q_q + Q_{q-1}) \tag{41}$$

$$E_{pq}^a = \frac{h_p h_q}{2} A_p^T (Q_q - Q_{q-1}) \tag{42}$$

$$R_{ds} = \begin{pmatrix} R_{ds11} & R_{ds12} & ... & R_{ds1N} \\ R_{ds21} & R_{ds22} & ... & R_{ds2N} \\ \vdots & \vdots & \ddots & \vdots \\ R_{dsN1} & R_{dsN2} & ... & R_{dsNN} \end{pmatrix} \tag{43}$$

$$\begin{aligned}
R_{dspq} &= \frac{1}{2}[(h_p + h_q)(R_{p-1,q-1} - R_{pq}) \\
&\quad + (h_p - h_q)(R_{p,q-1} - R_{p-1,q})],
\end{aligned} \tag{44}$$

$$R_{da} = \begin{pmatrix} R_{da11} & R_{da12} & ... & R_{da1N} \\ R_{da21} & R_{da22} & ... & R_{da2N} \\ \vdots & \vdots & \ddots & \vdots \\ R_{daN1} & R_{daN2} & ... & R_{daNN} \end{pmatrix} \tag{45}$$

$$\begin{aligned}
R_{dapq} &= \frac{1}{2}(h_p - h_q)(R_{p-1,q-1} \\
&\quad - R_{p-1,q} - R_{p,q-1} + R_{pq}),
\end{aligned} \tag{46}$$

and

$$\phi_{0r}^T = \begin{pmatrix} \phi^T(0) & \phi^T(-r) \end{pmatrix}$$

$$\hat{\phi}^T(\alpha) = \begin{pmatrix} \phi^{(1)T}(\alpha) & \phi^{(2)T}(\alpha) & ... & \phi^{(N)T}(\alpha) \end{pmatrix}$$

From the quadratic expression (31), we can obtain the Lyapunov-Krasovskii derivative condition as follows:

Proposition 3. The derivative of the Lyapunov-Krasovskii functional $V(\phi)$ expressed in (6) to (10), with Q, S and R piecewise linear, along the trajectory of the system expressed by (1) to (5), satisfies

$$\begin{aligned}&\dot{V}(\phi)\\ &\leq -\phi_{0r\alpha}^T Y \phi_{0r\alpha}\\ &\quad -\int_0^1 [\int_0^\alpha \phi_{\alpha\beta}^T \begin{pmatrix} W & R_{da} \\ R_{da}^T & W \end{pmatrix} \phi_{\alpha\beta} d\beta] d\alpha \quad (47)\end{aligned}$$

for arbitrary matrix function $U(A_0, A^1, A^2, ..., A^K)$ satisfying

$$\begin{pmatrix} U & -I \\ -I & S_d \end{pmatrix} > 0 \tag{48}$$

where

$$Y = \begin{pmatrix} -\frac{1}{3}D^a U D^{aT} + \Delta & -D^s - \frac{1}{3}D^a U E^{aT} \\ -D^{sT} - \frac{1}{3}E^a U D^{aT} & S_d - E^{sa} + R_{ds} - W \end{pmatrix}$$

$$E^{sa} = E^s + E^{sT} + \frac{1}{3}E^a U E^{aT}$$

$$\phi_{0r\alpha}^T = \begin{pmatrix} \phi_{0r}^T & \int_0^1 \hat{\phi}^T(\alpha) d\alpha \end{pmatrix}$$

$$\phi_{\alpha\beta}^T = \begin{pmatrix} \hat{\phi}^T(\alpha) & \hat{\phi}^T(\beta) \end{pmatrix}$$

Therefore, the Lyapunov-Krasovskii derivative condition (13) is satisfied for some sufficiently small $\varepsilon > 0$ if

$$\begin{pmatrix} \Delta & -D^s & -D^a \\ -D^{sT} & S_d + R_{ds} - (E^s + E^{sT}) - W & -E^a \\ -D^{aT} & -E^{aT} & 3S_d \end{pmatrix} > 0 \tag{49}$$

$$\begin{pmatrix} W & R_{da} \\ R_{da}^T & W \end{pmatrix} > 0 \tag{50}$$

are satisfied for all $(A_0, A^1, A^2, ..., A^K) \in \Omega$.

Proof. From (31), it can be verified that

$$\begin{aligned}&\dot{V}(\phi)\\ &= -\int_0^1 \begin{pmatrix} \phi_{DE}^T & \hat{\phi}^T(\alpha) \end{pmatrix} \begin{pmatrix} U & -I \\ -I & S_d \end{pmatrix} \begin{pmatrix} \phi_{DE} \\ \hat{\phi}(\alpha) \end{pmatrix} d\alpha\\ &\quad -\phi_{0r}^T \Delta \phi_{0r} + \phi_{0r}^T [D^s U D^{sT} + \frac{1}{3} D^a U D^{aT}] \phi_{0r}\\ &\quad +2\phi_{0r}^T [D^s U E^{sT} + \frac{1}{3} D^a U E^{aT}] \int_0^1 \hat{\phi}(\beta) d\beta\\ &\quad + \int_0^1 \hat{\phi}^T(\beta) d\beta [E^s U E^{sT} + \frac{1}{3} E^a U E^{aT}] \int_0^1 \hat{\phi}(\beta) d\beta\\ &\quad -(\int_0^1 \hat{\phi}(\alpha) d\alpha)^T R_{ds} (\int_0^1 \hat{\phi}(\alpha) d\alpha)\\ &\quad -\int_0^1 [\int_0^\alpha \phi_{\alpha\beta}^T \begin{pmatrix} 0 & R_{da} \\ R_{da}^T & 0 \end{pmatrix} \phi_{\alpha\beta} d\beta] d\alpha \quad (51)\end{aligned}$$

where

$$\begin{aligned}\phi_{DE}^T &= \phi_{0r}^T [D^s + (1-2\alpha) D^a]\\ &\quad + \int_0^1 \hat{\phi}^T(\beta) d\beta [E^s + (1-2\alpha) E^a]\end{aligned}$$

In view of (48), we can apply Jensen's inequality (Lemma 4 of Gu (2001), see also Shiryayev (1996)) in the first term above, and collecting terms to obtain

$$\begin{aligned}&\dot{V}(\phi)\\ &\leq -\phi_{0r\alpha}^T Y \phi_{0r\alpha}\\ &\quad -\int_0^1 [\int_0^\alpha \phi_{\alpha\beta}^T \begin{pmatrix} W & R_{da} \\ R_{da}^T & W \end{pmatrix} \phi_{\alpha\beta} d\beta] d\alpha\end{aligned}$$

for arbitrary matrix W. Therefore, the Lyapunov-Krasovskii derivative condition is satisfied if there exist a matrix W and a matrix function U such that (48), (50) and

$$Y > 0 \tag{52}$$

are satisfied. Finally, we can use Proposition 3 in Gu (2001) to eliminate the matrix function U from (48) and (52) to obtain (49). ■

The above result is much simpler than the corresponding result in Gu, *et. al.* (2001), and is in general much less conservative. Notice, we have constrained the parameters W^R and Z^R in Gu, *et. al.* (2001) to be idential and denoted as W. We can, of course, leave them independent, in which case there is a higher computational requirement but can be less conservative in nonuniform mesh.

5. STABILITY CRITERION AND EXAMPLES

From the above discussion, we can summarize

Corollary 4. The system described by (1) to (5) is asymptotically stable if there exist n by n matrices P, Q_p, $S_p = S_p^T$, $R_{pq} = R_{qp}^T$, $p, q = 0, 1, ..., N$; and Nn by Nn real matrix $W = W^T$, such that (27), (49) and (50) are satisfied, with notation defined in (18) to (23) and (32) to (46).

Proof. This follows from Proposition 2 and Proposition 3. Condition (17) is already implied by (49). ■

To illustrate the accuracy of discretized Lyapunov functional method for systems with distributed delay, the following examples are presented. The first example illustrates the improvements over Gu, et. al. (2001) regarding the speed of convergence to the analytical limit.

Example 5. Consider the system

$$\dot{x}(t) = -\int_{-r}^0 x(t+\theta) d\theta$$

It can be calculated analytically that the system is stable for $r < r_{\max}^{\text{analytical}} = \frac{\pi\sqrt{2}}{2} \approx 2.22$. The estimate of Kolmanovskii and Richard (1997) is $r < r_{\max}^{\text{kr}} = 1.41$. The calculation using the old method covered in Gu, et. al. (2001), requires $N = 10$ to reach $r_{\max}^{N=10} = 2.19$. Using the new method discussed here, for $N = 1, 2, 3$, the corresponding maximum time delay to retain stability can be calculated as $r_{\max,\text{new}}^{N=1} = 1.93$, $r_{\max,\text{new}}^{N=2} = 2.21$ and $r_{\max,\text{new}}^{N=3} = 2.22$.

The next example considers a system resulting from a model transformation from a system with single pointwise delay. This example not only shows the accuracy of the method, it also illustrates that the system treated here is more general.

Example 6. Consider the system

$$\dot{x}(t) = A_0 x(t) + \int_{-r/2}^{0} A^1 x(t+\xi)d\xi + \int_{-r}^{-r/2} A^2 x(t+\xi)d\xi \tag{53}$$

where

$$A_0 = \begin{pmatrix} -1.5 & 0 \\ 0.5 & -1 \end{pmatrix} \tag{54}$$

$$A^1 = \begin{pmatrix} 2 & 2.5 \\ 0 & -0.5 \end{pmatrix} \tag{55}$$

$$A^2 = \begin{pmatrix} -1 & 0 \\ 0 & -1 \end{pmatrix} \tag{56}$$

This system is the result of model transformation from the system with single delay

$$\dot{x}(t) = \begin{pmatrix} -3 & -2.5 \\ 1 & 0.5 \end{pmatrix} x(t) + \begin{pmatrix} 1.5 & 2.5 \\ -0.5 & -1.5 \end{pmatrix} x(t-r/2). \tag{57}$$

As was discussed in Gu and Niculescu (2000), although system (57) is asymptotically stable for $r < 4.8368$, the transformed system (53) is asymptotically stable only for $r < 2$ due to the presence of additional dynamics (here the delay is scaled to fit the convention in this this article). Such additional dynamics has also been discussed in Kharitonov and Melchor-Aguilar (2000). Using Corollary 4, the stability limit $r_{\max}$ such that system (53) is asymptotically stable for $r \leq r_{\max}$ are estimated. Let N_{d1} and N_{d2} be the number of divisions of the intervals $[-r/2, 0]$ and $[-r, -r/2]$, respectively. For $N_{d1} = N_{d2} = 1$ $(N = 2)$ we obtain $r_{\max} = 1.9725$ with only about 1.4% conservatism. The estimate is improved to $r_{\max} = 1.9999$ for $N_{d1} = N_{d2} = 2$ $(N = 4)$.

6. CONCLUSION

The discretized Lyapunov functional for systems with distributed delays are improved through variable elimination and Jensen's inequality. The resulting criterion converges to analytical solution faster, and is applicable to a wider class of systems.

REFERENCES

[1] Boyd, S., El Ghaoui, L., Feron, E. and Balakrishnan, V., 1994, *Linear Matrix Inequalities in System and Control Theory* (Philadelphia: SIAM).

[2] Gahinet, P., Nemirovski, A., Laub, A. J. and Chilali, M., 1995, *LMI Control Toolbox for use with MATLAB* (Natick, MA: MathWorks).

[3] Gu, K., Han, Q.-L., Luo, A. C. J. and Niculescu, S.-I., 2001, Discretized Lyapunov functional for systems with distributed delay and piecewise constant coefficients. *International Journal of Control*, **74**(7), 737-744.

[4] Gu, K., 2001, A further refinement of discretized Lyapunov functional method for the stability of time-delay systems. *International Journal of Control*, **74**, 967-976.

[5] Gu, K. and Niculescu, S.-I., 2000, Additional dynamics in transformed time-delay systems, *IEEE Transactions on Automatic Control*, **45**, 572-575.

[6] Hale, J. and Verduyn Lunel, S. M., 1993, *Introduction to Functional Differential Equations* (New York: Springer-Verlag).

[7] Kharitonov, V., 1998, Robust stability analysis of time delay systems: A survey. *Proc. IFAC Syst. Struct. Contr.*, Nantes, France.

[8] Kharitonov, V. L. and Melchor-Aguilar, D. A., 2000, On delay-dependent stability conditions. *System & Control Letters*, **40**, 71-76.

[9] Kolmanovskii, V. B., Niculescu, S.-I. and Gu, K., 1999, Delay effects on stability: a survey. *Proc. 38th IEEE Conf. Decision and Control*, Phoenix, AZ, Dec. 7-10, 1999, 1993-1998.

[10] Kolmanovskii, V., Niculescu, S.-I. and Richard, J.-P., 1999, On the Liapunov-Krasovskii functionals for stability analysis of linear delay systems, *International Journal of Control*, **72**, 374-382.

[11] Nesterov, Y. and Nemirovskii, A., 1994, *Interior-Point Polynomial Algorithms in Convex Programming* (Philadelphia: SIAM).

[12] Niculescu, S.-I., Verriest, E. I., Dugard, L. and Dion, J. M., 1997, Stability and robust stability of time-delay systems: A guided tour, in *Stability and Control of Time-Delay Systems* (L. Dugard and E. I. Verriest, Eds.), LNCIS, Springer-Verlag, London, 1-71.

[13] Shiryayev, A. N., 1996, *Probability*, Springer Verlag.

www.elsevier.com/locate/ifac

TRACKING AND DISTURBANCE ATTENUATION FOR TIME DELAY SYSTEMS IN A ROBUST SENSE

Roman Prokop, Petr Husták, Zdenka Prokopová

Institute of Information Technologies, Tomas Bata University in Zlin
Nam. TGM 275, 762 72 Zlin, Czech Republic
fax: ++420 67 754 3333
E-mail: prokop@ft.utb.cz, hustak@ft.utb.cz

Abstract: The paper is aimed to the design of linear continuous controllers. Final controllers ensure the internal stability, asymptotic tracking, disturbance rejection and attenuation. The control synthesis is based on general solutions of Diophantine equations in the ring of proper and Hurwitz stable rational functions. The control objectives are obtained through the divisibility conditions in this ring. Robustness of proposed algorithms can be studied through the infinity norm and sensitivity function. Several approximations of time delay terms and two control structures of controlled loop were considered. The methodology brings a scalar parameter for tuning and influencing of controller parameters. *Copyright © 2001 IFAC*

Keywords: Time Delay Systems, Robust Control, Disturbance Attenuation.

1. INTRODUCTION

The dynamics of many technological plants can be adequately approximated by first order transfer functions plus dead time. This transfer function can be stable or unstable one. The situation where linear systems have unstable poles may occur e.g. in a continuous-time stirred exothermic tank reactor, in polymerisation processes or in a class of biochemical processes where the processes must operate at an unstable steady state. Moreover, a time delay is an inherent part of many technological plants.

There are several available design and tuning methods for stable systems without or with time delay, see e.g. (Aström and Hägglund, 1995). The dead time can be treated in various ways. One of them is using the well-known Smith predictor improved by Watanabe and Ito (1981) and Aström, *et al.* (1994). However, Smith predictor type controllers suffer from several drawbacks, e.g. lack of robustness and difficulties in applying for unstable systems. The second way how to control time-delay systems is to extend Ziegler-Nichols rules for such systems. The unstable cases are studied e.g. in De Paor and O'Malley (1984), Venkatashankar and Chidambaram (1994). The third way tries to approximate the time-delay term $e^{-\tau s}$ by Pade or Taylor series and to design a linear but robust controller.

In this contribution, a general robust technique for first order time-delay systems is proposed. The control design is performed in the ring of proper and Hurwitz stable rational functions where the H_∞ norm serves as a tool for perturbation evaluation. Moreover, a scalar parameter $m > 0$ is defined for control and robust tuning.

2. DESCRIPTION OVER RINGS

Let $R_{PS}(s)$ denote a ring of Hurwitz stable and proper rational functions having no poles in the region $\mathbf{Re}\, s > -m,\ m \geq 0$.

Any transfer function G(s) of a (continuous-time) linear system has been traditionally expressed as a ratio of two polynomials in s. For the purposes of this contribution it is necessary to express the transfer functions as a ratio of two elements of $R_{PS}(s)$. It can be easily performed by dividing, both the polynomial denominator and numerator by the same stable polynomial of the order of the original denominator. Moreover, a scalar parameter m>0 seems to be a suitable „tuning parameter" influencing control behaviour as well as robustness of the closed loop system. Then all transfer functions could be described by

$$G(s)=\frac{b(s)}{a(s)}=\frac{\dfrac{b(s)}{(s+m)^n}}{\dfrac{a(s)}{(s+m)^n}}=\frac{B(s)}{A(s)}, \qquad (1)$$

$$n=\max(\deg(a),\deg(b)), \quad m>0$$

Reference and disturbance signals are expressed by

$$w=\frac{1}{s}=\frac{\dfrac{1}{s+m}}{\dfrac{s}{s+m}}=\frac{G_w(s)}{F_w(s)} \qquad (2)$$

$$v=\frac{\omega^2}{s^2+\omega^2}=\frac{\dfrac{\omega^2}{(s+m)^2}}{\dfrac{s^2+\omega^2}{(s+m)^2}}=\frac{G_v(s)}{F_v(s)}.$$

3. APPROXIMATION OF TIME DELAY TERMS

The simplest description for a time delay process can be expressed by a first order

$$G_1(s)=\frac{Ke^{-\tau s}}{s+\alpha} \qquad (3)$$

The value of the parameter $\alpha > 0$ represents stable systems, $\alpha < 0$ represents unstable ones and $\alpha = 0$ is an integrator. For linear control design, it is necessary to approximate the time delay term in (3). It can be done in several methods. The first one and the simplest case is to neglect the delay term $e^{-\tau s}$. Then the time delay is considered as a perturbation of a nominal transfer function. So the first nominal approximation is

$$G_2(s)=\frac{K}{s+\alpha} \qquad (4)$$

Next two approximations are based on the Taylor series approximation of $e^{-\tau s}$ in the numerator or in the denominator. Approximations $e^{-\tau s} \approx (1-\tau s) \approx (1+\tau s)^{-1}$ then give

$$G_3(s)=\frac{K(1-\tau s)}{s+\alpha}=\frac{b_1 s+b_0}{s+\alpha} \qquad (5)$$

$$G_4(s)=\frac{K}{(s+\alpha)}\frac{1}{(1+\tau s)}=\frac{b_0}{s^2+a_1 s+a_0} \qquad (6)$$

The last model can be obtained by the traditional Padé approximation

$$G_5(s)=\frac{K}{s+\alpha}\frac{1-\dfrac{\tau}{2}s}{1+\dfrac{\tau}{2}s}=\frac{b_1 s+b_0}{s^2+a_1 s+a_0} \qquad (7)$$

All approximated transfer function can be written in R_{ps} in the form

$$G(s)=\frac{b_1 s+b_0}{s^2+a_1 s+a_0}=\frac{\dfrac{b_1 s+b_0}{(s+m)^2}}{\dfrac{s^2+a_1 s+a_0}{(s+m)^2}}=\frac{B(s)}{A(s)} \qquad (8)$$

Figs. 1, 2 show step responses and Nyquist plots of the stable plant (3) and its approximated transfer functions (4) - (7).

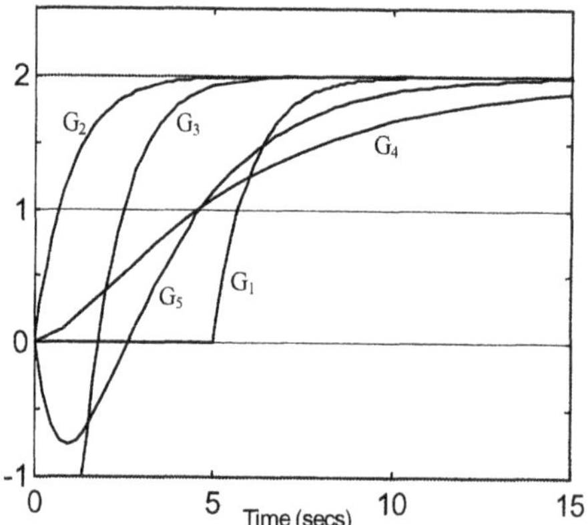

Fig. 1. Step responses (K=2,α=1,τ=5).

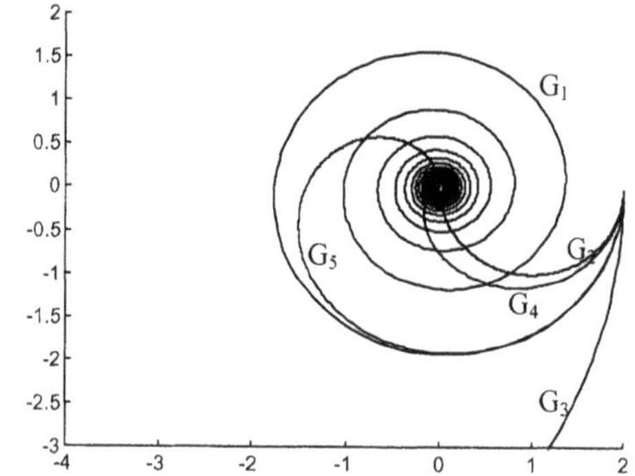

Fig. 2. Nyquist plots (K=2, α=1,τ=5).

4. CONTROL AND DISTURBANCE REJECTION DESIGN IN R_{PS}

Suppose a two degree-of-freedom control system depicted in Fig.3. Note that the traditional one degree-of-freedom system is obtained simply by R=Q. The first step of the control design is to stabilize the system by a proper feedback loop.

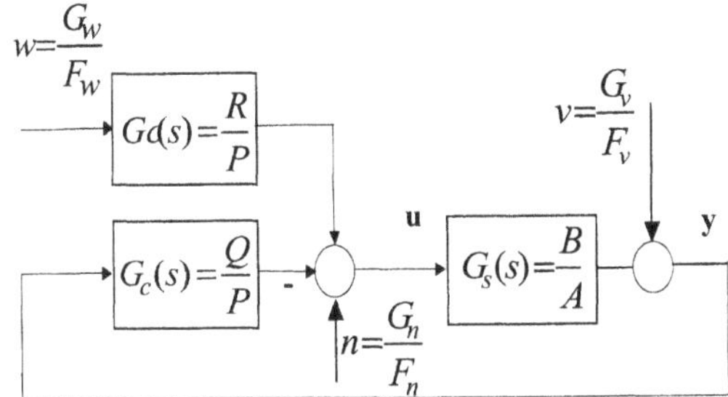

Fig. 3. Feedback feedforward closed loop system (FBFW).

The traditional engineering design approach of PID like controllers was performed either in the frequency domain or in polynomial representation (see e.g. Aström and Hägglund, 1995). However, the fractional approach developed in Vidyasagar (1985), Kučera (1993) enables a deeper insight into control tuning and a more elegant expression of all suitable controllers. The situation and details for time-delay free systems can be found in Prokop and Corriou (1995) or Prokop, *et al.*, 1998).

Suppose a two degrees-of-freedom system FBFW with the control law governed by

$$Pu = Rw - Qy \tag{9}$$

Basic relations following from Fig. 3 are

$$y = \frac{B}{A}u + v \qquad u = \frac{R}{P}w - \frac{Q}{P}y + n \tag{10}$$

and w, v, n are independent external inputs into the closed loop system.

Further, the following equations hold:

$$y = \frac{BR}{AP+BQ}\frac{G_w}{F_w} + \frac{AP}{AP+BQ}\frac{G_v}{F_v} + \frac{BP}{AP+BQ}\frac{G_n}{F_n} \tag{11}$$

For the structure FB (R=Q) the last relation gives the controlled error e=w-y:

$$e = \frac{AP}{AP+BQ}\frac{G_w}{F_w} + \frac{AP}{AP+BQ}\frac{G_v}{F_v} + \frac{BP}{AP+BQ}\frac{G_n}{F_n} \tag{12}$$

The first step of the control design is to stabilize the system by a proper feedback loop. It can be formulated in an elegant way in R_{PS} by the Diophantine equation:

$$AP + BQ = 1 \tag{13}$$

with a general solution for SISO systems $P-P_0+BT$, $Q=Q_0-AT$; where T is free in R_{PS} and P_0, Q_0 is a pair of particular solutions (Youla – Kučera parameterization of all stabilizing controllers). Details and proofs can be found e.g. in Vidyasagar (1985), Kučera (1993), Prokop and Corriou (1997). Then control error for FBFW structure:

$$e = (1 - BR)\frac{G_w}{F_w} + AP\frac{G_v}{F_v} + BP\frac{G_n}{F_n} \tag{14}$$

Now, it is necessary to solve both structures FB and FBFW separately. For asymptotic tracking and the FBFW structure, the second Diophantine equation gets the form:

$$F_w Z + BR = 1 \tag{15}$$

where $Z \in R_{PS}$ is not used in the control law.

The tracking error e tends to zero if

a) F_w divides AP for 1DOF (16)

b) F_w divides 1-BR for 2DOF (17)

Another control problem of practical importance is a disturbance rejection and disturbance attenuation. In both cases, the effect of disturbances v and n should be asymptotically eliminated from the plant output. Since the both disturbances are external inputs into the feedback part of the system, the effect must be processed by a feedback controller. It means that the second and third parts in (11) and (12) are

$$\frac{AP}{AP+BQ}\frac{G_v}{F_v} \tag{18}$$

$$\frac{BP}{AP+BQ}\frac{G_n}{F_n} \tag{19}$$

must belong to $R_{PS}(s)$, i.e. all AP+BQ, F_v, F_n should cancel. In other words, a multiple F_v, F_n must divide P. More precisely F_v, must divide the multiple AP and F_n the multiple BP. When define relatively prime elements A_0, F_{v0} and B_0, F_{n0} in $R_{PS}(s)$

$$\frac{A}{F_v} = \frac{A_0}{F_{v0}}, \qquad \frac{B}{F_n} = \frac{B_0}{F_{n0}} \tag{20}$$

then the problem of disturbance rejection and attenuation is solvable if and only if the pairs F_v, B and F_n, B are relatively prime and the feedback controller is given by

$$C_b = \frac{Q}{P} = \frac{Q}{P_1 F_{v0} F_{n0}} \tag{21}$$

where P_1, Q is any solution of the equation

$$A F_{v0} F_{n0} P_0 + BQ = 1 \tag{22}$$

5. ROBUST ANALYSIS AND TUNING

The fractional approach developed by Vidyasagar (1985) enables a deeper insight into control tuning and robustness. Let R_{PS} be a set of proper and Hurwitz stable rational functions. This set is a ring and the norm H_∞ can be easily defined through the frequency response in the sense

$$\|G\| = \sup_{Re s\geq 0}|G(s)| = \sup_{\omega\in E}|G(j\omega)| \qquad (23)$$

$$\|G_1\ G_2\| = \sup_{Re s\geq 0}\{|G_1(s)|^2 + |G_2(s)|^2\}^{\frac{1}{2}} \qquad (24)$$

Almost all models differ from a physical system. Let G(s) = B(s) / A(s) be a nominal plant and consider a family of perturbed systems G'(s) = B'(s) / A'(s) where

$$\begin{aligned} &\|A - A'\| \leq \varepsilon_1 \quad \|B - B'\| \leq \varepsilon_2 \\ &or \quad \|A - A' \quad B - B'\| \leq \varepsilon \end{aligned} \qquad (25)$$

where $\varepsilon_1, \varepsilon_2, \varepsilon$ are positive constants.
For robust control it is necessary to choose a part of stabilizing controllers P,Q which stabilize also all perturbed plants from the family given by (25). For perturbed plants choose such P,Q in (13) or (22) which fulfil the sufficient conditions

$$\varepsilon_1\|P_0 + BT\| + \varepsilon_2\|Q_0 - AT\| < 1 \qquad (26)$$

or the necessary and sufficient condition in the form

$$\varepsilon\left\|\begin{matrix} P_0 + BT \\ Q_0 - AT\end{matrix}\right\| < 1 \qquad (27)$$

In the case where the perturbations are not known, the notion of the sensitivity function

$$\in = \frac{y}{v} = \frac{e}{w} = A(P_0 + BT) \qquad (28)$$

can be used in the sense as in Doyle, Francis and Tannenbaum (1992). For above mentioned SISO systems sensitivity function $\in$ is a non-linear function of m>0 and the norm of the $\in$ can be minimized by a simple scalar optimization method. In this way, the "most robust" controller of given structure can be obtained.

6. ILLUSTRATIVE EXAMPLES

All approximated transfer function (4)-(7) can be considered as a special case of (8), i.e. the second order system with the relative degree 1. Further, the step-wise reference with $F_w = \dfrac{s}{s+m}$ is assumed and v is a harmonic signal with $F_v = \dfrac{s^2+\omega^2}{(s+m)^2}$.

Then equation (22) takes the form:

$$\begin{aligned} &(s^2 + a_1 s + a_0)(s^2 + \omega^2)(p_1 s + p_0) + \\ &(b_1 s + b_0)(q_3 s^3 + q_2 s^2 + q_1 s + q_0) = (s+m)^5 \end{aligned} \qquad (29)$$

After comparing left and right hand side terms of (29) the solution can be expressed by

$$\begin{aligned} P &= \frac{p_1 s + p_0}{s+m} + \frac{b_1 s + b_0}{(s+m)^2}T; \\ Q &= \frac{q_3 s^3 + q_2 s^2 + q_1 s + q_0}{(s+m)^3} - \frac{s^2 + a_1 s + a_0}{(s+m)^2}\frac{s^2+\omega^2}{(s+m)^2}T; \end{aligned} \qquad (30)$$

where T is an arbitrary element of R_{PS}.
The divisibility condition $F_w \backslash AP$ is achieved for $T = t_0 = -\dfrac{p_0 m}{b_0}$ and the final feedback part is:

$$\frac{Q}{F_v P} = \frac{\tilde{q}_4 s^4 + \tilde{q}_3 s^3 + \tilde{q}_2 s^2 + \tilde{q}_1 s + \tilde{q}_0}{(\tilde{p}_2 s^2 + \tilde{p}_1 s)(s^2+\omega^2)} \qquad (31)$$

where

$$\begin{aligned} &\tilde{p}_1 = p_0 + p_1 m + b_1 t_0; \quad \tilde{p}_2 = 1; \\ &\tilde{q}_4 = q_3 - t_0; \quad \tilde{q}_3 = q_2 + q_3 m - a_1 t_0; \\ &\tilde{q}_2 = q_1 + q_2 m - a_0 t_0 - \omega^2 t_0; \\ &\tilde{q}_1 = q_0 + q_1 m - a_1\omega^2 t_0; \quad \tilde{q}_0 = q_0 m - a_0\omega^2 t_0; \end{aligned} \qquad (32)$$

The feedforward part of the controller is then obtained in the following form

$$\frac{R}{F_v P} = \frac{r_0 (s+m)^4}{(\tilde{p}_2 s^2 + \tilde{p}_1 s)(s^2+\omega^2)} \qquad (33)$$

where $r_0 = \dfrac{m^2}{b_0}$ is the particular solution of (15)

$$s(z_1 s + z_0) + (b_1 s + b_0) r_0 = (s+m)^2 \qquad (34)$$

Example 1: Consider an integrator with time delay:

$$G_I(s) = \frac{e^{-\tau s}}{s} \qquad (35)$$

for τ=1. The aim of the controller is to stabilize the control loop, track a stepwise reference and to compensate a harmonic disturbance of a given frequency and arbitrary amplitude. The control law was constructed according to (31), (33) for both structures. All controller parameters are implicitly nonlinear functions of m>0.
Fig. 4 demonstrates the FBFW control response for Padé approximation (7) and for m = 0.95. Fig. 5

shows the control response for the same plant but in the FB closed loop. The load disturbances n=–0.1 were injected in all simulations in time t=30.

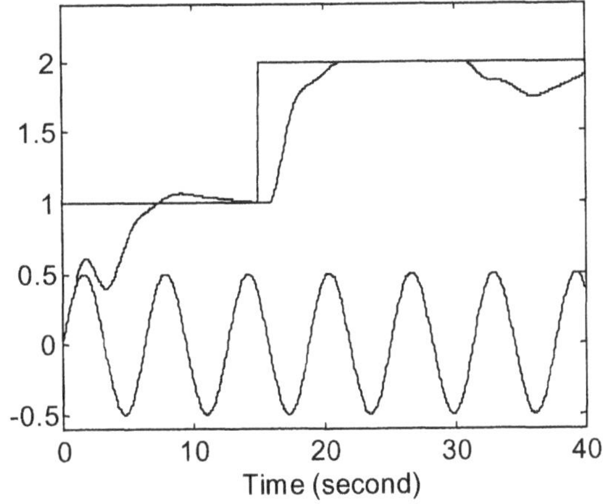

Fig. 4. Control response of the nominal plant $G(s) = \frac{1}{s}e^{-s}$ approximated by Pade approximation in an FBFW control loop for m=0.95, ω=1.

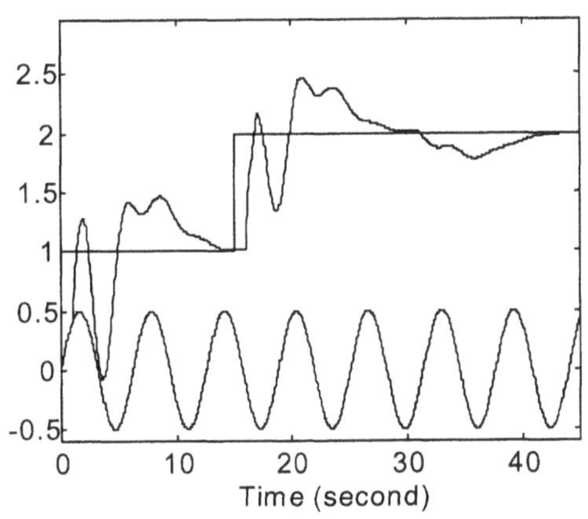

Fig. 5. Regulation response by an FB controller for m=1.01, ω=1.

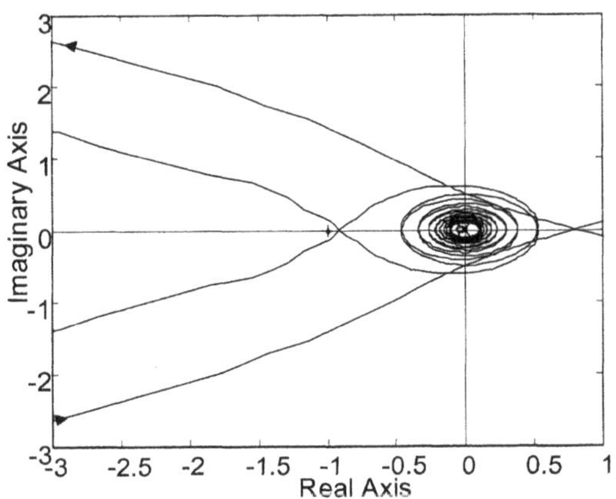

Fig. 6. Open loop Nyquist plots for integrator with time delay (35) and fourth order controller (31) with m= 1.2.

The effect of the tuning scalar parameter m>0 for robustness is illustrated in Fig. 6 and Fig. 7. There are shown the open loop Nyquist plots of integrator (35) and controller (31) for two values of m>0. The first value m=1.2 represents small robustness of the controlled system because of small distance of the critical point (-1; 0) from the Nyquist plot. However, the second value m= 0.8 represents the substantial increase of robustness since the distance from the critical point is fairly greater.

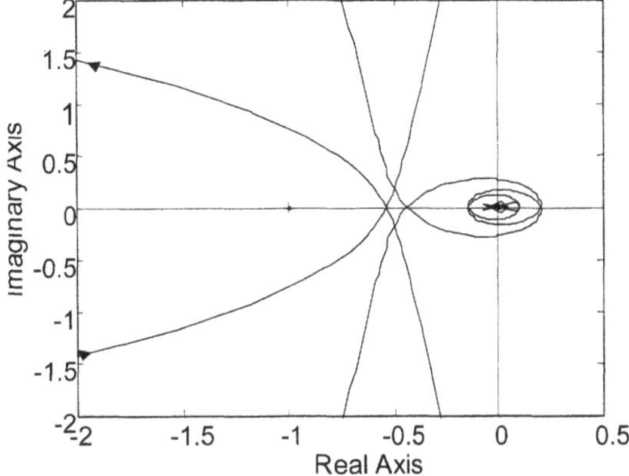

Fig. 7. Open loop Nyquist plots for integrator with time delay (35) and fourth order controller (31) with m= 0.8.

Example 2: Let the nominal stable transfer function be

$$G_0(s) = \frac{e^{-s}}{s^2 + s + 1} \tag{36}$$

The control aim and simulation conditions are the same as in Ex.1. The control law is obtained for the nominal plant (8) with the Taylor approximation in the numerator of (36). The control response depicted in Fig. 8 shows the behavior m=1, ω=1.

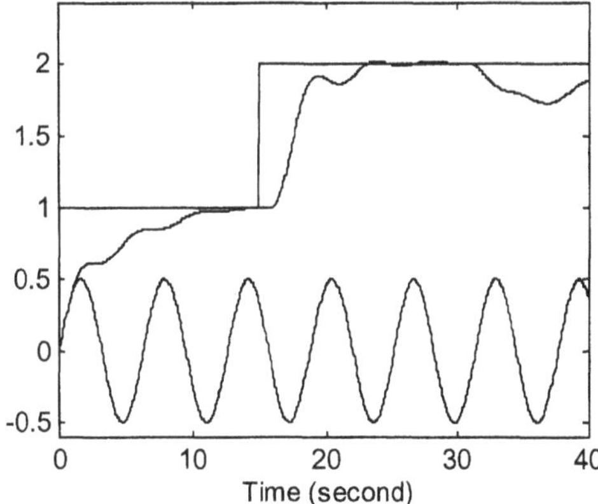

Fig. 8. Control responses of the second order system with Taylor numerator approximation.

Example 3: Consider an unstable transfer function with time delay

$$G_S(s) = \frac{e^{-0.5s}}{s-1} \tag{37}$$

The control responses for FB (m=1.5) and also FBFW (m=1.6) structures with plant (37) are shown in Fig. 9 and Fig. 10, respectively. Pade approximation was used.

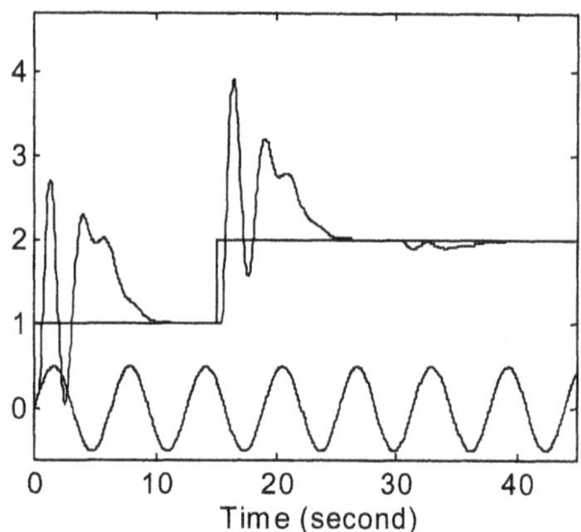

Fig. 9. FB response of unstable plant.

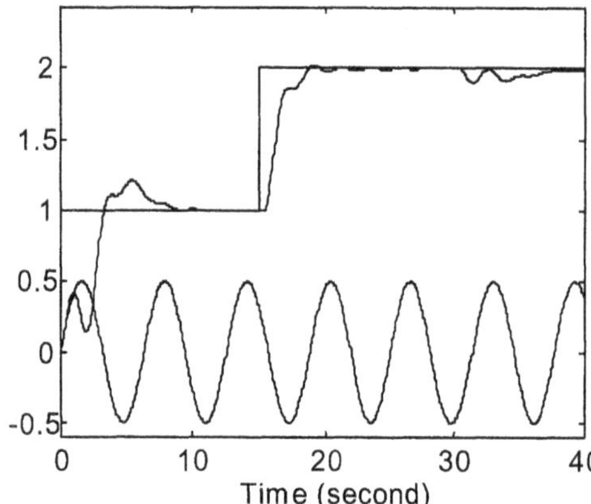

Fig.10. FBFW control response of unstable plant.

7. CONCLUSION

The task of simultaneous regulation and disturbance attenuation for a class of time delay systems is considered. A controller design methodology based on fractional representation was developed for stable, unstable and integrators with time delay. Resulting control laws give a class of generalized control structures. The proposed method enables to tune and influence the robustness and control behaviour by a single scalar parameter m>0. The tuning parameter can be chosen arbitrarily or it is a result of the robust and sensitivity optimization. The proposed methodology is supported by a Matlab + Simulink program system for automatic design and simulation.

ACKNOWLEDGEMENTS

This work was supported by the Europoly project INCO Copernicus No. CP 977 010.

REFERENCES

Aström, K.J., C.C. Hang, P. Persson, and W.K. Ho (1992). Towards intelligent PID control. *Automatica*, **Vol. 28, No.1**, pp. 1-9.

Aström, K.J., C.C. Hang and B.C. Lim (1994). A new Smith predictor for controlling a process with an integrator and long dead-time. *IEEE Trans. on AC*, **Vol. 39, No. 2**, pp. 343-345.

Aström, K.J. and T. Hägglund (1995). *PID Controllers: Theory, Design and Tuning.* Instrument Society of America, USA.

De Paor, A.M. and M.O'Malley (1989). Controllers of Ziegler Nichols type for unstable processes. *Int. J. Control*, **Vol. 49**, pp. 1273-84.

Doyle, C.D., B.A. Francis and A.R. Tannenbaum (1992). *Feedback Control Theory.* Macmillan, New York.

Grimble, M.J. and V. Kučera (1996). *Polynomial Methods for Control Systems Design.* Springer, Berlin.

Hwang, S.H., S.J. Shiu and M.L Lin (1997). PID control of unstable systems having time delay. In: *Prepr. IFAC ECC Bruxelles*, No.559.

Kučera, V. (1993). Diophantine equations in control - A survey, *Automatica*, **Vol. 29, No.6**, pp. 1361-75.

Persson, P. and K.J. Aström (1993). PID control revisited. In: *Prepr. 12th IFAC World Congress*, Sydney, **Vol. 8**, pp. 241-244.

Prokop, R. and A. Mészáros (1996). Design of robust controllers for SISO time delay systems. *Journal of Electrical Engineering*, **Vol. 47, No. 11-12**, pp. 287-294.

Prokop, R. and J.P. Corriou (1997). Design and analysis of simple robust controllers, *Int. J. Control*, **Vol. 66, No. 6**, pp. 905-921.

Prokop, R., V. Bobál and Z. Prokopová (1998). Design of robust controllers for time-delay systems. In: *Prepr. IFAC-IFIP-IMACS Conf. Control of Ind.Systems*, Belfort, France, pp. 359-364.

Venkatashankar, V. and M. Chidambaram (1994). Design of P and PI controllers for unstable firs-order plus time delay systems, *Int. J. Control*, **Vol. 60, No. 1**, pp. 137-144.

Vidyasagar, M. (1985). *Control system synthesis: a factorization approach.* MIT Press, Cambridge, M.A.

www.elsevier.com/locate/ifac

NECESSARY AND SUFFICIENT STABILITY CRITERIA FOR LINEAR SYSTEMS WITH DELAYS

Bugong Xu

Department of Automatic Control Engineering, South China University of Technology, Guangzhou 510641, People's Republic of China

Abstract: New necessary and sufficient stability criteria for linear systems with multiple uncertain delays are established by using the time-domain and the frequency-domain methods, respectively. The results are first established based on a new-type time-domain stability theorem for retarded dynamical systems and a new technique for estimating the derivative of Lyapunov function along the solution of system at some specific instants. Then, the same criteria are derived by the frequency-domain techniques. The established stability criteria are not only the least conservative but also involve the least tuning parameters. *Copyright © 2001 IFAC*

Keywords: stability criteria, delay, linear systems, time-domain method, frequency-domain method.

1. INTRODUCTION

It is well known that the main time-domain stability analysis methods for linear time-delay systems are Lyapunov methods including Lyapunov functional method and Lyapunov function method together with Razumikhin-type techniques (see for example, Kolmanovskii, and Nosov, 1986; Hale, and Lunel, 1993). For linear systems with time-varying delays, Lyapunov functional method needs generally the derivative conditions on delays, see for example, Trinh and Aldeen (1995), Jeung *et al.* (1996), Kim (1996), and Xu (1997b). But the derivative conditions on delays may introduce some conservative factors into the results, see the remarks given by Xu (1999, 2000a, 2001). On the other hand, Lyapunov function method together with Razumikhin-type techniques can be used to deal easily with the case of time-varying delays without the derivative conditions on delays. Unfortunately, the existing Razumikhin-type techniques in the literature also provides generally conservative results, see some results and remarks given in Xu and Liu (1994) and Xu (1994, 1995, 1997a, 1999, 2000a, 2001). Generally speaking, both Lyapunov functional method and Lyapunov function method provide sufficient stability conditions that include some free tuning scalar and/or matrix parameters. Therefore, the numerical schemes for determining these parameters so that less conservative or optimal results can be obtained have attracted many researchers, see Chen and Latchman (1994), Li and de Souza (1995), Trinh and Aldeen (1995), Niculescu *et al.* (1995), Luo and Bosch (1997), Kolmanovskii *et al.* (1999), Xu (1997b, 1999, 2000a, 2001), Xu and Lam (1999), Kolmanovskii and Ricard (1999), Gu (1999) and also see the recent results in Mahmoud (2000) and the book edited by Dugard and Verriest (1998). However, a more meaningful work for stability analysis of linear time-delay systems is to establish as least conservative stability conditions as possible and at the same time to reduce the number of tuning parameters as least as possible.

Consider the following linear system with multiple uncertain time-varying delays:

$$\left.\begin{aligned} &\dot{x}(t) = A_0 x(t) + \sum_{k=1}^{m} A_k x(t-\tau_k(t)) \\ &x_{t_0}(\theta) = x(t+\theta) = \phi(\theta) \\ &t_0+\theta \in \bigcup_{k=1}^{m}\left\{ t-\tau_k(t) \,\middle|\, \begin{matrix} t-\tau_k(t) \le t_0, \\ t \ge t_0 \end{matrix} \right\} \bigcup \{t_0\} \end{aligned}\right\} \quad (1)$$

and its special case:

$$\dot{x}(t) = A_0 x(t) + \sum_{k=1}^{m} A_k x(t - \tau_k), \quad t \geq 0 \qquad (2)$$

with multiple unknown constant delays $\tau_k < \infty$ for $k = 1, 2, \ldots, m$, where $\phi \in C_n$, $x \in R^n$, $A_k \in R^{n \times n}$ for $k = 0, 1, \ldots, m$ are constant matrices, $\tau_k(t) \leq \tau_{kM} < \infty$ and $\tau_k \leq \tau_{kM} < \infty$ for $k = 1, 2, \ldots, m$ with known constant $\tau_{kM} > 0$ for all $k = 1, 2, \ldots, m$ or $\tau_k(t) \leq \tau < \infty$ for $k = 1, 2, \ldots, m$ with unknown constant $\tau > 0$ are the *uncertain* delays. In this paper, new necessary and sufficient delay-dependent and delay-independent uniformly asymptotically (or uniformly) stability conditions for linear systems with multiple *uncertain* delays (1) and (2) are first established based on the new-type stability theorem for retarded dynamical systems and the new technique for estimating the derivative of Lyapunov function along the solution of system at some specific instants (Xu, 1999, 2000a, 2001). Then, the same criteria are derived by the frequency-domain techniques.

The notations used in this paper are as follows. R^n is the real vector space of dimension n; $R^{n \times n}$ is the real matrix space of dimension $n \times n$; R_+ denotes the set of nonnegative real numbers; C^n is the complex vector space of dimension n; C_n denotes the Banach space of continuous functions mapping $[-\tau, 0]$ into R^n, where $\tau > 0$; $y_t(\theta) \in R^n$ denotes $y(t+\theta) \in R^n$ for $t \in R$ and $\theta \in R$ so that $y(t) = y_t(0)$; $\|\cdot\|$ denotes the Euclidean norm in R^n; $\|\phi\|_\tau = \sup_{-\tau \leq \theta \leq 0} \|\phi(\theta)\|$ with $\phi(\theta) \in R^n$ for given $\phi \in C_n$; A^T is the transpose of A; A^* is the conjugate transpose of A, i.e. $\overline{A}^T$; $\lambda(\cdot)$ denotes any eigenvalue of a matrix; $\lambda_{\max}(A)$ and $\lambda_{\min}(A)$ are the maximum and minimum eigenvalues of $A^T = A \in R^{n \times n}$, respectively; $A > 0$ (or < 0) denotes a positive definite (or negative definite) matrix; $A \leq B$ means that $A - B \leq 0$ is negative semi-definite; $\mathrm{Re}\, s$ is the real part of $s \in C$; $|z|$ is the modulus of $z \in C$ or the absolute value of $z \in R$; I_n is the $n \times n$ identity matrix; and finally, $j = \sqrt{-1}$.

2. PRELIMINARIES

Since all the stability properties discussed in this paper are global ones in the solution space of system (1) or (2), the following stability definitions (Xu, 2001, 2000b) are used.

Definition 1. System (1) or (2) is said to be exponentially (uniformly asymptotically) stable if the equilibrium $x^* \equiv 0$ of system (1) or (2) is globally exponentially (uniformly asymptotically) stable and said to be uniformly stable if the equilibrium $x^* \equiv 0$ of system (1) or (2) is globally uniformly stable.

Definition 2. Let $\alpha > 0$ be constant, $\Pi: R^n \to R_+$ be continuous, $\|x\| \leq \Pi(x)$ for $x \in R^n$, and $\overline{\Pi}_t = \sup_{-\tau \leq \theta \leq 0} \{\Pi(\phi(\theta))\}$ for given $\phi \in C_n$. System (1) is exponentially stable with respect to the constant decay degree $\alpha > 0$ if along the solution $x(t_0, \phi)(t)$ of system (1) through any $(t_0, \phi) \in R \times C_n$, one has

$$\|x(t_0, \phi)(t)\| \leq \overline{\Pi}_{t_0} \exp\{-\alpha(t - t_0)\}, \; \forall t \geq t_0. \quad (3)$$

Definition 3. System (2) is said to be α-stable locally in the delays if there are positive $\tau_{kM} > 0$ for $k = 1, 2, \ldots, m$ such that all the roots of the characteristic equation:

$$\det\left[sI_n - \left(A_0 + \sum_{k=1}^{m} e^{-\tau_k s} A_k \right) \right] = 0 \qquad (4)$$

lie in the region of $\mathrm{Re}\, s \leq -\alpha < 0$ in the complex plane for arbitrary values of $\tau_k \in [0, \tau_{kM}]$, $k = 1, 2, \ldots, m$.

Definition 4. System (2) is said to be ε-stable independently of the delays if for any positive $\tau_{kM} > 0$ for $k = 1, 2, \ldots, m$, there is a sufficiently small positive number $\varepsilon = \varepsilon(\tau_{kM}) > 0$ such that all the roots of the characteristic equation (4) lie in the region of $\mathrm{Re}\, s < -\varepsilon < 0$ in the complex plane for arbitrary values of $\tau_k \in [0, \tau_{kM})$, $k = 1, 2, \ldots, m$, where $\varepsilon \to 0$ as $\tau_{kM} \to \infty$ for any $k \in \{1, 2, \ldots, m\}$.

Lemma 1. Let $P > 0 \in R^{n \times n}$ and $D \in R^{n \times n}$ be constant matrices. Then, for any $x, y \in C^n$ and any $\eta = \rho e^{-j(\theta/2)} \in C$ with $\rho > 0$ and $\theta \in R$, one has

$$x^* PDy e^{-j\theta} + e^{j\theta} y^* D^T Px \leq \frac{1}{\rho} x^* PDP^{-1} D^T Px + \rho y^* Py \qquad (5)$$

and the equality holds if and only if

$e^{j(\theta/2)} D^T Px = \eta Py$.

Proof: It is easy to see that for any $\eta = \rho e^{-j(\theta/2)} \in C$ and any $x, y \in C^n$, $(e^{j(\theta/2)} D^T Px - \eta Py)^* P^{-1} (e^{j(\theta/2)} D^T Px - \eta Py) \geq 0$ and the equality holds if and only if $e^{j(\theta/2)} D^T Px = \eta Py$. Q.E.D.

Lemma 2: Let $P > 0 \in R^{n \times n}$ and $D \in R^{n \times n}$ be constant matrices, $x, y \in C^n$, and $K > 0$. Then, there exists a positive $\rho_M > 0$ such that

$$\max_{\substack{x^* Px = K \\ y^* Py = K}} \{2x^* PDy\} = \max_{x^* Px = K} \left\{ \begin{array}{c} \frac{1}{\rho_M} x^* PDP^{-1} D^T Px \\ + \rho_M x^* Px \end{array} \right\} = 2\rho_M K \tag{6}$$

and

$$\max_{x^* Px = K} \left\{ \frac{1}{\rho_M} x^* PDP^{-1} D^T Px + \rho_M x^* Px \right\} \leq \max_{x^* Px = K} \left\{ \frac{1}{\rho} x^* PDP^{-1} D^T Px + \rho x^* Px \right\} \tag{7}$$

for any $\rho > 0$.

Proof: Since $K > 0$, there must exist a positive $\rho_M > 0$ such that $\rho_M^2 K = \max_{x^* Px = K} \{x^* PDP^{-1} D^T Px\}$. Let $X(K) = \{x \in C^n \mid x^* Px = K\}$. For any given $\tilde{x} \in X(K)$, it is not difficult to see that there exists a positive $\tilde{\rho} > 0$ and a vector $z \in X(K)$ satisfying $D^T P\tilde{x} = \tilde{\rho} e^{j\theta} Pz$ with a $\theta \in R$ such that

$$\tilde{\rho}^2 K = \tilde{x}^* PDP^{-1} D^T P\tilde{x} = \tilde{\rho}^2 z^* Pz = \tilde{\rho}^2 x^* Px \tag{8}$$

for any $x \in X(K)$. Note that $\rho_M P^{1/2} z$ depends linearly on $P^{1/2} y$ when $y = z \in X(K)$. This together with (8) implies that

$$\rho_M K = \max_{\substack{x^* Px = K \\ y^* Py = K}} \left\{ \left(y^* Pyx^* PDP^{-1} D^T Px \right)^{1/2} \right\} = \max_{\substack{x^* Px = K \\ y^* Py = K}} \{x^* PDy\} \tag{9}$$

From (8) and (9), it is easy to obtain (6), i.e.

$$\max_{\substack{x^* Px = K \\ y^* Py = K}} \left\{ \frac{1}{\rho_M} x^* PDP^{-1} D^T Px + \rho_M y^* Py \right\} = \max_{x^* Px = K} \left\{ \frac{1}{\rho_M} x^* PDP^{-1} D^T Px + \rho_M x^* Px \right\} = 2\rho_M K = \max_{\substack{x^* Px = K \\ y^* Py = K}} \{2x^* PDy\} \tag{10}$$

By (6) and Lemma 1, (7) is obvious. Q.E.D.

Remark 1: Obviously, all the results established in Lemma 1 and Lemma 2 still hold for the case of $x, y \in R^n$ and $\eta = \rho > 0 \in R_+$.

3. TIME-DOMAIN METHOD

In this section, the necessary and sufficient stability criteria for system (1) are derived by using the time-domain method (Xu, 2001).

Theorem 1. Let $\tau_{kM} > 0$ for $k = 1, 2, \ldots, m$ and $\gamma > 0$. System (1) is exponentially stable with respect to the constant decay degree $\alpha = \gamma/2 > 0$ if and only if there is a positive definite matrix $P > 0 \in R^{n \times n}$ and m positive scalars $\rho_k > 0$ for $k = 1, 2, \ldots, m$ such that

$$T_0 + \sum_{k=1}^{m} e^{\frac{1}{2}\gamma\tau_{kM}} \left(\frac{PA_k P^{-1} A_k^T P}{\rho_k} + \rho_k P \right) \leq -\gamma P \tag{11}$$

where $T_0 = PA_0 + A_0^T P$.

Proof. Let $V(x) = x^T Px$, where $P > 0 \in R^{n \times n}$. Then, by Lemma 1, Lemma 2, Remark 1, and Theorem A.1 in Appendix, one obtains

$$\begin{aligned}
\dot{V}(y_t(0)) &= y_t^T(0) T_0 y_t(0) + \sum_{k=1}^{m} 2e^{\frac{1}{2}\gamma(\tau_k(t) + \theta x_k)} y_t^T(0) PA_k y_t(0_k) \\
&\leq \max_{\substack{V(y_t(0)) \\ = V(y_t(0_k)) \\ = V(x_t(0))}} \left\{ \begin{array}{l} y_t^T(0) T_0 y_t(0) \\ + \sum_{k=1}^{m} 2e^{\frac{1}{2}\gamma\tau_{kM}} \left| y_t^T(0) PA_k y_t(0_k) \right| \end{array} \right\} \\
&\leq \max_{\substack{V(y_t(0)) \\ = V(x_t(0))}} \left\{ \begin{array}{l} y_t^T(0) T_0 y_t(0) + \sum_{k=1}^{m} e^{\frac{1}{2}\gamma\tau_{kM}} \rho_k y_t^T(0) P y_t(0) \\ + \sum_{k=1}^{m} e^{\frac{1}{2}\gamma\tau_{kM}} y_t^T(0) \left(\frac{PA_k P^{-1} A_k^T P}{\rho_k} \right) y_t(0) \end{array} \right\} \\
&= -\gamma V(y_t(0)), \quad \forall y_t \in S_{a.sta.}(L_t(\theta))
\end{aligned} \tag{12}$$

whenever (i.e. if and only if) $V(x_t(0)) = \overline{V}_{t0} \exp\{-\gamma(t-t_0)\}$ on $t \geq t_0 \in R$, where $S_{a.sta.}(L_t(\theta))$ is defined as (24) in Appendix, $y_t(-\tau_k(t)) = y_t(0_k)e^{\frac{1}{2}\gamma(\tau_k(t)+\theta_{Xk})}$ and $V(y_t(0_k)) = y_t^T(0_k)Py_t(0_k) = V(x_t(0))$ with $\theta_{Xk} \in (-\infty, 0]$ for all $k = 1, 2, \ldots, m$. By Theorem A.1 in Appendix and Definition 2 with $\lambda_{\min}(P)x^T x \leq V(x) \leq \lambda_{\max}(P)x^T x$ and $\Pi(x) = V(x)$, the proof is completed. Q.E.D.

Theorem 2. System (1) is uniformly asymptotically stable if and only if there is a positive definite matrix $P > 0 \in R^{n\times n}$ and m positive scalars $\rho_k > 0$ for $k = 1, 2, \ldots, m$ such that

$$PA_0 + A_0^T P + \sum_{k=1}^{m}\left(\frac{PA_k P^{-1} A_k^T P}{\rho_k} + \rho_k P\right) < 0 \quad (13)$$

and uniformly stable if and only if there is a positive definite matrix $P > 0 \in R^{n\times n}$ and m positive scalars $\rho_k > 0$ for $k = 1, 2, \ldots, m$ such that

$$PA_0 + A_0^T P + \sum_{k=1}^{m}\left(\frac{PA_k P^{-1} A_k^T P}{\rho_k} + \rho_k P\right) \leq 0 \,. \quad (14)$$

Proof. According to the standard definitions of uniform stability (Hale, and Lunel, 1993, Xu, 2001) and Definition 2, it is not difficult to see that the uniform asymptotical stability (or uniform stability) for the zero solution of system (1) is equivalent to the corresponding exponential stability with respect to a *unknown* positive constant decay degree (or zero decay degree). Therefore, this theorem can be proved similarly to Theorem 1 based on Theorem A.1 in Appendix so that the proof is omitted here. Q.E.D.

Remark 2. According to Lemma 1, Lemma 2, Remark 1, Definition 1, Definition 2, and the proof of Theorem 1, it is easy to see that the established stability criteria above are not only the least conservative but also involve the least tuning parameters.

4. FREQUENCY-DOMAIN METHOD

It should be pointed out that the results established in Section 3 are also suitable to system (2). However, in order to compare the time-domain method with the frequency-domain method, let us derive the corresponding stability criteria by using the frequency-domain method in this section.

Theorem 3. Let $\alpha > 0$ and $\tau_{kM} > 0$ for $k = 1, 2, \ldots, m$. System (2) is α-stable locally in the delays if and only if there is a positive definite matrix $P > 0 \in R^{n\times n}$ and m positive scalars $\rho_k > 0$ for $k = 1, 2, \ldots, m$ such that

$$T_0 + \sum_{k=1}^{m} e^{\alpha\tau_{kM}}\left(\frac{PA_k P^{-1} A_k^T P}{\rho_k} + \rho_k P\right) \leq -2\alpha P \quad (15)$$

where $T_0 = PA_0 + A_0^T P$.

Proof. Let $\tau_k \in [0, \tau_{kM}]$ for $k = 1, 2, \ldots, m$, $s = \sigma + j\omega \in C$ with $\sigma \in R$ and $\omega \in R$, $W(\tau_k, s) = A_0 + \sum_{k=1}^{m} e^{-\tau_k s} A_k = W(\tau_k, \sigma, \omega)$ and $\tilde{y} := \tilde{y}(\tau_k, \tilde{\sigma}, \tilde{\omega}) \in C^n$ satisfy $W(\tau_k, \tilde{\sigma}, \tilde{\omega})\tilde{y} = \lambda(W(\tau_k, \tilde{\sigma}, \tilde{\omega}))\tilde{y}$ at $s = \tilde{s} = \tilde{\sigma} + j\tilde{\omega} \in C$, where

$$\begin{aligned}\tilde{\sigma} &= \operatorname{Re}\lambda(W(\tau_k, \tilde{\sigma}, \tilde{\omega})) \\ &= \frac{\tilde{y}^*\left(\tilde{P}W(\tau_k, \tilde{\sigma}, \tilde{\omega}) + W^*(\tau_k, \tilde{\sigma}, \tilde{\omega})\tilde{P}\right)\tilde{y}}{2\tilde{y}^*\tilde{P}\tilde{y}} \\ &= \frac{\left[\begin{array}{l}\tilde{y}^*(\tilde{P}A_0 + A_0^T\tilde{P})\tilde{y} \\ +\sum_{k=1}^{m} e^{-\tau_k\tilde{\sigma}}\tilde{y}^*(\tilde{P}A_k e^{-j\tau_k\tilde{\omega}} + A_k^T\tilde{P}e^{j\tau_k\tilde{\omega}})\tilde{y}\end{array}\right]}{2\tilde{y}^*\tilde{P}\tilde{y}}\end{aligned} \quad (16)$$

for any $\tilde{P} > 0 \in R^{n\times n}$. Let us prove the sufficiency by contradiction. Assume that condition (15) holds. If there is a root $\tilde{s} = \tilde{\sigma} + j\tilde{\omega} \in C$ of the corresponding characteristic equation (4) for $\tau_k \in [0, \tau_{kM}]$, $k = 1, 2, \ldots, m$, such that $\tilde{\sigma} > -\alpha$, then by Lemma 1, Lemma 2, and equation (16), one has

$$\begin{aligned}-\alpha &< \tilde{\sigma} \\ &\leq \frac{\left[\begin{array}{l}\tilde{y}^*(PA_0 + A_0^T P)\tilde{y} \\ +\sum_{k=1}^{m} e^{-\tilde{\sigma}\tau_{kM}}\tilde{y}^*\left(\frac{PA_k P^{-1}A_k^T P}{\rho_k} + \rho_k P\right)\tilde{y}\end{array}\right]}{2\tilde{y}^* P\tilde{y}} \\ &< \frac{\left[\begin{array}{l}\tilde{y}^*(PA_0 + A_0^T P)y \\ +\sum_{k=1}^{m} e^{\alpha\tau_{kM}}\tilde{y}^*\left(\frac{PA_k P^{-1}A_k^T P}{\rho_k} + \rho_k P\right)\tilde{y}\end{array}\right]}{2\tilde{y}^* P\tilde{y}} \\ &\leq -\alpha\end{aligned} \quad (17)$$

which is a contradiction. The proof for the

sufficiency is completed. For the necessity, since $\tilde{\sigma} \le -\alpha$ for all $\tau_k \in [0, \tau_{kM}]$, $k = 1, 2, \ldots, m$, by Lemma 2 and equation (16), it is easy to see that there must exist a $P > 0 \in R^{n\times n}$ and m positive scalars $\rho_k > 0$ for $k = 1, 2, \ldots, m$ such that

$$\max_{\tau_k \in [0, \tau_{kM}]} \{\tilde{\sigma}\} = \max_{v^* P v = \frac{1}{2}} \left\{ \begin{array}{l} v^*(PA_0 + A_0^T P)v \\ + \sum_{k=1}^{m} e^{\alpha\tau_{kM}} v^* \left(\dfrac{PA_k P^{-1} A_k^T P}{\rho_k} + \rho_k P \right) v \end{array} \right\} = -\alpha. \tag{18}$$

Otherwise, by Lemma 2, equation (16) and the last equality in (18), $-\alpha > \max_{\tau_k \in [0, \tau_{kM}]} \{\tilde{\sigma}\}$ which is a contradiction. Q.E.D.

Theorem 4. System (2) is ε-stable independently of the delays if and only if there is a constant matrix $P > 0 \in R^{n\times n}$ and m positive scalars $\rho_k > 0$ for $k = 1, 2, \ldots, m$ such that (13) holds.

Proof. Since the eigenvalues of a matrix depend continuously on its elements, condition (13) implies that for any positive $\tau_{kM} > 0$, $k = 1, 2, \ldots, m$, there is a sufficiently small number $\varepsilon > 0$ such that

$$T_0 + \sum_{k=1}^{m} e^{\varepsilon\tau_{kM}} \left(\frac{PA_k P^{-1} A_k^T P}{\rho_k} + \rho_k P \right) \le -2\varepsilon P \tag{19}$$

where $T_0 = PA_0 + A_0^T P$. Note that $\varepsilon \to 0$ as $\tau_{kM} \to \infty$ for any $k \in \{1, 2, \ldots, m\}$. By Lemma 2 and Theorem 3, it is easy to see that all the roots of equation (4) lie in the region of $\mathrm{Re}\, s < -\varepsilon < 0$ in the complex plane for arbitrary values of $\tau_k \in [0, \tau_{kM})$ for $k = 1, 2, \ldots, m$ if and only if inequality (19) (i.e. (13)) holds. According to Definition 4, the proof is completed. Q.E.D.

5. CONCLUSION

The new necessary and sufficient stability criteria for linear systems with multiple uncertain time-varying/constant delays have been established by using both time-domain and frequency-domain methods. The results are derived based on the new stability theorems and the new technique established and used in Xu (2001, 2000). It has been remarked that the established stability criteria for the systems under consideration are not only the least conservative but also involve the least tuning parameters.

ACKNOWLEGEMENTS

This work was supported by NSFC Project 60074026 and Guangdong Province Natural Science Foundation of China Project 000409.

APPENDIX: A NEW-TYPE STABILITY THEOREM FOR RETARDED DYNAMICAL SYSTEMS

Consider a retarded dynamical system

$$\dot{x}(t) = f(t, x_t) \tag{20}$$

where "•" denotes the right-hand derivative, $f : R \times C_n \to R^n$, and $f(t, \phi)$ is continuous and Lipschitzian in ϕ so that for an initial function $x_{t0} = \phi \in C_n$ at $t = t_0 \in R$, system (20) has a unique solution $x(t_0, \phi)(t)$ on $[t_0 - \tau, \infty)$. Suppose that $f(t, 0) \equiv 0$ for all $t \in R$ so that $x^* \equiv 0$ is the equilibrium of system (20). For simplicity, we denote the value of the solution $x(t_0, \phi)(t) \in R^n$ by $x(t)$ and the solution segment $x(t + \theta) = x_t(\theta) \in R^n$ for all $\theta \in [-\tau, 0]$ by $x_t \in C_n$ at a given $t \ge t_0$.

Theorem A.1. Let $V(x) = x^T P x$, where $x \in R^n$ and $P > 0 \in R^{n\times n}$. Then, (i) The equilibrium $x^* \equiv 0$ of system (20) is globally uniformly stable if along the solution $x(t_0, \phi)(t)$ of system (20) through any $(t_0, \phi) \in R \times C_n$,

$$\dot{V}(y_t(0)) \le 0, \quad \forall y_t \in S_{sta.}(\bar{V}_{t0}) \tag{21}$$

whenever $V(x_t(0)) = \bar{V}_{t0}$ on $t \ge t_0$, where $S_{sta.}(\bar{V}_{t0})$ is defined as

$$S_{sta.}(\bar{V}_{t0}) = \left\{ y_t \in C_n \;\middle|\; \begin{array}{l} V(y_t(0)) = \bar{V}_{t0} \\ \left\| P^{1/2} y_t(\theta) \right\|^2 = \left\| P^{1/2} y_t(0) e^{\frac{1}{2}\theta_X} \right\|^2 \\ \theta \in [-\tau, 0], \quad \theta_X \in (-\infty, 0] \end{array} \right\}; \tag{22}$$

(ii) The equilibrium $x^* \equiv 0$ of system (20) is globally exponentially (uniformly asymptotically) stable with respect to the constant decay degree $\gamma/2 > 0$ if along the solution $x(t_0, \phi)(t)$ of system (20) through any $(t_0, \phi) \in R \times C_n$,

$$\dot{V}(y_t(0)) \le -\gamma V(y_t(0)), \quad \forall y_t \in S_{a.sta.}(L_t(\theta)) \quad (23)$$

whenever $V(x_t(0)) = \overline{V}_{t0} \exp\{-\gamma(t-t_0)\}$ on $t \ge t_0$, where $\gamma > 0$ and $S_{a.sta.}(L_t(\theta))$ is defined as

$$S_{a.sta.}(L_t(\theta)) = \left\{ \begin{array}{l} y_t \in C_n \\ \left| \begin{array}{l} L_t(\theta) = L(t+\theta) = \overline{V}_{t0} e^{-\gamma(t+\theta-t_0)} \\ V(y_t(0)) = L_t(0) \\ \left\| P^{1/2} y_t(\theta) \right\|^2 = \left\| P^{1/2} y_t(0) e^{-\frac{1}{2}\gamma(\theta-\theta_X)} \right\|^2 \\ \theta \in [-\tau, 0], \quad \theta_X \in (-\infty, 0] \end{array} \right. \end{array} \right\}. \quad (24)$$

Proof. For the proof, see Xu (2001).

REFERENCES

Chen, J. and H.A. Latchamn (1994). Asymptotic stability independent of delays: simple necessary and sufficient conditions. In *Proceedings of 1994 American Control Conference,* Baltimore, Maryland, **1**, 1027-1031.

Dugard, L. and E.I. Verriest. (Eds). (1998). *Stability and Control of Time-delay Systems.* Springer, London.

Gu, K. (1999). Discretized Lyapunov functional for uncertain systems with multiple time-delay. *International Journal of Control,* **72**, 1436-1445.

Hale, J.K. and S.M.V. Lunel (1993). *Introduction to Functional Differential Equations.* Springer-Verlag, New York.

Jeung, E.T., D.C. Oh, J.H. Kim and H.B. Park (1996). Robust controller design for uncertain systems with time delays: LMI approach. *Automatica,* **32,** 1229-1231.

Kim, J.H. (1996). Robust stability of linear systems with delayed perturbations. *IEEE Transactions on Automatic Control,* **41**, 1820-1822.

Kolmanovskii, V.B. and V.R., Nosov (1986). *Stability of Functional Differential Equations.* Academic Press, London.

Kolmanovskii, V.B., S.I. Niculescu and J.P. Richard (1999). On the Liapunov-Krasovskii functional for stability analysis of linear delay systems. *International Journal of Control,* **72**, 374-384.

Kolmanovskii, V.B. and J.P. Richard (1999). Stability of some linear systems with delays. *IEEE Transactions on Automatic Control,* **44**, 984-989.

Li, X. and C.E. de Souza (1995). LMI approach to delay-dependent robust stability and stabization of uncertain liner delay systems. In *Proceedings of the 34th Conference on Decision and Control,* New Orleans, Louisiana, USA, **4**, 3614-3619.

Luo, J.S. and P.P.J. van den Bosch (1997). Independent of delay stability criteria for uncertain linear state space models. *Automatica,* **33**, 171-179.

Mahmoud, M.S. (2000). *Robust Control and Filtering for Time-Delay Systems.* Marcel Dekker, Inc., New York.

Niculescu, S.I., A.T. Neto, J.M. Dion and L. Dugard (1995). Delay-dependent stability of linear systems with delayed state: a LMI approach. In *Proceedings of the 34th Conference on Decision and Control,* New Orleans, Louisiana, USA, **2**, 1495-1496.

Trinh, H. and M. Aldeen (1995). Stability robustness bounds for linear systems with delayed perturbations. *IEE Proceedings Control Theory and Applications*, **142**, 345-350.

Xu, B. (1994). Comments on 'Robust stability of delay dependence for linear uncertain systems'. *IEEE Transactions on Automatic Control,* **39**, 2365.

Xu, B. (1995). On delay-independent stability of large-scale systems with time delays. *IEEE Transactions on Automatic Control,* **40**, 930-933.

Xu, B. (1997a). An improved Razumikhin-type theorem and its applications—Author's reply. *IEEE Transactions on Automatic Control,* **42**, 430.

Xu, B. (1997b). Stability robustness bounds for linear systems with multiple time-varying delayed perturbations. *International Journal of Systems Science*, **28**, 1311-1317.

Xu, B. (1999). Decay estimates for retarded dynamic systems. *International Journal of Systems Science,* **30**, 427-439;

Xu, B. (2000a). Decentralized stabilization of large-scale linear continuous systems with *N*×*N* time-varying delays. *International Journal of Systems Science,* **31**, 489-496.

Xu, B. (2000b). Stability criteria for linear time-invariant systems with multiple delays. *Journal of Mathematical Analysis and Applications*, **252**, 484-494.

Xu, B. (2001). Stability of Retarded Dynamical Systems: A Lyapunov Function Approach. *Journal of Mathematical Analysis and Applications*, **253**, 590-615.

Xu, B., Y. Fu and L. Bai (1996). Further results on robust bounds for large-scale time-delay systems with structured and unstructured uncertainties. *International Journal of Systems Science,* **27**, 1491-1495.

Xu, B. and J. Lam (1999). Decentralized stabilization of large-scale interconnected time-delay systems. *Journal of Optimization Theory and Applications,* **103**, 231-240.

Xu, B. and Y. Liu (1994). An improved Razumikhin-type theorem and its applications. *IEEE Transactions on Automatic Control,* **39**, 839-841.

www.elsevier.com/locate/ifac

A REVISIT OF SOME DELAY-DEPENDENT STABILITY CRITERIA FOR UNCERTAIN TIME-DELAY SYSTEMS *

Keqin Gu and Qing-Long Han

Department of Mechanical & Industrial Engineering
Southern Illinois University at Edwardsville
Edwardsville, Illinois 62026, USA, E-mail: kgu@siue.edu

Abstract: This article revisits some delay-dependent stability criteria previous discussed in the literature. A more systematic approach is taken, and different forms of stability criteria are developed. The form is more convenient to treat related uncertainty in different system matrices. The relationship between the stability criteria based on simple Lyapunov-Krasovskii functional method and on Razumikhin Theorem are more explicitly shown. *Copyright © 2001 IFAC*

Keywords: Stability, Time-delay, Linear Matrix Inequality

1. INTRODUCTION

Time-delay systems are frequently encountered in engineering, biology, economy, and other areas (Hale and Verduyn Lunel, 1993). In the wake of intensive research on the robust stability and control theory, the stability and control of time-delay systems received renewed interests. The development of efficient computational algorithm for nonsmooth convex optimization problem, which made it possible to efficiently solve Linear Matrix Inequalities (LMI) (Boyd, et. al., 1994), inspired intensive activities to formulate such problems in an LMI form. We will only mention some more recent activities on delay-dependent stability using the time-domain approaches. For a more comprehensive survey, see Niculescu, *et. al.* (1997), Kolmanovskii, *et. al.* (1999) and Kharitonov (1998).

For systems with small delay, a model transformation technique is often used to transform a system with a point-wise delay into one with a distributed delay, and a simple Lyapunov-Krasovskii or Razumikhin stability criterion is used to the resulting system. See, for example, Su and Huang (1992), De Souza and Li (1999), Li and De Souza (1997), Niculescu (1999). We will call such results as "simple delay-dependent stability criteria". Some intrinsic limitations due to model transformation (known as "additional dynamics") have been discussed in Gu and Niculescu (2000) and Kharitonov and Melchor-Aguilar (2000). Some recent studies in Gu (1997) and Gu (2001) are free from such transformations, and proved to be much less conservative. In spite of these, due to its simplicity, different variations of simple delay-dependent stability criteria are still actively pursued in recent publications.

In this article, the ideas of such simple delay-dependent stability criteria are revisited. A more systematic approach is used. As a result, it is much easier to consider systems with related uncertainties. The relationship between the Lyapunov-Krasovskii method and Razumikhin method is more easily shown.

* This work is partially supported by National Science Foundation Grant INT-9818312

2. LYAPUNOV-KRASOVSKII FUNCTIONAL BASED STABILITY CONDITIONS

Consider the system

$$\dot{x}(t) = A_0(t)x(t) + A_1(t)x(t-r) \quad (1)$$

where $A_0 \in \mathsf{R}^{n\times n}$, $A_1 \in \mathsf{R}^{n\times n}$ are uncertain matrices not known completely, except that they are within a compact set Ω which we will refer to as the uncertainty set,

$$(A_0(t), A_1(t)) \in \Omega \text{ for all } t \geq 0 \quad (2)$$

Use model transformation, we can transform the above system to the following system with distributed delays

$$\dot{x}(t) = \bar{A}_0(t)x(t) + \int_{-2r}^{0} \bar{A}(t,\theta)x(t+\theta)d\theta \quad (3)$$

$$(\bar{A}_0(t), \bar{A}(t,\cdot)) \in \bar{\Omega},\ t \geq 0 \quad (4)$$

where $\bar{\Omega}$ represents the set

$$\left\{ (\bar{A}_0, \bar{A}(\cdot)) \left| \begin{array}{l} \bar{A}_0 = A_0 + A_1,\ \bar{A}(\theta) = -A_1 A_{0\theta}, \\ \bar{A}(-r+\theta) = -A_1 A_{1\theta},\ \theta \in [-r, 0] \\ (A_0, A_1) \in \Omega \text{ and } (A_{0\theta}, A_{1\theta}) \in \Omega \end{array} \right. \right\} \quad (5)$$

For the sake of convenience, we have omitted explicit notation for time-dependence of $\bar{A}$ and write, for example, $\bar{A}(\theta)$ for $\bar{A}(t,\theta)$. As is described in Gu and Niculescu (2000) and Kharitonov and Melchor-Aguilar (2000), the stability of the system described by (3) to (5) implies the stability of the system described by (1) and (2), but not vice versa.

To discuss the stability of a system with distributed delays described by (3) and (4), use the Lyapunov-Krasovskii functional

$$V(\phi) = \phi^T(0)P\phi(0) + \int_{-2r}^{0}\int_{\theta}^{0} \phi^T(\xi)S(\theta)\phi(\xi)d\xi d\theta$$

where $P > 0$ and $S(\theta) > 0$ for $-2r \leq \theta \leq 0$. The derivative of $V(x_t)$ along the system trajectory can be calculated as

$$\begin{aligned} \dot{V}(\phi) &= \dot{V}(x_t)|_{x_t=\phi} \\ &= \phi^T(0)[P\bar{A}_0 + \bar{A}_0^T P + \int_{-2r}^{0} S(\theta)d\theta]\phi(0) \\ &\quad + 2\phi^T(0)\int_{-2r}^{0} P\bar{A}(\theta)\phi(\theta)d\theta \\ &\quad - \int_{-2r}^{0} \phi^T(\theta)S(\theta)\phi(\theta)d\theta \end{aligned}$$

where x_t represents the segment of system trajectory in the interval $[t-2r, t]$:

$$x_t(\theta) = x(t+\theta),\ -2r \leq \theta \leq 0$$

For arbitrary $R(\theta, A_0, A(\theta))$ satisfying

$$\int_{-2r}^{0} R(\theta, \bar{A}_0, \bar{A}(\theta))d\theta = 0 \quad (6)$$

we have

$$\dot{V}(\phi) = \int_{-2r}^{0} \phi_{0\theta}^T \begin{pmatrix} M(\theta) & P\bar{A}(\theta) \\ \bar{A}^T(\theta)P & -S(\theta) \end{pmatrix} \phi_{0\theta} d\theta \quad (7)$$

where $\phi_{0\theta}^T = \left(\phi^T(0)\ \phi^T(\theta) \right)$ and

$$M(\theta) = \frac{1}{2r}(P\bar{A}_0 + \bar{A}_0^T P) + S(\theta) + R(\theta, \bar{A}_0, \bar{A}(\theta)) \quad (8)$$

We can therefore conclude that

Lemma 1. The system with distributed delays described by (3) and (4) is asymptotically stable if there exist a symmetric matrix

$$P > 0, \quad (9)$$

a symmetric matrix function $R(\theta, \bar{A}_0, \bar{A}(\theta))$ satisfying the constraint (6), and a symmetric matrix function $S(\theta)$ such that

$$\begin{pmatrix} M(\theta) & P\bar{A}(\theta) \\ \bar{A}^T(\theta)P & -S(\theta) \end{pmatrix} < 0,\ -2r \leq \theta \leq 0 \quad (10)$$

for all $(\bar{A}_0, \bar{A}(\cdot)) \in \bar{\Omega}$ where $M(\theta)$ is defined in (8).

Proof. Use the Lyapunov-Krasovskii Stability Theorem (Theorem 2.1, Chapter 5 of Hale and Verduyn Lunel, 1993). ■

Apply Lemma 1 to the system with the specific uncertainty described by (5), we can conclude:

Lemma 2. The system described by (3) to (5) is asymptotically stable if there exist symmetric matrices $P > 0$, S_0, S_1, and matrix function $R = R(A_0, A_1, A_{0\theta}, A_{1\theta})$ such that

$$\begin{pmatrix} N + S_0 + R & -PA_1A_{0\theta} \\ -(PA_1A_{0\theta})^T & -S_0 \end{pmatrix} < 0 \quad (11)$$

$$\begin{pmatrix} N + S_1 - R & -PA_1A_{1\theta} \\ -(PA_1A_{1\theta})^T & -S_1 \end{pmatrix} < 0 \quad (12)$$

for all $(A_0, A_1) \in \Omega$ and $(A_{0\theta}, A_{1\theta}) \in \Omega$, where

$$N = \frac{1}{2r}(P(A_0 + A_1) + (A_0 + A_1)^T P)$$

Proof. Apply Lemma 1 with the choices

$$S(\theta) = \begin{cases} S_0, & -r < \theta \leq 0 \\ S_1, & -2r < \theta \leq -r \end{cases}$$

and for $-r < \theta \leq 0$

$$R(\theta) = R(A_0, A_1, A_{0\theta}, A_{1\theta})$$
$$R(\theta - r) = -R(A_0, A_1, A_{0\theta}, A_{1\theta})$$

■

From the above, we can further conclude

Proposition 3. The system described by (1) and (2) is asymptotically stable if there exist symmetric matrices $P > 0$, S_0 and S_1 such that

$$\begin{pmatrix} M & -PA_1A_{0\theta} & -PA_1A_{1\theta} \\ -A_{0\theta}^TA_1^TP & -S_0 & 0 \\ -A_{1\theta}A_1P & 0 & -S_1 \end{pmatrix} < 0 \quad (13)$$

for all $(A_0, A_1) \in \Omega$ and $(A_{0\theta}, A_{1\theta}) \in \Omega$, where

$$M = \frac{1}{r}[P(A_0 + A_1) + (A_0 + A_1)^T P] + S_0 + S_1$$

Proof. We only need to prove the stability of the system described by (3) to (5). The sufficient condition is given by Lemma 2. Use Proposition 3 of Gu (1999) to eliminate the matrix variable R, we can see that the conditions in this Proposition is equivalent to those of Lemma 2. ■

3. STABILITY OF SYSTEMS WITH NORM BOUNDED UNCERTAINTY

Now we will consider the norm bounded uncertainty described by

$$A_0(t) = A_{0n} + \Delta A_0(t) \quad (14)$$

$$A_1(t) = A_{1n} + \Delta A_1(t) \quad (15)$$

where

$$\begin{pmatrix} \Delta A_0(t) & \Delta A_1(t) \end{pmatrix} = EF(t) \begin{pmatrix} G_0 & G_1 \end{pmatrix} \quad (16)$$

and

$$||F|| \leq 1 \quad (17)$$

We can state

Proposition 4. The system described by (1) and (2), with uncertainty described by (14) to (17) is asymptotically stable if there exist symmetric matrices $P > 0$, S_0, S_1 and scalar μ such that

$$\begin{pmatrix} M_n & -PA_{1n}A_{0n} & -PA_{1n}A_{1n} & PE & \frac{1}{r}(G_0+G_1)^T & -PA_{1n}E \\ -A_{0n}^TA_{1n}^TP & -S_0 + \mu G_0^TG_0 & \mu G_0^TG_1 & 0 & -(G_1A_{0n})^T & 0 \\ -A_{1n}A_{1n}P & \mu G_1^TG_0 & -S_1 + \mu G_1^TG_1 & 0 & -(G_1A_{1n})^T & 0 \\ E^TP & 0 & 0 & -I & 0 & 0 \\ \frac{1}{r}(G_0+G_1) & -G_1A_{0n} & -G_1A_{1n} & 0 & -I & -G_1E \\ -(PA_{1n}E)^T & 0 & 0 & 0 & -(G_1E)^T & -\mu I \end{pmatrix} < 0 \quad (18)$$

where

$$M_n = \frac{1}{r}[P(A_{0n}+A_{1n})+(A_{0n}+A_{1n})^TP]+S_0+S_1$$

Proof. According Proposition 3, the system is asymptotically stable if (13) is satisfied for arbitrary (A_0, A_1) satisfying (14) to (17), and $(A_{0\theta}, A_{1\theta})$ similarly satisfies

$$A_{0\theta}(t) = A_{0n} + \Delta A_{0\theta}(t) \quad (19)$$

$$A_{1\theta}(t) = A_{1n} + \Delta A_{1\theta}(t) \quad (20)$$

where

$$\begin{pmatrix} \Delta A_{0\theta}(t) & \Delta A_{1\theta}(t) \end{pmatrix} = EF_\theta(t) \begin{pmatrix} G_0 & G_1 \end{pmatrix} \quad (21)$$

and

$$||F_\theta|| \leq 1 \quad (22)$$

Let

$$\bar{P}_\theta = \begin{pmatrix} M_n & -PA_{1n}A_{0\theta} & -PA_{1n}A_{1\theta} \\ -A_{0\theta}^TA_{1n}^TP & -S_0 & 0 \\ -A_{1\theta}A_{1n}P & 0 & -S_1 \end{pmatrix}$$

$$\bar{E}_\theta = \begin{pmatrix} PE \\ 0 \\ 0 \end{pmatrix}$$

$$\bar{G}_\theta = \begin{pmatrix} \frac{1}{r}(G_0 + G_1) & -G_1A_{0\theta} & -G_1A_{1\theta} \end{pmatrix}$$

Then (13) can be written as

$$\bar{P}_\theta + \bar{E}_\theta F(t)\bar{G}_\theta + (\bar{E}_\theta F(t)\bar{G}_\theta)^T < 0$$

A sufficient condition for the above is

$$\bar{P}_\theta + \lambda \bar{E}_\theta \bar{E}_\theta^T + \frac{1}{\lambda}\bar{G}_\theta^T\bar{G}_\theta < 0$$

for some $\lambda > 0$, or equivalently (multiplying both sides by λ and use Shur's complement twice)

$$\begin{pmatrix} \lambda\bar{P}_\theta & \lambda\bar{E}_\theta & \bar{G}_\theta^T \\ \lambda\bar{E}_\theta^T & -I & 0 \\ \bar{G}_\theta & 0 & -I \end{pmatrix} < 0$$

Redefine λP as P, λS_0 as S_0 and λS_1 as S_1. In the resulting matrix inequality, considering the uncertainty $(A_{0\theta}, A_{1\theta})$, the above can again be written as

$$\bar{P} + \bar{E}F_\theta(t)\bar{G} + (\bar{E}F_\theta(t)\bar{G})^T < 0$$

for appropriately defined $\bar{P}$, $\bar{E}$ and $\bar{G}$. A sufficient condition for the above is

$$\bar{P} + \frac{1}{\mu}\bar{E}\bar{E}^T + \mu\bar{G}^T\bar{G} < 0$$

for some $\mu > 0$. Use Schur's complement to reach (18). ■

It is interesting to compare the above stability condition with Theorem 3.1 of de Souza and Li (1999). First, the uncertainty treated is different: the above considers the uncertainty entries in A_0 and A_1 to be related, while de Souza and Li (1999) considers the case of unrelated uncertainty. It is possible to use the above idea to consider unrelated uncertainty, with resulting LMI of larger dimention than above. Second, the dimension of LMI here is smaller. A dual form of the above stability condition is possible as discussed in the following Corollary, which can be used for the purpose of state feedback stabilization.

Corollary 5. The system described by (1) and (2), with uncertainty described by (14) to (17) is asymptotically stable if there exist scalar λ, symmetric matrices $U > 0$, V_0 and V_1, such that

$$\begin{pmatrix} M_{nd} & -A_{1n}A_{0n}U & -A_{1n}A_{1n}U & E & \frac{1}{r}U(G_0+G_1)^T & -\lambda A_{1n}E & 0 \\ -UA_{0n}^TA_{1n}^T & -V_0 & 0 & 0 & -U(G_1A_{0n})^T & 0 & UG_0^T \\ -UA_{1n}A_{1n} & 0 & -V_1 & 0 & -U(G_1A_{1n})^T & 0 & UG_1^T \\ E^T & 0 & 0 & -I & 0 & 0 & 0 \\ \frac{1}{r}(G_0+G_1)U & -G_1A_{0n}U & -G_1A_{1n}U & 0 & -I & -\lambda G_1E & 0 \\ -\lambda(A_{1n}E)^T & 0 & 0 & 0 & -\lambda(G_1E)^T & -\lambda I & 0 \\ 0 & G_0U & G_1U & 0 & 0 & 0 & -\lambda I \end{pmatrix} < 0$$

where

$$M_{nd} = \frac{1}{r}[(A_{0n}+A_{1n})U+U(A_{0n}+A_{1n})^T]+V_0+V_1$$

Proof. Left multiply the first three rows of (18) and right multiply the first three colums by P^{-1}, multiply the last row and column by $1/\mu$, and define $U = P^{-1}$, $V_i = P^{-1}S_iP^{-1}$, $i = 0, 1$, $\lambda = 1/\mu$. Then use Schur's complement. ■

4. RAZUMIKHIN THEOREM BASED RESULTS

Now consider the system with time-varying delay

$$\dot{x}(t) = A_0(t)x(t) + A_1(t)x(t - r(t)) \quad (23)$$

with the uncertain system matrices and time-delay described by

$$(A_0(t), A_1(t)) \in \Omega \text{ for all } t \geq 0 \quad (24)$$

and

$$0 < r(t) \leq r_{\max} \text{ for all } t \geq 0 \quad (25)$$

Using model transformation, one obtains

$$\dot{x}(t) = \bar{A}_0(t)x(t) + \int_{-2r(t)}^{0} \bar{A}(t,\theta)x(t+\phi(t,\theta))d\theta \quad (26)$$

where

$$\bar{A}_0(t) = A_0(t) + A_1(t)$$

and for $r(t) < \theta \leq 0$

$$\begin{aligned} \bar{A}(t,\theta) &= -A_1(t)A_0(t+\theta) \\ \bar{A}(t,\theta - r(t)) &= -A_1(t)A_1(t+\theta) \\ \phi(t,\theta) &= \theta \\ \phi(t,\theta - r(t)) &= \theta - r(t+\theta) \end{aligned}$$

Therefore, we can write

$$(\bar{A}_0(t), \bar{A}(t,\cdot)) \in \bar{\Omega},\ t \geq 0 \quad (27)$$

where $\bar{\Omega}$ represents the set

$$\left\{ \bar{A}_0, \bar{A}(\cdot) \left| \begin{array}{l} \bar{A}_0 = A_0 + A_1,\ \bar{A}(\theta) = -A_1A_{0\theta} \\ \bar{A}(-r+\theta) = -A_1A_{1\theta}, -r \leq \theta < 0 \\ \text{for } (A_0, A_1) \in \Omega,\ (A_{0\theta}, A_{1\theta}) \in \Omega \end{array} \right. \right\} \quad (28)$$

Using Razumikhin Theorem to system (26), we obtain

Lemma 6. The system described by (26) to (27) is asymptotically stable if there exist a scalar function $\alpha(t,\theta) \geq 0$, a matrix $P > 0$ and a matrix function $R(\theta, A_0, A(t,\theta))$ satisfying

$$\int_{-2r(t)}^{0} R(\theta, A_0, A(t,\theta))d\theta = 0 \quad (29)$$

such that

$$\begin{pmatrix} N(\theta) & P\bar{A}(t,\theta) \\ \bar{A}^T(t,\theta)P & -\alpha(t,\theta)P \end{pmatrix} < 0 \quad (30)$$

for all $(\bar{A}_0(t), \bar{A}(t,\cdot)) \in \bar{\Omega}$, $t \geq 0$, where

$$N = \frac{1}{2r_{\max}}[P\bar{A}_0 + \bar{A}_0^TP] + \alpha(\theta)P + R(\theta, \bar{A}_0, \bar{A}(\theta))$$

Proof. Choose Lyapunov function

$$V(x(t)) = x^T(t)Px(t) \quad (31)$$

The Razumikhin Theorem (Theorem 4.2 of Chapter 5 in Hale and Verduyn Lunel (1993)) indicates that the system is asymptotically stable if there exists a $p > 1$ such that the derivative of the Lyapunov function along the system trajectory satisfies

$$\dot{V}(x(t)) \leq -\varepsilon||x(t)||^2 \quad (32)$$

whenever

$$V(x(t+\theta)) < pV(x(t)) \quad (33)$$

We can calculate, under condition (33), that

$$\begin{aligned} \dot{V}(x(t)) &\leq 2x^T(t)P[\bar{A}_0(t)x(t) \\ &\quad + \int_{-2r(t)}^{0} \bar{A}(t,\theta)x(t+\phi(\theta))d\theta] \\ &\quad + \int_{-2r(t)}^{0} \alpha(t,\theta)[px^T(t)Px(t) \\ &\quad - x^T(t+\phi(\theta))Px(t+\phi(\theta))]d\theta \\ &= \int_{-2r(t)}^{0} x_{0\theta}^T \begin{pmatrix} N_{p,r}(\theta) & P\bar{A}(t,\theta) \\ \bar{A}^T(t,\theta)P & -\alpha(t,\theta)P \end{pmatrix} x_{0\theta}d\theta \end{aligned}$$

for $\alpha(t,\theta) \geq 0$, where $x_{0\theta}^T = \left(x^T(t) \; x^T(t+\phi(\theta)) \right)$ and

$$N_{p,r} = \frac{1}{2r}[P\bar{A}_0 + \bar{A}_0^T P] + \alpha(\theta)pP + R(\theta, A_0, A(\theta))$$

Since (30) implies $N(\theta) < 0$, integrating on the interval $[-2r(t), 0]$, using (29), we can conclude

$$P\bar{A}_0 + \bar{A}_0^T P < 0$$

Therefore,

$$N_{p,r}(\theta) \leq N_{p,r_{\max}}(\theta) \quad (34)$$

As a consequence, if $N(\theta) = N_{1,r_{\max}}(\theta)$ in (30) is replaced by $N_{p,r_{\max}}(\theta)$, then (30) would guarantee (32). As p can be arbitrarily close to 1, the asymptotic stability of the system is proven. ■

It is interesting to compare Lemma 6 and Lemma 1. It is noticed that if we replace r by $r_{\max}$ and restrict $S(\theta)$ to have the expression $\alpha(\theta)P$ in the conditions of Lemma 1, we arrive at the conditions in Lemma 6. Clearly, Lemma 6 is more restrictive if applied to the time-invariant delay case.

The observation above allows us to directly obtain, for generic uncertainty, the following Proposition from Proposition 3 (replacing r by $r_{\max}$ and replacing S_i by $\alpha_i P$):

Proposition 7. The system described by (23) to (24) is asymptotically stable if there exist symmetric matrix $P > 0$ and scalars $\alpha_0 \geq 0$, $\alpha_1 \geq 0$, such that

$$\begin{pmatrix} N & -PA_1A_{0\theta} & -PA_1A_{1\theta} \\ -A_{0\theta}^T A_1^T P & -\alpha_0 P & 0 \\ -A_{1\theta}A_1 P & 0 & -\alpha_1 P \end{pmatrix} < 0 \quad (35)$$

for all $(A_0, A_1) \in \Omega$ and $(A_{0\theta}, A_{1\theta}) \in \Omega$, where

$$N = \frac{1}{r_{\max}}[P(A_0+A_1)+(A_0+A_1)^T P]+\alpha_0 P+\alpha_1 P$$

Same procedure also allows us to obtain the stability condition for norm-bounded uncertainty and time-varying delay parallel to Proposition 4.

It is important to notice that the conditions in the above Proposition and its derived form for norm bounded uncertainty are no longer in LMI form due to the multiplicative terms $\alpha_0 P$ and $\alpha_1 P$. Due to this reason, some previous works tend to avoid such scaling factors. For example, it can be shown that the conditions proposed in Li and de Souza (1997) effectively choose $\alpha_0 = \alpha_1 = 1$. This can be seen by eliminating the variables P_1 and P_2 using the technique proposed in Gu (1999) among the Schur's complement of (3.17) and (3.22) in Li and de Souza (1997).

However, in practice it is important to include, or at least choose reasonable values of the scaling factors to minimize the conservatism. To understand this, consider a system without uncertainty, for which (35) is used. If A_0 and A_1 are multiplied by 60 and r is divided by 60. This corresponds to the system evolving 60 times faster, or simply due to using different units to measure time (say, using minutes instead of seconds). Clearly, the stability is not affected. One need to choose α_0 and α_1 3600 times as large in order to accommodate this change.

To obtain the least conservative result using this type of matrix inequality, one need to conduct a nonconvex search. As a compromise, one can choose a reasonable fixed set of α_0 and α_1, and accept the additional conservatism. For example, in using (35), let $\bar{\alpha}$ be the solution of the generalized eigenvalue problem

$$\bar{\alpha} = \sup \alpha$$

subject to, for all $(A_0, A_1) \in \Omega$,

$$U + \alpha P < 0 \quad (36)$$

$$P > 0 \quad (37)$$

where

$$U = \frac{1}{r_{\max}}[P(A_0 + A_1) + (A_0 + A_1)^T P]$$

It is then clear that any feasible (α_0, α_1) needs to fall in the triangular region $\alpha_0 > 0$, $\alpha_1 > 0$, $\alpha_0 + \alpha_1 < \bar{\alpha}$. With norm bounded uncertainty, (36) can be written in a LMI form. Let

$$\beta_0 = \max_{(A_0, A_1) \in \Omega} ||A_0|| \quad (38)$$

$$\beta_1 = \max_{(A_0, A_1) \in \Omega} ||A_1|| \quad (39)$$

A reasonable choice seems to be

$$\alpha_0 = \frac{\beta_0 \bar{\alpha}}{2(\beta_0 + \beta_1)} \quad (40)$$

$$\alpha_1 = \frac{\beta_1 \bar{\alpha}}{2(\beta_0 + \beta_1)} \quad (41)$$

For norm bounded uncertainty, the calculation of β_0 and β_1 can also be formulated as a generalized eigenvalue problem. A more ideal choice is to let β_0 and β_1 to assume the maximum of $||A_1 A_{0\theta}||$ and $||A_1 A_{1\theta}||$ within the uncertainty set, respectively, which requires more computation.

5. NUMERICAL EXAMPLES

Consider the uncertain time-delay system (1), with uncertain matrices A_0 and A_1 described by (14) to (17), where the nominal matrices are

$$A_{0n} = \begin{pmatrix} -2 & 0 \\ 0 & -0.9 \end{pmatrix}, \quad A_{1n} = \begin{pmatrix} -1 & 0 \\ -1 & -1 \end{pmatrix} \quad (42)$$

First consider the case without uncertainty, *i.e.*,

$$E = 0, \qquad G_0 = G_1 = I \tag{43}$$

Using simple Lyapunov-Krasovskii method in either Proposition 3 or Proposition 4, it is calculated that the system is asymptotically stable for $r < \bar{r} = 1$.

Now consider the uncertain system with the following parameters

$$E = 0.2I, \qquad G_0 = G_1 = I \tag{44}$$

Using Proposition 4, it can be calculated that the system is asymptotically stable for $r < \bar{r} = 0.5936$.

Next, we consider the system with time-varying delay (23) using Razumikhin Theorem. First consider systems without uncertainty described by (14) to (17) and (42) and (43). Using Proposition 7, choose α_0 and α_1 as (38) to (41), it can be calculated that the system is asymptotically stable for $r_{\max} < \bar{r} = 0.9042$. For system with uncertainty, E, G_0 and G_1 as described in (44), using the parallel of Proposition 4 (replacing S_k by $\alpha_k P$, $k = 0, 1$, and replacing r by $r_{\max}$), choosing α_0 and α_1 in the same way, it can be calculated that the system is asymptotically stable for $r_{\max} < \bar{r} = 0.5597$.

As compared to the corresponding results presented in Li and de Souza (1997) and de Souza and Li (1999), the above results clearly demonstrated that the new formulation allows us to take advantage of the related uncertainty to reduce conservatism.

For time-invariant delay, the results are still much more conservative than the discretized Lyapunov functional method discussed in Han and Gu (2001) (for most coarse discretization of $N = 1$, stability is assured for $r < \bar{r} = 5.3$ for systems without uncertainty, and $\bar{r} = 2.78$ with uncertainty). The result can be further improved by using the new formulation in Gu (2001).

6. CONCLUSIONS

Some delay-dependent stability criteria using simple Lyapunov-Krasovskii functional method and Razumikhin Theorem have been revisited with a more systematic formulation.

ACKNOWLEDGEMENT

Part of this work was conducted while the first author was visiting CINVESTAV-IPN, Mexico. He would like to thank Professor Vladimir L. Kharitonov for various assistance and discussions on related topics.

REFERENCES

[1] Boyd, S., El Ghaoui, L., Feron, E. and Balakrishnan, V., 1994, *Linear Matrix Inequalities in System and Control Theory* (Philadelphia: SIAM).

[2] De Souza, C. E. and Li, X., 1999, Delay-dependent robust H_∞ control of uncertain linear state-delayed systems. *Automatica*, **35**, 1313-1321.

[3] Gu, K., 1997, Discretized LMI set in the stability problem of linear uncertain time-delay systems. *International Journal of Control*, **68**, 923-934.

[4] Gu, K., 1999, Partial solution of LMI in stability problem of time-delay systems. *Proc. 38th IEEE Conf. on Decision and Control*, Phoenix, AZ, December 7–10, 1999, 227-232.

[5] Gu, K., 2001, A further refinement of discretized Lyapunov functional method for the stability of time-delay systems. *International Journal of Control*, **74**, 967-976.

[6] Gu, K. and Niculescu, S.-I., 2000, Additional dynamics in transformed time-delay systems, *IEEE Transactions on Automatic Control*, **45**, 572-575.

[7] Hale, J. and Verduyn Lunel, S. M., 1993, *Introduction to Functional Differential Equations* (New York: Springer-Verlag).

[8] Han, Q.-L. and Gu, K., 2001, On robust stability of time-delay systems with norm-bounded uncertainty. *IEEE Transactions on Automatic Control*, **46**.

[9] Kharitonov, V. L. and Melchor-Aguilar, D. A., 2000, On delay-dependent stability conditions. *System & Control Letters*, **40**, 71-76.

[10] Kharitonov, V., 1998, Robust stability analysis of time delay systems: A survey. *Proc. IFAC Syst. Struct. Contr.*, Nantes, France.

[11] Kolmanovskii, V. B., Niculescu, S.-I. and Gu, K., 1999, Delay effects on stability: a survey. *Proc. 38th IEEE Conf. Decision and Control*, Phoenix, AZ, Dec. 7-10, 1999, 1993-1998.

[12] Li, X. and de Souza, C. E., 1997, Criteria for robust stability and stabilization of uncertain linear systems with state delay. *Automatica*, **33**, 1657-1662.

[13] Niculescu, S.-I., 1999, A model transformation class for delay-dependent stability analysis, *1999 American Control Conference*, 314-318.

[14] Niculescu, S.-I., Verriest, E. I., Dugard, L. and Dion, J. M., 1997, Stability and robust stability of time-delay systems: A guided tour, in *Stability and Control of Time-Delay Systems* (L. Dugard and E. I. Verriest, Eds.), LNCIS, Springer-Verlag, London, 1-71.

[15] Su, T.-J. and Huang, C.-G., 1992, Robust stability of delay dependence for linear uncertain systems. *IEEE Transactions on Automatic Control*, **37**, 1656-1659.

www.elsevier.com/locate/ifac

DELAY DEPENDENT STABILITY OF SCALAR DISCRETE DELAY SYSTEMS

Erik I. Verriest

School of Electrical and Computer Engineering, Georgia Institute of Technology, Atlanta, Georgia, USA
erik.verriest@ece.gatech.edu

Abstract: Necessary and sufficient conditions for the delay dependent stability of a linear difference-delay equation are derived. In particular, the delay dependent stability of a scalar delay delay system is related to the existence of a positive Markov parameter sequence for an associated fixed *second* order system.

Keywords: Difference-Delay Systems, Stability, Positive Markov Sequence

1. INTRODUCTION

Discrete delay systems of the form

$$x_{k+1} = Ax_k + Bx_{k-n}, \tag{1.1}$$

have recently been studied with renewed vigor. Sufficient conditions for robust stability, i.e., stability independent of the delay where obtained by Mori *et al.* (1982), Brierley (1986) and Ivanov and Verriest (1994), and Verriest and Ivanov (1995). Further work appeared in Ivanov and Verriest (1998). In independent work of Kolmanovski *et al.* (1999) similar results were presented. These results are similar to the continuous time delay case in their use of a Lyapunov-Krasovskii functional, and lead to Riccati equation conditions (Kolmanovski, 1995; Verriest and Ivanov, 1991;1994). Testing of these can be done via the LMI-methods. Further extensions with frequency domain criteria are obtained in (Verriest, 2001b; Xu *et al.*, 2001). See also Lee *et al.* (1992), Kuruklis (1994) and Rodionov (1996).

In this paper exact criteria (i.e., necessary and sufficient conditions) are obtained for the delay dependent stability. The idea is to use the standard Jury test for a delay-difference system. This test involves the construction of a finite sequence. If all terms in this sequence are positive, the system is stable. A linear time invariant system is associated with this test itself.

2. SCALAR DELAY DIFFERENCE EQUATION

In this section we consider the scalar system

$$x_{k+1} = ax_k + bx_{k-n} \tag{2.2}$$

Note that this in fact corresponds to a discrete system of order $n+1$. Its characteristic polynomial is

$$p_n(z) = z^{n+1} - az^n - b. \tag{2.3}$$

In testing the delay dependent stability, an obvious but tedious approach consists in finding the roots of $p_n(z)$ in (2.3) for successively larger values of n, and test if their modulus exceeds one. This is of course computationally quite intensive and therefore lacks elegance. A computationally efficient method is derived below.

2.1 *Jury Test*

It is assumed that the reader is familiar with the mechanics of the Jury stability test. See for instance (Franklin *et al.*, 1990). The celebrated Jury stability test, for this characteristic polynomial gives an array starting with

$$\begin{array}{cccccc} 1 & -a & 0\cdots 0 & 0 & -b \\ -b & 0 & 0\cdots 0 & -a & 1 \end{array}$$

There are $n+1$ entries in these rows, and both rows have $n-3$ positions in common having the entry 0. When the next pair of rows in the array is formed, it has the form

$$\begin{array}{cccccc} 1-b^2 & -a & 0\cdots 0 & 0 & ab \\ ab & 0 & 0\cdots 0 & -a & 1-b^2 \end{array}$$

Now there are n entries per row, and $n-4$ of these are 0 in both rows. One can easily follow this pattern down almost to the end. In fact, one can generalize this by relabeling the first pair of rows as

$$\begin{array}{cccccc} \alpha_1 & \beta_1 & 0\cdots 0 & 0 & \gamma_1 \\ \gamma_1 & 0 & 0\cdots 0 & \beta_1 & \alpha_1 \end{array}$$

The following rows have the same structure

$$\begin{array}{ccccc} \alpha_2 & \beta_2 & 0\cdots & 0 & \gamma_2 \\ \gamma_2 & 0 & 0\cdots & \beta_2 & \alpha_2 \end{array}$$

and so on up to the $(n-2)$nd pair, which is

$$\begin{array}{cccc} \alpha_{n-2} & \beta_{n-2} & 0 & \gamma_{n-2} \\ \gamma_{n-2} & 0 & \beta_{n-2} & \alpha_{n-2} \end{array}$$

where the iteration is given by the nonlinear recursions

$$\alpha_{k+1} = \alpha_k - \frac{\gamma_k^2}{\alpha_k} \tag{2.4}$$

$$\beta_{k+1} = \beta_k \tag{2.5}$$

$$\gamma_{k+1} = -\frac{\beta_k \gamma_k}{\alpha_k}, \tag{2.6}$$

for $k = 1, \ldots n-3$. The initialization is

$$\alpha_1 = 1$$

$$\beta_1 = -a$$

$$\gamma_1 = -b.$$

Continuing he Jury table, the $(n-1)$st pair of rows is

$$\begin{array}{ccc} \alpha_{n-1} & \beta_{n-1} & \gamma_{n-1} \\ \gamma_{n-1} & \beta_{n-1} & \alpha_{n-1}, \end{array}$$

where thus still

$$\alpha_{n-1} = \alpha_{n-2} - \frac{\gamma_{n-2}^2}{\alpha_{n-2}} \tag{2.7}$$

$$\beta_{n-1} = \beta_{n-2} \tag{2.8}$$

$$\gamma_{n-1} = -\frac{\beta_{n-2}\gamma_{n-2}}{\alpha_{n-2}}. \tag{2.9}$$

The nth pair in the Jury table is

$$\begin{array}{cc} \alpha_n & \gamma_n \\ \gamma_n & \alpha_n, \end{array}$$

where now

$$\alpha_n = \alpha_{n-1} - \frac{\gamma_{n-1}^2}{\alpha_{n-1}} \tag{2.10}$$

$$\gamma_n = \frac{\alpha_{n-1} - \gamma_{n-1}}{\alpha_{n-1}} \beta_{n-1}. \tag{2.11}$$

The table finally terminates with

$$\alpha_{n+1} = \alpha_n - \frac{\gamma_n^2}{\alpha_n}.$$

According to the Jury test for stability the characteristic polynomial $p_n(z)$ will have all its zeroes inside the unit circle if and only if all $\alpha_k; k = 1\ldots, n+1$, are positive.

Let us now analyse these recursions. It is obvious that $\beta_1 = \beta_2 = \cdots = \beta_{n-1} = -a$. Hence the three component recursion reduces to

$$\alpha_{k+1} = \alpha_k - \frac{\gamma_k^2}{\alpha_k} \tag{2.12}$$

$$\gamma_{k+1} = a\frac{\gamma_k}{\alpha_k}, \tag{2.13}$$

for $k = 1, \ldots, n-1$, with termination

$$\alpha_n = \alpha_{n-1} - \frac{\gamma_{n-1}^2}{\alpha_{n-1}}$$

$$\gamma_n = a\left(\frac{\gamma_{n-1}}{\alpha_{n-1}} - 1\right)$$

Elimination of γ gives

$$\alpha_{k+2} = \alpha_{k+1} - \frac{a^2}{\alpha_k \alpha_{k+1}}(\alpha_k - \alpha_{k+1})$$

for $k = 1$ to $k = n-2$. This means that

$$\alpha_{k+2} + \frac{a^2}{\alpha_{k+1}} = \alpha_{k+1} + \frac{a^2}{\alpha_k}$$

and thus that the quantity $\alpha_{k+1} + a^2/\alpha_k$ is a constant of the (parameter) motion for $k = 1, \ldots, n-2$. The value of this constant follows from the initial conditions:

$$\alpha_1 = 1 \quad , \quad \alpha_2 = \alpha_1 - \frac{\gamma_1^2}{\alpha_1} = 1 - b^2$$

Hence:

$$\alpha_{k+1} + \frac{a^2}{\alpha_k} = 1 + a^2 - b^2. \tag{2.14}$$

As γ_{n-1} is required in the terminal step, we must also propagate the γ variable. By multiplying the equations

$$\gamma_{k+1} = \frac{a}{\alpha_k}\gamma_k \tag{2.15}$$

for $k = 1$ to $k = n-2$, one gets

$$\prod_{i=1}^{n-2} \gamma_{i+1} = a^{n-2}\frac{\prod_{i=1}^{n-2}\gamma_i}{\prod_{i=1}^{n-2}\alpha_i}.$$

from which

$$\gamma_{n-1} = \frac{-a^{n-2}b}{\alpha_1\cdots\alpha_{n-2}} \tag{2.16}$$

2.2 *Solution of Nonlinear Recursions*

The nonlinear recursions

$$\alpha_{k+1} = (1+a^2-b^2) - \frac{a^2}{\alpha_k} \tag{2.17}$$

$$\gamma_{k+1} = \frac{a\gamma_k}{\alpha_k} \tag{2.18}$$

with initialization $\alpha_1 = 1, \gamma_1 = -b$, can be linearized (*exactly*) by letting

$$\alpha_k = \frac{N_k}{D_k}.$$

A deeper introduction to such exact linearization is given in Verriest (2001a). A linear recursion is

$$\begin{bmatrix} N_{k+1} \\ D_{k+1} \end{bmatrix} = \begin{bmatrix} 1+a^2-b^2 & -a^2 \\ 1 & 0 \end{bmatrix} \begin{bmatrix} N_k \\ D_k \end{bmatrix}, \tag{2.19}$$

with initialization

$$\begin{bmatrix} N_1 \\ D_1 \end{bmatrix} = \begin{bmatrix} 1 \\ 1 \end{bmatrix}. \tag{2.20}$$

Note that this is not a unique possible choice. Also:

$$\gamma_k = \frac{a}{\alpha_{k-1}}\frac{a}{\alpha_{k-2}}\cdots\frac{a}{\alpha_1}\gamma_1 = \frac{-a^{k-1}b}{N_{k-1}}D_1 = \frac{-a^{k-1}b}{N_{k-1}},$$

terminating with

$$\gamma_n^{\mathrm{Ter}} = -a\left(1+\frac{a^{n-2}b}{N_{n-1}}\right)$$

and thus finally giving the last α in the 'Jury'-sequence:

$$\alpha_{n+1} = \alpha_n - \frac{\gamma_n^2}{\alpha_n}$$

Testing proceeds now as follows: Iterate the linear model for N_k, i.e., the state $[N_k, D_k]'$, and compute from it the sequence

$$\begin{bmatrix} \alpha_k \\ \gamma_k \end{bmatrix} = \begin{bmatrix} \frac{N_k}{D_k} \\ \frac{-a^{k-1}b}{D_k} \end{bmatrix}.$$

Compute also for each k, a *potential* terminating element

$$\gamma_k^{\mathrm{Ter}} = \gamma_k - a$$

and

$$\omega_k \stackrel{\mathrm{def}}{=} \alpha_{k+1}^{\mathrm{Ter}} = \alpha_k - \frac{(\gamma_k^{\mathrm{Ter}})^2}{\alpha_k} = \alpha_k - \frac{a^2}{\alpha_k}\left(1+\frac{a^{n-2}b}{D_k}\right)^2.$$

If N is the largest integer such that $\alpha_\ell > 0$ and $\omega_\ell > 0$ for all $\ell \le N+1$, then the characteristic polynomial $p_n(z)$ passes the Jury test, and the delay system is stable for all delays $n \le N$.

We conclude then with the following statement:

Theorem 2.1. Compute the sequences $\{\alpha_k\}$ and $\{\omega_k\}$ as indicated from the linear recursion of (N_k, D_k). If N is the largest integer such that $\forall n \le N, \alpha_n > 0, \omega_n > 0$, then the delay difference system is asymptotically stable for all delays $n < N$.

Note that this algorithmic test can be represented by a *tree*: The diamond indicates a nonpositive element, while the 'o' denotes a positive quantity.

$$\begin{array}{lccccc} n: & 1 & 2 & 3 & 4 & 5 \\ \alpha: \longrightarrow & o \longrightarrow & o \longrightarrow & o \longrightarrow & o \longrightarrow & \diamond \\ & | & | & | & | & \\ \omega: & o & o & o & o & \end{array}$$

In this example, $N = 4$, indicating that the system is delay stable for all $n < N = 4$.

Note however that, $\operatorname{sgn} D_{k+1} = \operatorname{sgn} N_k$, and thus

$$\operatorname{sgn}\alpha_k = \operatorname{sgn} N_k \operatorname{sgn} D_k = \operatorname{sgn} N_k \operatorname{sgn} N_{k-1}.$$

Thus, if $\operatorname{sgn} N_n = 1$ for all n, then $\operatorname{sgn}\alpha_n = 1$ for all n. If the N-sequence changes sign, so thus the α-sequence. To perform Jury's stability test, it suffices therefore to check the sign of the N_n sequence and its (nonlinear) terminator, ω_n. But a close inspection of the terminator leads now to the following observation:

$$\omega_k = \frac{N_k}{D_k} - \frac{a^2}{N_k D_k}\left(D_k + a^{n-2}b\right)^2 = \frac{N_k^2 - a^2\left(D_k + a^{n-2}b\right)^2}{N_k D_k}.$$

Since again $N_k D_k = N_k N_{k-1}$ is positive if α_k is positive, it turns out that ω_k will be positive if

$$N_k^2 > a^2\left(D_k + a^{k-2}b\right)^2,$$

or, again with $N_k > 0$,

$$N_k > |a|\left|D_k + a^{k-2}b\right| = |a|\left|N_{k-1} + a^{k-2}b\right|.$$

We conclude that determining the maximum delay for which the first order delay system is globally asymptotically stable is equivalent to testing the positivity of the response N_k of some fixed second order system. If N_k is positive, it should be larger than $|a|\left|N_{k-1} + a^{k-2}b\right|$ to guarantee stability for delays up to and including $k-1$.

An example of the tree for the test is:

$$\begin{array}{lccccccccc}
n: & 1 & & 2 & & 3 & & 4 & & 5 \\
N_n: & 1 = N_1 & \longrightarrow & N_2 & \longrightarrow & N_3 & \longrightarrow & N_4 & \longrightarrow & \diamond \\
 & & \searrow & \vee & \searrow & \vee & \searrow & \vee & & \\
f_n(N_{n-1}): & & & f_2(N_1) & & f_3(N_2) & & f_4(N_3) & &
\end{array}$$

---

where $f_n(N) = |a||N + a^{n-2}b|$. In the above tree, $N_n > 0$ and $N_n > f_n(N_{n-1})$ for $n = 1, \ldots, 4$, but $N_5 < 0$. Hence one concludes that the corresponding delay system is asymptotically stable for $n = 1, 2$, and 3.

The problem boils down to the following:
Given the recursion

$$\begin{bmatrix} N_{k+1} \\ D_{k+1} \end{bmatrix} = \begin{bmatrix} 1 + a^2 - b^2 & -a^2 \\ 1 & 0 \end{bmatrix}\begin{bmatrix} N_k \\ D_k \end{bmatrix}, \tag{2.21}$$

with initialization

$$\begin{bmatrix} N_1 \\ D_1 \end{bmatrix} = \begin{bmatrix} 1 \\ 1 \end{bmatrix}, \tag{2.22}$$

when is the sequence $\{N_k\}$ positive?
Note that this sequence is precisely the *Markov parameter* sequence (or pulse response) for the *fixed* second order linear system:

$$A = \begin{bmatrix} 1 + a^2 - b^2 & -a^2 \\ 1 & 0 \end{bmatrix}$$

$$B = \begin{bmatrix} 1 \\ 1 \end{bmatrix}, \quad C = [1\ 0]$$

In addition, the inequality (involving a simple nonlinearity) $N_n > f_n(N_{n-1})$ must hold. In turn, a further simplification of this condition gives:

$$\frac{N_k}{|a|^k} > \left|\frac{N_{k-1}}{|a|^{k-1}} + \frac{b}{|a|}\frac{a^{k-2}}{|a|^{k-2}}\right|,$$

or

$$\widehat{N}_k > \left|\widehat{N}_{k-1} + \frac{b}{|a|}(\operatorname{sgn} a)^k\right|,$$

where $\widehat{N}_k = \frac{N_k}{|a|^k}$ is a normalized version of N_k. Let us depict this in the $(\widehat{N}_{k-1}, \widehat{N}_k)$-plane: If $a > 0$, this means $\widehat{N}_k > \widehat{N}_{k-1} + \frac{b}{a}$, while for $a < 0$, one must test $\widehat{N}_k > |\widehat{N}_{k-1} + \frac{b}{|a|}|$. For k even, this means that $\widehat{N}_k$ should lie in the wedge bounded by the $\widehat{N}_k$ axis and the line with slope 1 through $(0, \frac{b}{|a|})$, while for odd k, it must be in the region bounded below by the $\widehat{N}_k$ axis, the line connecting $(0, \frac{b}{|a|})$ and $(\frac{b}{|a|}, 0)$ and the line through $(\frac{b}{|a|}, 0)$ with slope 1.

The explicit solution is readily found, e.g. using standard Z-transform techniques: Let ξ_1 and ξ_2 be the roots of the characteristic equation $z^2 - (1 + a^2 - b^2)z + a^2$.

Then in the generic case of disjoint roots, one finds

$$\begin{bmatrix} N_k \\ D_k \end{bmatrix} = \frac{\xi_1^k - \xi_2^k}{\xi_1 - \xi_2}\begin{bmatrix} 1 \\ 1 \end{bmatrix} + \frac{\xi_1^{k-1} - \xi_2^{k-1}}{\xi_1 - \xi_2}\begin{bmatrix} -a^2 \\ b^2 - a^2 \end{bmatrix} \tag{2.23}$$

Note that a sufficient condition for the N-sequence to be positive, is that the poles of the second order system are real and positive. These poles are the zeros of the characteristic equation $z^2 - (1 + a^2 - b^2)z + a^2$. It suffices therefore that $(1 + a^2 - b^2)^2 > 4a^2$, i.e., $(1 + a + b)(1 + a - b)(1 - a + b)(1 - a - b) > 0$. The region satisfying this is the diamond with vertices (0,1), (1,0), (-1,0) and (0,-1). This region in parameter space is actually known to be the region guaranteeing stability *independent* of delay (Verriest and Ivanov, 1995)

3. EXAMPLE

The sufficient condition for stability independent of the delay fails for the delay system $x_{k+1} = 0.4x_k - 0.61x_{k-n}$. Computation of the $\{\alpha_k\}$ and $\{\omega_k\}$ sequences yields

n	α_n	ω_n	n	α_n	ω_n
1	1	1	10	.3863	.1039
2	.6279	.5891	11	.3737	.0861
3	.5331	.4209	12	.3597	.0702
4	.4878	.3231	13	.3431	.0557
5	.4599	.2584	14	.3216	.0421
6	.4400	.2119	15	.2904	.0292
7	.4242	.1762	16	.2369	.0168
8	.4108	.1477	17	.1125	.0047
9	.3984	.1241	18	$-.6349$	$-.0074$

The α and ω are positive up to index 17. This means that this delay system is stable for all delays up to 16. For $n = 17$, the system becomes unstable. Indeed, a direct check of the eigenvalues of the 18-th order system shows two eigenvalues with modulus 1.000056476.

4. CONCLUSIONS

Exact necessary and sufficient conditions for the asymptotic stability of a scalar delay system, dependent on the delay, were derived. With the delay time as a variable parameter, it was shown, that with increased delay instability can result. It was shown that a fixed second order system determines the stability of the delay system, through the positivity of its Markov parameter sequence, and a related sequence. This allows to 'paint' the parameter plain in colors according to the maximum value of the delay time before instability sets in. Future work will look at scalar systems with multiple delays. Although the situation is more complex, it is expected that the behavior of a higher order system will determine the delay dependent stability in this case.

Acknowledgement

The support of the NSF-CNRS collaborative grant INT-9818312 is gratefully acknowledged.

5. REFERENCES

Brierley, S.D. (1986). Delay-independent stability criterion for discrete systems using the Lyapunov equation, *Proceedings of the 25th Conference on Decision and Control*, Athens, Greece, 910-911.

Franklin, G.F., J.D. Powell and M. Workman (1990). *Digital Control of Dynamic Systems*, 2nd Ed., Addison-Wesley.

Ivanov, A.F. and E.I. Verriest (1994). Robust stability of delay-difference equations, in *Systems and Networks: Mathematical Theory and Applications*, (U. Helmke, R. Mennicken and J. Saurer, eds.), Univ. of Regensburg Press, 725-726, 1994.

Ivanov A.F. and E.I. Verriest (1998). Stability and Existence of Periodic Solutions in Discrete Delay Systems, Research Report 32/98 School of Information Technology and Mathematical Sciences, University of Ballarat, Victoria, Australia.

V.B. Kolmanovskii, V.B. (1995). On the application of the Lyapunov second method to the Volterra difference equations," *Automation and Remote Control*, **56**, No. 11, 1545-1556.

Kolmanovskii, V.B., J.-F. Lafay and J.-P. Richard (1999). Riccati equations in stability theory of diference equations with memory, *Proc. 5-th European Control Conference*, Karlsrühe, Germany, August 31- Sept. 3, DA-12,2.

Kuruklis, S. A. (1994). The asymptotic stability of $x_{n+1} - ax_n + bx_{n-k} = 0$," *J. Math Anal. Appl.*,**188**, 719-731.

Lee, C.-H., T.-H. S. Li and F.-C. Kung (1992). D-stability analysis for discrete systems with a time delay," *Systems & Control Letters*, **19**, 213-219.

Mori, T., N. Fukuma and M.M. Kuwahara (1982). Delay-independent stability criteria for discrete-delay systems, *IEEE Transactions on Automatic Control*, **27**, No. 4, 964-966.

Rodionov, A.M. (1996). A method for investigating the stability of differential and discrete equations with delay, *Automation and Remote Control*, **57**, No. 12, 1720-1727.

Verriest, E.I. (2001a), Systems over projective fields, *Proceedings of the 5th IFAC Symposium Nonlinear Control Systems*, (NOLCOS'01), Saint-Petersburg, Russia, 207-212.

Verriest, E.I. (2001b), Frequency domain stability criterion for discrete delay systems, *Proceedings IFAC Workschop on Time Delay Systems (TDS3)* (this volume).

Verriest, E.I. and A.F. Ivanov (1991). Robust stabilization of systems with delayed feedback, *Proceedings of the second Int'l Symposium on Implicit and Robust Systems*, Warsaw, Poland, 190-193.

Verriest, E.I. and A.F. Ivanov (1994). Robust stability of systems with delayed feedback, *Circuits, Systems, and Signal Processing*,**13**, No. 2-3, 213-222.

Verriest, E.I. and A.F. Ivanov (1995). Robust stability of delay-difference equations, *Proceedings of the 34th IEEE CDC*, New Orleans, LA, 386-391.

Xu, S., J. Lam, and C. Yang (2001). Quadratic stability and stabilization of uncertain linear discrte-time systems with state delay," *Systems and Control Letters*, **43**, 77-84.

www.elsevier.com/locate/ifac

OSCILLATIONS IN UNCERTAIN LOSSLESS PROPAGATION MODELS

Vladimir Răsvan * **Silviu-Iulian Niculescu** **

* *Dept. of Automatic Control, University of Craiova, A.I.Cuza, 13, Craiova, RO-1100, Romania*
E-mail: vrasvan@automation.ucv.ro
** *HEUDIASYC (UMR CNRS 6599), Université de Technologie de Compiègne, Centre de Recherche de Royallieu, 60205, Compiègne, France.*
E-mail: silviu@hds.utc.fr

Abstract: This paper focuses on forced oscillations (periodic and almost periodic) in some dynamical models described by coupled differential and difference equations including time-varying uncertainties. The existence and stability result is based on a theorem on invariant manifolds Banach spaces due to Kurzweil and Halanay. A Liapunov-Krasovskii functional approach combined with the $\mathcal{S}$-procedure is considered. *Copyright © 2001 IFAC*

Keywords: lossless propagation; oscillations; uncertainty; Krasovskii functionals.

1. INTRODUCTION

A quite well-established fact is that *oscillatory systems* with some interconnections between them display what is called *propagation phenomena* (see, e.g. Hale and Verduyn Lunel, 1993). *Lossless propagation* is associated to transmission lines without lossess: such lines correspond in engineering to LC electrical lines or ot lossless stream, water or gas pipes. In engineering, the propagation problem is strongly related to *electric* and *electronic applications*, e.g. circuit structures consisting of multipoles connected through LC transmission lines Brayton an Miranker (1964) (see also Marinov and Neittaanmäki, 1991). Some propagation phenomena may be also met in *power distribution* systems if the distribution area is quite large. We shall note that the lossless propagation occurs also for *non-electric* 'signals' as water, steam or gas flows and pressures.

We shall not insist on the modeling part, but we just point out that a long list of references that were published along the time may be found in the above cited references. The mathematical model is described in all these cases by a *mixed initial* and *boundary value problem* for hyperbolic partial differential equations modeling the lossless propagation. The boundary conditions are of special type beeing in "feedback connection" with some system described by ordinary differential equations. This sends to the so-called "derivative boundary conditions" considered by Cooke (1970) (see also Cooke and Krumme, 1968) but also to the even more general boundary conditions of Abolinia and Myshkis (1960), described by Volterra operators. Integration along characteristics of the hyperbolic partial differential equation (which is in fact the method of d'Alembert) mentioned in the cited references allows the association of a certain system of functional equations to the mixed problem; more precisely, a one-to-one correspondence may be established and proved between the solutions of the mixed problem for hyperbolic partial differential equations and the initial value (Cauchy) problem for the associated system of functional equations.

In certain cases, some of them considered in Hale and Martinez-Amores (1977), Halanay and Răsvan (1997), Cooke (1970), Cooke and Krumme (1968), this system of functional differential equations reads as follows:

$$\begin{cases} \dot{x}_1(t) = Ax_1(t) + Bx_2(t-\tau) \\ \qquad +\tilde{f}_1(t, x_1(t), x_2(t), x_2(t-\tau)) \\ x_2(t) = Cx_1(t) + Dx_2(t-\tau) \\ \qquad +\tilde{f}_2(t, x_1(t), x_2(t), x_2(t-\tau)) \end{cases}, \quad (1)$$

This system has been treated by Hale and Martinez-Amores (1977) by writing the second equation as:

$$\frac{d}{dt}\Big[x_2(t) - Cx_1(t) - Dx_2(t-\tau) \quad (2)$$
$$-\tilde{f}_2(x_1(t), x_2(t), x_2(t-\tau))\Big] = 0,$$

and applying the general results for neutral systems presented in Hale and Verduyn Lunel (1993).

An earlier approach (Răsvan, 1975) suggested the treatment of (1) as a special case of neutral systems by letting $x_2(t) = \dot{z}(t)$. This last approach was used in absolute stability (Răsvan, 1975), forced nonlinear oscillations (Halanay and Răsvan, 1977) and approximation by ODEs (Halanay and Răsvan, 1981) (which "projected back" on the partial differential equations gave the method of lines).

Throughout this paper, we shall deal with the problem of *forced oscillations* (periodic or almost periodic) for a special version of (1), namely:

$$\begin{cases} \dot{x}_1(t) = (A + \Delta A(t))x_1(t) + Bx_2(t-\tau) \\ \qquad -b_1\phi(t, \sigma(t)) + f_1(t) \\ x_2(t) = Cx_1(t) + Dx_2(t-\tau) \\ \qquad -b_2\phi(t, \sigma(t)) + f_2(t) \\ \sigma(t) = c^T x_1(t). \end{cases} \quad (3)$$

Here $x_1 \in \mathbf{R}^{n_1}$ and $x_2 \in \mathbf{R}^{n_2}$, $\tau > 0$ is the propagation time (here, the delay), A, B, C, D are matrices of appropriate dimensions and b_1, b_2, c are vectors. The bounded vector functions $f_1(\cdot)$, $f_2(\cdot)$ might be periodic or almost periodic. Here $\Delta A(\cdot)$ is an unknown real norm-bounded matrix function which represent a time-varying parameter uncertainty. The admissible uncertainty are assumed to be of the form:

$$\Delta A(t) = DF(t)E, \quad F \in \mathbf{R}^{i \times j}, \quad (4)$$

where $F(t)$ is unknown time-varying matrix with measurable elements satisfying:

$$F^T(t)F(t) \leq I, \quad (5)$$

and D, E are known real matrices which characterize how the uncertain parameters in $F(t)$ enter the nominal matrix A.

Some (delay-independent) exponential stability results for the linear case with constant matrix coefficients and no uncertainties are proposed in Niculescu and Răsvan (2000a) (a frequency-domain approach: appropriate finite-dimensional transfer functions) and in Niculescu and Răsvan (2000b) (a time-domain approach: appropriate Liapunov-Krasovskii functionals). The corresponding numerical schemes use norm properties of some continuous- and discrete-time transfer functions that can be easily "translated" in solving algebraic Riccati equations or appropriate linear matrix inequalities.

This note completes and extends into a time-domain framework the contributions of Halanay and Răsvan (1977), and it can be interpreted or seen as the Liapunov-Krasovskii counterpart. The main interest lies in the simplicity of the corresponding numerical schemes.

We shall, therefore, consider the general system (3) and the following Liapunov-Krasovskii type functional candidate: $V : \mathbf{R}^n \times \mathcal{L}^2(-\tau, 0; \mathbf{R}^n) \mapsto \mathbf{R}_+$:

$$V(x_1, x_2) = x_1^T P x_1 + \int_{-\tau}^{0} x_2(\theta)^T R x_2(\theta) d\theta \quad (6)$$

searching for the positive definite matrices $P > 0$ and $R > 0$ that should satisfy the requirements that will be deduced in the sequel.

Note that the functional $V(x_1, x_2)$ is similar to the one proposed in Niculescu and Răsvan (2000b) for the asymptotic stability analysis case study, and represents a natural extension of the form proposed in Niculescu *et al.* (1997) (retarded case).

The proposed analysis is *constructive*. In fact, we shall compute some basic estimates necessary for applying the Kurzweil-Halanay type results concerning invariant manifolds on Banach spaces.

2. BASIC ESTIMATES

At this end, we differentiate along the solutions of (3) the function $V^* : \mathbf{R} \mapsto \mathbf{R}_+$ defined by:

$$V^*(t) = V(e^{\alpha t}x_1(t), e^{\alpha t}x_{2t}) = e^{2\alpha t}x_1(t)^T P x_1(t)$$
$$+ \int_{t-\tau}^{t} e^{2\alpha\theta} x_2(\theta) R x_2(\theta) d\theta, \quad (7)$$

where $\alpha > 0$ is subject to choice, $(x_1(t), x_2(t))$ is some solution of (3) starting from $t_0 \in \mathbf{R}$ and $x_{2t} \equiv x_2(t+\theta)$, $-\tau \leq \theta \leq 0$. We deduce:

$$\dot{V}^*(t) = e^{2\alpha t}\left\{2\alpha x_1(t)^T P x_1(t)\right.$$
$$+2x_1(t)^T P\left[(A + DF(t)E)x_1(t) + \right.$$

$$+Bx_2(t-\tau) - b_1\phi_1(t, c_1^T x_1(t)) + f_1(t)\big]$$
$$+x_2(t)^T R x_2(t) - e^{-2\alpha\tau} x_2(t-\tau)^T R x_2(t-\tau)\big\}.$$

We substitute $x_2(t)$ by its expression prescribed from the difference equation in (3), and using some standard matrix inequalities, we obtain:

$$-\dot{V}^*(t) \leq e^{2\alpha t}\left[x_1(t)^T \quad x_2(t-\tau)^T \quad \phi(t, c^T x_1(t))\right] \cdot$$
$$\begin{bmatrix} \begin{pmatrix} -(A+\alpha I)^T P \\ -P(A+\alpha I) \\ -C^T RC \\ -PDD^T P \\ -E^T E \end{pmatrix} & -(PB + C^T RD) & \begin{pmatrix} Pb_1 + \\ C^T Rb_2 \end{pmatrix} \\ -(B^T P + D^T RC) & \begin{pmatrix} e^{-2\alpha\tau} R \\ -D^T RD \end{pmatrix} & D^T Rb_2 \\ \begin{pmatrix} b_1^T P + \\ b_2^T RC \end{pmatrix} & b_2^T RD & -b_2^T Rb_2 \end{bmatrix} \cdot$$
$$\cdot \begin{bmatrix} x_1(t) \\ x_2(t-\tau) \\ \phi(t, c^T x_1(t)) \end{bmatrix}$$
$$-e^{2\alpha t}\left[x_1(t)^T P f_1(t) + f_1(t)^T P x_1(t) + f_2(t)^T RCx(t)\right.$$
$$+x_1(t)^T C^T R f_2(t) + f_2(t)^T RD x_2(t-\tau)$$
$$+x_2(t-\tau)^T D^T R f_2(t) + f_2(t)^T R b_1 \phi(t, c^T x_1(t))$$
$$\left.+\phi(t, c^T x_1(t)) b_1^T R f_2(t) + f_2(t)^T R f_2(t)\right]. \quad (8)$$

We seek for the quadratic form $-\dot{V}^*(t)$ from the above expression to be nonnegative definite but positive on the (n_1+n_2)-dimensional subspace of the variables x_1 and x_2. But nonnegative definiteness of $-\dot{V}^*$ with $P > 0, R > 0$ is not possible since $-b_2^T Rb_2 < 0$.

This situation clearly indicates that we have to apply the *S-procedure* (Yakubovich, 1962): taking into account (8) we add and substract from $-\dot{V}^*$ the quantity $\theta_0\left(\sigma - \frac{1}{\bar{\phi}}\phi(\sigma)\right)\phi(\sigma)$, where $\theta_0 > 0$. Our requirement will be now that the quadratic form:

$$-S(x_1, x_2(t-\tau), \phi) = -\dot{V}^*(x_1(t), x_2(t-\tau), \phi)$$
$$-\theta_0\left(c^T x_1(t) - \frac{1}{\bar{\phi}}\phi(t, c^T x_1(t))\right)\phi(t, c^T x_1(t))$$

has to be *nonnegative definite*, but *positive-definite* on the (n_1+n_2)-dimensional subspace of the first two variables.

If this is true, then the following estimate is obtained from (8):

$$\dot{V}^*(t) \leq -\delta e^{2\alpha t}\left(|\, x_1(t)\,|^2 + |\, x_2(t-\tau)\,|^2\right)$$
$$+e^{2\alpha t}\left[\gamma_1\left(|\, x_1(t)\,| + |\, x_2(t-\tau)\,|\right) + \gamma_2\right] \quad (9)$$

for some $\delta > 0$, and $\gamma_1, \gamma_2 > 0$.

In obtaining (9), we took into account boundedness of f_1 and f_2. If we recall the obvious inequality $2\sqrt{\alpha^2+\beta^2} \geq \alpha + \beta$, with $\alpha > 0$, $\beta > 0$, then (9) becomes:

$$\dot{V}^*(t) \leq -\delta e^{2\alpha t}\left(|\, x_1(t)\,|^2 + |\, x_2(t-\tau)\,|^2\right)$$
$$+e^{2\alpha t}\left[2\gamma_1\left(|\, x_1(t)\,|^2 + |\, x_2(t-\tau)\,|^2\right)^{\frac{1}{2}} + \gamma_2\right]. \quad (10)$$

The RHS of the above inequality will be nonpositive on the set defined by:

$$|\, x_1\,|^2 + |\, x_2\,|^2 \geq \rho^2, \quad (11)$$

where ρ is the positive root of the equation:

$$\rho X^2 - 2\gamma_1 X - \gamma_2 = 0.$$

On the set defined by (11), the Liapunov functional is non-increasing. Since $P > 0$, $R > 0$, we shall have on this set (i.e. sufficiently "far away from the origin"):

$$e^{2\alpha t}\lambda_0\left(|\, x_1(t)\,|^2 + \int_{-\tau}^{0} |\, x_2(t+\theta)^2\,|\, d\theta\right) \leq$$
$$\leq V(x_1(t), x_{2t}) \leq V(x_1(\tilde{t}), x_{2\tilde{t}}) \leq \Lambda_0\left(|\, x_1(\tilde{t})\,|^2\right.$$
$$\left.+\int_{-\tau}^{0} |\, x_2(\tilde{t}+\theta)^2\,|\, d\theta\right) e^{2\alpha\tilde{t}}, \quad t \geq \tilde{t}, \quad (12)$$

where $\lambda_0 > 0$, and $\Lambda_0 > 0$ are respectively the smallest and the largest among the eigenvalues of P and R (considered together in a single finite set of n_1+n_2 positive real numbers). Outside the set defined by (11), i.e. inside the ball $|\, x_1\,|^2 + |\, x_2\,|^2 \leq \rho^2$, the solutions of the system are bounded and we have:

$$|\, x_1(t)\,|^2 + \int_{-\tau}^{0} |\, x_2(t+\theta)\,|^2\, d\theta \leq \rho^2(1+\tau). \quad (13)$$

From here and (12), we deduce:

$$\left(|\, x_1(t)\,|^2 + \int_{-\tau}^{0} |\, x_2(t+\theta)\,|^2\, d\theta\right)^{\frac{1}{2}} \leq$$
$$\sqrt{\frac{\Lambda_0}{\lambda_0}} e^{-\alpha(t-\tilde{t})}\left(|\, x_1(\tilde{t})\,|^2 + \int_{-\tau}^{0} |\, x_2(\tilde{t}+\theta)\,|^2\, d\theta\right)$$
$$+\rho\sqrt{1+\tau}, \quad t \geq \tilde{t}. \quad (14)$$

Remark 1. This is the first of the estimates of the type used by Halanay (1969), and Halanay and Răsvan (1977) for a simpler system. Note that they used a different argument to prove the corresponding inequality.

We shall consider, further, two solutions $(x_1^i(t), x_2^i(t))$ $i = \overline{1,2}$, of (3), define their difference $(z_1(t), z_2(t))$ as:

$$z_1(t) = x_1^1(t) - x_1^2(t), \; z_2(t) = x_2^1(t) - x_2^2(t) \quad (15)$$

satisfying

$$\begin{cases} \dot{z}_1(t) = (A + \Delta A(t))z_1(t) + Bz_2(t-\tau) \\ \quad -b_1[\phi(t, c^T x_1^1(t)) - \phi(t, c^T x_1^2(t))] \\ z_2(t) = Cz_1(t) + Dz_2(t-\tau) \\ \quad -b_2[\phi(t, c^T x_1^1(t)) - \phi(t, c^T x_1^2(t))], \end{cases} \quad (16)$$

and differentiate along (16) the Liapunov functional (6) written along (15):

$$\tilde{V}^*(t) = V(e^{\alpha t} z_1(t), e^{\alpha t} z_2(t)).$$

A simple comparison of (16) and (3) will give the estimate

$$\left(| z_1(t) |^2 + \int_{-\tau}^{0} | z_2(t+\theta) |^2 \, d\theta \right)^{\frac{1}{2}}$$
$$\leq \gamma_4 e^{-\alpha(t-\tilde{t})} \left(| x_1^1(\tilde{t}) - x_1^2(\tilde{t}) |^2 \right.$$
$$\left. + \int_{-\tau}^{0} | x_2^1(\tilde{t}+\theta) - x_2^2(\tilde{t}+\theta) |^2 \, d\theta \right)^{\frac{1}{2}}, \quad t \geq \tilde{t}. (17)$$

Remark 2. We have thus obtained the second estimate (see below).

3. MAIN RESULTS

With the estimates computed above we are able to apply the following results of Kurzweil (1967) and Halanay (1967) concerning invariant manifolds on Banach spaces:

Theorem 1. (Kurzweil-Halanay). Suppose for a general flow in the Banach space $\mathcal{B}$ there exists real positive constants l, T, $\alpha < 1$, k_1, k_2 such that:
i) $||\tilde{c}|| \leq l$, $t \in [\tilde{t} + T, \tilde{t} + 2T]$ imply $||c(t; \tilde{t}, \tilde{c})|| \leq l$;
ii) $||\tilde{c}_1|| \leq l$, $||\tilde{c}_2|| \leq l$, $t \in [\tilde{t} + T, \tilde{t} + 2T]$ imply

$$||c(t; \tilde{t}, \tilde{c}_1) - c(t; \tilde{t}, \tilde{c}_2)|| \leq \alpha ||\tilde{c}_1 - \tilde{c}_2||;$$

iii) the following condition holds:

$$||c(t; \tilde{t}, \tilde{c}_1) - c(t; \tilde{t}, \tilde{c}_2)|| \leq k_1 e^{k_2(t-\tilde{t})} ||\tilde{c}_1 - \tilde{c}_2||, \forall t \geq \tilde{t}.$$

Then there exists $p : \mathbf{R} \mapsto \mathcal{B}$ and the positive constants k, ν such that:
a) $||\tilde{p}(t)|| \leq l$;
b) $p(t) = c(t; t_1, p(t_1))$, $\forall t \geq t_1$, $t_1 \in \mathbf{R}$;
c) $||c(t; \tilde{t}, \tilde{c}) - p(t)|| \leq k e^{-\nu(t-\tilde{t})} ||\tilde{c} - p(\tilde{t})||$, $||\tilde{c}|| \leq l$, $t \geq \tilde{t}$;
d) If $c(t + \Omega, \tilde{t} + \Omega, \tilde{c}) \equiv c(t; \tilde{t}, \tilde{c}) \equiv c(t; \tilde{t}, \tilde{c})$ then $p(t + \Omega) = p(t)$.
e) If every sequence $\{h_n\}_n$, $h_n \to +\infty$ contains a subsequence h_{n_k} such that for every bounded $c(t; \tilde{t}, \tilde{c})$, the sequence $c(t + h_{n_k}, \tilde{t} + h_{n_k}, \tilde{c})$ converges uniformly with respect to $\tilde{t}$, $\tilde{c}$ for $\tilde{t} \in \mathbf{R}$, $||\tilde{c}|| \leq l$ then p is almost periodic.
Remark that the condition in **d)** is satisfied if the flow is defined by a periodic system and the condition in **e)** if the flow is defined by a system which is almost periodic with respect to t, uniformly with respect to the state variables.

The Banach space in our case is the Hilbert space $\mathbf{R}^{n_1} \times \mathcal{L}^2(-\tau, 0; \mathbf{R}^{n_2})$ with the natural norm occuring in (12) (which coincides up to a multiplicative constant with the norm induced by the positive-definite Liapunov functional (6)).

The simple application of the general result will ensure *existence* and *global exponential stability* of a unique periodic or almost periodic solution for (3) according to the possible situations described in the conclusion of the Theorem 1.

We may summarize by stating:

Theorem 2. Consider the system (3) under the following assumptions:
i) There exist positive-definite matrices P and R and the real numbers α and θ_v such that the quadratic form $-S(x_1, x_2, \phi)$ defined above should be nonnegative-definite but positive-definite on $\mathbf{R}^{n_1} \times \mathbf{R}^{n_2}$.
ii) The nonlinear function $\phi(t, \sigma)$ satisfies:

$$0 \leq \frac{\phi(t, \sigma_1) - \phi(t, \sigma_2)}{\sigma_1 - \sigma_2} \leq \bar{\phi},$$

for all $\sigma_1, \sigma_2 \in \mathbf{R}$, $\sigma_1 \neq \sigma_2$.
iii) $| f_1(t) | + | f_2(t) | \leq M$, $\forall t \in \mathbf{R}$.

Then there exists a solution of (3) which is bounded on $\mathbf{R}$ and periodic if $\phi(\cdot, \sigma)$ and $f_i(t)$ are periodic with commensurable periods (their periods are in rational ratio) or almost periodic if $\phi(\cdot, \sigma)$ and $f_i(t)$ are periodic with non-commensurable periods or almost periodic. Moreover, this solution is exponentially stable.

4. CONCLUDING REMARKS AND FURTHER COMMENTS

It has been pointed out throughout the paper that proving existence and stability of forced oscillations requires application of some fixed-point type theorem. In our case this theorem is given by Kurzweil-Halanay theorem which allows to obtain not only existence but also exponential stability of forced oscillations.

Application of this fundamental result requires at its term some specific estimates of the solutions. Two are the ways of obtaining these estimates - both emerging from the stability theory of the systems with sector restricted nonlinearities. The first one is that based on Popov-like frequency domain inequalities, most suitable for infinite-dimensional systems but leading to computational complications. The second one, based on Liapunov Krasovskii functionals seems more restrictive(theoretically) but better from computational point of view; moreover it leads to linear matrix inequalities which may be solved by high-quality software.

For this reasons we took the second approach which allowed also to obtain some robustness with respect to coefficient perturbations satisfying some quadratic inequalities. Worth mentioning that existence of the solution of the involved matrix inequalities is also equivalent to some Popov-like inequality which is now *delay-independent* and *finite-dimensional*.

REFERENCES

Abolinia V. E. and Myshkis, A. D. (1960). Mixed problem for an almost linear hyperbolic system in the plane (in Russian), *Matem. sbornik*, vol.**12**, no.4, pp. 423-442.

R. K. Brayton and W. L. Miranker (1964). Oscillations in a distributed network. *Arch. Rat. Mech. Anal.*, **17**, 358-376.

Cooke, K. L. (1970). A linear mixed problem with derivative boundary conditions. in *Seminar on differential equations and dynamical systems (III)*, Lecture Series, No. 51, University of Maryland, College Park, 11-17.

Cooke, K. L. and Krumme, D. W. (1968). Differential-Difference Equations and Nonlinear Partial-Boundary Value Problems for Linear Hyperbolic Partial Differential Equations. *J. Math. Anal. Appl.*, vol.**24**, no.2, 372-387.

Halanay, A. (1967). Invariant manifolds for systems with time lags. in *Differential equations and dynamical systems* (Eds. Hale & LaSalle), Academic Press: NY, 399-413.

Halanay, A. (1969). "Almost periodic solutions for a class of nonlinear systems with time lags," *Revue Roumaine Math. Pures* **14**, 1269-1276.

Halanay, A. and Răsvan, Vl. (1977). "Almost periodic solutions for a class of systems described by delay-differential and difference equations," in *J. Nonlinear Analysis. Theory, Methods & Applications*, vol. **1**, no. 1, pp. 197-206.

Halanay, A. and Răsvan, Vl. (1981). Approximation of delays by differential equations. in *Recent Advances in Differential Equations* (R. CONTI, Ed.), Academic Press, pp. 155-197.

Halanay, A. and Răsvan, Vl. (1997). Stability radii for some propagation models. in *IMA J. Math. Contr. Information*, vol. **14**, 95-107.

Hale, J. K. and Verduyn Lunel, S. M. (1993). *Introduction to Functional Differential Equations*. Springer-Verlag, New York.

Hale, J. K. and Martinez Amores, P. (1977). Stability in neutral equations. *J. Nonlinear Analysis. Theory, Methods & Applications*, vol. **1**, no. 1, pp. 161-172.

Kurzweil, J. (1967). Invariant manifolds for flows. in *Differential equations and dynamical systems* (Eds. Hale & LaSalle), Academic Press: NY, 431-468.

Marinov, C. and Neittaanmäki, P. (1991). *Mathematical Models in Electrical Circuits: Theory and Applications*, Kluwer Academic Press.

Niculescu, S. -I. and Răsvan, Vl (2000a). Delay-independent stability in lossless propagation models with applications (I): A complex-domain approach. in *Proc. MTNS 2000*, Perpignan, France (June 2000).

Niculescu, S. -I. and Răsvan, Vl (2000b). Delay-independent stability in lossless propagation models with applications (II): A Lyapunov based approach. in *Proc. MTNS 2000*, Perpignan, France (June 2000).

Niculescu, S. -I., Verriest E. I., Dugard, L. and Dion, J. -M. (1997). Stability and robust stability of time-delay systems: A guided tour. in *Stability and Control of Time-Delay Systems* (L. Dugard and E. I. Verriest, Eds.), LNCIS, Springer-Verlag, London, **228**, pp. 1-71.

Răsvan, Vl. (1975). *Absolute stability of time lag control systems* (in Romanian), Ed. Academiei, Bucharest, 1975 (Russian revised edition by Nauka, Moscow, 1983).

Yakubovich (1962). Solution of some matrix inequalities that are met in automotic control theory. (in Russian) *Dokl. Akad. Nauk SSSR* **143**, 1304-1307.

www.elsevier.com/locate/ifac

MATRIX CONVEX DIRECTIONS FOR TIME DELAY SYSTEMS [1]

J. Santos [2], S. Mondié and V. Kharitonov

Departamento de Control Automático CINVESTAV-IPN, Av. IPN 2508, A.P. 14-740, 07300 México, D.F., México, [khar,smondie,jsantos]@ctrl.cinvestav.mx

Abstract: The characterization of convex direction for real Hurwitz matrices is extended to the case of multivariable time delay linear systems. The above is achieved using the results of convex directions for quasipolynomials. *Copyright © 2001 IFAC*

Keywords: time delay systems, convex directions.

1. INTRODUCTION

Many actual systems have the property of aftereffect, i.e., the future states depend not only on the present but also on the past history. Delay is believed to occur in such areas as mechanics, chemistry, biology, medicine, economics, atomic energy, information theory, etc.

For an accurate modelling of such systems, the effect that the delays may have on the behavior of the system response must indeed be taken into account.

Stability is one of the important characteristics of dynamical systems. The concept of convex directions for polynomials and quasipolynomials proves to be a useful tool in stability and robust stability study, (see Rantzer (1992), Kharitonov and Zhabko (1993)). Recently some interesting extention were made to the case of matrices, see Kokame et all. (1994). There, a complete characterization of convex directions for real Hurwitz matrices has been given. Similar results for discrete time systems were reported in Xie (1997).

The aim of this paper is to characterize matrix convex directions for time delay systems. Since this characterization must be valid for all values of delay, it must also be valid when the delay is equal to zero, therefore, matrix convex directions of Kokame et all. (1994) are the only matrix candidates for our characterization. It worths to be mentioned the fact that the caracterization does not depend of delay does not imply that we are talking about stability independent of delay. The characteristic function of the time delay system corresponding to each candidate, which is a quasipolynomial of the retarded type (Bellman and Cooke, 1963), is then analyzed with the help of results on convex directions for quasipolynomials (Kharitonov and Zhabko, 1993).

New proofs of the basic results on matrix convex directions obtained in Kokame et all. (1994) are also given. They are essentially based on the ideas of the proofs developed for delayed systems and on results about convex directions for polynomials.

2. PRELIMINARIES

We introduce first some background on convex directions.

Convex Directions of Polynomials

A notion of convex direction has been proposed in Rantzer (1992) and is defined as follows:

Definition 1. A polynomial $g(s)$ is a convex direction if for any stable polynomial $p(s)$, the stability of $p(s)+g(s)$ implies that of

$$p(s)+\mu g(s), \qquad \mu\in[0,1] \tag{1}$$

where $\deg(p(s))>\deg(g(s))$.

In the following we will use the following two results.

Remark 1. Any first or zero degree polynomial $g(s)$ with real coefficients is a convex direction for real Hurwitz stable polynomials.

Remark 2. Any constant polynomial $g(s)$ is a convex direction for complex Hurwitz stable polynomials.

[1] Supported by Conacyt, Mexico 31951-A.
[2] Financed by Conacyt 130255.

Matrix convex directions

Consider a linear time-invariant system of the form

$$\dot{x}(t) = Ax(t) + \mu Dx(t) \quad (2)$$

$$x(t) \in R^n, \; A, D \in R^{n\times n}, \; \mu \in R.$$

where matrix A is Hurwitz and μ is a scalar parameter.

Definition 2. (Kokame et all., 1994) Matrix $D \in R^{n\times n}$ is a convex direction if for any Hurwitz matrix $A \in R^{n\times n}$, the stability of

$$\dot{x}(t) = Ax(t) + Dx(t)$$

implies that of (2) for all $\mu \in [0, 1]$.

Geometric interpretation: the previous definition means that system (2) has only one open interval of stability on the real axis μ. The size and location of the interval depends on matrices A and D, but the fact that (2) has only one stability interval depends on matrix D. When system (2) is stable for $\mu = 0$ and $\mu = 1$ the stability interval contains segment $[0, 1]$. We can say even more: if D is a convex direction then for any (not necessarily Hurwitz) real square matrix A there exist no more than one stability interval on the μ-axis. In other words, the definition implies that if matrix D is a convex direction, then for any Hurwitz matrix A the system (2) has only one stability interval and vice versa, if for every Hurwitz matrix A the system (2) has only one stability interval, then matrix D is a convex direction.

Initial Problem: *Characterize the set of matrix convex directions for system (2).*

The characterization has been given in Kokame et all., (1994), where the following auxiliary results were also stated.

Lemma 1. (Kokame et all., 1994) The identity matrix is a matrix convex direction.

Lemma 2. (Kokame et all., 1994) If $D \in R^{n\times n}$ is a matrix convex direction, so is kD for an arbitrary real scalar k.

Lemma 3. (Kokame et all., 1994) If $D \in R^{n\times n}$ is a matrix convex direction, so is $S^{-1}DS$, where S is a real nonsingular matrix.

Lemma 4. (Kokame et all., 1994) Let D be a block triangular matrix

$$D = \begin{bmatrix} D_{11} & D_{12} \\ 0 & D_{22} \end{bmatrix},$$

where $D_{ij} \in R^{n_i\times n_j}$, for $i, j = 1, 2$. If D is a matrix convex direction, so are both D_{11} and D_{22}.

The next theorems describe main results from Kokame et all., (1994). We supply them with new proofs, because we will use later on the same arguments for the case of time delay systems.

Theorem 5. (Kokame et all., 1994) For $n = 2$ a matrix $D \in R^{2\times 2}$ is a matrix convex direction if and only if D, up to similarity transformations, has one of the following three forms:

$$D_1 = \begin{bmatrix} 0 & d \\ 0 & 0 \end{bmatrix}, \; D_2 = \begin{bmatrix} d & 0 \\ 0 & d \end{bmatrix}, \; D_3 = \begin{bmatrix} x_1 & 0 \\ 0 & x_2 \end{bmatrix}, \quad (3)$$

where d, x_1, x_2 are real numbers and $x_1 x_2 \leq 0$.

Proof. 1. For D_1 the characteristic polynomial of the system (2) is

$$p(s) = s^2 + s(-\alpha_{22} - \alpha_{11}) + (\alpha_{11}\alpha_{22} - \alpha_{21}\alpha_{12}) - \mu d\alpha_{21},$$

and can be written as $p(s) = p_0(s) + \mu g(s)$, where $g(s) = -d\alpha_{21}$. Since $g(s)$ is a constant polynomial it is a convex direction, then it follows from remark 1 that D_1 is a matrix convex direction.

2. For D_2, the statement follows directly from lemma 1 and lemma 2.

3. For D_3, the characteristic polynomial of system (2) has the form

$$\begin{aligned} p(s) = & s^2 + s(-\alpha_{22} - \alpha_{11}) + (\alpha_{11}\alpha_{22} - \alpha_{12}\alpha_{21}) \\ & + \mu(\alpha_{11}x_2 + \alpha_{22}x_1 - (x_1 + x_2)s) + \mu^2 x_1 x_2 \\ = & p_0(s) + \mu g_1(s) + \mu^2 g_2(s), \; \mu \in [0, 1]. \quad (4) \end{aligned}$$

It is not possible to apply here directly arguments based on polynomial convex directions because (4) depends quadratically on μ. Therefore we imbed first the family into the bigger one

$$p(s) = p_0(s) + \mu_1 g_1(s) + \mu_2 g_2(s) \quad (5)$$

where new parameters μ_1, μ_2, vary in the triangle ABC (see Figure 1). This new family contains the original one for $\mu_1 = \mu$, $\mu_2 = \mu^2$, $\mu \in [0, 1]$ (see curve η on Figure 2). On the other hand (5) is the convex hull of three polynomials: $p_0(s)$, $p_0(s)+g_1(s)$, $p_0(s)+g_1(s)+g_2(s)$, and therefore we may apply the edge theorem to check stability of the new family. Indeed, this theorem says that the stability of the three edges l_1, l_2, l_3 implies that of (5).

Polynomials, corresponding to extreme points A and B are stable because (4) is stable for $\mu = 0$ and $\mu = 1$. We need first to verify stability of polynomial $p_0(s)+g_1(s)$, which corresponds to the third extreme point, C. Comparing coefficients of $p_0(s) + g_1(s) + g_2(s)$ and $p_0(s) + g_1(s)$ we see that the only difference is in the free coefficient. This coefficient of $p_0(s)+g_1(s)$ is bigger on $-x_1x_2$ (remember that $x_1x_2 \leq 0$) than the corresponding coefficient of $p_0(s) + g_1(s) + g_2(s)$. Both polynomials are of degree two, therefore stability of $p_0(s) + g_1(s) + g_2(s)$ implies that of $p_0(s) + g_1(s)$.

Now the stability of the three edges is verified with the help of the concept of convex direction.

For the edge l_1 we have to verify the stability of $p_0(s) + \mu[g_1(s) + g_2(s)]$ for all $\mu \in [0,1]$. Since $g_1(s) + g_2(s)$ is of the first degree, then it is a convex direction, and stability of polynomials corresponding to extreme points A, B, imply that of the edge l_1.

For the edge l_2, we have to verify the stability of $p_0(s) + \mu_1 g_1(s)$ for all $\mu_1 \in [0,1]$. Again $g_1(s)$ is of the first degree, so it is a convex direction, and edge l_2 is stable.

For the edge l_3, we have to verify the stability of the segment $p_0(s) + g_1(s) + \mu_2 g_2(s)$ for all $\mu_2 \in [0,1]$. Polynomial $g_2(s)$ is constant and therefore it is a convex direction, so edge l_3 is stable. Since all three edges, l_1, l_2, l_3, are stables, this implies the stability of the family (5) and, of course, that of (4). Therefore matrix D_3 is a matrix convex direction. ■

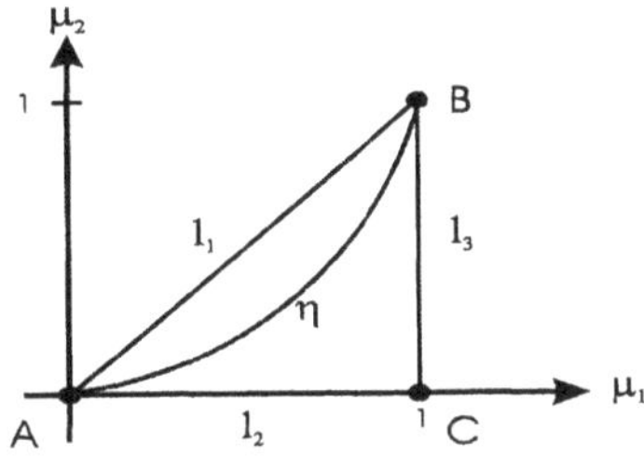

Fig. 1

Theorem 6. (Kokame et all., 1994) A matrix $D \in R^{3\times 3}$ is a matrix convex direction if and only if D coincides up to similarity transformation with one of the following two matrices:

$$D_4 = \begin{bmatrix} 0 & d & 0 \\ 0 & 0 & 0 \\ 0 & 0 & 0 \end{bmatrix}, \quad D_5 = \begin{bmatrix} d & 0 & 0 \\ 0 & d & 0 \\ 0 & 0 & d \end{bmatrix},$$

where d is a real number.

Proof. 1. For matrix D_4, the characteristic polynomial of the system (2) is

$$p(s) = p_0(s) + \mu g(s), \tag{6}$$

where

$$p_0(s) = \det(sI^{3\times 3} - A),$$
$$g(s) = d(-\alpha_{21}s + \alpha_{21}\alpha_{33} - \alpha_{31}\alpha_{23}).$$

Since $g(s)$ is a polynomial of the first degree, then it is a convex direction, therefore D_4 is a matrix convex direction.

2. For matrix D_5, we can observe that $D_5 = dI^{3\times 3}$ is the identity matrix multiplied by a real scalar d. By lemma 1 and lemma 2 one can conclude that matrix D_5 is a matrix convex direction. ■

Theorem 7. (Kokame et all., 1994) A matrix $D \in R^{n\times n}$, $n \geq 4$, is a matrix convex direction if a only if D is the identity matrix multiplied a real scalar d.

Proof. It follows directly from lemma 1 and lemma 2. ■

3. MATRIX CONVEX DIRECTIONS FOR TIME DELAY SYSTEMS

Given a linear delay system of the form

$$\dot{x}(t) = Ax(t) + \mu Dx(t-h) \tag{7}$$

where $x(t) \in R^n$, $A, D \in R^{n\times n}$, $h \geq 0$, $\mu \in R$ is the system parameter.

The following definition of matrix convex directions for time delay systems, is motivated by our geometric interpretation of matrix convex directions for systems without time delay.

Definition 3. A matrix $D \in R^{n\times n}$ is a convex direction for time delay systems if for any Hurwitz matrix $A \in R^{n\times n}$, and for any $h \geq 0$ the system (7) has only one interval of stability on the real axis of μ.

It is important to emphasize that this definition introduces matrix convex directions which do not depend on the value of the delay. But as it has been mentioned already in the introduction it does not imply that stability should be independent of delay.

Problem Statement: Characterize the set of matrix convex directions for (7).

The following remark is a direct consequence of definition 3.

Remark 3. If matrix $D \in R^{n\times n}$ is a convex direction, so is kD for a real scalar k.

Machinery for the characterization

Since we look for a characterization of matrix convex direction independent of the value of h, they must also be matrix convex directions for $h = 0$. Therefore, the only candidates to be matrix convex direction in this sense are those matrices which are convex directions for systems without delay. This observation reduces drastically the set of possible candidates.

For each candidate, our approach is based on the fact that it is possible to factorize the characteristic function of (7) in order to apply to each factor stability conditions based on the concept of convex direction for quasipolynomials.

Convex directions for quasipolynomials

It follows from the above arguments that the characterization of matrix convex directions for linear systems with delays and for quasipolynomials are intimately related. For this reason, we recall now important results on convex directions of quasipolynomials (Kharitonov and Zhabko, 1993).

Lemma 8. Let

$$\mathcal{H} = \left\{ \begin{array}{c} p(s) = s^n + a_1 s^{n-1} + \cdots + a_{n-1}s + a_n \\ p(s) = 0 \Rightarrow \mathrm{Re}(s) < 0 \end{array} \right\}$$

be the set of Hurwitz stable polynomials with real coefficients, and $q(s)$ be a real polynomial of degree less or equal to n. Then

$$g(s) = q(s)e^{\tau s}$$

is a convex direction for $\mathcal{H}$ if

$$\tau \leq 0, \tag{8}$$

and the inequality

$$\frac{\partial \arg(q(jw))}{\partial w} \leq \left| \frac{\sin(2\arg(q(jw)))}{2w} \right| \tag{9}$$

holds for all positive frequencies w where $q(jw) \neq 0$.

Lemma 9. Let

$$\mathcal{H} = \left\{ \begin{array}{c} p(s) = s^n + c_1 s^{n-1} + \cdots + c_{n-1}s + c_n \\ p(s) = 0 \Rightarrow \mathrm{Re}(s) < 0 \end{array} \right\}$$

be the set of Hurwitz stable polynomials with complex coefficients, and $q(s)$ be a complex polynomial of degree less or equal n. Then

$$g(s) = q(s)e^{\tau s}$$

is a convex direction for $\mathcal{H}$ if

$$\tau \leq 0, \tag{10}$$

and the inequality

$$\frac{\partial \arg(q(jw))}{\partial w} \leq 0 \tag{11}$$

holds for all real w where $q(jw) \neq 0$.

4. BASIC LEMMAS

We present now some basic statements which will be used later on for deriving matrix convex directions.

Lemma 10. The identity matrix $I \in R^{n \times n}$, is a matrix convex direction.

Proof. When $D = I \in R^{n \times n}$ system (7) has the following characteristic function

$$f(s) = \det(sI - A - \mu I e^{-sh}). \tag{12}$$

There exists a nonsingular matrix S such that

$$S^{-1}AS = \begin{bmatrix} \lambda_1 & * & \cdots & 0 \\ 0 & \lambda_2 & \ddots & \vdots \\ \vdots & \ddots & \ddots & * \\ 0 & \cdots & 0 & \lambda_n \end{bmatrix}$$

where λ_i, $i = 1, 2, ..., n$, are eigenvalues of A and $f(s)$ can be factorized as

$$\begin{aligned} f(s) &= (s - \lambda_1 - \mu e^{-sh})(s - \lambda_2 - \mu e^{-sh}) \\ &\quad \cdots (s - \lambda_n - \mu e^{-sh}). \end{aligned}$$

Each factor is of the form $f_i(s) = p_i(s) + \mu g_i(s)$, where $p_i(s) = s - \lambda_i$, and $g_i(s) = q_i(s)e^{-sh}$, $q_i(s) = -1$. It follows from lemma 8 and lemma 9 that $g_i(s)$ is a convex direction (real or complex, depending of λ_i). Therefore stability of $p_i(s)$ and $p_i(s) + g_i(s)$ implies that of $p_i(s) + \mu g_i(s)$ for all $\mu \in [0, 1]$. This means that $I^{n \times n}$ is a matrix convex direction. ■

Lemma 11. If $D \in R^{n \times n}$ is a convex direction, so is $S^{-1}DS$, where S is a real nonsingular matrix.

Proof. Since $D \in R^{n \times n}$ is a convex direction, then system (7) has one interval of stability on the real axis μ for all Hurwitz matrix A, so that, applying the similarity transformation to the system we obtain the new one

$$\begin{aligned} \dot{x}(t) &= S^{-1}\left(Ax(t) + \mu Dx(t-h)\right)S \\ &= \hat{A}x(t) + \mu S^{-1}DSx(t-h) \end{aligned} \tag{13}$$

which is stable if and only if the original one is stable. It is clear that the new system has only one stability interval on the axis μ when it is the case for system (7). In other words $S^{-1}DS$ is a matrix convex direction if and only if D is a matrix convex direction. ■

Lemma 12. Given a complex number z, then $\varphi(s) = s + \lambda - ze^{-sh}$ is stable if $\lambda > |z|$.

Proof. Suppose, by contradiction that $\varphi(s)$ has a root s_0 such that $\mathrm{Re}(s_0) \geq 0$. Then $s_0 + \lambda - ze^{-s_0 h} = 0$ and $|s_0+\lambda| = |z|\left|e^{-s_0 h}\right|$. Since $\left|e^{-s_0 h}\right| \leq 1$, we can write that $|s_0+\lambda| \leq |z|$. By our assumption $\mathrm{Re}(s_0) \geq 0$, $\lambda \leq |s_0+\lambda|$. Therefore $\lambda \leq |z|$. This inequality leads indeed to a contradiction. ■

Lemma 13. Let D be a block triangular matrix:

$$D = \begin{bmatrix} D_{11} & D_{12} \\ 0 & D_{22} \end{bmatrix}$$

where $D_{ij} \in R^{n_i \times n_j}$, for $i,j = 1,2$. If D is a matrix convex direction, so are both D_{11} and D_{22}.

Proof. If D is a matrix convex direction then the system

$$\dot{x}(t) = Ax(t) + \mu Dx(t-h) \qquad (14)$$

has only one stability interval for every Hurwitz stable matrix A. Let us assume by contradiction that D_{11} is not a matrix convex direction, i.e., there exists matrix A_{11}^* such that the system

$$\dot{y}(t) = A_{11}^* y(t) + \mu D_{11} y(t-h) \qquad (15)$$

is stable for $\mu = 0$ and $\mu = \alpha > 0$, but it is not stable for some $\mu_0 \in (0,\alpha)$. Now, let us choose a matrix A_{22}^* such that the system

$$\dot{z}(t) = A_{22}^* z(t) + \mu D_{22} z(t-h) \qquad (16)$$

is stable for all $\mu \in [0,\alpha]$.

Let $A_{22}^* = -\lambda I$ where $\lambda > 0$, then the characteristic equation of system (16) is given by

$$f(s) = \det(sI + \lambda I - \mu D_{22} e^{-sh}). \qquad (17)$$

Matrix D_{22} can be written as $D_{22} = T^{-1}JT$, where

$$J = \begin{bmatrix} z_1 & * & 0 & 0 \\ 0 & z_2 & \ddots & \vdots \\ \vdots & 0 & \ddots & * \\ 0 & \cdots & 0 & z_n \end{bmatrix}$$

is the canonical form of matrix D_{22}. We can apply the last transformation to (17) in order to obtain

$$f(s) = \Pi_{k=1}^{n_2}(s + \lambda - \mu z_k e^{-sh})$$

It follows from lemma 12 that system (16) is stable for all $\mu \in [0,\alpha]$ if $\lambda > \alpha \max_{1 \leq k \leq n_2}\{|z_k|\} = \lambda_0$.

Now let $\lambda > \lambda_0$. Consider matrix

$$A^* = \begin{bmatrix} A_{11}^* & 0 \\ 0 & -\lambda I \end{bmatrix}.$$

Then system (14) has the form

$$\dot{x}(t) = \begin{bmatrix} A_{11}^* & 0 \\ 0 & -\lambda I \end{bmatrix} x(t) + \mu \begin{bmatrix} D_{11} & D_{12} \\ 0 & D_{22} \end{bmatrix} x(t-h). (18)$$

The characteristic equation of (18) is

$$f(s) = \det(sI_1 - A_{11}^* - \mu D_{11} e^{-sh}) \cdot \det(sI_2 + \lambda I_2 - \mu D_{22} e^{-sh}). \qquad (19)$$

We know that $f(s)$ is stable for $\mu = 0$ and for $\mu = \alpha$ but is unstable for some $\mu_0 \in (0,\alpha)$. However D is a matrix convex direction, therefore the system (18) should be stable for all $\mu \in [0,\alpha]$. The contradiction proves the lemma. ■

5. CHARACTERIZATION OF MATRIX CONVEX DIRECTIONS

We are now able to verify if the candidates, namely, the matrix convex directions for linear systems, are also matrix convex directions for time delay systems.

Matrix convex directions in $R^{2\times 2}$

For 2×2 matrices we have to analyze the following three cases:

Case 1: Consider the following matrix candidate

$$D_\alpha = \begin{bmatrix} 0 & d \\ 0 & 0 \end{bmatrix}, \quad d \in R.$$

The characteristic equation of system (7) for this matrix is $f(s) = p(s) + \mu g(s)$ where $f(s) = s^2 + s(-\alpha_{22} - \alpha_{11}) + (\alpha_{11}\alpha_{22} - \alpha_{21}\alpha_{12})$, $g(s) = d\alpha_{21}e^{-sh} = q(s)e^{\tau s}$ with $q(s) = d\alpha_{21}$, and $\tau = -h$. It is clear that $\tau = -h < 0$, and

$$\frac{\partial \arg(q(jw))}{\partial w} = \frac{\sin(2\arg(q(jw))}{2w} = 0.$$

So, $g(s)$ satisfies lemma 8 and D_α is a matrix convex direction.

Case 2: Consider the matrix candidate

$$D_\beta = \begin{bmatrix} d & 0 \\ 0 & d \end{bmatrix}, \quad d \in R$$

We notice that $D_\beta = dI^{2\times 2}$ is the identity matrix multiplied by a real scalar d. Due to lemma 10 and remark 3 we infer that matrix D_β is a matrix convex direction.

Case 3: Consider the matrix candidate

$$D_\gamma = \begin{bmatrix} x_1 & 0 \\ 0 & x_2 \end{bmatrix}, \quad x_1, x_2 \in R, \quad x_1 x_2 \leq 0.$$

We are able to analyze only the subcase $x_1 x_2 = 0$. The analysis of the general case, $x_1 x_2 < 0$, seems to be rather involved. Let us assume that $x_1 = 0$ while $x_2 \neq 0$. The characteristic equation of system (7) is then

$$f(s) = s^2 - (\alpha_{22} + \alpha_{11})s + (\alpha_{11}\alpha_{22} - \alpha_{12}\alpha_{21}) \\ +\mu(\alpha_{11}x_2 - x_2 s)e^{-sh},$$

where $g(s) = q(s)e^{\tau s}$, $q(s) = x_2(\alpha_{11} - s)$, and $\tau = -h$. Since $f(s)$ has real coefficients it follows from 8 that it is a convex direction, and therefore in this particular case ($x_1 = 0$, $x_2 \neq 0$) D_γ is matrix convex direction.

The analysis of the subcase $x_2 = 0$, $x_1 \neq 0$ is similar to the previous one.

Matrix convex directions in $R^{3\times 3}$

For 3×3 matrices, we have the following possibilities.

Case 4: The matrix candidate

$$D_\delta = \begin{bmatrix} d & 0 & 0 \\ 0 & d & 0 \\ 0 & 0 & d \end{bmatrix}, \quad d \in R,$$

is a matrix convex direction. This is a direct consequence of lemma 10 and remark 3.

Case 5: Consider the candidate

$$D_\epsilon = \begin{bmatrix} 0 & d & 0 \\ 0 & 0 & 0 \\ 0 & 0 & 0 \end{bmatrix}, \quad d \in R.$$

The characteristic equation of system (7) for this matrix is

$$f(s) = \det(sI^{3\times 3} - A) + \\ +\mu d(-\alpha_{21}s + (\alpha_{21}\alpha_{33} - \alpha_{31}\alpha_{23}))e^{-sh}.$$

It can be written as $f(s) = p(s) + \mu g(s)$, where $g(s) = q(s)e^{\tau s}$. Using lemma 8 it is easy to verify that quasipolynomial $f(s)$ is a convex direction. Therefore D_ϵ is a matrix convex direction.

Matrix convex directions in $R^{n\times n}$, $n \geq 4$

Consider the candidate

$$D_\phi = dI^{n\times n}, \quad n \geq 4,$$

where $I^{n\times n}$ is the matrix identity and d is a real scalar. From lemma 10 and remark 3 we conclude that matrix D_ϕ is a matrix convex direction.

6. SUMMARY AND FURTHER DEVELOPMENTS

Finally, we are able to state the theorems resulting from the analysis of convex directions for time delay systems performed in this paper.

Theorem 14. A matrix D is a convex direction for time delay systems in $R^{2\times 2}$, if D coincides, up to similarity transformations, with one of the following three matrices:

$$D_\alpha = \begin{bmatrix} 0 & d \\ 0 & 0 \end{bmatrix}, \quad D_\beta = \begin{bmatrix} d & 0 \\ 0 & d \end{bmatrix}, \quad D_\gamma = \begin{bmatrix} x_1 & 0 \\ 0 & x_2 \end{bmatrix},$$

where d, x_1, x_2 are real numbers and $x_1 x_2 = 0$.

Theorem 15. A matrix D is a convex direction for time delay systems in $R^{3\times 3}$, if and only if D coincides, up to similarity transformations, with one of the following two matrices:

$$D_\delta = \begin{bmatrix} 0 & d & 0 \\ 0 & 0 & 0 \\ 0 & 0 & 0 \end{bmatrix}, \quad D_\epsilon = \begin{bmatrix} d & 0 & 0 \\ 0 & d & 0 \\ 0 & 0 & d \end{bmatrix},$$

where d is a real number.

Theorem 16. A matrix D is a convex direction for time delay systems in $R^{n\times n}$, $n \geq 4$, if and only if D is the identity multiplied by a real scalar d.

The analysis of Case 3 when $x_1 x_2 < 0$ is open due to the complexity of the corresponding quasipolynomial. Further developments of the above characterization are, for instance: the delay dependent matrix convex directions characterization, the analysis of matrix convex directions for two or more delays, and, given the system $\dot{x}(t) = Ax(t) + \mu[D_1 x(t - h_1) + D_2 x(t - h_2)]$, where A is Hurwitz and h_1, h_2 are delays, the characterization of matrices D_1, D_2 such that the segment $D_1 x(t - h_1) + D_2 x(t - h_2)$ is a convex direction.

REFERENCES

H. Kokame, H. Ito and T. Mori (1994). Entire Family of Convex Directions for Real Hurwitz Matrices, Proc. 33rd Conference on Decision and Control, Lake Buena Vista, pp. 3835-3836.

L. Xie (1997). Issues on Robustness Analysis of Linear Control Systems with Parametric Uncertainties, Ph. D. Thesis, Department of Electrical and Computer Engineering, The University of Newcastle.

R. Bellman and K. L. Cooke (1963). *Differential-Difference Equations*, Academic Press, New York.

V. L. Kharitonov and A. P. Zhabko (1994). Robust Stability of Time Delay Systems, *IEEE Trans. Automat. Contr.*, vol 39, pp. 2388-2397.

A. Rantzer (1992). Stability Conditions for Polytopes of Polynomials, *IEEE Trans. Automatic Control*, vol.37, pp. 79-89.

www.elsevier.com/locate/ifac

ON DELAY ROBUSTNESS ANALYSIS OF A SIMPLE CONTROL ALGORITHM IN HIGH-SPEED NETWORKS

Silviu-Iulian Niculescu *

* *HEUDIASYC (UMR CNRS 6599)*
Université de Technologie de Compiègne,
Centre de Recherche de Royallieu, BP 20529,
60205, Compiègne, France.
E-mail: silviu@hds.utc.fr

Abstract: This paper focuses on the robustness analysis of a simple (second-order) control algorithm for data transfer in high-speed networks. The model under consideration (derived using a fluid approximation technique) can be found in Izmailov (1996). The novelty of the approach lies in characterizing the stability of the scheme in the delay-parameter space (where the delays correspond to the control-time interval, and to the round-trip time, respectively), and its interest is due to the computation of some *delay-insensitive* measures for the algorithm which give upper and lower bounds for the uncertainty in the knowledge of the round-trip times. *Copyright © 2001 IFAC*

Keywords: delays; stability; upper/lower bounds; control algorithm; robustness.

1. INTRODUCTION

The growing interest on improving performances and quality-of-service in high-speed networks (Walrand and Varaiya, 1996) is at the origin of the development of various control algorithms using deterministic or stochastic, continuous- or discrete-time model representations of the network. Independently of the representation, all the models take in consideration a large variety of delays: round-trip or/and control-time intervals, propagation, transport, action delays, total and averaged network-induced delays, to cite only a few.

In the sequel, we shall focus on a simple continuous-time deterministic model of a *single connection* between a source controlled by an access regulator and a distant node served with a transmission *constant* capacity μ, model derived in Izmailov (1996) using a fluid approximation method (Bolot and Shankat, 1992) for controlling a congestion node. In fact, it is described by a second-order delay-differential equation of the form:

$$\begin{cases} \dot{x}_1(t) = x_2(t-\tau_1) - \mu \\ \dot{x}_2(t) = -a(x_1(t-\tau_2) - \bar{X}) \\ \qquad\quad -b(x_1(t-\tau_2-r) - \bar{X}), \end{cases} \quad (1)$$

where x_1 represents the buffer contents, x_2 the current input rate, and $\bar{X}$ the target value, respectively. Using the new variable $y(t) = x_1(t) - \bar{X}$, (1) leads to the following second-order delay equation including two discrete delays τ and r:

$$\ddot{y}(t) + ay(t-\tau) + by(t-\tau-r) = 0, \quad (2)$$

The 'total' delay $\tau = \tau_1 + \tau_2$ represents the *round-trip time*, and r is the *control time-interval*. Further discussions (analysis, optimization) on such a model can be found in Izmailov (1995, 1996). This paper focuses on describing some of the 'delay limitations' (upper/lower bounds, stability/instability characterization) of the closed-loop system (2), that is the *robustness* of the scheme in terms of round-trip times and control-time intervals.

Assume now that the control algorithm 'gains' (a,b) satisfy the constraints: $b < 0$, $a >\mid b \mid$. The conditions $a > 0$ and $b < 0$ are already encountered in the algorithm proposed by Izamilov (1996) for some optimal choices A^* and B^* function of the ratio $\frac{r}{\tau}$. The constraint $a >\mid b \mid$ will be discussed later. In view of delay robustness analysis, assume further that both delays (τ, r) are free parameters of the system, and that the gains (a,b) are fixed. The problem considered here is to analyze the stability of (2) in the *delay-parameter space* $O\tau r$, that is to find intervals of the form $(\underline{r}_i, \overline{r}_i)$ and $(\underline{\tau}_i, \overline{\tau}_i)$ function of the parameters (a,b) such that the second-order system (2) is asymptotically stable. In fact, these *upper* and *lower bounds* will define some *delay-insensitivity* measures, or equivalently, the robustness of the control algorithm in terms of delays.

The notion of delay-parameter space and similar analysis in the scalar case can be found in Hale and Huang (1993), where they show that the geometry of the stability regions is complex, and that the computations of the corresponding boundaries is a difficult problem (see also Cooke and van den Driessche, 1986; Hale, 1991). Furthermore, as seen in Toker and Özbay (1996), the non-scalar case turns to be a $\mathcal{NP}$-hard problem if the delays are not commensurate (that is, rationally-independent).

If the ratio between the delays is rational, an algorithm for computing the optimal first-delay interval (including the origin) can be found in Chen *et al* (1994). For more general stability/instability delay intervals, see for instance Niculescu (1997). Both approaches need the construction of some appropriate matrix pencils of high dimensions, but the computed bounds are exact. Further comments as well as algorithms simplification are proposed in Niculescu (2001). Note also the complex stability behavior (ill-possedness: Louisell, 1995, interference: MacDonald, 1987, robustness loss: Datko, 1994) if one 'changes' from commensurate to incommensurate delays even for first- or second-order delay-difference equations. To the best of author's knowledge, there does not exist a complete stability characterization in the delay-parameter space for a second-order system involving two discrete delays.

The remaining paper is organized as follows: the stability analysis as well as various comments and interpretations are proposed in the next section. A simple numerical example is presented in Section 3 and some concluding remarks end the paper.

2. STABILITY ANALYSIS

As specified above, the behavior of (1) or (2) is *complex*. Indeed, if, for example, one takes $\tau = r = 0$ (the origin of the delay-parameter space), the corresponding equation has an *oscillatory* behavior, since the characteristic equation has the roots $s = \pm j\sqrt{a+b}$ on the imaginary axis ($a + b > 0$ by hypothesis; see the Introduction).

Assume now that $\tau = 0$, but $r \neq 0$, that is we shall consider the behavior of the system on the delay axis Or, if the control-time interval is 'seen' as a 'free' parameter. The corresponding characteristic equation becomes transcendental:

$$s^2 + a + be^{-sr} = 0, \tag{3}$$

and is similar to the closed-loop system of an oscillator with a delayed output feedback of 'gain' $-b > 0$ (Niculescu and Abdallah, 2000, see also Abdallah *et al.* 1993). Then, one can prove that:

Proposition 1. The system (2) with $\tau = 0$, $b < 0$, $a >\mid b \mid$ is asymptotically stable for all the triples (a, b, r) satisfying simultaneously:

$$\underline{r}_i(a,b) = \frac{2i\pi}{\sqrt{a - \mid b \mid}} < r < \frac{(2i+1)\pi}{\sqrt{a + \mid b \mid}} = \overline{r}_i(a,b), \tag{4}$$

for $i = 0, 1, \ldots$. If $r = \underline{r}_i(a,b)$ or $r = \overline{r}_i(a,b)$, the associated characteristic equation in closed-loop has at least one eigenvalue on the imaginary axis. Furthermore, for each pair (a,b) there exists a value $r^*(a,b)$, such that for any $r > r^*(a,b)$ the considered system is unstable.

Remark 1. It is quite evident that for $i = 0$, the first stability delay-interval is:

$$0 < r < \overline{r}_0(a,b) = \frac{\pi}{\sqrt{a + \mid b \mid}},$$

that is the system becomes stable for any sufficiently small positive delay $r = \varepsilon > 0$. The choice of the 'gains' (a,b) such that $a >\mid b \mid$ becomes clear: stability guaranteed for small control-time intervals although the stability at the origin ($r = 0$) is not ensured.

Note however that even if $\mid b \mid > a > 0$, one can define some delay-intervals guaranteeing stability. Further comments on stabilizing oscillations using delay terms can be found in Niculescu (2001).

Remark 2. Using a continuity argument (Els'golts' and Norkin, 1973; Neimark, 1949) in terms of the solutions of the characteristic equation of (2) with respect to the delay τ, the result above proves that the *control time-interval* could have a *stabilizing* or *destabilizing* effect for small *round trip-times* τ. Indeed, for $\tau = 0$, Proposition 1 shows a *sequence* of *stability* and *instability* control-time intervals. Further arguments are presented below.

The next step consists in analyzing the behavior of the characteristic equation of (2) in the hypothesis that r satisfies the constraints given in Proposition 1, and with τ as a free parameter, that is r and τ 'decoupled' variables. Since the stability is guaranteed for $\tau = 0$ and, by continuity (see, for instance, the arguments in Els'golts' and Norkin, 1973), for sufficiently small positive values $\tau = \varepsilon > 0$, we are interested to compute the *optimal* $\tau_i = \tau_i(a, b, r)$ for a control-time interval r given in $(\underline{r}_i(a, b), \overline{r}_i(a, b))$, and guaranteeing the above property.

For such a r, the characteristic equation of (2) can be rewritten as:

$$s^2 + (a + be^{-sr})e^{-s\tau} = 0. \qquad (5)$$

Using the continuity principle argument with respect to the delay τ, it follows that one only needs to characterize the solutions of the characteristic equation with respect to the imaginary axis. Thus, one can have two different situations: no solutions and at least one solution on the imaginary axis, respectively. Indeed, if there does not exist any solution of (5) on $j\mathbf{R}$, then the system will be stable for *any* delay value (delay-independent stability, see, for instance, Niculescu, 2001). If there exists at least one root, then one can compute the bound on the delay τ^* for which the system will loss its stability, and such that for any $\tau \in [0, \tau^*)$, the system will be asymptotically stable (for some r appropriately chosen).

We have the following results:

Proposition 2. (Delay bounds). The system (2) with $b < 0$, $a >| b |$ is stable for all delays r and τ satisfying the constraints: small

$$\begin{cases} \underline{r}_i(a,b) = \dfrac{2i\pi}{\sqrt{a-\mid b\mid}} < r < \\ \quad < \dfrac{(2i+1)\pi}{\sqrt{a+\mid b\mid}} = \overline{r}_i(a,b), \quad i = 0, 1, \ldots \\ 0 \leq \tau < \tau(a,b,r) = \tau_i(a,b,r) \\ \quad = \min\limits_{\omega_s} \left\{ \dfrac{1}{\omega_s} arctan \left(\dfrac{-bsin(\omega_s r)}{a + bcos(\omega_s r)} \right) \right\}, \end{cases} \qquad (6)$$

where ω_s belongs to the set of positive solutions of the equation:

$$\omega^4 = a^2 + b^2 + 2abcos(\omega r). \qquad (7)$$

Furthermore, assume that the chosen delay $r \in (\underline{r}_i(a, b), \overline{r}_i(a, b)))$ and the solution $\tilde{\omega}_s$ defining the corresponding upper bound $\tau(a, b, r)$ in (6) satisfy the condition:

$$\tilde{\omega}_s^3 > \frac{1}{2} a \mid b \mid rsin(\tilde{\omega}_s r). \qquad (8)$$

Then:

i) if $\tau = \tau_i(a, b, r) + \varepsilon$, with $\varepsilon > 0$ sufficiently small, the system (2) is unstable.
ii) if the equation (7) has only one positive solution, then there does not exist any positive value of the round-trip time $\tau > \tau_i(a, b, r)$ such that the system (2) is asymptotically stable.

Remark 3. The range of i guaranteeing the inequalities in (6) can be computed as follows:

$$0 \leq i < \frac{1}{2} \cdot \frac{1}{\sqrt{\frac{a+|b|}{a-|b|}} - 1}.$$

Proposition 3. (Instability). If for a pair (a, b) with $b < 0$ and $a >| b |$, the delay r does not satisfy the constraints (4), then the closed-loop system is unstable for any delay τ satisfying:

$$0 \leq \tau < \tau(a,b,r) = \tau_i(a,b,r) = \min_{\omega_s} \left\{ \frac{1}{\omega_s} arctan \left(\frac{-bsin(\omega_s r)}{a + bcos(\omega_s r)} \right) \right\}, \qquad (9)$$

where ω_s belongs to the set of solutions of the equation:

$$\omega^4 = a^2 + b^2 + 2abcos(\omega r). \qquad (10)$$

Furthermore, assume that the delay r and the solution $\tilde{\omega}_s$ defining the corresponding upper bound $\tau_i(a, b, r)$ in (9) satisfy the condition:

$$\tilde{\omega}_s^3 > \frac{1}{2} a \mid b \mid rsin(\tilde{\omega}_s r). \qquad (11)$$

Then:

i) if $\tau = \tau_i(a, b, r) + \varepsilon$, with $\varepsilon > 0$ sufficiently small, the system (2) is unstable.
ii) if the equation (10) has only one positive solution, then there does not exist any positive round-trip time value such that the closed-loop system (2) is asymptotically stable.

Remark 4. First, note that, if any, the bound on τ in (6) is correctly defined. Indeed, the equation (7) has always at least one solution $| \omega_s | \in (\sqrt{a+b}, \sqrt{a-b})$. Furthermore, $-bsin(\omega_s r) > 0$ since $\omega_s r \in (2i\pi, (2i+1)\pi)$, and thus $arctan \left(\frac{-bsin(\omega_s r)}{a+bcos(\omega_s r)} \right)$ is positive ($a >| b |$, by hypothesis).

Remark 5. If the equation (7) has only one positive solution, the stability/instability scheme becomes simpler. In fact, in the best case, we can have only one *round-trip time* interval guaranteeing stability, that is only one switch from stability to instability and no reversal. In the worst case, we can conclude on the *delay-independent* instability in terms of round-trip times.

Proof [Proposition 2]: Using the stability argument presented above, it follows that the characteristic equation (5) on the imaginary axis becomes:

$$\omega^2 = (a + be^{-j\omega r})e^{-j\omega\tau}. \qquad (12)$$

If one takes the modulus, we shall have:

$$\omega^4 = (a + bcos(\omega r))^2 + b^2 sin(\omega r)^2,$$

which is equivalent to (7). Since that equation has always at least one positive solution ω_s, it follows that there exists a positive maximal delay value τ^*, such that the system will be stable for all $\tau \in [0, \tau^*)$ (see arguments in Niculescu, 2001). If, for example, one takes the imaginary part of (12), it is easy to see that taking for τ the value:

$$\tau^* = \min_{\omega_s} \frac{1}{\omega_s} arctan\left(\frac{-bsin(\omega_s r)}{a + bcos(\omega_s r)}\right) > 0,$$

the equation (12) is satisfied by the corresponding 'frequency' ω_s. The next step is to prove that the bound in (6) is the optimal one.

Optimality: Assume that there exists a smaller $0 < \tilde{\tau} < \tau^*$ such that the system is unstable. Then, based on the continuity argument (see also Proposition 3.1 in Niculescu, 2001, p. 89), it follows that there exists some delay $\bar{\tau}$ such that $0 < \bar{\tau} < \tilde{\tau}$, and if $\tau = \bar{\tau}$, the characteristic equation has at least one solution $\bar{\omega}$ on the imaginary axis, and no roots with strictly positive real parts. This solution should satisfy the equation:

$$\bar{\omega}^2 = (a + be^{-j\bar{\omega}r})e^{-j\bar{\omega}\bar{\tau}}. \qquad (13)$$

In conclusion, $\bar{\omega}$ should satisfy (modulus, imaginary part constraints):

$$\begin{cases} \bar{\omega}^4 = (a + bcos(\bar{\omega}r))^2 + b^2 sin(\bar{\omega}r)^2 \\ (a + bcos(\bar{\omega}r))sin(\bar{\omega}\bar{\tau}) = -bsin(\bar{\omega}r)cos(\bar{\omega}\bar{\tau}), \end{cases}$$

which leads to the equality:

$$\bar{\tau} = \frac{1}{\bar{\omega}} arctan\left(\frac{-bsin(\bar{\omega}r)}{a + bcos(\bar{\omega}r)}\right).$$

From the definition of τ^* in (6), it follows that $\bar{\tau} \geq \tau^*$, which contradicts $\bar{\tau} < \tilde{\tau} < \tau^*$. In conclusion, the bound on τ in (6) is the *optimal* one.

Instability: The second step is to analyze the *sign* of the real part of the derivative of $\frac{ds}{d\tau}$ in $s = j\omega_s$, with $\omega_s > 0$. Some simple but tedious computations lead to the following:

$$Re\left[\left(\frac{ds}{d\tau}\right)\right]_{s=j\omega_s} = \frac{2\omega_s^5}{2\omega_s^3 - a \mid b \mid sin(\omega_s r)} \qquad (14)$$

If the constraint (8) is satisfied it follows that the right-hand term is positive, that means the instability of the considered system for some delay $\tau = \tau_i(a, b, r) + \varepsilon$, with $\varepsilon > 0$ sufficiently small. If the equation (7) has only one positive solution $\omega_s > 0$, then, due to (14), it follows that at each crossing, the number of roots with positive real parts is strictly positive. Thus, the proof is complete. ∇∇∇

Remark 6. The proof of Proposition 3 can be developed using similar arguments and it is omitted.

Remark 7. If the variables are 'coupled', that is, we define some $\alpha > 0$, such that $r = \alpha\tau$, the analysis is still valid, but based on a different argument.

Remark 8. Proposition 2 says that the 'choice' of the control time-interval becomes an important factor to *induce* stability or instability in the associated fluid model for small round-trip times.

Note that the analysis proposed in Izmailov (1996) (which extends the technique proposed in Izmailov, 1995 if $r = \tau$) does not allow to conclude on such type of property. The advantage of this analysis is to give *explicitly* upper and lower stability/instability bounds in terms of control-time intervals and round-trip times. Furthermore, for a given triplet (a, b, r), we are able to compute the *upper* limit of the round-trip time guaranteeing the stability of the corresponding scheme, that is the behavior for small round-trip times.

Remark 9. As it is proved by Proposition 2, decreasing the duration of control-time intervals is not always the best solution to improve the performances of the algorithm in terms of robustness (see also the example proposed in the next section). A different argument for such a conclusion was presented in Izmailov (1996).

As a natural consequence of Propositions 2 and 3 and, we have the following stability/instability results:

Corollary 1. (Stability robustness). The above algorithm with the control-time interval defined by $(\underline{r}_i(a, b), \overline{r}_i(a, b))$ in (4) guarantees the stability of the fluid model if the round-trip delay τ takes any positive value in the interval $[0, \tau_i(a, b, r))$, where $\tau_i(a, b, r)$ is defined in (9).

Corollary 2. (Instability robustness). If for a pair (a, b), the control-time interval r is chosen such that it does not satisfy the conditions in (4), then for any round-trip time $\tau \in [0, \tau(a, b, r))$, with $\tau(a, b, r)$ defined in (6), the fluid model is unstable.

Remark 10. The maximal delay bound τ_i guaranteeing stability can be *optimized* function of $r \in (\underline{r}_i(a,b), \overline{r}_i(a,b))$, and we shall have a *'max-min'* problem. For the sake of brevity, such aspects are not detalied here. However, some comments on the proposed example are considered.

Note also that the performances of the algorithm are not discussed.

3. NUMERICAL EXAMPLE

Consider the closed-loop system (2) with $a = 2$ and $b = -1.75$.

Then, applying the results above, it follows that if $\tau = 0$, the stability is guaranteed for all $r \in (0, r_{max})$ with $r_{max} = 1.6223$, and there does not exist other control-times intervals guaranteeing stability (Proposition 1).

Assume now that $r = 0.3 < r_{max}$. Then the maximum round-trip time value guaranteeing the stability of the scheme is $\tau_{max}(0.3) = 1.3198$, and there does not exist other delay intervals guaranteeing stability (Proposition 2). In fact, the *optimal* control-time interval is $r = 0.322$, and the corresponding *maximal* round-trip time is $\tau_{max}(0.322) = 1.3234$.

Note that if r becomes smaller, the maximal round-trip time value guaranteeing stability decreases. Thus, for $r = 0.2$, it becomes $\tau_{max}(0.2) = 1.1408$, $r = 0.1$, $\tau_{max}(0.05) = 0.3457$. The same property holds if the control-time interval is greater: $\tau_{max}(0.4) = 1.2717$ if $r = 0.4$, and $\tau_{max}(0.75) = 0.6955$ if $r = 0.75$ (see also the graphics included).

4. CONCLUDING REMARKS

In this paper, we considered a simple control algorithm in the literature for data transfer in high-speed networks, and we have analyzed some stability regions in terms of delays: round-trip time and control-time interval respectively. Some upper and lower bounds for guaranteeing stability/instability are porposed. Our interest was focused on the effects of choosing some control-time interval on the *maximal* allowable value on the round-trip time ensuring stability/instability of the corresponding control algorithm.

5. ACKNOWLEDGEMENTS

The author wish thank anonymous reviewers for their useful comments.

REFERENCES

Abdallah, C., Dorato, P., Benitez-Read and Byrne, R. (1993). Delayed positive feedback can stabilize oscillatory systems. *Proc. American Contr. Conf.* 3106-3107.

Bolot, J. -C. and Shankat, A. U. (1992). Analysis of a fluid control approximation to flow control dynamics. *Proc. IEEE Infocom'92*, Florence, Italy, 2398-2407.

Chen, J., Gu, C. and Nett, C. N. (1994). A new method for computing delay margins for stability of linear delay systems, in *Proc. 33rd IEEE Conf. Dec. Contr.*, Lake Buena Vista, Florida, US, 433-437.

Cooke, K. L. and van den Driessche, P. (1986). On zeroes of some transcendental equations. in *Funkcialaj Ekvacioj* **29**, 77-90.

Datko, R. (1994). Time delay perturbations and robust stability. in *Differential Equations, Dynamical Systems, and Control Science*, LNM, Marcel Dekker: New York, vol. **152**, 457-468.

El'sgol'ts, L. E. and Norkin, S. B. (1973). *Introduction to the theory and applications of differential equations with deviating arguments* (Mathematics in Science and Eng., **105**, Academic Press: New York).

Hale, J. K. (1991). Dynamics and delays. in *Delay differential equations and dynamical systems* (S. BUSENBERG, M. MARTELLI, Eds.), Springer-Verlag: Berlin, LNM, 16-30.

Hale, J. K. and Huang, W. (1993). Global geometry of the stable regions for two delay differential equations. *J. Math. Anal. Appl.*, vol. **178**, 344-362.

Izmailov, R. (1995). Adaptive feedback control algorithms for large data transfer in high-speed networks. *IEEE Trans. Automat. Contr.*, vol. **40**, 1469-1471.

Izmailov, R. (1996). Analysis and optimization of feedback control algorithms for data transfers in high-speed networks. in *SIAM J. Contr. Optimiz.*, vol. **34**, 1767-1780.

Louisell, J. (1995). Absolute stability in linear delay-differential systems: Ill-possedness and robustness. *IEEE Trans. Automat. Contr.*, vol. **40**, 1288-1291.

MacDonald, N. (1987). An interference effect of independent delays. in *IEE Proc. Contr. Theory & Appl.*, Pt. **D 134**, 38-42.

Neimark, J. (1949). D-subdivisions and spaces of quasi-polynomials. *Prikl. Math. Mech.*, vol. **13**, 349-380.

Niculescu, S. -I. (2001). *Delay effects on stability: A robust control approach* (Springer-Verlag: Heidelberg, Germany, LNCIS, May 2001, to appear).

Niculescu, S. -I. (1997). Delay-interval stability and hyperbolicity of linear time-delay systems:

A matrix pencil approach, in *Proc. 5th European Contr. Conf.*, Brussels, Belgium, 1997.

Niculescu, S. -I. and Abdallah, C. T. (2000). Delay effects on static output feedback stabilization. in *Proc. 39th IEEE Conf. Dec. Contr.*, Sydney, Australia.

Toker, O. and Özbay, H. (1996). Complexity issues in robust stability of linear delay-differential systems. *Math., Contr., Signals, Syst.*, vol. **9**, 386-400.

Walrand, J. and Varaiya, P. (1996). *High-performance communication networks* (Morgan Kaufmann Publ., San Francisco, CA).

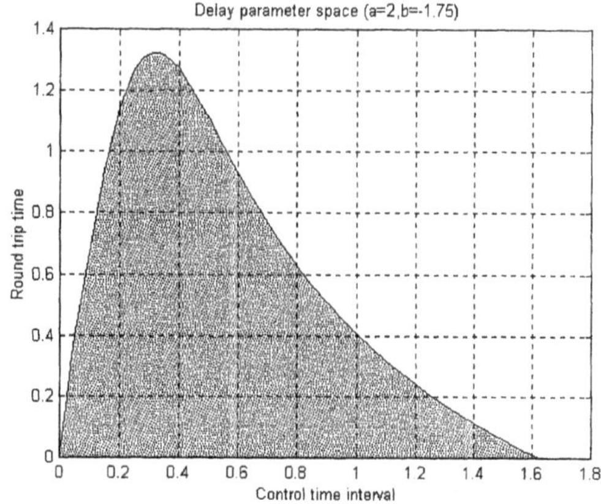

www.elsevier.com/locate/ifac

TIME DELAY AND POWER CONTROL IN SPREAD SPECTRUM WIRELESS NETWORKS

Aly I. El-Osery [*] **Chaouki Abdallah** [**]
Mo Jamshidi [*]

[*] *Autonomous Control Engineering Center (ACE)*
The University of New Mexico
Albuquerque, NM 87131, USA
{elosery,jamshidi}@eece.unm.edu
[**] *Electrical & Computer Engineering Department*
The University of New Mexico
Albuquerque, NM 87131, USA
chaouki@eece.unm.edu

Abstract: Delay in the computation of the signal-to-interference ratio in communication systems is unavoidable. In the the case of mobile communication, delay is a very critical problem due the fast variation of the communication channel and the need for effective, fast and accurate power control. In this paper we present our approach to dealing with the delay in mobile communication systems as well as our controller to achieve and maintain the desired signal quality. We will concentrate on the *code division multiple access* (CDMA) systems. *Copyright © 2001 IFAC*

Keywords: Power control, CDMA

1. INTRODUCTION

The popularity of the CDMA in commercial wireless systems is growing rapidly, and has been recognized as a viable alternative to the more traditional methods such as *frequency division multiple access* (FDMA) and *time division multiple access* (TDMA). Although there are different types of spread spectrum, we will concentrate on direct sequence (DS) spread spectrum. DS-CDMA has many advantages: i) universal one-cell frequency reuse, ii) narrow band interference rejection, iii) inherent multipath diversity, iv) soft hand-off capability, and v) soft capacity limit. Unlike FDMA and TDMA, in CDMA the users are separated by pseudorandom codes rather than in frequency or time, and the transmitted signal is spread to occupy a much larger bandwidth. In this case the central mechanism for interference management is *power control*. The role of power control is to keep the received power of all users at an equal level so as to avoid problems such as the *near-far problem*. This problem occurs due to the lack of power control: If all mobiles were to transmit at a fixed power the mobile closest to the base station will overpower all others. The other role of power control is to account for the fast varying changes in the mobile radio channel. One of the main issues in power control is thus to address the inherent delay in the system due to the averaging process used to estimate the *signal-to-interference ratio* which is an accepted measure for the quality-of-service for each user, and the fact that the control command is send at fixed intervals. This delay becomes critical in the case of small-scale fading.

As is well known, the mobile channel is best modeled statistically usually as Rayleigh or Ricean channel (Rappaport, 1996). These models are called *small-scale* fading models and the parameters of the Rayleigh or Ricean distribution are themselves modeled as random variables that change at a much slower rate, according to the so called *large-scale* fading characteristics. Large-scale fading has usually a log-normal probability density function whose parameters depend on the location of the mobile relative to the base station. In particular, the average power is determined under the assumption that electromagnetic waves will experience a path loss inversely proportional to the distance traveled raised to some power. An accurate model of the wireless channel and its state is usually unobtainable. Any power control algorithms developed should be able to adjust the power levels of each mobile using local measurements only, so that in reasonable time, all users will maintain the desired signal-to-interference ratio. In this paper, only the uplink (mobile-to-base), also known as *reverse link*, control will be studied. The reverse link, is more challenging due to the fact that mobile stations have more computation and power limitation than the base station.

Some of the early work in power control was provided by (Zander, 1992). In (Kim, 1999; Wu, 1999; Grandhi *et al.*, 1993) centralized power control was studied, and due to the complexity of the system, centralized power control was suggested to be used only for providing theoretical limits. When all users could be accommodated with acceptable signal-to-interference ratio, (Foschini and Miljanic, 1993) suggested a distributed power control algorithm that guarantees convergence and computes the required transmission power of each mobile station. In (Jeantti and Kim, 2000) a second-order constrained power control (CSOPC) algorithm was presented. This approach uses the current and past power values to determine the necessary transmission power of each mobile. CSOPC was compared with the algorithm presented in (Foschini and Miljanic, 1993) and was shown to converge at a faster rate. Convergence analysis of distributed power control algorithms is investigated in (Huang and Yates, 1998). In (Yates, 1995) a framework for uplink power control in cellular radio systems was presented. Our approach to solving the power control problem will be within such framework. Many of the above mentioned approaches did not use knowledge already established in the area of control theory, and did not address the delay problem.

In this paper we present our approach to dealing the delay in the estimation of the signal-to-interference ratio and our design of the power control algorithm. The paper is organized as follows. In Section 2 we discuss mobile radio propagation model, which will help understand the nature of the received signal power. In Section 3 our approach to predicting the changes in the channel due to flat fading will be presented. Section 4 discusses power control and presents our algorithms. Simulation and results are given in Section 5, and our conclusion is given in Section 6.

2. MOBILE RADIO PROPAGATION

Since signal-to-interference ratio is the measure used to determine the quality of the signal, and consequently, the power adjustments that need to be made, accurately estimating the power of the received signal is a critical part of power control. In a typical mobile-radio situation, a moving mobile station is in communication with a base station at a fixed position. This movement is usually in such a way that the direct line between the mobile station and the base station is obstructed by buildings and other obstacles. Therefore, the mode of propagation of the electromagnetic energy from the transmitter to the receiver will be largely by way of scattering, either by reflection from flat sides of the buildings or by diffraction around such buildings.

2.1 Channel Model

For wideband fading channel, the baseband impulse response of the channel having L resolvable taps is given by (Proakis, 1983)

$$h(t) = \sum_{l=1}^{L} \xi(d, \alpha, \eta)\beta_l \delta(t - \tau_l) \exp(j\theta_l), \qquad (1)$$

where d is the distance between the transmitter and the receiver, α is the path loss exponent and η is the shadowing variable which is a normally distributed variable with zero mean and variance σ_s^2. The factor $\xi(d, \alpha, \eta)$ is due to large-scale fading over the link between the ith base station and the jth mobile station, β_l is the magnitude of the lth path of the small-scale fading channel response at time τ_l, and θ_l is phase shift of the lth path relative to the 0th path and is uniformly distributed between 0 and 2π. $\xi(d, \alpha, \eta)$ is defined as

$$\xi^2(d, \alpha, \eta) = d^{-\alpha} 10^{\eta/10}. \qquad (2)$$

Assuming the communication channel is of the form given in Equation (1) the average received power is given by

$$\overline{p}_r(n) = Gp_t, \qquad (3)$$

where G is the communication link and is defined as $\overline{\xi}^2(d, \alpha, \eta) \sum_{l=1}^{L} \overline{\beta}_l^2$, $\overline{\beta}_l$ is the average of the small-scale fading component, and p_t is the transmission power. As seen in Equation (3), in order to computed the signal-to-interference ratio, we need to have an averaging window. Also, in CDMA the power command is send at fixed intervals based on previous estimation of the signal quality, e.g. in the IS-95 standard the power command is send every 1.25msec. These factors cause an unavoidable delay in the system, and introduces power control errors due to the fact that the power command is based on previous estimates. Consequently, it is important to predict such changes in order to minimize the power control error. In the next section, we discuss in detail Clark's model for flat fading which shows the spatial correlation in the received signal power due to scattering. Using this property, a prediction algorithm will used to accurately track changes in the communications channel.

2.2 Clarke's Model for Flat Fading

In a typical radio-communication scenario, the mobile station is in motion and is trying to communicate with a base station that is at a fixed location. In this situation, doppler shift is introduced. Clarke's model deduced the statistical characteristics of the fields and signals in the reception of radio frequencies by a moving vehicle from a scattering propagation model (Clarke, 1968).

Assuming that the total field is vertically polarized and is composed of the superposition of N waves, the E-field component can be described as (Clarke, 1968):

$$E_z = E_0 \sum_{n=1}^{N} C_n \exp\{j\vartheta_n + 2\pi f_{md} t \cos\alpha_n\}, \quad (4)$$

where $f_{md} = v f_0/c$ is the maximum doppler shift, v is the velocity of the mobile station, f_0 is the carrier frequency, c is the speed of light, E_0 is the common amplitude of the N waves (assumed constant), η is the intrinsic impedance of free space, the time variations is of the form $\exp\{j2\pi f_0 t\}$, ϑ_n is the random phase (uniformally distributed throughout 0 to 2π) of the nth wave arriving at any angle α_n to the x-axis (Figure 1), and C_n is a random variable describing the amplitude of the nth wave. The ensemble average of the C_n terms is assumed to satisfy

$$\sum_{n=1}^{N} \overline{C_n^2} = 1. \quad (5)$$

The random received signal envelope has a Rayleigh distribution given by

$$p(r) = \begin{cases} \dfrac{r}{\sigma_r} \exp\left(-\dfrac{r^2}{2\sigma_r^2}\right) & (0 \le r \le \infty), \\ 0 & (r < 0), \end{cases} \quad (6)$$

where σ_r is the rms value of the received voltage signal, and $r(t)$ is the complex envelope of the received signal. Figure 2 shows a Rayleigh distribution signal envelope at 845MHz with receiver speed of 100km/h as a function of time.

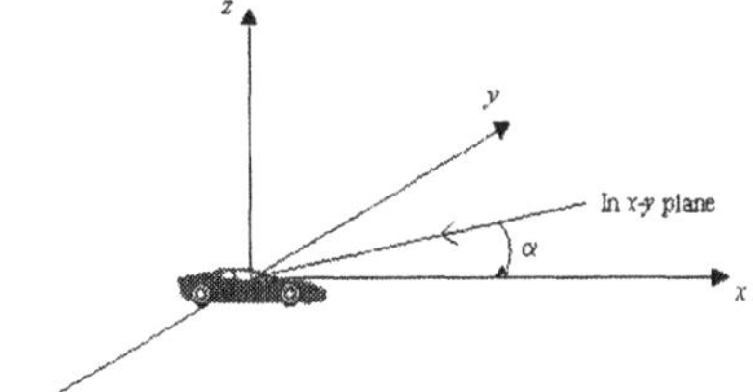

Fig. 1. Plane wave

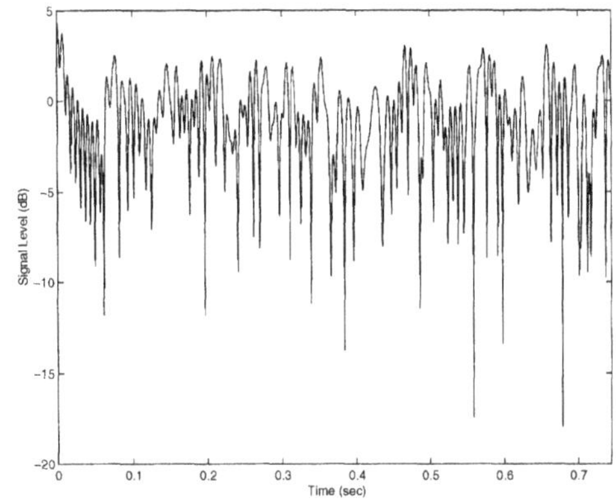

Fig. 2. A typical Rayleigh fading envelope at 845MHz and receiver speed of 100km/h

3. RECEIVED POWER PREDICTION

The fast changes in the received power are dominated by the small-scale fading. Using Equation (4), the spatial correlation function of the power is derived (see Appendix). Taking advantage of this case we can derive a linear predictor that uses the past values to predict the future change of the channel. Since variation of $\bar{\beta}$ in Equation (3) is much faster compared to the other terms, we will concentrate on estimating it.

Assuming that the we have N observations of $\bar{\beta}^2$, in the mean-square-sense, its estimate at $n+1$, given N measured values, is given by

$$\hat{\beta}^2(n+1) = \sum_{i=1}^{N} h_i \bar{\beta}^2(n-i), \quad (7)$$

where $\hat{\beta}^2$ is the prediction of $\bar{\beta}^2$, h_i is computed as follows (Shanmugan and Breipohl, 1988)

$$\begin{bmatrix} h_1 \\ h_2 \\ \vdots \\ h_N \end{bmatrix} = \begin{bmatrix} R_{\beta^2}(0) & \cdots & R_{\beta^2}(N-1) \\ R_{\beta^2}(1) & \cdots & R_{\beta^2}(N-2) \\ \vdots & \ddots & \vdots \\ R_{\beta^2}(N-1) & \cdots & R_{\beta^2}(0) \end{bmatrix}^{-1} \begin{bmatrix} R_{\beta^2}(1) \\ R_{\beta^2}(2) \\ \vdots \\ R_{\beta^2}(N) \end{bmatrix}, \quad (8)$$

and

$$R_{\beta^2}(\tau) = 1 + J_0^2(2\pi f_{md}\tau), \quad (9)$$

where $J_0(\cdot)$ is the zero-order Bessel function of the first kind.

4. POWER CONTROL

The goal of the power control algorithms is to find the transmission power of each mobile such that the following inequality is satisfied

$$\gamma_i = \frac{G_{ki}p_i}{\sum_{j\neq i}^{Q} p_j G_{kj} + n_i} \geq \gamma_i^*, \qquad (10)$$

where γ_i is the signal-to-interference ratio for mobile i, G_{ki} is the communication gain between mobile station i and its assigned base station k, γ_i^* is the desired signal-to-interference for the ith mobile station with receiver noise n_i and transmission power p_i which is constraint as follows

$$0 \leq p_i \leq p_{i\_\max}, \qquad (11)$$

where $p_{i\_\max}$ is the maximum transmission power of mobile i.

Since it is desirable that the mobile station transmits the signal at the minimum power possible to maintain the quality-of-service required, Inequality (10) becomes equality. Consequently, the transmission power is written as

$$p_i(n+1) = \min\{p_{i\_\max}, \frac{\gamma_i^*}{\gamma_i(n)} p_i(n)\}, \qquad (12)$$

where $\gamma_i(n)$ is the signal-to-interference ratio of mobile i at iteration n. It is important to note that unlike centralized power control, only the total interference is needed to compute the power levels. Convergence of the iterative algorithm given by Equation (12) is studied in (Yates, 1995). Along the same lines, different power control algorithms can be designed to have faster convergence rate. In the next subsections we will present the CSOPC approach followed by our own.

4.1 Constraint Second-Order Power Control

The constraint second-order power control, denoted CSOPC, presented in (Jeantti and Kim, 2000) uses current and past values in order to provide a faster convergence of the power command. In this section a brief overview of the CSOPC approach will be given. Equation (10) could be written as a set of linear equations as follows

$$\mathbf{XP} = \Xi, \qquad (13)$$

where $\mathbf{P} = \begin{bmatrix} p_1 & \ldots & p_Q \end{bmatrix}^T$, Q is the total number of mobile stations in the system,

$$\mathbf{X} = \mathcal{I} - \mathbf{A}, \qquad (14)$$

$$\mathbf{A} = \{a_{ij}\}, \qquad a_{ij} = \gamma_i^* w_{ij}, \qquad (15)$$

$$\Xi = \{b_i\}, \qquad b_i = \gamma_i^* \frac{n_i}{G_{ik}}, \qquad (16)$$

$\mathcal{I}$ is a $Q \times Q$ identity matrix, and w_{ij} is defined as

$$w_{ij} = \begin{cases} \frac{G_{kj}}{G_{ki}} & i \neq j, \\ 0 & i = j. \end{cases} \qquad (17)$$

Thus Equation (10) has been converted to a set of linear equations, that could be iteratively solved for $\mathbf{P}$ (Jeantti and Kim, 2000).

CSOPC is developed by applying the *successive overrelaxation method* (SOR) (Young, 1971) to Equation (13). The CSPOC results in (Jeantti and Kim, 2000) were compared with the *distributed constraint power control* (DCPC) in (Foschini and Miljanic, 1993). CSOPC was proven to be more effective; consequently, later in this paper the CSOPC algorithm will be used as the comparison benchmark. Through some manipulations the following iterative algorithm was obtained (Jeantti and Kim, 2000).

$$\begin{aligned} p_i(n+1) = \min\{&p_{i\_\max}, \max\{0, \\ &a(n)\frac{\gamma_i^*}{\gamma_i(n)} p_i(n) + (1-a(n))p_i(n-1)\}\}, \end{aligned} \qquad (18)$$

where as described earlier, $p_{i\_\max}$ is the maximum allowable power for mobile i, $\gamma_i(n)$ is the signal-to-interference ratio of mobile i at iteration n, $p_i(0)$ is chosen randomly between 0 and $p_{i\_\max}$, and $a(n)$ is a decreasing sequence such that $\lim_{n\to\infty} a(n) = 1$. As an example, the following $a(n)$ sequence was used in (Jeantti and Kim, 2000):

$$a(n) = 1 + \frac{1}{1.5^n}, \quad n = 1, 2, \ldots, l, \qquad (19)$$

where l is the total number of iterations. Equation (18) determines the necessary power using the current and the past power values, which accounts for the terminology of "second-order". Note that if $a(n) = 1$, Equation (18) reduces to Equation (12) and that the *min* and *max* operators are used to guarantee that the power will be within the allowable range based on Equation (11).

4.2 Predictive Power Control

In this section we present our power control algorithm. Our approach is to view each mobile-to-base station connection as a separate subsystem as follows (El-Osery and Abdallah, 2000)

$$s_i(n+1) = \frac{\xi^2 \hat{\beta}^2(n+1)(p_i(n) + u_i(n))}{I_i(n)}, \qquad (20)$$

where ξ^2 is defined by Equation (2), $\hat{\beta}^2$ is defined in Equation (7), $I_i(n) = \sum_{j\neq i}^{Q} p_i(n)G_{kj}(n) + n_i$ is the interference experience by mobile i due to other users (assumed constant in short periods of time), and $u_i(n)$ is the increment by which the power of mobile i should be increased. Using the linear predictor described in Equation (7), we modify Equation (20) as follows

$$\begin{aligned} s_i(n+1) &= \alpha(n)(p_i(n) + u_i(n)) \\ &= \alpha(n)p_i(n) + v_i(n), \end{aligned} \tag{21}$$

where $\alpha(n) = (\xi^2/I_i(n))\sum_{i=1}^{N} h_i\beta^2(n-i)$, and h_i is defined in Equation (8). Now if we choose $v_i(n) = -\alpha(n)p(n) + z_i(n)$ then Equation (21) becomes

$$s_i(n+1) = z_i(n), \tag{22}$$

where $z_i(n)$ is the control law. The goal is to find the right control command that will make each s_i follow a desired signal to interference ratio γ_i^* by using a discrete-time integrator defined as follows

$$\zeta_i(n+1) = \zeta_i(n) + e_i(n) = \zeta_i(n) + s_i(n) - \gamma_i^*. \tag{23}$$

Defining $x_i(n)$ as

$$x_i(n) = \begin{pmatrix} \zeta_i(n) \\ e_i(n) \end{pmatrix}. \tag{24}$$

Using the above notation the system can now be expressed as a second-order linear state-space system by

$$\begin{aligned} x_i(n+1) = \begin{pmatrix} \zeta_i(n+1) \\ e_i(n+1) \end{pmatrix} &= \begin{pmatrix} 1 & 1 \\ 0 & 0 \end{pmatrix} x_i(n) + \\ \begin{pmatrix} 0 \\ 1 \end{pmatrix} z_i(n) &+ \begin{pmatrix} 0 \\ -1 \end{pmatrix} \gamma_i^*(n). \end{aligned} \tag{25}$$

Now introducing the new variable z_i' defined as

$$z_i'(n) = z_i(n) - \gamma_i^*, \tag{26}$$

and consequently, Equation (25) becomes

$$x_i(n+1) = \begin{pmatrix} 1 & 1 \\ 0 & 0 \end{pmatrix} x_i(n) + \begin{pmatrix} 0 \\ 1 \end{pmatrix} z_i'(n). \tag{27}$$

We then choose the feedback controller

$$z_i'(n) = -\begin{pmatrix} k_\zeta & k_e \end{pmatrix} x_i(n). \tag{28}$$

If we choose the appropriate feedback gains k_ζ and k_e, then Equation (25) will be stable. Finally, the transmission power at iteration $n+1$ is

$$\begin{aligned} p_i(n+1) &= \min\{p_{i\_\max}, \max\{0, p_i(n) + u_i(n)\} \\ &= \min\{p_{i\_\max}, \max[0, \frac{-1}{\alpha(n)}\{(k_\zeta \; k_e)x_i(n) + \gamma_i^*\}]\}. \end{aligned} \tag{29}$$

Equation (29) is a non-linear power control command representing a predictive power control (PPC). One of the states used to obtain this control command is an integrator which is very useful in suppressing noise that occurs during the computation of the signal-to-interference ratio.

5. SIMULATION AND RESULTS

For comparison purposes, we have simulated our power control algorithms as well as the CSOPC. We have set the desired energy per bit, E_b, to interference, I_0, ratio to be 7dB. Based on the IS-95 standard, we use the bit rate to be 9.6kbps, the channel bandwidth to be 1.2288MHz, and the power control command is updated every 1.25msec. Using these parameters, $\gamma_i^* = (E_b/I_0)(R_b/B_c)$ where $i = 1, \ldots, Q$, R_b is the bit rate, and B_c is radio channel bandwidth. The system is then simulated at different mobile speeds. The gains k_ζ and k_e are set to be 0.9995. Figure 3 shows a comparison between the CSOPC algorithm and our nonlinear predictive power control (PPC) introduced in this paper for a mobile speed of 100km/h. As seen in the figure the out performs CSOPC. The fluctuations in the plot is due to the errors in the prediction algorithm. The two power control algorithms are also tested with a mobile speed of 20km/h, see Figure 4, in this case the advantage of the new control is much more obvious. By using CSOPC there are high fluctuations due to the delay in computing the signal-to-interference ratio.

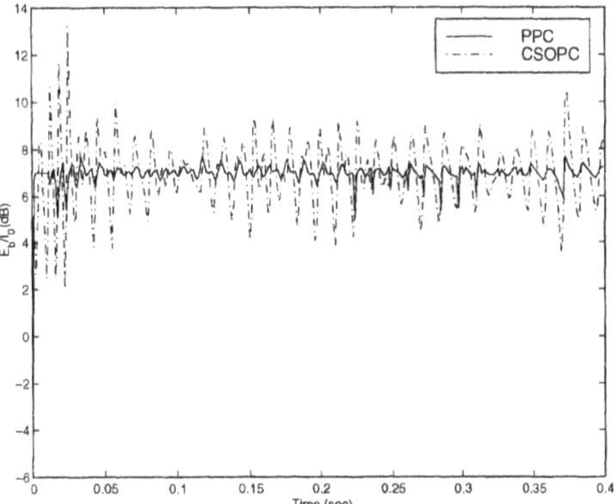

Fig. 3. PPC versus CSOPC with receiver speed of 100km/h

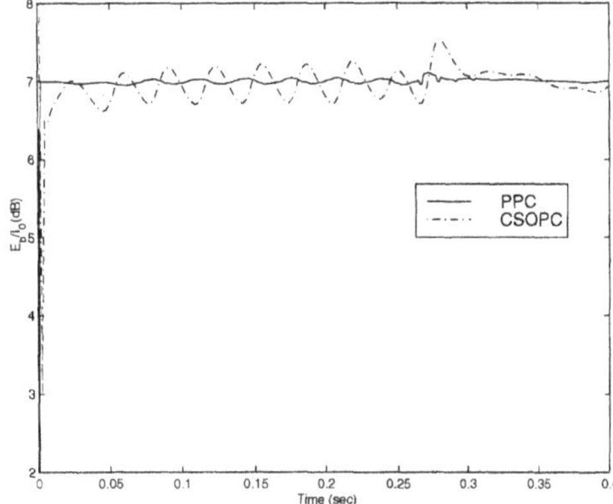

Fig. 4. PPC versus CSOPC with receiver speed of 20km/h

6. CONCLUSION

Delay in wireless communication systems is a serious problem. Due to the fast variation in the communication channel, the need to overcome this the delay problem becomes evident. Controlling the transmission power in CDMA wireless networks is very critical, and as shown in the simulation section, it is greatly affected by the delay. In this paper we have presented our approach in dealing with the delay in spread spectrum wireless networks. As shown in the the results section by predicting the changes in the channel and taking the delay into account while designing the power control algorithms we can improve the performance of our power control algorithm with a minimum increase in computational expense.

Appendix A. AUTOCORRELATION FUNCTION

Assuming that N is sufficiently large and the phases ϑ_n are independent, and as a consequent of the Central Limit Theorem, the E component, Equation (4), is complex Gaussian random variable. The field component E_z, may be expressed by

$$E_z(t) = x(t) + jy(t), \quad (A.1)$$

where $\mathbb{E}\{x(t)\} = \mathbb{E}\{y(t)\} = 0$ and $\mathbb{E}\{x^2(t)\} = \mathbb{E}\{y^2(t)\} = \sigma_r^2 = E_0^2/2$.

Since both x and y are Gaussian distributed and are independent, then $\mathbb{E}\{x(t)y(t)\} = 0$. Using the above properties, we start deriving the autocorrelation function of the received power.

$$\begin{aligned}
&R_{\|E_z\|^2}(\tau) = \\
&\mathbb{E}\{(x^2(t) + y^2(t))(x^2(t+\tau) + y^2(t+\tau)\} = \\
&\mathbb{E}\{x^2(t)x^2(t+\tau) + \mathbb{E}\{y^2(t)y^2(t+\tau)\} + \\
&\mathbb{E}\{x^2(t)y^2(t+\tau)\} + \mathbb{E}\{y^2(t)x^2(t+\tau)\},
\end{aligned} \quad (A.2)$$

Using the following property for Gaussian variables

$$\begin{aligned}
\mathbb{E}\{x_1x_2x_3x_4\} = \mathbb{E}\{x_1x_2\}\mathbb{E}\{x_3x_4\} + \\
\mathbb{E}\{x_1x_3\}\mathbb{E}\{x_2x_4\} + \mathbb{E}\{x_1x_4\}\mathbb{E}\{x_2x_3\},
\end{aligned} \quad (A.3)$$

the autocorrelation function $R_{\|E_z\|^2}$ becomes

$$R_{\|E_z\|^2} = 2R_{x^2x^2}(\tau) + 4\sigma_r, \quad (A.4)$$

where

$$\begin{aligned}
&R_{x^2x^2}(\tau) = \mathbb{E}\{x^2(t)x^2(t+\tau)\} \\
&= \sigma_r^4 + 2\mathbb{E}^2\{x(t)x(t+\tau)\} \\
&= \sigma_r^4 + 2E_0^4\mathbb{E}^2\{\sum_{n=1}^{N} C_n \cos\vartheta_n \sum_{m=1}^{N} C_n \cos(\vartheta_m + \\
&\quad 2\pi f_{md}\tau \cos\alpha_m)\} \\
&= \sigma_r^4 + \frac{1}{2}E_0^4\{\sum_{n=1}^{N} \mathbb{E}\{C_n^2 \cos(2\pi f_{md}\tau \cos\alpha_n)\}^2 \\
&= \sigma_r^4 + \frac{1}{2}E_0^4\{\int_{-\pi}^{+\pi} f_\alpha(\alpha)\cos(2\pi f_{md}\tau \cos\alpha_n)d\alpha\}^2 \\
&= \sigma_r^4 + \frac{1}{2}E_0^4 J_0^2(2\pi f_{md}\tau).
\end{aligned} \quad (A.5)$$

Finally,

$$R_{\|E_z\|^2}(\tau) = 4\sigma_r^4 + E_0^4 J_0^2(2\pi f_{md}\tau). \quad (A.6)$$

Now the autocorrelation of the small-scale fading becomes

$$R_{\beta^2}(\tau) = 1 + J_0^2(2\pi f_{md}\tau). \quad (A.7)$$

REFERENCES

Clarke, R.H. (1968). A statistical Theory of Mobile-Radio Reception. *Bell Sys. Tech. Journal* **47**, 957–1000.

El-Osery, A.I. and C.T. Abdallah (2000). Distributed Power Control in CDMA Cellular Systems. *IEEE Antennas and Propagation Magazine* **42**(4), 152–159.

Foschini, G.J. and Z. Miljanic (1993). A Simple Distributed Autonomous Power Control Algorithm and its Convergence. *IEEE Transactions on Vehicular Technology* **42**(4), 641–646.

Grandhi, S.A., R. Vijayan and D.J. Goodman (1993). Centralized Power Control in Cellular Radio Systems. *IEEE Trans. on Vehicular Technology* **42**(4), 466–468.

Huang, C.-Y. and R.D. Yates (1998). Rate of convergence for minimuim power assignment algorithms in cellular radio systems. *Wireless Networks* **4**, 223–231.

Jeantti, R. and S.-L. Kim (2000). Second-Order Power Control with Asymptotically Fast Convergence. *IEEE Journal on Selected Areas in Communications* **18**(3), 447–457.

Kim, D. (1999). A simple Algorithm for Adjusting Cell-Site Transmitter Power in CDMA Cellular Systems. *IEEE Trans. on Vehicular Technology* **48**(4), 1092–1098.

Proakis, J.G. (1983). *Digital Communications.* McGraw-Hill. New York.

Rappaport, T.S. (1996). *Wireless Communications: Principles and Practice.* Prentice Hall PTR. New Jersey.

Shanmugan, K.S. and A.M. Breipohl (1988). *Random Signals: Detection, Estimation and Data Analysis.* John Wiley & Sons. New York.

Wu, Q. (1999). Performance of Optimum Transmitter Power Control in CDMA Cellular Mobile Systems. *IEEE Trans. on Vehicular Technology* **48**(2), 571–575.

Yates, R.D. (1995). A Framework for Uplink Power Control in Cellular Radio Systems. *IEEE Journal on Selected Areas in Communications* **13**(7), 1341–1347.

Young, D.M. (1971). *Iterative Solution of Large Linear Systems.* Academic Press. New York.

Zander, J. (1992). Performance of Optimum Transmitter Power Control in Cellular Radio Systems. *IEEE Transations on Vehicular Technology* **41**(1), 57–62.

www.elsevier.com/locate/ifac

RESULTS ON FLUID MODELLING PACKET SWITCHED NETWORKS

Fabien Chatté [*] Bertrand Ducourthial [*]
Dritan Nace [*] Silviu-Iulian Niculescu [*]

[*] *Laboratoire Heudiasyc, UMR CNRS 6599, Université de Technologie de Compiègne, Centre de Recherche de Royallieu, BP. 20529, 60205 Compiègne* CEDEX, *France.*

Abstract: This paper focuses on the use of fluid modelling of packet switched networks. A simplified network protocol as well as the corresponding exact (continuous-time) fluid modelling are simulated. Fluid and discrete simulations are compared in different situations and with different parameters, in order to study the influence of the protocol and some network parameters on the matching of the results. Conditions that allows using fluid modelling of discrete packet switching protocols are then identified. *Copyright © 2001 IFAC*

Keywords: Networks, Models, Continuous systems, Discrete systems, Simulation.

1. INTRODUCTION

The actual development of Internet and communication networks with the need to support interactive applications (which are time constrained), makes the *congestion* an important and critical problem. There exists a lot of methods in the literature for controlling this phenomenon, and we shall not insist on them (see *e.g.* Jacobson (1988); Chiu and Jain (1989); Walrand and Varaiya (1996); Mathis et al. (1997); Alman et al. (1999); Floyd et al. (2000)). The main idea consists in controlling the sending rate on the base of a feedback from the destination, in order to avoid congestion that could appear somewhere in the network between the source and the destination.

To obtain simpler models and analysis techniques, we are interesting by a *fluid modelling* of the networks packets model (see, for instance, Bolot and Shakar (1992); Mascolo (1999)). Indeed, the interest of such a continuous network model lies in its simplicity. One of the simplest fluid modelling is given by a chain of integrators with an *uncertain* delay in the input for a source-destination pair of nodes. The number of integrators in the chain corresponds to the number of congestion nodes in the corresponding path (source to destination). Although the uncertainty in the delay makes the problem more complicated, the advances in robust analysis and control of uncertain system including or not delays Niculescu (2001); Zhou et al. (1995) allow to be optimistic for improving the congestion control in packets switching networks.

Using a continuous framework for the study and the control of a packet-switched network leads to two main problems. The first one is the *accuracy* of the fluid modelling. The second one is the *equivalence* of discrete and fluid results, depending on network and protocol parameters.

This paper focuses on the second problem, and deals with the *limitations* of the use of the fluid modelling of a discrete protocol with congestion control. To avoid any problem due to some imperfections of the modelling, a simplified protocol is used, for which a continuous fluid model is relatively easy to implement. Simulations using the packet simulator *ns* and *matlab* are discussed for one, two and three sources one destination. Still more sources have been studied in Chatté et al. (2001). A simple qualitative criterion (*closeness*) is proposed to compare fluid and discrete results. Conditions of the use of the fluid modelling are then given.

The protocol is presented in Section 2, as well as its fluid modelling. Discrete and fluid simulations are discussed in Section 3. Concluding remarks end the paper.

2. PROTOCOL

In this section we present the protocol used in this study.

2.1 Discrete protocol

A source node sends data packets to a destination node. The destination node sends back to the source some acknowledgement packets. It is assumed that, at the beginning of the communication, the network is empty (there is no packet in the network) and there is no congestion. Each data packet sent by the source contains a time-stamp in its header (sending date). The destination node copies this time-stamp into the acknowledgment packet before sending it back to the source node. The source can then compute the *round-trip-time* of the ith packet, denoted as $\mathtt{rtt}_i$ by:

$$\mathtt{rtt_i} = \text{receive time} \;-\; \text{time stamp in ack}$$

The first round-trip-time, computed when the network was still empty, denoted as $\mathtt{rtt\_min}$, is used as a *reference delay*. We have $\mathtt{rtt_i} \geq \mathtt{rtt\_min}$ for all further packets i. The source rate increases linearly if

$$\mathtt{rtt_i} \leq \mathtt{rtt\_min} + \mathtt{T}$$

where T is a fixed time delay, called *tolerance*. The source rate decreases exponentially if

$$\mathtt{rtt_i} > \mathtt{rtt\_min} + \mathtt{T}$$

Notice that congestion is due to limit bandwidth of links and not to processing power of nodes or routers. It is assumed that there is exactly *one bottleneck link* in the path from the source to the destination and no bottleneck on the way back to the source. The propagation delay Df between the source and the bottleneck link, and the delay Dr from the bottleneck back to the source are assumed constant (see Figure 1). We call *congestion node* the initial extremity of the bottleneck link (node C). The bottleneck has a fixed rate of μ *kbits/s*, and the size of the queue is infinite (to avoid the packets loss). All data packets sent by the source have the size s, fixed to 1000 bytes.

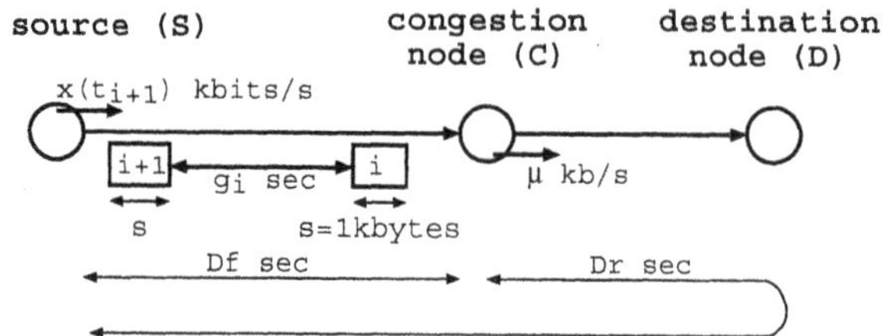

Fig. 1. Protocol overview

The source controls its sending rate $x(t_i)$ by adjusting the inter-packet interval (gap), called g_i. Let t_i be the date at which the packet i is sent. We have:

$$g_i = t_{i+1} - t_i = \frac{s}{x(t_i)}$$

The protocol has a minimal rate of 3000 *bits/s* in order to avoid a dead locked situation:

$$x(t_i) \geq 3kbits/s, \forall t_i$$

Indeed, if the rate falls close to 0 *bit/s*, an infinite gap will appear. In this case, the source will not be advertised on the end of the congestion because no packet will be sent and thus no acknowledgements received.

The tolerance T can be used to manage the average queue size. For example, if $\mathtt{T} = s/\mu$, the presence of one packet in the queue of the bottleneck node C is allowed. Hence, the source rate will enter later in the exponentially decreasing phase.

The simulations of the protocol have been done with the network simulator ns (2000) version 2.1b7a. The algorithm is based on two main primitives: send and receive a packet (see Chatté et al. (2001) for the algorithm and the *ns* code). It adjusts the inter-packet gap to obtain the wanted rate, depending on the increasing or decreasing phase. Three parameters are necessary: linear increasing factor α, exponential decreasing factor β, and the tolerance T. The larger α is, the faster the rate increases (in an increasing phase). The smaller β is, the faster the rate decreases (in a decreasing phase).

2.2 Fluid modelling

The continuous simulations have been done with *matlab*.

The fluid modelling is composed of two continuous functions of the time $q(t)$ and $x(t)$: $q(t)$ denotes the bottleneck queue size at time t and $x(t)$ gives the source sending rate at time t. When a data arrives at the bottleneck at time t, its sending rate is equal to $x(t - \texttt{Df})$. Indeed, it was sent by the source at time $t - \texttt{Df}$ (see Figure 1). Data are sent from the bottleneck node with a sending rate equal to μ if the queue is not empty, and a sending rate equal to $x(t - \texttt{Df})$ if the queue is empty and $x(t - \texttt{Df}) < \mu$.

Hence we have:

- if $q(t) = 0$ and $x(t - \texttt{Df}) \leq \mu$ then $q(t) = 0$
- else, $q(t) = \int_{t_{\text{queue not empty}}}^{t} (x(t - \texttt{Df}) - \mu)dt$

where $t_{\text{queue not empty}}$ is the date at which the queue becomes not empty:

$$x(t_{\text{queue not empty}} - \texttt{Df}) = \mu$$

A data spends $q(t)/\mu$ time into the queue, where t is the time at which the data reaches the queue. We denotes as **queuing_time** the function that gives the time spent by a packet in the queue in function of its leaving date: $\texttt{queuing\_time}(t + q(t)/\mu) = q(t)/\mu$. If an acknowledgement is received at time t by the source, it means that the corresponding packet left the queue at date $t - \texttt{Dr}$, which means that it arrived in the queue at date $t - \texttt{Dr} - \texttt{queuing\_time(t} - \texttt{Dr)}$.

Then, at the arrival of an acknowledgment at date t, if $q(t - \texttt{Dr} - \texttt{queuing\_time(t} - \texttt{Dr)})/\mu > \texttt{T}$, the sending rate of the source must be reduced exponentially, else it must increase linearly. Hence we have:

- if $(q(t - \texttt{Dr} - \texttt{queuing\_time(t} - \texttt{Dr)})/\mu) \leq \texttt{T}$ then $x(t) = x_{\texttt{min}} + \alpha \times (t - t_{\textbf{update min}})$
- else, $x(t) = max\left(x_{max} \times exp(\frac{t - t_{\text{update max}}}{\beta}), 3000\right)$

with $t_{\text{update min}} = t_{\text{waiting}\leq\texttt{T}} + \texttt{Dr}$ and $x_{\texttt{min}} = x(t_{\text{waiting}\leq\texttt{T}} + \texttt{Dr})$, $t_{\text{update max}} = t_{\text{waiting}>\texttt{T}} + \texttt{Dr}$ and $x_{\texttt{max}} = x(t_{\text{waiting}>\texttt{T}} + \texttt{Dr})$, where $t_{\text{waiting}\leq\texttt{T}}$ and $t_{\text{waiting}>\texttt{T}}$ are respectively defined by:

$$\texttt{queuing\_time}(t_{\text{waiting}\leq\texttt{T}}) \leq \texttt{T} \quad \text{and} \quad \texttt{queuing\_time}(t_{\text{waiting}>\texttt{T}}) > \texttt{T}$$

Initially $x_{\texttt{min}} = x_{\texttt{max}} = 3000$, $t_1 = t_{\text{update min}} = t_{\text{update max}} = 0$, and the queue is empty ($q(0) = 0$).

2.3 Comparison of the fluid and discrete results

The source rate is periodic in the fluid model, while it is quasi-periodic in the discrete case. So, we chose to compare the periods of the two models. In order to compare the periods, an average discrete period has been computed over a simulation of 300 s:

$$\text{discrete\_period} = \frac{\sum \text{period\_duration}}{\text{nb\_periods}}$$

Note that the comparison of the periods is a good indicator of the closeness of the discrete and fluid curves, while it does not take care about phases. Hence, if the periods are close, the only difference that can exist between the models is the phase. In particular, when the periods are close each other, $x_{\texttt{min}}$ and $x_{\texttt{max}}$ are also close each other (there is a relation between those parameters). We then define the closeness indicator as follows:

$$\text{closeness} = \left| 100 - \left(\frac{\text{fluid\_period}}{\text{discrete\_period}} \times 100 \right) \right|$$

3. SIMULATIONS

Discrete and fluid simulations have been done on networks with one two and three sources.

3.1 One source network

3.1.1. Influence of coefficients α and β

3.1.1.1. Hypotheses. This set of simulations has been done on the network of Figure 1. The throughput of the links (S, C) and (C, S) were varying between $1Mbits/s$ and $5Mbits/s$ in order to avoid a congestion at the source node S. The throughput of C was fixed to $\mu = 750kbits/s$. The maximal throughput of the link (D, C) has been fixed to $1Mbits/s$ (asymmetric connection). The propagation delay of the link (S, C) is $\texttt{Df} = 330\ ms$. The delay of the connections (C, D) and (D, C) are both equal to 100 ms, and the delay of (C, S) is 130 ms. Hence, $\texttt{Df} = \texttt{Dr} = 330\ ms$. The tolerance T was fixed to 0 packets authotized in the queue. Several values of α and β have been tested. The coefficient α was varying between 20 and 3000, and β between 0.25 and 6. In those simulations, the larger α is (*ie.*, more the rate increases quickly), the smaller β is (*ie.* more the rate descreases quickly).

3.1.1.2. Results. For a large range of value $(20 \leq \alpha \leq 1400, 0.5 \leq \beta \leq 6)$the closeness is quite good (under 6%). For the other values tested ($\alpha >$ 1400, and $\beta < 0.5$), the discrete model was not periodic wheras the fluid model was still periodic, so the fluid and discrete models do not match. Note that when the discrete protocol becomes aperiodic, the traffic load (evaluated using the total number of bits received by the destination D and the duration of the simulation) of the congestion node becomes low, and the protocol is inefficient. The corresponding α and β values should not be used in practice.

Figure 2 display results of discrete (up) and fluid (down) simulations, with $\alpha = 60$ and $\beta = 2$. Other values of these parameters give similar results (Chatté et al. (2001)).

As a conclusion, the coefficients α and β should be choosen to obtain a periodic discrete protocol. In the opposite case, the fluid modelling does not match with the discrete protocol.

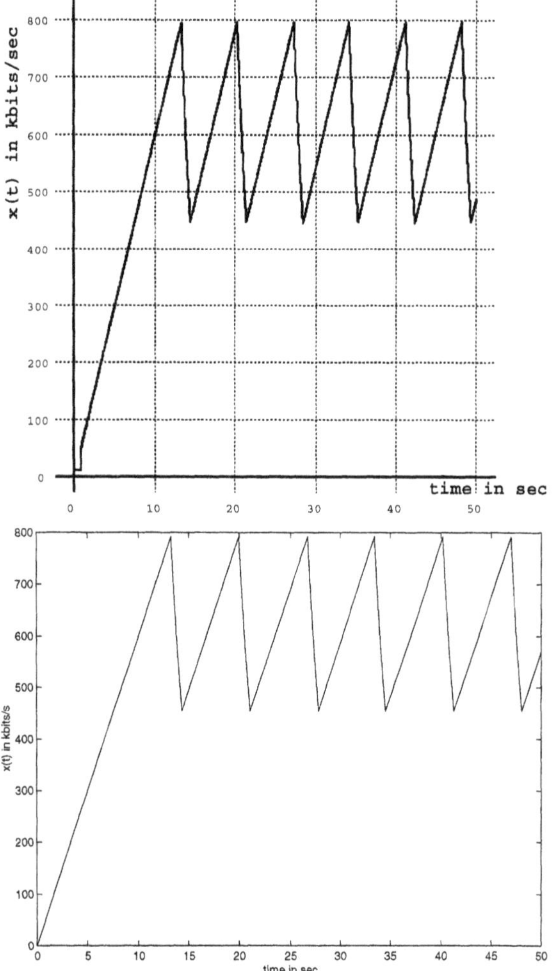

Fig. 2. Discrete and fluid simulations, T = 0 packets, $\mu = 750$ *kbits/s*

3.1.2. Influence of the tolerance T

3.1.2.1. Hypotheses. This two sets of simulations have been done on the same network than before, the propagation delays (Df and Dr), and the bandwidth of the link were the same. In the first set of simulations, the coefficient α was fixed to 800 and β to 0.75. In the second set, α was fixed to 375 and β to 2. Several values of the tolerance T have been tested. The tolerance was varying between 0 and 35 packets authorized in the queue for the first set and between 0 and 30 for the second one.

3.1.3. Results In both sets of simulation, the best results ($0\% \leq$ closeness $\leq 3.15\%$) are obtained for a tolerance of zero and two packets. For other values of T (1, 3, 4, ..., 30, 35), the closeness becomes bad, in general over 10%. Hence, the value of the tolerance T as an important influence (which is difficult to analyse) on the pertinence of the results obtained by the fluid modelling of the packet switching protocol.

3.2 Two sources network

3.2.1. General hypotheses These sets of simulations have been done on the network of Figure 3. The rate of the links between the source nodes and the congestion node are symmetric and varied between 1 $Mbits/s$ and 5 $Mbits/s$ (in order to avoid a congestion on the source node). The link between the congestion node and the receiver is asymmetric: on the going way μ was fixed to 750 $kbits/s$, and to 1 $Mbits/s$ on the way back.

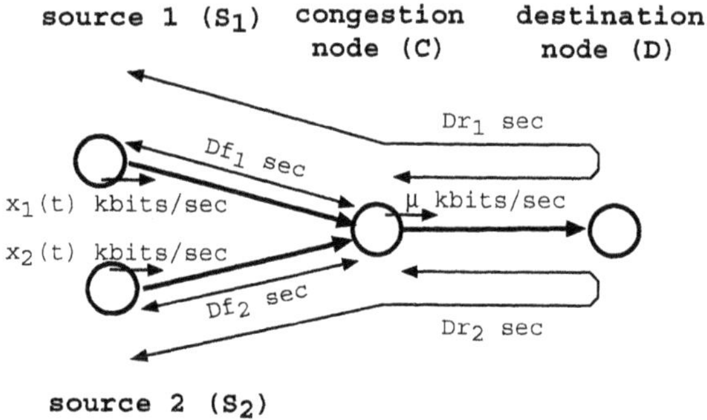

Fig. 3. Two sources network overview.

3.2.2. Influence of the tolerance

3.2.2.1. Hypotheses. In this set of simulations, the coefficients α_1 and α_2 are both fixed to 60, while β_1 and β_2 are both fixed to 2. The propagation delays $\mathtt{Df}_1$ and $\mathtt{Df}_2$ are both fixed to 330 ms, and $\mathtt{Dr}_1$ and $\mathtt{Dr}_2$ are also fixed to 330 ms. In order to verify if the tolerance T has an influence on the closeness of the two models, the tolerance was varying between 0 to 30 packets.

3.2.2.2. Results. The best results ($1.4\% \leq$ closeness $\leq 2.5\%$) are obtained for a tolerance of three packets or situated between ten and fifteen packets. In a network with two sources, two packets can arrive at the same time to the congestion node. In this case, if there is a tolerance of zero packet, one of the two sources could conclude that there is a beginning of congestion because one of its packets has waited in the queue. This source will reduce its rate, whereas in fact, there is not any congestion in the network. This problem is avoided if the tolerance has a minimum value of one packet. The results obtained with two sources network confirms those obtained with a single source network: the closeness of the results is good if the tolerance is chosen in one of the intervals defined before. The tolerance of the protocol still has an important influence on the matching of discrete and fluid results.

3.2.3. Influence of the coefficients α and β

3.2.3.1. Hypotheses for two identical sources. The simplest case consists in using two identical sources: $\alpha_1 = \alpha_2$, $\beta_1 = \beta_2$. To manage this set of simulations, a tolerance of three packets were used. The propagation delay is symmetric on the link between the sources and the congestion node (100 ms) ($\mathtt{Df}_1 = \mathtt{Df}_2 = 100$). The propagation delay on the link between the congestion node and the receiver is symmetric too and fixed to 100 ms ($\mathtt{Dr}_1 = \mathtt{Dr}_2 = 300$). In order to verify if the coefficients have an influence on the closeness, the coefficients α_1 and α_2 was vrying between 20 and 1600, and the coefficients β_1 and β_2 between 0.25 and 6.

3.2.3.2. Results for two identical sources. With two identical sources, the range of value with which, the two models are close, is smaller than the range obtained for one source network ($20 \leq \alpha \leq 400$ and $1 \leq \beta \leq 6$). Conditions to obtain a periodic behavior of the discrete protocol with two sources are stronger than for one sources; the range of acceptable values is then smaller.

3.2.3.3. Hypotheses for two different sources. In order to observe the influence of two different sources, a set of simulations has been performed, in which $\alpha_1 \neq \alpha_2$ and $\beta_1 \neq \beta_2$. A tolerance of two packets has been used in this set of simulation. The propagation delays were the same than those used for two identical sources.

3.2.3.4. Results for two different sources. Figure 4 displays fluid model (up), and discrete model (down), a result among others. The closeness between the two models does not seemed to be influenced by the difference between the sources. If α_1, α_2, β_1, β_2 are chosen in there respective range defined in 3.2.3.2, the closeness is always under 5%.

3.2.4. Influence of the propagation delays In this set of simulations, the coefficients α_1 and α_2 are both fixed to 60, while $\beta_1 = \beta_2$ are both fixed to 2. The tolerance was fixed to 2 packets authorized in the queue. The results obtained

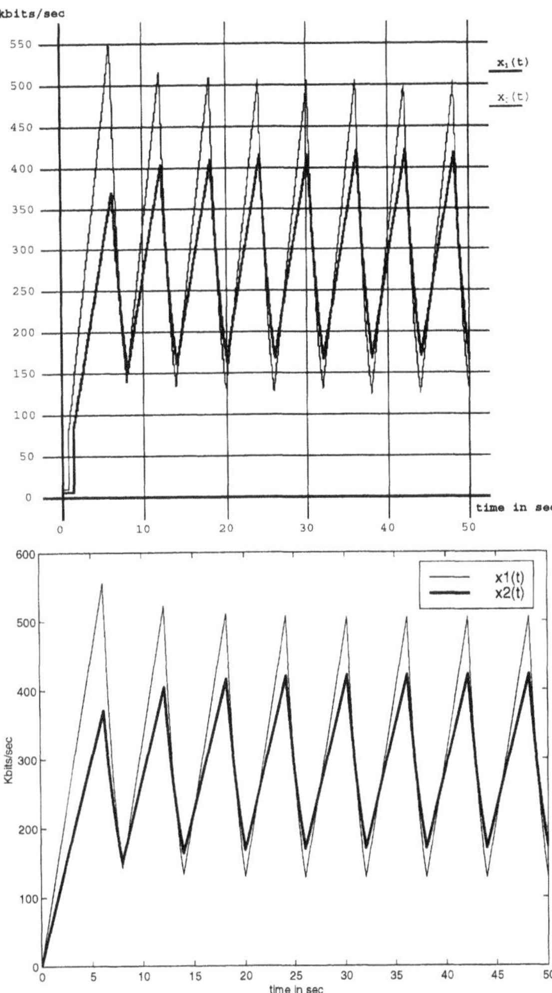

Fig. 4. Discrete and fluid simulation, $\alpha_1 = 90$, $\alpha_2 = 60$, $\beta_1 = 1.32$, $\beta_2 = 2$, $\mathtt{Df}_1 = \mathtt{Df}_2 = \mathtt{Dr}_1 = \mathtt{Dr}_2 = 330$ ms.

indicate that the propagation delays do not have a real influence on the closeness of the models.

3.3 Three sources network

In this set of simulations we focus on the influence of the tolerance on the closeness.

3.3.1. Hypotheses This set of simulations has been done in a network similar to the precedent (see Figure 3): the network was composed of three source nodes, one congestion node and one receiver node. The bandwidth of the links between the source nodes and the congestion node are symmetric and fixed to 1 $Mbits/s$. The link between the congestion node and the receiver one is asymmetric: on the going way μ was fixed to 1.1 $Mbits/s$, and on the way back the bandwidth was fixed to 2.1 $Mbits/s$. The propagation delay on the link between the sources and the congestion node is asymmetric: on the going way it was fixed to 330 ms, and to 130 ms on the way back. The propagation delay on the link between the congestion node and the receiver one is symmetric and fixed to 100 ms. The propagation delays $\mathtt{Df}_1$, $\mathtt{Df}_2$, $\mathtt{Df}_3$, $\mathtt{Dr}_1$, $\mathtt{Dr}_2$, $\mathtt{Dr}_3$ are fixed to 330 ms. All the sources were identical: $\alpha_1 = \alpha_2 = \alpha_3 = 60$ and $\beta_1 = \beta_2 = \beta_3 = 2$.

3.3.2. Results One more time, the closeness is good (under 5 %) only if the tolerance is chosen in a certain range of values ($4 \leq \mathtt{T} \leq 12$). For a tolerance greater than 12 packets, the closeness becomes very bad, and the bigger the tolerance is, the worse the closeness is. In this set of simulation, contrarily to the precedents, there is only one range of values in which the two models are close. Equally, we remarked that for a large tolerance ($\mathtt{T} > 20$), the discrete protocol has not the same behavior. Indeed, the period increases with the tolerance T. But, it is no more the case when T is greater than 20. The continuous modelling does not report this phenomena, and then its results become very bads.

4. CONCLUSION

In this work, the limitation in the use of a fluid modelling of a discrete congestion control protocol has been investigated. A simple protocol and a simple network has been used in order to obtain an exact fluid modelling. The results obtained by comparison of both discrete and fluid periods indicate that a fluid modelling can be a pertinent approximation of a discrete packet switched protocol. For simple topologies (one, two, three sources), the fluid modelling gives good results, and we extend here results of Bolot and Shakar (1992).

However, there is some conditions, depending on the protocol or the network parameters, for which the continuous modelling is not pertinent. Indeed, to obtain a good closeness, the discrete protocol should have a periodic behavior (in our case, α and β must be chosen in a specific range). Note that the propagation delays do not seemed to have any influence on the closeness. But, the sensibility of the protocol to the detection of the begining of a congestion seems to have a great (and still unexplained) influence on the pertinence of the results of the continuous model. In our case, only some specific ranges or values of the tolerance T ensure that the fluid simulations gave good results.

Other simulations with five or ten sources have been done Chatté et al. (2001). Some similar results have been obtained. In particular, for a tolerance greater than ten packets, the fluid modelling does not match anymore with the discrete protocol.

In future works, we plan to improve our results with more complete simulations, and more configurations. The comparison of the discrete and fluid curves could take care of the phase. Such a work could lead to the determination of restrictive conditions on the protocol and network parameters that ensure accurate continuous simulations of packet switched networks. These fluid modellings of discrete protocols should allow to design efficient control congestion systems for telecommunication networks.

REFERENCES

M. Alman, V. Paxson, and W. Stevens. Tcp congestion control. Request For Comment 2581, RFC editor, http://www.rfc-editor.org/, March 1999.

J.-C. Bolot and A. U. Shakar. Analysis of a fluid approximation to flow control dynamics. In *Proc of IEEE INFOCOM'92*, pages 2398–2407, 1992.

F. Chatté, B. Ducourthial, D. Nace, and S.-I. Niculescu. Results about fluid approximation of a packet switched network (extended version). Technical report, Heudiasyc UMR CNRS 6599, May 2001.

D. Chiu and R. Jain. Analysis of the increase and decrease algorithms for congestion avoidance in computer networks. *Computer Networks and ISDN Systems*, 17:1–14, 1989.

S. Floyd, J. Padhye, and J. Widmer. Equation-based congestion control for unicast applications. In *Proc. of SIGCOMM 2000*. ACM, May 2000.

V. Jacobson. Congestion avoidance and control. In *Proc. of ACM SIGCOMM'88*, pages 314–329, 1988.

S. Mascolo. Congestion control in high-speed communication networks using the Smith principle. *Automatica*, 35(1999):1921–1935, 1999.

M. Mathis, J. Semke, J. Mahdavi, and T. Ott. The macroscopic behavior of the tcp congestion avoidance algorithm. *Computer Communications Review*, 27(3), July 1997.

S.-I. Niculescu. *Delay effects on stability: A robust control approach.* LNCIS. Springer-Verlag, Heidelberg, May 2001. 383 pages, à paraître.

The ns manual. The ns project, http://www.isi.edu/nsnam/ns/ns-documentation.html, August 2000.

J. Walrand and P. Varaiya. *High-performance communication networks.* Morgan Kaufmann, San Francisco, CA, USA, 1996.

K. Zhou, J. Doyle, and K. Glover. *Robust and optimal control.* Prentice Hall, New Jersey, USA, 1995.

www.elsevier.com/locate/ifac

STABILITY OF COMMUNICATIONS NETWORKS IN THE PRESENCE OF DELAYS

Chaouki Abdallah [*,1] **John Chiasson** [**]

[*] *ECE Dept, University of New Mexico, Alburquerque NM 87131-1356, USA, chaouki@eece.unm.edu*
[**] *ECE Dept, University of Tennessee, Knoxville TN 37996, USA, chiasson@utk.edu*

Abstract: In this paper, we present a study of the destabilizing effects of delay in communications networks. We the apply results from a companion paper to find the range of delays for which a Smith predictor controller guarantees the stability of an ATM/ABR network. *Copyright © 2001 IFAC*

Keywords: Time Delay, Stability, Smith Predictor

1. THE ATM/ABR EXAMPLE

Congestion control in the Available Bit Rate (ABR) class of Asynchronous Transfer Mode (ATM) networks poses interesting challenges due to the presence of multiple-delays, magnitude and rate constraints on the inputs, amplitude limitation on the state, and additive disturbances. The ABR class is designed as a best-effort class for applications such as file transfer and email. Thus, no service guarantees are required, but the source of data packets controls its data rate, using a feedback signal provided by switches downstream that measure the congestion of the network. Due to the presence of this feedback, many classical and advanced control theory concepts have been suggested to deal with the congestion control problem in the ATM/ABR case.

We will focus our modeling on one particular queue in the network which is associated with a link shared by many virtual circuits D. Cavendish and Gerla (1996). A virtual circuit is established for any two stations wishing to communicate, by informing all switches between them of their requirements. Assuming that the flow of packets is conserved, the queue level model for each buffer in the ATM network (as proposed in D. Cavendish and Gerla (1996)) is

$$\dot{x}(t) = \sum_{i=1}^{n} u_i(t - T_i) - d(t) \tag{1}$$

where

- $x(t)$ is the queue level associated with the considered link;
- n is the number of *virtual circuits* sharing the queue associated with the considered link;
- $u_i(t)$ is the inflow cell rate caused by the ith *virtual circuit*;
- T_i is the propagation delay from the ith source to the queue, and is usually uncertain;
- $d(t)$ is the rate of packets leaving the queue.

More details on this model and its physical interpretation is available from Mascolo (2000) (see also Nicolescu (2001) p. 60-63). Equation (1) appears to be linear since the saturation effect due to the limited buffer capacity has been neglected. Our control design however, must ensure that this condition is actually satisfied, so that the controller is not saturated. The author in Mascolo (2000) proposes the Smith predictor control

[1] The research of C.T. Abdallah is partially supported by NSF INT-9818312

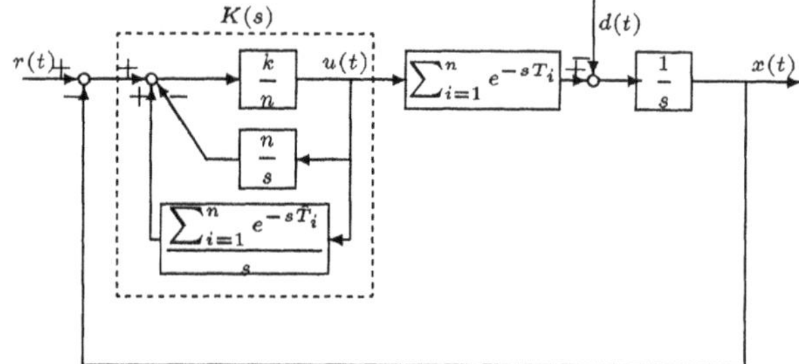

Fig. 1. The controlled system.

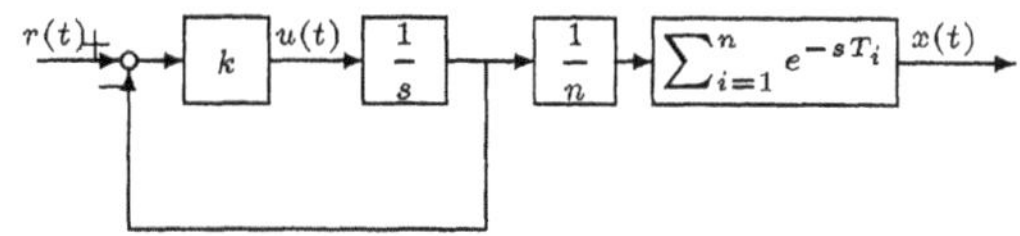

Fig. 2. The desired closed-loop system.

scheme of Figure 1. The objectives of the control law proposed in Mascolo (2000) are to guarantee

- *Stability*[2]:

$$x(t) \leq r^0 \quad t \geq 0$$

where r^0 is the queue capacity. This condition guarantees no cell loss and is not the usual stability requirement;

- *Full Link Utilization*:

$$x(t) > 0 \qquad t > T_{tr} \tag{2}$$

The time T_{tr} in (2) mainly accounts for the transient time of the dynamics. Let $\hat{T}_i$ denote the estimated values of the delays T_i. Assuming that $\hat{T}_i = T_i$ and disregarding for the moment the presence of disturbance $d(t)$, the scheme in Figure 1 can be shown to be equivalent to the one in Figure 2Mascolo (2000). In this case, the designer knows exactly the value of all delays $T_i; i = 1, \cdots, n$, and the closed-loop system exhibits the following nice properties.

- With $r(t) = r^0 \cdot 1(t)$ ($1(t)$ is the unit step) denoting the desired reference level of the queue, and $d(t) = a \cdot 1(t)$, with $0 \leq a \leq 1$ the *stability property* (boundedness) is satisfied.
- Further, with $d(t) = a \cdot 1(t)$, with $0 \leq a \leq 1$ the *full utilization property* is satisfied, provided that Mascolo (2000)

$$r^0 > a \left(\frac{1}{k} + \frac{1}{n} \sum_{i=1}^{n} T_i \right).$$

The technique which is used in order to obtain the desired closed-loop system is the well-known Smith's principle Smith (1957)Power and Simpson (1978).

The main drawback of the approach of Mascolo (2000) is that it assumes that the propagation delays T_i are *exactly* known. When this is not the case, even stability, *in the sense defined above*[3], can be lost. In order to illustrate this limitation and the potential of losing stability (boundedness), a simulation was performed. Assuming that $n = 4$, $T_1 = 10$ sec, $T_2 = 30$ sec, $T_3 = 60$ sec, $T_4 = 120$ sec, $k = 0.1$, and $r^0 = 40$, the stability condition is satisfied (see Figure 3) if $\hat{T}_i = T_i \; i = 1, \ldots 4$.

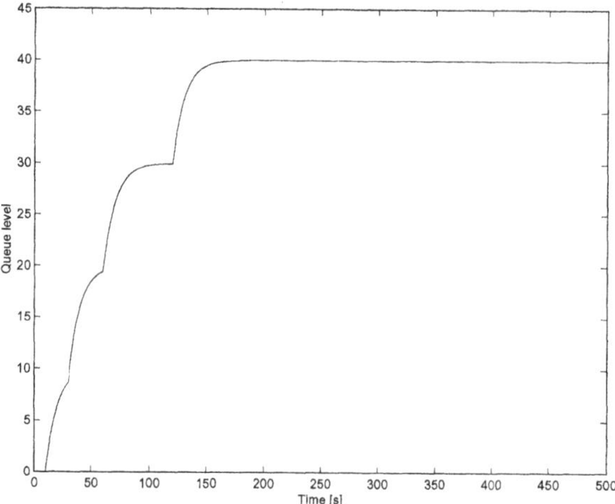

Fig. 3. Queue level when the propagation delays are assumed to be known.

Next, the actual delays $T_i; i = 1, \cdots, n$ are perturbed by 5%, without modifying their assumed values $\hat{T}_i; i = 1, \cdots, n$. As shown in Figure 4, stability is lost since $x(t)$ is greater than r^0 during some time intervals. Stability is regained however if the controller gain k is changed to $k = 0.01$ (see Figure 5) at the expense of a slower response. When the T_is however are not known, the designer cannot apriori design a Smith-predictor-type controller in order to guarantee the desired degree of stability and performance. Moreover, even BIBO stability may be lost if one does not take the effects of the uncertainty in the delays.

2. DELAYS IN THE ATM MODEL

The transfer function of the ATM system is

$$X(s) = \frac{\frac{k}{n}\sum_{i=1}^{n} e^{-sT_i}}{s + k - \frac{k}{n}\left(\sum_{i=1}^{n} e^{-sT_i} - e^{-s\hat{T}_i}\right)} R(s) - \frac{s + \frac{k}{n}(n - \sum_{i=1}^{n} e^{-s\hat{T}_i})}{s\left(s + k - \frac{k}{n}\left(\sum_{i=1}^{n} e^{-sT_i} - e^{-s\hat{T}_i}\right)\right)} d$$

First of note that the transfer function

$$\frac{s + \frac{k}{n}(n - \sum_{i=1}^{n} e^{-s\hat{T}_i})}{s}$$

[2] The network community refers to this as a "stability" condition. Stability in the dynamic systems sense will guarantee BIBO stability and then care must be taken to ensure the bound r^0 is not violated.

[3] That is, $x(t) \leq r^0$ is violated.

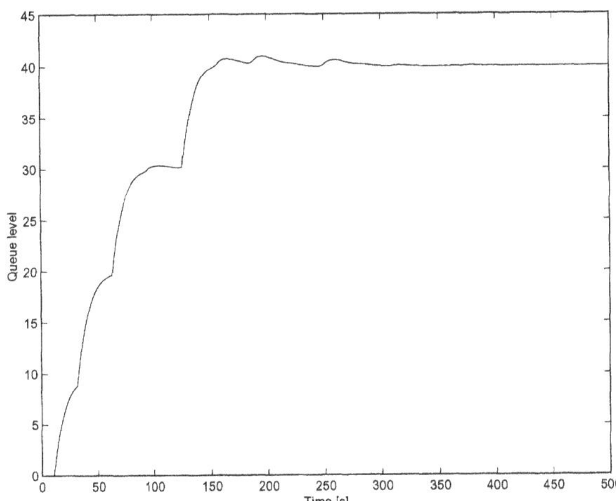

Fig. 4. Queue level when the propagation delays are perturbed by 5%.

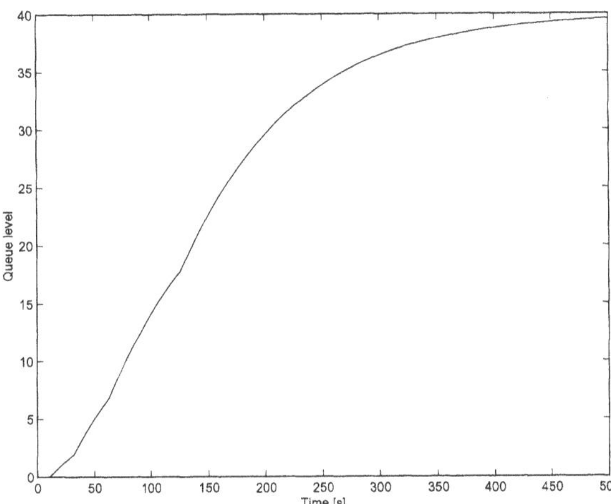

Fig. 5. Queue level when the propagation delays are perturbed by 5% and $k = 0.01$.

is stable since the only possible pole location is $s = 0$ and there it follows that

$$\left.\frac{s + \frac{k}{n}(n - \sum_{i=1}^{n} e^{-s\hat{T}_i})}{s}\right|_{s=0}$$
$$= \lim_{s \longrightarrow 0} \frac{s + \frac{k}{n}(n - \sum_{i=1}^{n} e^{-s\hat{T}_i})}{s}$$
$$= \lim_{s \longrightarrow 0} \frac{s + \frac{k}{n}(n \sum_{i=1}^{n} \hat{T}_i)s}{s}$$
$$= 1 + k \sum_{i=1}^{n} \hat{T}_i \neq \infty.$$

Consequently, the stability of the ATM system is reduced to studying the stability of $s + k - \frac{k}{n}\left(\sum_{i=1}^{n} e^{-sT_i} - e^{-s\hat{T}_i}\right)$. In this section, being stable means

$$s + k - \frac{k}{n}\left(\sum_{i=1}^{n} e^{-sT_i} - e^{-s\hat{T}_i}\right) \neq 0 \text{ for } \mathrm{Re}(s) \geq 0 \quad (3)$$

In order to analyze the stability of this system, some further refinements in the model are now made. Let $T_i = m_i h_0, \hat{T}_i = m_i h$ for $i = 1, ..., n$ where the m_i are positive integers[4]. Define $c(z) \triangleq \sum_{i=1}^{n} z^{m_i}$ so that $c(e^{-sh_0}) = \sum_{i=1}^{n} e^{-sT_i}, c(e^{-sh}) = \sum_{i=1}^{n} e^{-s\hat{T}_i}$. The characteristic equation of the ATM system (3) may be rewritten as

$$a(s, e^{-h_0 s}, e^{-hs}) \triangleq a_0(s) + a_1(s)\left[c(e^{-sh_0}) - c(e^{-sh})\right] \quad (4)$$

where $a_0(s) = s + k, a_1(s) = -k/n$. Here the delays h_0, h are noncommensurate in general. By design, the compensator $G_c(s) = k$ is chosen so that (4) is stable for $h = h_0$, that is, $a_0(s)$ is stable. The interest here is to determine the range of values of h about h_0 for which (4) remains stable. As h is decreased (or decreased) in value from h_0, there is a first value h^* for which the roots of (4) are on the $j\omega$ axis, that is,

$$\begin{aligned} &a(s, e^{-h_0 s}, e^{-h^* s}) \\ &\triangleq a_0(s) + a_1(s)\left(c(e^{-sh_0}) - c(e^{-sh^*})\right) \\ &= 0 \end{aligned} \quad (5)$$

for some $s^* = j\omega^*$.

To exploit this idea ($z_0 = e^{-sh_0}, z = e^{-sh}$), define ($m = \max_i\{m_i\}$)

$$\begin{aligned} &a(s, z_0, z) \\ &= a_0(s) + a_1(s)\left(c(z_0) - c(z)\right), \ c(z) = \sum_{i=1}^{n} z^{m_i} \\ &\tilde{a}(s, z_0, z) \\ &\triangleq z_0^m z^m a(-s, 1/z_0, 1/z) \\ &= z_0^m z^m \left[a_0(-s) + a_1(-s)\left(c(1/z_0) - c(1/z)\right)\right] \\ &= z_0^m z^m a_0(-s) + a_1(-s)\left(z^m c(z_0) - z_0^m c(z)\right) \end{aligned}$$

Solving $a(s, z_0, z) = 0$ to get $c(z) = \left(a_0(s) + a_1(s)c(z_0)\right)/a_1(s)$ and substituting this into $\tilde{a}(s, z_0, z) = 0$ gives

$$\begin{aligned} 0 = &z_0^m z^m a_0(-s) + \\ &a_1(-s)\left(z^m c(z_0) - z_0^m \frac{a_0(s) + a_1(s)c(z_0)}{a_1(s)}\right) \\ z^m = &\frac{z_0^m\left(a_1(-s)a_0(s) + a_1(s)a_1(-s)c(z_0)\right)}{\left(z_0^m a_1(s)a_0(-s) + a_1(-s)a_1(s)c(z_0)\right)} \\ = &\frac{a_1(-s)a_0(s) + a_1(s)a_1(-s)c(z_0)}{a_1(s)a_0(-s) + a_1(-s)a_1(s)c(1/z_0)} \end{aligned}$$

where $z_0^{-m}c(z_0) \equiv c(1/z_0)$. With $s = j\omega, z = e^{-jh\omega}, z_0 = e^{-jh_0\omega}$ this becomes

[4] For all practical purposes, there will always be h_0, h such that the delays can be written as integer multiples of these values.

$$
\begin{aligned}
z^m &= e^{-jmh\omega} \\
&= \frac{a_1(-j\omega)a_0(j\omega) + a_1(j\omega)a_1(-j\omega)c(e^{-jh_0\omega})}{a_1(j\omega)a_0(-j\omega) + a_1(-j\omega)a_1(j\omega)c(e^{jh_0\omega})} \\
&= \left|\frac{a_1(-j\omega)a_0(j\omega) + a_1(j\omega)a_1(-j\omega)c(e^{-jh_0\omega})}{a_1(j\omega)a_0(-j\omega) + a_1(-j\omega)a_1(j\omega)c(1/e^{-jh_0\omega})}\right| \\
&\times e^{j\angle \frac{a_1(-j\omega)a_0(j\omega)+a_1(j\omega)a_1(-j\omega)c(e^{-jh_0\omega})}{a_1(j\omega)a_0(-j\omega)+a_1(-j\omega)a_1(j\omega)c(e^{jh_0\omega})}} \\
&= e^{2j\angle\left(a_1(-j\omega)a_0(j\omega)+a_1(j\omega)a_1(-j\omega)c(e^{-jh_0\omega})\right)}
\end{aligned}
$$

and solving for z

$$
\begin{aligned}
z &= e^{-jh\omega} \\
&= e^{j\frac{2\angle\left(a_1(-j\omega)a_0(j\omega)+a_1(j\omega)a_1(-j\omega)c(e^{-jh_0\omega})\right)+2\pi k}{m}}
\end{aligned} \qquad (6)
$$

for $k = 0, 1, ..., m-1$. By definition, for $h = h^*$ corresponds to $a(s, e^{-h_0 j\omega}, e^{-h^* j\omega}) = 0$ for some $s^* = j\omega^*$ where $e^{-h^* j\omega}$ is given by the right-hand side of (6) for $k = 0, 1, ..., m-1$. Thus, the righ-hand side of (6) is substituted into (5) to get

$$
\begin{aligned}
&0 = \\
&a_0(j\omega) + a_1(j\omega)\left[c(e^{-jh_0\omega}) - c(e^{-jh^*\omega})\right]\Big|_{e^{-jh^*\omega} = \text{ rhs of (6)}}
\end{aligned}
$$

For each $k = 0, 1, ..., m-1$ this equation is solved for the corresponding ω_{ki}, $i = 1, 2, ...$ which are then substituted back into (6) so for $k = 0, 1, ..., m-1$ it follows that

$$
\begin{aligned}
&e^{-jh^*\omega_{ki}} = \\
&e^{j\left(2\angle\left(a_1(-j\omega_{ki})a_0(j\omega_{ki})+a_1(j\omega_{ki})a_1(-j\omega_{ki})c(e^{-jh_0\omega_{ki}})\right)+2\pi k\ \right)/n}
\end{aligned}
$$

Finally, for each ω_{ki} this is solved h^*_{ki}. Then

$$
\begin{aligned}
h_1^* &= \max_{k,i}\{h^*_{ki} : h^*_{ki} < h_0\} \\
h_2^* &= \min_{k,i}\{h^*_{ki} : h^*_{ki} > h_0\}.
\end{aligned}
$$

Finally, for $h_1^* < h < h_2^*$

$$
\begin{aligned}
&a(s, e^{-h_0 s}, e^{-hs}) \triangleq \\
&a_0(s) + a_1(s)\left[c(e^{-sh_0}) - c(e^{-sh})\right] \neq 0 \text{ for } \operatorname{Re}\{s\} \geq 0
\end{aligned}
$$

and

$$
\begin{aligned}
a(s, e^{-h_0 s}, e^{-h_1^* s}) &= 0 \text{ for some } s = j\omega \\
a(s, e^{-h_0 s}, e^{-h_2^* s}) &= 0 \text{ for some } s = j\omega
\end{aligned}
$$

3. CONCLUSIONS

In this paper, we have applied results form our companion paper Chiasson and Abdallah (2001) in order to exactly determine the range of delays for which Smith predictor controllers remain stabilizing. The results are specialized to ATM/ABR networks, but can also be applied in many areas of queuing theory and communications.

REFERENCES

J. Chiasson and C.T. Abdallah. A test for robust stability of time delay systems. December 2001. Sante Fe, NM.

S. Mascolo D. Cavendish and M. Gerla. Sp-eprca: an atm rate based congestion control scheme based on a smith predictor. 1996. preprint.

S. Mascolo. Smith's principle for congestion control in high-speed data network. *IEEE Transactions on Automatic Control*, 45(2):358–364, 2000.

S.I. Nicolescu. *Delay effects on Stability*. Springer-Verlag, 2001. Berlin.

H.M. Power and R.J. Simpson. *Introduction to Dynamics and Control*. McGraw-Hill, 1978.

O.J.M. Smith. Closed control of loops with dead time. *Chemical Engineering Progress*, 53(5): 217–219, 1957.

LOAD BALANCING INSTABILITIES DUE TO TIME DELAYS IN PARALLEL COMPUTATIONS

Chaouki Abdallah * **J. Douglas Birdwell** **
John Chiasson ** **Victor Chupryna** ** **Zhong Tang** **
Tsewei Wang ***

* *ECE Dept, University of NewMexico, Alburquerque NM 87131-1356, USA, chaouki@eece.unm.edu*
** *ECE Dept, University of Tennessee, Knoxville TN 37996, USA, birdwell@utk.edu, chiasson@utk.edu, tang@hickory.engr.utk.edu*
*** *ChEDept, University of Tennessee, Knoxville TN 37996, USA, twang@utk.edu*

Abstract: A deterministic dynamic linear time-delay model is presented to model load balancing in a cluster of nodes used for parallel computations. The model is analyzed for stability in terms of the delays in the transfer of information between nodes and the gains in the load balancing algorithm. *Copyright © 2001 IFAC*

Keywords: Time Delay, Stability, Load Balancing, Parallel Computation, Cluster Computing

1. INTRODUCTION

Parallel computer architectures utilize a set of computational elements (CE) to achieve performance that is not attainable on a single processor, or CE, computer. A common architecture is the cluster of otherwise independent computers communicating through a shared network. To make use of parallel computing resources, problems must be broken down into smaller units that can be solved individually by each CE while exchanging information with CEs solving other problems.

The Federal Bureau of Investigation (FBI) National DNA Indexing System (NDIS) and Combined DNA Indexing System (CODIS) software are candidates for parallelization. New methods developed by Wang et al (Wang, 2001)(Wang, 1999)(Birdwell, 2000)(Birdwell, 1999)(Birdwell, 2001) lead naturally to a parallel decomposition of the DNA database search problem while providing orders of magnitude improvements in performance over the current release of the CODIS software. The projected growth of the NDIS database and in the demand for searches of the database necessitates migration to a parallel computing platform.

Effective utilization of a parallel computer architecture requires the computational load to be distributed more or less evenly over the available CEs. The qualifier "more or less" is used because the communications required to distribute the load consumes both computational resources and network bandwidth. A point of diminishing returns exists.

Distribution of computational load across available resources is referred to as the *load balancing* problem in the literature. Various taxonomies of load balancing algorithms exist. Direct methods examine the global distribution of computational load and assign portions of the workload to resources before processing begins. Iterative methods examine the progress of the computation and the expected utilization of resources, and adjust the workload assignments periodically as computation progresses. Assignment may be either deterministic, as with the dimension exchange/diffusion (Corradi and Zambonelli, 1999) and gradient methods, stochastic, or optimization based. A comparison of several deterministic methods is provided by Willeback-LeMain and Reeves (Willebeek-LeMair and Reeves, 1993).

To adequately model load balancing problems, several features of the parallel computation environment should be captured: (1) The workload awaiting processing at each CE; (2) the relative performances of the CEs; (3) the computational requirements of each workload component; (4) the delays and bandwidth constraints of CEs and network components involved in the exchange of workloads, and (5) the delays imposed by CEs and the network on the exchange of measurements. A queuing theory (Kleinrock, 1975) approach is well-suited to the modeling requirements and has been used in the literature by Spies (Spies, 1996) and others. However, whereas Spies assumes a homogeneous network of CEs and models the queues in detail, the present work generalizes queue length to an expected waiting time, normalizing to account for differences among CEs, and aggregates the behavior of each queue using a continuous state model. The present work focuses upon the effects of delays in the exchange of information among CEs, and the constraints these effects impose on the design of a load balancing strategy.

2. MODEL

The mathematical model of a given computing node is given by

$$\begin{aligned}\frac{dx_i(t)}{dt} &= \lambda_i - \mu_i + u_i(t) - \sum_{j\neq i} p_{ij}u_j(t)\\ y_i(t) &= x_i(t) - \frac{\sum_{j=1}^{n} x_j(t-\tau_{ij})}{n}\\ u_i(t) &= -K_i y_i(t-h_i)\end{aligned} \tag{1}$$

where: $x_i(t)$ is the expected waiting time experienced by a task inserted into the queue of the ith node, q_i is the number of tasks in the ith node, t_{p_i} is average time needed to process a task on the ith node ($x_i(t) = q_i t_{p_i}$), λ_i is the rate of generation of waiting time on the ith node caused by the addition of tasks (rate of increase in x_i), μ_i is the rate of reduction in waiting time caused by the service of tasks at the ith node ($\mu_i \equiv (1 \times t_{p_i})/t_{p_i} = 1$ for all i), $u_i(t)$ is the rate of removal (transfer) of the tasks from node i at time t by load balancing at node i, p_{ij} is the fraction of $u_j(t)$ that node j allocates to node i at time t; $\sum_{i=1}^{n} p_{ij} = 1$, ($p_{jj} = 0$), $p_{ij}u_j(t-h_{ij})$ is the rate of removal (transfer) of tasks at time t from node j by (to) node i, h_{ij} is the time delay for task transfer from node j to node i. ($h_{ii} = 0$), τ_{ij} is the time delay for communicating the node waiting time x_j to node i ($\tau_{ii} = 0$), and n is the number of nodes. All rates are in units of the rate of change of expected waiting time (time/time, which is dimensionless).

Here $u_i(t) < 0$ means tasks are being sent to other nodes while $u_i(t) > 0$ means the ith node is receiving tasks from other nodes. A delay is experienced by transmitted tasks before they are received at the other node. The control law $u_i(t) = -K_i y_i(t-h_i)$ states that if the ith node output $x_i(t)$ is above the local average $\left(\sum_{j=1}^{n} x_j(t-\tau_{ij})\right)/n$ then it sends data to the other nodes, while if it is less than the local average, it accepts data from the other nodes. Here h_i is the delay in sending the tasks $-K_i y$ to the other nodes.

Often, the p_{ij} are functions of the state x_i so as to send a higher fraction of the data to those nodes that have less tasks. However, this is left out in this model to retain linearity of the system so that a stability analysis can be carried out.

3. STABILITY ANALYSIS

To study the stability of the model, three nodes ($n = 3$) are considered with $K_1 = K_2 = K_3 = K$, $p_{ij} = 1/2$, for all $i \neq j$, $p_{ii} = 0$, $\tau_{ij} = h$ for $i \neq j$, $\tau_{ii} = 0$, $h_i = 2h$ for all $i = 1, 2, 3$. The transfer function from the inputs $d_1 = \lambda_1 - \mu_1$, $d_2 = \lambda_2 - \mu_2$, and $d_3 = \lambda_3 - \mu_3$ to the output $y_1(s) \triangleq x_1(s) - \left(x_1(s) + e^{-hs}x_2(s) + e^{-hs}x_3(s)\right)/3$ is ($z = e^{-hs}$)

$$\begin{aligned}y_1(s) &= \\ &\frac{-\frac{1}{18}(6s + K(2z^2 - z^3 - z^4))(2s + K(2z^2 + z^3))}{-\frac{1}{4}s(2s + K(2z^2 + z^3)^2}d_1\\ &+ \frac{\frac{1}{18}z\left(3s + K(-2z + z^2 + z^3)\right)\left(2s + K(2z^2 + z^3)\right)}{-\frac{1}{4}s(2s + K(2z^2 + z^3)^2}\times\\ &(d_2 + d_3)\end{aligned}$$

$$\begin{aligned}&= \frac{2}{9}\frac{6s + K(2z^2 - z^3 - z^4)}{s(2s + K(2z^2 + z^3)}d_1\\ &- \frac{2}{9}\frac{z\left(3s + K(-2z + z^2 + z^3)\right)}{s(2s + K(2z^2 + z^3))}(d_2 + d_3)\end{aligned}$$

$$= \frac{2}{9}\frac{b_1(s,z)}{sa(s,z)}d_1 - \frac{2}{9}\frac{zb_2(s,z)}{sa(s,z)}(d_2 + d_3)$$

where $b_1(s,z) = 6s + K(2z^2 - z^3 - z^4)$, $b_2(s,z) = 3s + K(-2z + z^2 + z^3)$ and $a(s,z) = 2s + K(2z^2 + z^3)$.

The stability of this time delay system is analyzed using the techniques developed in (Bellman and

Cooke, 1963)(Chiasson and Lee, 1985)(Hertz and Zeheb, 1984)(Kamen, 1982)(Chiasson, 1988)(Chiasson and Abdallah, 2001) (see especially (Chiasson, 1988)(Chiasson and Abdallah, 2001)). The problem is broken into the three parts where it is first shown that $a(s,e^{-hs})$ for stable for $0 \le h < \frac{2}{K}\frac{0.674889}{2.85}$. Then it is shown that $b_1(s,e^{-hs})/s$ and $b_2(s,e^{-hs})/s$ are both stable independent of delay.

3.1 Stability of $a(s,e^{-hs})$

First $a(s,z)$ is considered. With $K>0$, the polynomial $a(s,z)$ is stable for $h=0$ since $a(s,1)=2s+3K$. Next, define

$$a(s,z) = 2s + K(2z^2+z^3)$$
$$\tilde{a}(s,z) \triangleq z^3 a(-s,1/z) = z^3(-2s) + K(2z+1)$$

The idea here (see (Chiasson, 1988)(Chiasson and Abdallah, 2001)) is to simply note that the polynomial is stable for $h=0$ and then to increase h until there are zeros of the polynomial on the $j\omega$ axis. At this point, there are zeros of the form $s=j\omega, z=e^{-j\omega h}$ which must satisfy $a(s,z)=\tilde{a}(s,z)=0$ as $-j\omega$ and $1/e^{-j\omega h}$ are simply the conjugates of $s=j\omega, z=e^{-j\omega h}$, respectively. To find the first value of h that results in the system being unstable, the common zeros of $\{a(s,z),\tilde{a}(s,z)\}$ on the $j\omega$ axis are found. Specifically, the variable s is eliminated from $a(s,z)$ and $\tilde{a}(s,z)$ resulting in

$$R(z) = z^6+2z^5+2z+1$$
$$= (z^2+2z+1)(z^4+z^3-2z^2+z+1)=0.$$

These roots are

$$z_{1,2} = -\frac{1}{2} \pm j\frac{\sqrt{3}}{2} = e^{j2\pi/3}, e^{j4\pi/3}$$
$$z_{3,4} = \frac{-1+\sqrt{17}}{4} \pm j\frac{1}{\sqrt{2}}\sqrt{\frac{-1+\sqrt{17}}{4}}$$
$$= 0.780776 \pm j0.624811$$
$$z_{5,6} = \frac{-1-\sqrt{17}}{4} \pm \frac{1}{\sqrt{2}}\sqrt{\frac{1+\sqrt{17}}{4}}$$

Only the first four roots $z_{1,2}$,$z_{3,4}$ have magnitude one and the corresponding values of s are

$$s_{1,2} = -\frac{K}{2}(2z^2+z^3)\Big|_{z=-\frac{1}{2}\pm j\frac{\sqrt{3}}{2}} = -\frac{K}{2}\left(\frac{1}{2} \mp j\frac{\sqrt{3}}{2}\right)$$
$$s_{3,4} = -\frac{K}{2}(2z^2+z^3)\Big|_{0.780776\pm j0.624811} = \mp\frac{K}{2}2.85j$$

As $\mathrm{Re}(s_{1,2}) \neq 0$, these do not correspond to zeros of $a(s,e^{-hs})$ on the $j\omega$ axis. Setting $z=e^{-sh}$ with $s=\mp\frac{K}{2}2.85j$, $z=0.780776 \pm j0.624811 = e^{\pm j0.674889}$ gives

$$e^{\pm j0.674889} = e^{\pm\frac{K}{2}2.85jh}$$
$$\Longrightarrow h = \frac{2}{K}\frac{0.674889}{2.85}$$

That is, for a given K, this gives the smallest positive value of the delay for which $a(s,e^{-hs})$ has a zero in the right-half plane (specifically, on the $j\omega$ axis). Or, in terms of the gain K, $a(s,e^{-hs})$ has no zeros in the right-half plane for

$$K < K_{\max} \triangleq \frac{2}{h}\frac{0.674889}{2.85}$$

and further has zeros on the $j\omega$ axis for $K=K_{\max}$.

3.2 Stability of $b_1(s,e^{-hs})/s$ *and* $b_2(s,e^{-hs})/s$

Consider next transfer function

$$\frac{b_1(s,e^{-hs})}{s} = \frac{6s+K(2e^{-2hs}-e^{-3hs}-e^{-4hs})}{s}$$

The denominator has a single root at $s=0$. However, this is not a pole since

$$\left.\frac{b_1(s,e^{-hs})}{s}\right|_{s=0} = \lim_{s\to 0}\frac{b_1(s,e^{-hs})}{s}$$
$$= \lim_{s\to 0}\frac{6s+K(2e^{-2hs}-e^{-3hs}-e^{-4hs})}{s}$$
$$= \lim_{s\to 0}\frac{6s+3Khs}{s}$$
$$= 6+3K \neq \infty$$

Similarly, for the transfer function

$$\frac{b_2(s,e^{-hs})}{s} = \frac{3s+K(-2e^{-hs}+e^{-2hs}+e^{-3hs})}{s}$$

the denominator has a single root at $s=0$. As before, this is not a pole since

$$\left.\frac{b_2(s,e^{-hs})}{s}\right|_{s=0} = \lim_{s\to 0}\frac{b_2(s,e^{-hs})}{s}$$
$$= \lim_{s\to 0}\frac{3s+K(-2e^{-hs}+e^{-3hs}+e^{-3hs})}{s}$$
$$= \lim_{s\to 0}\frac{3s-4Khs}{s}$$
$$= 3-4Kh \neq \infty$$

Consequently, the transfer functions $b_1(s,e^{-hs})/s$, $b_2(s,e^{-hs})/s$ are both stable independent of delay.

4. SIMULATIONS

Experimental procedures to determine the delay values are given in the thesis (Dasgupta, 2001*a*) and summarized in the paper (Dasgupta, 2001*b*). These give representative values for a Fast Ethernet network with three nodes of $\tau_{ij} = 400\ \mu\mathrm{sec}$ for $i \neq j, \tau_{ii}=0$, and so with $h_i = 2\times 400\ \mu\mathrm{sec}$, the maximum value for the gain is

$$K_{\max} = \frac{2}{h}\frac{0.674889}{2.85} = \frac{2}{400\times 10^{-6}}\frac{0.674889}{2.85} = 1184$$

The simulation was performed with three nodes ($n=3$), $K_1=K_2=K_3=K$, $p_{ij}=1/2$, for all i,j, $\tau_{ij}=h$ for $i\neq j$, $\tau_{ii}=0$, $h_i=2h$ for $i=$

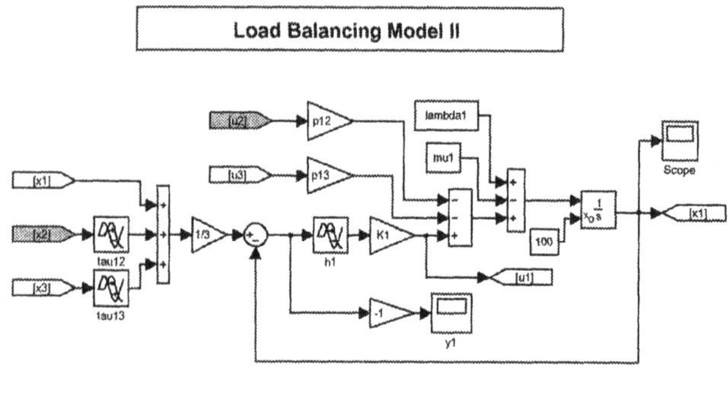

$1, 2, 3$ and $h = 400$ μsec. The inputs were set as $\lambda_1 = 2\mu_1, \lambda_2 = 0, \lambda_3 = 0$ and the initial conditions were $x_1(0) = 100$, $x_2(0) = 5$ and $x_3(0) = 3$. The figure below is a block diagram of one node of the system where $K_1 = K$, $\tau_{12} = \tau_{13} = 400$ μsec and $h_1 = 800$ μsec.

The figures below are plots of $y_1(t)$ from the simulation for three runs where the gain is set as $K = 0.99K_{\max}, K = K_{\max}$, and $K = 1.01K_{\max}$, respectively.

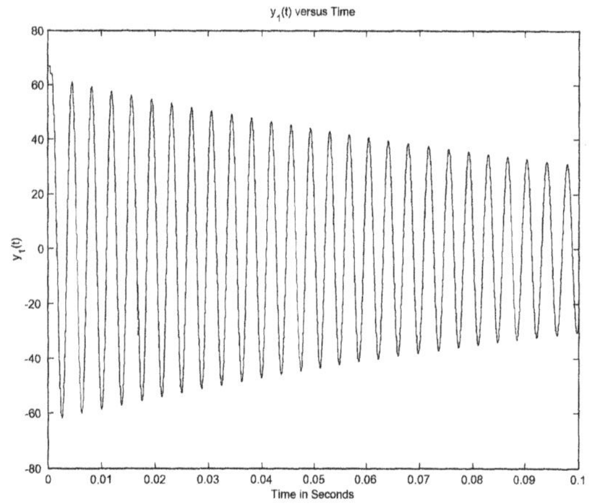

Plot of $y_1(t)$ versus time for $K = 0.99K_{\max}$

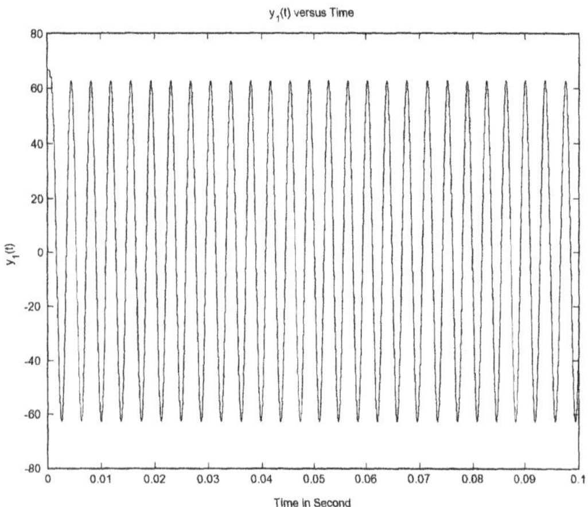

Plot of $y_1(t)$ versus time for $K = K_{\max}$

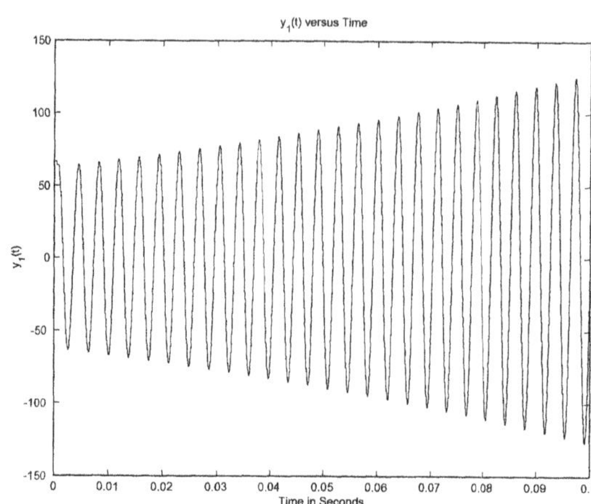

Plot of $y_1(t)$ versus time for $K = 1.01K_{\max}$

5. CONCLUSIONS

In this work, a load balancing algorithm was modeled as a linear time-delay system. Under the assumption of symmetric nodes and controllers (all intercommunication delays are identical and the controller gains identical) a systematic procedure was presented to determine the stability of the system by an explicit relationship between the delay values and the control gain. The delays create a limit on the size of the controller gains in order to ensure stability.

6. ACKNOWLEDGEMENTS

The work of J.D. Birdwell, V. Chupryna, Z. Tang, and T.W. Wang was supported by U.S. Department of Justice, Federal Bureau of Investigation under contract J-FBI-98-083. The work of C.T. Abdallah was supported in part by the National Science Foundation through the grant INT-9818312. The views and conclusions contained in this document are those of the authors and should not be interpreted as necessarily representing the official policies, either expressed or implied, of the U.S. Government.

REFERENCES

Bellman, R. and K.L. Cooke (1963). *Differential-Difference Equations.* New York: Academic.

Birdwell, J.D., R. D. Horn D. J. Icove T. W. Wang P. Yadav S. Niezgoda (1999). A hierarchical database design and search method for codis. In: *Tenth International Symposium on Human Identification.* Orlando, FL.

Birdwell, J.D., T-W. Wang M. Rader (2001). The university of tennessee's new search engine for codis. In: *6th CODIS Users Conference.* Arlington, VA.

Birdwell, J.D., T. W. Wang R. D. Horn P. Yadav D. J. Icove (2000). Method of indexed storage and retrieval of multidimensional information. U. S. Patent Application 09/671,304.

Chiasson, J. (1988). A method for computing the interval of delay values for which a differential-delay system. *IEEE Transactions on Automatic Control* **33**(12), 1176–1178.

Chiasson, J. and C.T. Abdallah (2001). A test for robust stability of time delay systems. Sante Fe, NM.

Chiasson, J.N., S.D. Brierley and E.B. Lee (1985). A simplified derivation of the zeheb-walach 2-d stability test with applications to time-delay systems. *IEEE Transactions on Automatic Control.*

Corradi, A., L. Leonardi and F. Zambonelli (1999). Diffusive load-balancing polices for dynamic applications. *IEEE Concurrency* **22**(31), 979–993.

Dasgupta, P. (2001*a*). *Performance Evaluation of Fast Ethernet, ATM and Myrinet under PVM, MS Thesis.* University of Tennesse.

Dasgupta, P., J. D. Birdwell T. W. Wang (2001*b*). Timing and congestion studies under pvm. In: *Tenth SIAM Conference on Parallel Processing for Scientific Computation.* Portsmouth, VA.

Hertz, D., E.I. Jury and E. Zeheb (1984). Simplified analytic stability test for systems with commensurate time delays. *IEE Proceedings.* part D.

Kamen, E.W. (1982). Linear systems with commensurate time delays: Stability and stabilization independent of delay. *IEEE Transactions on Automatic Control* pp. 367–375.

Kleinrock, L. (1975). *Queuing Systems Vol I : Theory.* John Wiley & Sons. New York.

Spies, F. (1996). Modeling of optimal load balancing strategy using queuing theory. *Microprocessors and Microprogramming* **41**, 555–570.

Wang, T.W., J. D. Birdwell P. Yadav D. J. Icove S. Niezgoda S. Jones (1999). Natural clustering of dna/str profiles. In: *Tenth International Symposium on Human Identification.* Orlando, FL.

Wang, T.W., P. Yadav J. D. Birdwell (2001). Method of index storage and retrieval of dna profiles stored in a large database. In: *American Institute of Chemical Engineering Annual Conference.* Reno Nevada.

Willebeek-LeMair, M.H. and A.P. Reeves (1993). Strategies for dynamic load balancing on highly parallel computers. *IEEE Transactions on Parallel and Distributed Systems* **4**(9), 979–993.

www.elsevier.com/locate/ifac

AN ASYMPTOTICALLY STABLE APPROACH IN THE NUMERICAL SOLUTION OF DELAY DIFFERENTIAL EQUATIONS

S. Maset *,[1]

* *Dipartimento di Scienze Matematiche*
Università di Trieste
Via Valerio 12
34127 Trieste, Italy

Keywords: Time delay Systems, Numerical methods, Asymptotic stability.

1. THE ABSTRACT IMPLICIT EULER METHOD

It is known (see [1]) that there do not exist Runge-Kutta (RK) methods unconditionally preserving the asymptotic stability when they are applied to linear systems of Delay Differential Equations (DDEs)

$$y'(t) = Ly(t) + My(t-1) \tag{1}$$

where $L, M \in \mathbb{C}^{2\times 2}$. In other words, given a RK method and a stepsize it is possible to find an asymptotically stable system (1) which numerical solution is not asymptotically stable. Since 2×2 complex systems can be restated as 4×4 real systems, it is impossible to find RK methods unconditionally preserving the asymptitic behaviour for real systems of dimension four.

In view of such a very negative result we now present an approch for solving DDE which unconditionally preserves the asymptotic stability on all systems (1) with L, M matrices of arbitrary dimension. This is the first example of a numerical method with such a nice property.

Let us consider the DDE initial value problem

$$\begin{cases} y'(t) = Ly(t) + My(t-1), \ t \geq 0 \\ y(t) = \varphi(t) \quad t \in [-1, 0] \end{cases} \tag{2}$$

where $L, M \in \mathbb{C}^m$ and $\varphi \in C([-1,0], \mathbb{C}^m)$. The method we are proposing for numerical solving (2) is the classic implicit Euler method as applied to the abstract Cauchy problem corresponding to (2). For this reason we call it "the abstract implicit Euler method".

The abstract Cauchy problem is stated in the Banach space $C = C([-1,0], \mathbb{C}^m)$ equipped by the maximum norm, and is

$$\begin{cases} \frac{d}{dt} y_t = \mathcal{A} y_t, \ t > 0 \\ y_0 = \varphi \end{cases} \tag{3}$$

where $\mathcal{A} : D(\mathcal{A}) \to C$ is the unbounded linear operator defined by

$$\begin{aligned} D(\mathcal{A}) = \{ & \phi \in C : \phi' \in C \ and \\ & \phi'(0) = L\phi(0) + M\phi(-1)\}; \\ & \mathcal{A}\phi = \phi'. \end{aligned} \tag{4}$$

and y_t is given by

$$y_t(\theta) = y(t+\theta), \ \theta \in [-1, 0]. \tag{5}$$

When applied with constant stepsize Δt, the abstract implicit Euler method produces a sequence u^k, $k = 0, 1, 2, ...$, of functions in the space C, where u^k is an approximation of y_{t_k}, $t_k = k\Delta t$. We start with $u^0 = \varphi$ and, given u^k, the next function u^{k+1} is the solution of the boundary value problem

---

[1] Partially supported by MURST and INdAM-GNCS

$$\begin{cases} u^{k+1}(\theta) = u^k(\theta) + \Delta t \left(u^{k+1}\right)'(\theta) \\ \theta \in [-1,0], \\ \left(u^{k+1}\right)'(0) = Lu^{k+1}(0) + Mu^{k+1}(-1). \end{cases} \tag{6}$$

The method can be restated as follows

$$u^0 = \varphi,\ u^{k+1} = E\left(u^k\right),\ k = 0,1,2,... \tag{7}$$

where $E : C \to C$ is defined as follows. Given $\phi \in C$, the function $\chi = E(\phi) \in C$ is computed through the following steps:

1) Compute the function $\psi \in C$,

$$\psi(\theta) = e^{\frac{\theta}{\Delta t}}\phi(0) + \int_{\theta}^{0} e^{\frac{\theta-\xi}{\Delta t}} \frac{1}{\Delta t}\phi(\xi)\, d\xi,$$

$$\theta \in [-1,0]. \tag{8}$$

2) Compute the solution $K \in \mathbb{C}^m$ of the $m \times m$ linear system algebraic equation

$$\left(I_m - \Delta t\left(L + e^{-\frac{1}{\Delta t}}M\right)\right)K = L\psi(0) + M\psi(-1). \tag{9}$$

3) Compute the function $\chi \in C$,

$$\chi(\theta) = \psi(\theta) + \Delta t e^{\frac{1}{\Delta t}}K,\ \theta \in [-1,0]. \tag{10}$$

It is clear that the objects ϕ, ψ and χ are infinite dimensional and so the abstract implicit Euler method can't be realized on a computer. Below in the section "Using the abstract implicit Euler method on a computer" we show that if the initial function of the DDE (2) belongs to a particular and significant finite dimensional space, then we can realize on a computer the method because now we can work with finite dimensional objects.

2. PROPERTIES OF THE METHOD

Here we state the convergence of the method and the preservation of the asymptotic behaviour (see [2] for details).

Theorem (Convergence). If φ is of class C^2 and $\varphi'(0) = L\varphi(0) + M\varphi(-1)$, then, for all $T > 0$,

$$\max_{k=0,1,...,k_{max}} \max_{\theta\in[-1,0]} \left|u^k(\theta) - y_{t_k}(\theta)\right| = O(\Delta t); \tag{11}$$

with $k_{max}\Delta t = T$ and $||$ arbitrary norm on $\mathbb{C}^m$.

Theorem (Preservation of asymptotic stability). The abstract implicit method unconditionally preserves the asymptotic stability for all DDE (2), i.e. if the solution y vanishes for all initial functions φ, then the sequence u^k defined in (7) vanishes for all initial functions φ and for all stepsizes Δt.

3. USING THE ABSTRACT IMPLICIT EULER METHOD ON A COMPUTER

Let C_N, with $N = 0,1,2,...$, be the finite dimensional subspace of C

$$C_{N,N_1} = \left\{\sum_{n=0}^{N} \frac{(-\theta)^n}{n!}a_n : a_n \in \mathbb{C}^m\right\} \tag{12}$$

and let C_{N,N_1}, with $N, N_1 = 0,1,2,...$, be the finite dimensional subspace of C

$$C_{N,N_1} = \left\{\sum_{n=0}^{N} \frac{(-\theta)^n}{n!}a_n + \sum_{n=0}^{N_1} \frac{1}{n!}e^{\frac{\theta}{\Delta t}}\left(-\frac{\theta}{\Delta t}\right)^n b_n \right.$$
$$\left. : a_n \in \mathbb{C}^m,\ b_n \in \mathbb{C}^m\right\} \tag{13}$$

where $\frac{(-\theta)^n}{n!}$, $n = 0,...,N$, and $e^{\frac{\theta}{\Delta t}}\left(-\frac{\theta}{\Delta t}\right)^n$, $n = 0,...,N_1$, denotes the functions $\theta \mapsto \frac{(-\theta)^n}{n!}$ and $\theta \mapsto e^{\frac{\theta}{\Delta t}}\left(-\frac{\theta}{\Delta t}\right)^n$, $\theta \in [-1,0]$. We set $C_{N,-1} = C_N$.

A function $\phi \in C_{N,N_1}$, $\phi = \sum_{n=0}^{N} \frac{(-\theta)^n}{n!}a_n + \sum_{n=0}^{N_1} \frac{1}{n!}e^{\frac{\theta}{\Delta t}}\left(-\frac{\theta}{\Delta t}\right)^n b_n$, is uniquely defined by the couple of matrices (A,B) where $A \in \mathbb{C}^{m\times(N+1)}$ has columns $a_0, ..., a_N$ and $B \in \mathbb{C}^{m\times(N_1+1)}$ has columns $b_0, ..., b_{N_1}$. We will write $\phi = (A,B)$.

For the map E defining the abstract implicit Euler method we have

$$E : C_{N,N_1} \to C_{N,N_1+1} \tag{14}$$

(see [2] for details) and, given $\phi = (A,B) \in C_{N,N_1}$, the function $\chi = E(\phi) = (A^*, B^*) \in C_{N,N_1+1}$ is computed through the following steps (compare with the steps 1), 2) and 3) of section "The abstract implicit Euler method"):

1) Compute the function $\psi = (A^+, B^+) \in C_{N,N_1+1}$,

$$A^+ = A - AD(\Delta t),\ b_0^+ = b_0 + AD_1(\Delta t),$$
$$b_n^+ = b_{n-1},\ n = 1, ..., N_1 + 1 \tag{15}$$

where

$$D(\Delta t) = \begin{pmatrix} 0 & & & & \\ \Delta t & 0 & & & \\ -\Delta t^2 & \Delta t & 0 & & \\ \vdots & & \ddots & \ddots & \\ (-1)^{N+1}\Delta t^N & \cdots & -\Delta t^2 & \Delta t & 0 \end{pmatrix}$$

and $D_1(\Delta t)$ is the first column of $D(\Delta t)$.

2) Compute the solution $K \in \mathbb{C}^m$ of the algebraic equation

$$\left(I_m - \Delta t\left(L + e^{-\frac{1}{\Delta t}}M\right)\right)K = L\psi(0) + M\psi(-\tau) \tag{17}$$

where

$$\psi(0) = A_1^+ + B_1^+, \tag{18}$$

with A_1 and B_1 first columns of A and B, and

$$\psi(-1) = A^+ \left(\frac{1}{0!}, \dots, \frac{1}{N!}\right)^T +$$

$$B^+ \left(e^{-\frac{1}{\Delta t}} \frac{\left(\frac{1}{\Delta t}\right)^0}{0!}, \dots, e^{-\frac{1}{\Delta t}} \frac{\left(\frac{1}{\Delta t}\right)^{N_1+1}}{(N_1+1)!}\right)^T \tag{19}$$

3) Compute the function $\chi = (A^*, B^*) \in C_{N,N_1+1}$,

$$A^* = A^+,\ b_0^* = b_0^+ + \Delta t K,$$
$$b_n^* = b_n^+,\ n = 1, \dots, N_1. \tag{20}$$

Therefore by taking the initial function φ of (2) in the space C_N the function u^k, $k = 0, 1, 2...$ of the sequence (7) belongs to $C_{N,k-1}$. Given the finite dimensional object $u^0 = \varphi$, we can compute the sequence of finite dimensional object u^k, $k = 1, 2, \dots$, as showed above.

REFERENCES

[1] S. Maset. Instability of Runge-Kutta methods when applied to linear systems of delay differential equations. Numer. Math., in press

[2] S. Maset. The abstract Implicit Euler method for Retarded Functional Differential Equations. Quaderno matematico del Dipartimento di Scienze Matematiche dell'Università di Trieste, Serie II n.480.

www.elsevier.com/locate/ifac

A FREE-STEPSIZE IMPLEMENTATION OF SECOND ORDER STABLE METHODS FOR NEUTRAL DELAY DIFFERENTIAL EQUATIONS

Alfredo Bellen *,[1] **Marino Zennaro** *,[1]

* *Dipartimento di Scienze Matematiche, Universita' di Trieste, Italy*

Keywords: Delay Differential Equations, Neutral Equation, Contractivity, Asymptotic Stability, Numerical Analysis.

It is known that the solution of the general Delay Differential Equation (DDE)

$$\begin{cases} y'(t) = f\big(t, y(t), y(t-\tau(t))\big), & t_0 \le t \le t_f, \\ y(t) = \phi(t), & t \le t_0, \end{cases}$$

where $\tau(t) \ge 0$, undergoes the *smoothing* effect, that is the solution, which is not differentiable at the point t_0 even if the functions f, τ and ϕ are of class C^∞, becomes smoother and smoother at subsequent points $\xi_0 = t_0 < \xi_1 < \ldots$ (called *breaking points*) recursively generated as follows. If ξ_k is a breaking point where the function $y(t)$ is of class C^p, any ξ_j such that $\xi_k = \xi_j - \tau(\xi_j)$ is a subsequent breaking point where $y(t)$ is of class at least C^{p+1}. On the contrary, the smoothing effect does not apply to the DDE of *neutral* type (NDDE)

$$\begin{cases} y'(t) = f\big(t, y(t), y(t-\tau(t)), y'(t-\tau(t))\big), \\ \qquad\qquad\qquad t_0 \le t \le t_f, \\ y(t) = \phi(t), \quad t \le t_0, \end{cases}$$

where $y'(t)$ has a jump discontinuiuty at any breaking point. This prevents the convergence of any numerical step-by-step method to exceed the order one, unless the breaking points are all included as mesh points. This is a severe restriction, in particular for equations with many non monotonic delay functions $t - \tau(t)$ where the breaking points may proliferate in a chaotic manner as t increases.

Nevertheless, when the neutral equation may be transformed into the form (see Hale [4])

$$\begin{cases} \frac{d}{dt}\Big(y(t) - G\big(t, y(t-\tau(t))\big)\Big) = \\ \quad H\Big(t, y(t), y(t-\tau(t))\Big), \quad t_0 \le t \le t_f, \\ y(t) = \phi(t), \quad t \le t_0, \end{cases} \tag{1}$$

it is natural to solve it for the function

$$\Phi(t) = y(t) - G\big(t, y(t-\tau(t))\big)$$

that, unlike $y(t)$, is of class $C^1[t_0, t_f]$ provided that the function $H(t, u, v)$ is continuous, as usually occurs. This is the case, for instance, of the linear equation

$$y'(t) = Ly(t) + My(t-\tau) + Ny'(t-\tau), \tag{2}$$

where the delay τ is constant, which may be written as

$$\frac{d}{dt}\Big(y(t) - Ny(t-\tau)\Big) = L\Big(y(t) - Ny(t-\tau)\Big) + (M + LN)y(t-\tau).$$

In these cases the equation (1) takes the form of a Delay Differential Algebraic Equation (DDAE)

$$\begin{cases} \Phi'(t) = K\Big(t, \Phi(t), y\big(t-\tau(t)\big)\Big), \\ y(t) = \Phi(t) + G\Big(t, y\big(t-\tau(t)\big)\Big), \\ \qquad\qquad\qquad t_0 \le t \le t_f, \\ \Phi(t_0) = \phi(t_0) - G\Big(t_0, \phi\big(t_0 - \tau(t_0)\big)\Big), \\ y(t) = \phi(t), \quad t \le t_0, \end{cases} \tag{3}$$

[1] Partially supported by MURST and INdAM-GNCS

to be solved for $\Phi(t)$ and $y(t)$, where

$$K(t,u,v) = H\Big(t, u + G(t,v), v\Big).$$

For non vanishing delays, equation (3) is basically an ODE and may be solved step-by-step by a continuous method. In fact, for sufficiently small stepsizes, the algebraic component y may be separately computed, by recursion, after the differential component Φ .

Once we have integrated equation (3) till the point t_n by a *continuous* numerical method, the subsequent step for approximating Φ and y consists in solving, for $\Lambda(t)$ and $w(t)$, the local problem

$$\begin{cases} \Lambda'(t) = K\Big(t, \Lambda(t), \eta\big(t-\tau(t)\big)\Big), \\ w(t) = \Lambda(t) + G\Big(t, \eta\big(t-\tau(t)\big)\Big), \\ \qquad\qquad t_n \le t \le t_{n+1}, \\ \Lambda(t_n) = \eta(t_n) - G\Big(t_n, \eta\big(t_n - \tau(t_n)\big)\Big). \end{cases} \tag{4}$$

where $\eta(t)$ is the numerical continuous approximation of $y(t)$ previously computed in the past intervals. The system (4) is first solved for $\Lambda(t)$ with some continuous approximation $\lambda(t)$ and then for $w(t)$ with continuous approximation $\eta(t)$ given by

$$\eta(t) = \lambda(t) + G\Big(t, \eta\big(t-\tau(t)\big)\Big)$$

or, in order to avoid tracing back the solution until $\phi(t)$, by

$$\eta(t) = \mathcal{P}\Big(\lambda(\cdot) + G\Big(\cdot, \eta\big(\cdot - \tau(\cdot)\big)\Big)\Big)(t), \tag{5}$$

where $\mathcal{P}\big(u(\cdot)\big)$ is the interpolation of u in a suitable polynomial space and nodes in $[t_n, t_{n+1}]$.

Since the right-hand side function of the differential equation that defines the auxiliary function $\Phi(t)$ is Lipschitz continuous, any method of (global) uniform order $p \ge 2$ has a local error which is an $O(h^2)$ on any step interval including a discontinuity point, and an $O(h^{p+1})$ elsewhere (see [1], [3]). Assuming that the number of discontinuties is finite in $[t_0, t_f]$, we conclude that the method performs to the order 2 with free stepsizes, not necessarily including the discontinuity points into the mesh Δ.

Nevertheless, there still are some troubles with discontinuities for the continuous approximation $\eta(t)$ of the solution $y(t)$. In fact, if the interpolation (5) is used for the computation of $G\big(t, \eta(t - \tau(t))\big)$ in (4), the lack of smoothness of y at the discontinuity points causes a local loss of accuracy to the interpolation process down to an $O(h)$ whenever the delayed argument $t - \tau(t)$ crosses one of such points. However, this unpleasant phenomenon takes place only locally at those mesh points close to the discontinuity points, unless they are all included in the mesh Δ, in which case no loss of accuracy occurs. The global order of accuracy of $\eta(t)$ at all the other mesh points remains an $O(h^2)$.

We conclude this part by stressing the remarkable and, at the first sight, surprising property that, by handling the NDDE in the form (1), the loss of accuracy at some nodal values y_{n_i} does not affect the accuracy at subsequent points of the approximate trajectory where the expected accuracy is recovered. On the other hand, if interpolation is not used and $\eta(t)$ is computed by tracing it back until the initial function $\phi(t)$, not even these local losses of accuracy occur.

The occurencies mentioned above are illustrated by the following example.

Example Let us consider the equation

$$\begin{cases} y'(t) = y(t) + y'(t-1), \quad 0 \le t \le 3, \\ y(t) = 1, \quad t \le 0, \end{cases} \tag{6}$$

whose solution is the continuous function

$$y(t) = \begin{cases} e^t & \text{in } [0,1] \\ e^t + (t-1)e^{t-1} & \text{in } [1,2] \\ e^t + e^{t-1} + \dfrac{1}{2}(t-2)(t+2e)e^{t-2} & \text{in } [2,3] \end{cases}$$

We first apply the Heun method of order 2 to the equation in the form (6) with linear interpolation and constant stepsize $h = 3/N$, where N is the number of steps used to cover $[0,3]$. In Table 1 we display the results obtained by including the discontinuity points $\xi_1 = 1$ and $\xi_2 = 2$ in the mesh, whereas in Table 2 we display the results obtained by excluding them from the mesh. The first point $t_0 = 0$ and the final point $t_f = 3$ (which is a discontinuity point) are always included in the mesh.

We denote the continuous numerical solution by $\eta(t)$ and the nodal values by $y_n = \eta(t_n)$.

As expected, in Table 1 the maximum global order is an $O(h^2)$, whereas in Table 2 it is an $O(h)$.

N	$\max_{1 \le n \le N} \lvert y_n - y(t_n) \rvert$
30	$2.48 \cdot 10^{-1}$
300	$2.69 \cdot 10^{-3}$
3000	$2.71 \cdot 10^{-5}$

Table 1. Numerical results for equation (6) with meshes including $\xi_1 = 1$ and $\xi_2 = 2$.

N	$\max_{1\leq n\leq N} \lvert y_n - y(t_n)\rvert$
31	$4.01 \cdot 10^{-1}$
301	$2.35 \cdot 10^{-2}$
3001	$2.16 \cdot 10^{-3}$

Table 2. Numerical results for equation (6) without including $\xi_1 = 1$ and $\xi_2 = 2$ in the mesh.

Now let us rewrite equation (6) in the form (3), that is

$$\begin{cases} \Phi'(t) = \Phi(t) + y(t-1), & 0 \leq t \leq 3, \\ y(t) = \Phi(t) + y(t-1), \\ \Phi(0) = 0, \\ y(t) = 1, & t \leq 0. \end{cases} \tag{7}$$

The auxiliary function turns out to be

$$\Phi(t) = \begin{cases} e^t - 1 & \text{in } [0,1] \\ e^t + (t-2)e^{t-1} & \text{in } [1,2] \\ e^t + \frac{1}{2}(t-2)(t+2e-2)e^{t-2} & \text{in } [2,3] \end{cases}$$

We apply again the Heun method with linear interpolation and constant stepsize $h = 3/N$ to approximate the auxiliary function $\Phi(t)$, and we use linear interpolation of nodal values to compute the numerical continuous solution $\eta(t)$ from the recursive formula which defines $y(t)$. We use the same meshes as before and report the relevant results in Table 3 and Table 4 where the nodal approximations of the auxiliary function Φ are denoted by Φ_n . We call n_1 and n_2 the indices of the two mesh points at which the loss of accuracy occurs, due to the interpolation process. The mesh point t_{n_1} is close to $\xi_2 = 2$, whereas $t_{n_2} = t_N = 3$. This is illustrated in Figure ??.

Moreover, in Table 5 we display the results obtained avoiding interpolation for the recursive formula and not including the discontinuities $\xi_1 = 1$ and $\xi_2 = 2$ in the mesh.

N	$\max\limits_{1\leq n\leq N} \lvert y_n - y(t_n)\rvert$	$\max\limits_{1\leq n\leq N} \lvert \Phi_n - \Phi(t_n)\rvert$
30	$1.90 \cdot 10^{-1}$	$1.57 \cdot 10^{-1}$
300	$2.05 \cdot 10^{-3}$	$1.70 \cdot 10^{-3}$
3000	$2.07 \cdot 10^{-5}$	$1.71 \cdot 10^{-5}$

Table 3. Numerical results for equation (7) with meshes including $\xi_1 = 1$ and $\xi_2 = 2$.

N	$\max\limits_{1\leq n\leq N} \lvert y_n - y(t_n)\rvert$	$\max\limits_{n\neq n_1,n_2} \lvert y_n - y(t_n)\rvert$
31	$9.46 \cdot 10^{-2}$	$9.46 \cdot 10^{-2}$
301	$1.69 \cdot 10^{-3}$	$1.30 \cdot 10^{-3}$
3001	$2.83 \cdot 10^{-4}$	$1.34 \cdot 10^{-5}$

N	$\max_{1\leq n\leq N} \lvert \Phi_n - \Phi(t_n)\rvert$
31	$1.04 \cdot 10^{-1}$
301	$1.22 \cdot 10^{-3}$
3001	$1.25 \cdot 10^{-5}$

Table 4. Numerical results for equation (7) without including $\xi_1 = 1$ and $\xi_2 = 2$ in the mesh and with interpolation.

N	$\max\limits_{1\leq n\leq N} \lvert y_n - y(t_n)\rvert$	$\max\limits_{1\leq n\leq N} \lvert \Phi_n - \Phi(t_n)\rvert$
31	$1.18 \cdot 10^{-1}$	$1.11 \cdot 10^{-1}$
301	$1.39 \cdot 10^{-3}$	$1.29 \cdot 10^{-3}$
3001	$1.43 \cdot 10^{-5}$	$1.31 \cdot 10^{-5}$

Table 5. Numerical results for equation (7) without including $\xi_1 = 1$ and $\xi_2 = 2$ in the mesh and without interpolation.

All the results are consistent with our expectations.

Representation (3) also allows to perform a delay-independent stability analysis for the solution $y(t)$ as well as for the numerical approximation $\eta(t)$ (see Torelli and Vermiglio [3]). In particular, by assuming $\mu[L] < 0$, one can prove that, for the linear equation (2), the condition

$$\frac{||LN + M||}{-\mu[L]} + ||N|| \leq 1$$

implies the boundedness of the solution, while the stronger condition

$$\frac{||LN(s) + M||}{-\mu[L]} + ||N|| \leq \rho < 1$$

implies the asymptotic stability. The term $\mu[L]$, called *the logarithmic norm* of L, depends on the

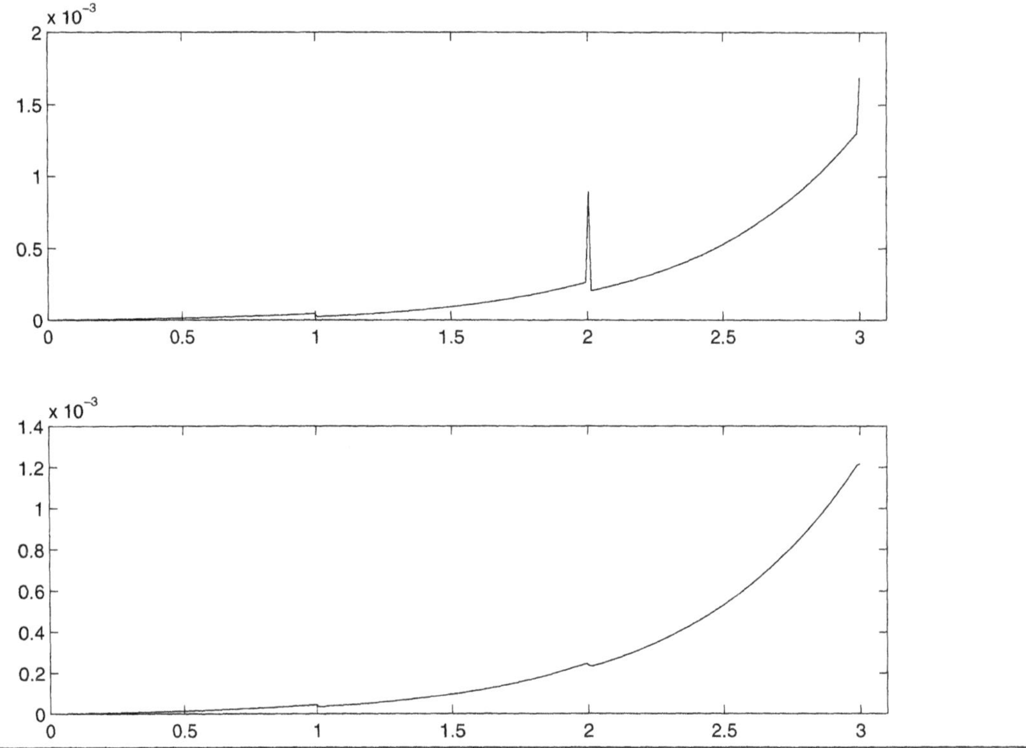

Fig. 1. Errors for the approximations of y (top) and Φ (bottom) with $N = 301$ for equation (7) by using interpolation (see Table 4).

norm involved and, for the inner product norm $\|\cdot\| = \sqrt{\langle\cdot|\cdot\rangle}$, it is

$$\mu[L] = \frac{\|L + L^T\|}{2}.$$

The approach is based on viewing the DDE as an ODE with forcing term and naturally extends to any kind of delay τ, even "state-dependent" $\tau = \tau(t, y(t))$. The same arguments used for the stability analysis of the solution $y(t)$ apply to the numerical approximation $\eta(t)$ and allow to select some second order continuous ODE-methods which preserve the asymptotic stability for any constant integration stepsize. A prominent method in this class is the two-stage Lobatto III-C method with linear interpolation (see Bellen and Zennaro [2]).

REFERENCES

[1] A. Bellen and M. Zennaro: *Numerical Methods for Delay Differential Equations*, Oxford University Press (in preparation).

[2] A. Bellen and M. Zennaro: *Strong contractivity properies of numerical methods for ordinary and delay differential equations*, Appl. Numer. Math. 9 (1992), 321-346.

[3] L. Torelli and R. Vermiglio: *Stability analysis of neutral delay differential equations in Hale's form*, Technical report RR UDMI/04/98, University of Udine, 1998

[4] J.K. Hale and S.M. Verduyn-Lunel: *Introduction to functional differential equations*, (1991), Springer Verlag, New York.

www.elsevier.com/locate/ifac

NUMERICAL COMPUTATION OF CHARACTERISTIC ROOTS FOR DELAY DIFFERENTIAL EQUATIONS

D. Breda [*,1] **S. Maset** [**,1] **R. Vermiglio** [*,1]

[*] *Dipartimento di Matematica e Informatica,*
Universita' di Udine,
Via delle Scienze 208, I-33100 Udine, Italy
dbreda@dimi.uniud.it
rossana@dimi.uniud.it
[**] *Dipartimento di Scienze Matematiche,*
Universita' di Trieste,
Via Valerio 12, I-34127 Trieste, Italy
maset@univ.trieste.it

Abstract: In this paper methods for computing characteristic roots of Delay Differential Equations (DDEs) with fixed discrete delay are investigated. Two different approaches are presented, based on different schemes of approximation applied to the equation restated as an abstract Cauchy problem, see (Maset, 2000). A number of numerical tests illustrates the convergence properties of both methods. *Copyright © 2001 IFAC*

Keywords: Delay Differential Equations, Characteristic Roots, Large Eigenvalues Problems.

1. INTRODUCTION

In this paper two approaches for computing characteristic roots of Delay Differential Equations (DDEs) with fixed discrete delay are investigated. The first one is based on the approximation of the solution operator, the second one on the approximation of the infinitesimal generator of the semigroup.

Consider the following system of DDEs:

$$y'(t) = Ly(t) + My(t-\tau),\ t \geq 0 \qquad (1)$$

where $L, M \in \mathbb{C}^{m\times m}$ and $\tau > 0$.

Let X be the Banach space $X := C([-\tau, 0], \mathbb{C}^m)$ equipped by the maximun norm. It is well-known that the family $\{T(t)\}_{t\geq 0}$ of linear bounded operators on X defined by

$$T(t)\varphi = y_t, \quad \varphi \in X \qquad (2)$$

is a C_0-semigroup, see (Diekmann, *et al.*, 1995). Here y_t is the function

$$y_t(\theta) = y(t+\theta),\ \theta \in [-\tau, 0] \qquad (3)$$

where y is the solution of (1) with initial data $\varphi \in X$. The infinitesimal generator $\mathcal{A} : D(\mathcal{A}) \to X$ of the semigroup is given by:

$$D(\mathcal{A}) = \{\varphi \in X \mid \varphi' \in X \text{ and } \varphi'(0) = L\varphi(0) + M\varphi(-\tau)\} \qquad (4)$$

$$\mathcal{A}\varphi = \varphi'. \qquad (5)$$

[1] Partially supported by MURST and INdAM-GNCS

Since the characteristic roots of (1) constitute the spectrum of $\mathcal{A}$, the idea is to approximate $\mathcal{A}$ by a finite dimensional matrix and then compute its eigenvalues. This is the second approach. On the other hand, the spectrum $\sigma(T(t))$ of the solution operator $T(t)$ is given by

$$\sigma(T(t)) \setminus \{0\} = e^{t\sigma(\mathcal{A})} \qquad (6)$$

and so the characteristic roots can be approximated by computing the eigenvalues of a numerical approximation of $T(\tau)$ and then taking the logarithm. This is the first approach. In Engelborghs and Roose (pre-print) a numerical approximation of the solution operator is obtained by using multistep methods. In this approach Runge-Kutta methods are used.

Detailed description of both approaches is presented in section 2 and section 3, while section 4 is dedicated to numerical tests carried out on a case studied in Engelborghs and Roose (pre-print).

In order to avoid cumbersome notation the scalar version of (1) is considered in the rest of the paper.

2. APPROXIMATION OF THE SOLUTION OPERATOR $T(\tau)$

The approach consists of applying Runge-Kutta (RK) methods to (1) and approximate in this way the operator $T(t)$, obtaining an analogous of (2) in a discrete form. For fixed N, N positive integer, consider an s-stage RK method (A, b, c) such that

$$\begin{array}{c} 0 < c_1 < c_2 < \cdots < c_s = 1 \\ b = (a_{s1}, a_{s2}, \ldots, a_{ss}) \end{array} \qquad (7)$$

as applied to (1) with stepsize $h = \frac{\tau}{N}$ (past values are approximated by past stage values):

$$\begin{cases} Y^{(n+1)} = 1_s y_n + hA\left(LY^{(n+1)} + \right. \\ \qquad\qquad \left. + MY^{(n+1-N)}\right) \\ y_{n+1} = Y_s^{(n+1)} \end{cases} \qquad (8)$$

where $Y^{(k+1)} = \left(Y_1^{(k+1)}, \ldots, Y_s^{(k+1)}\right)^T \in \mathbb{C}^s$ is the stage vector at the k-th step and $1_s = (1, 1, \ldots, 1)^T \in \mathbb{R}^s$. Combining equations (8) leads to:

$$\begin{aligned} Y^{(n+1)} &= (I_s - hAL)^{-1} 1_s e_s^T Y^{(n)} + \\ &+ h(I_s - hAL)^{-1} AMY^{(n+1-N)} \end{aligned} \qquad (9)$$

where $e_s = (0, \ldots, 0, 1)^T \in \mathbb{R}^s$. By setting:

$$X_n = \begin{pmatrix} Y^{(n)} \\ \vdots \\ Y^{(n+1-N)} \end{pmatrix} \in \mathbb{C}^{sN} \qquad (10)$$

it follows:

$$X_{n+1} = \Phi X_n, \; n = 0, 1, 2, \ldots \qquad (11)$$

where Φ is the matrix:

$$\Phi = \begin{pmatrix} P & 0 & \cdots & & 0 & Q \\ I_s & 0 & & & & 0 \\ & I_s & 0 & & & \cdot \\ & & \ddots & & & \cdot \\ & & & \ddots & & \cdot \\ & & & I_s & 0 & \cdot \\ & & & & I_s & 0 \end{pmatrix} \in \mathbb{C}^{sN \times sN} \qquad (12)$$

with $P = (I_s - hAL)^{-1} 1_s e_s^T$ and $Q = h(I_s - hAL)^{-1} AM$.

Applying recursively (11) leads to:

$$X_n = \Phi^n X_0, \; n = 0, 1, 2, \ldots \qquad (13)$$

where X_0 is defined by the initial data. Therefore Φ^N is an approximation of $T(\tau)$. Finally, the approximated characteristic roots λ^* are given by the logarithmic transformation of the eigenvalues μ of the matrix Φ:

$$\lambda^* = \frac{1}{h} \ln \mu, \; \mu \in \sigma(\Phi). \qquad (14)$$

3. APPROXIMATION OF THE INFINITESIMAL GENERATOR $\mathcal{A}$

For fixed N, N positive integer, consider the mesh on the interval $[-\tau, 0]$:

$$\begin{cases} \Omega_N = \{0\} \cup \\ \qquad \cup \bigcup_{n=0}^{N-1} \{\theta_n - c_i h \mid i = 1, \ldots, s\} \\ \theta_n = -nh, \quad n = 0, \ldots, N-1 \\ h = \dfrac{\tau}{N} \\ 0 < c_1 < c_2 < \cdots < c_s = 1 \end{cases} \qquad (15)$$

and replace the continuous space X by the discrete space $X_N = \mathbb{C}^{\Omega_N} = \mathbb{C}^{1+sN}$. Particular schemes of discretization for the infinitesimal generator can be obtained by applying s-stage RK methods (A, b, c) such that (7) holds, see (Maset, 2000). The infinitesimal generator $\mathcal{A}$ is approximated by the $(1 + sN) \times (1 + sN)$ complex matrix

$$\mathcal{A}_N = \begin{pmatrix} L & 0 & \cdots & 0 & M \\ & & \mathcal{B}_N & & \end{pmatrix} \qquad (16)$$

where:

$$\mathcal{B}_N = \frac{1}{h} \begin{pmatrix} w & W & & & \\ & w & W & & \\ & & & \ddots & \\ & & & w & W \end{pmatrix} \qquad (17)$$

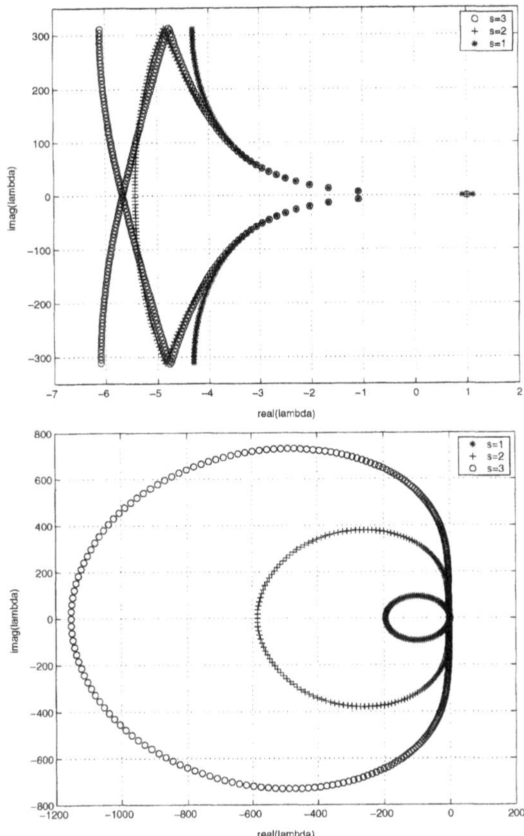

Fig. 1. Approximated characteristic roots of (19) by $T(\tau)$ approximation method (top) and $\mathcal{A}$ approximation method (bottom) for $N = 100$.

is a $sN \times (1 + sN)$ real matrix with $w = A^{-1}1_s \in \mathbb{R}^s$, $W = -A^{-1} \in \mathbb{R}^{s\times s}$ and the vector w is aligned with the s-th column of W. Given $\varphi \in X_N$, $\mathcal{A}_N\varphi$ is a discrete derivative of the vector φ. The first row corresponds to the splicing condition

$$\varphi'(0) = L\varphi(0) + M\varphi(-\tau). \tag{18}$$

The other lines are finite differences. In this way $\sigma(\mathcal{A}_N)$ is a direct approximation of the characteristic roots.

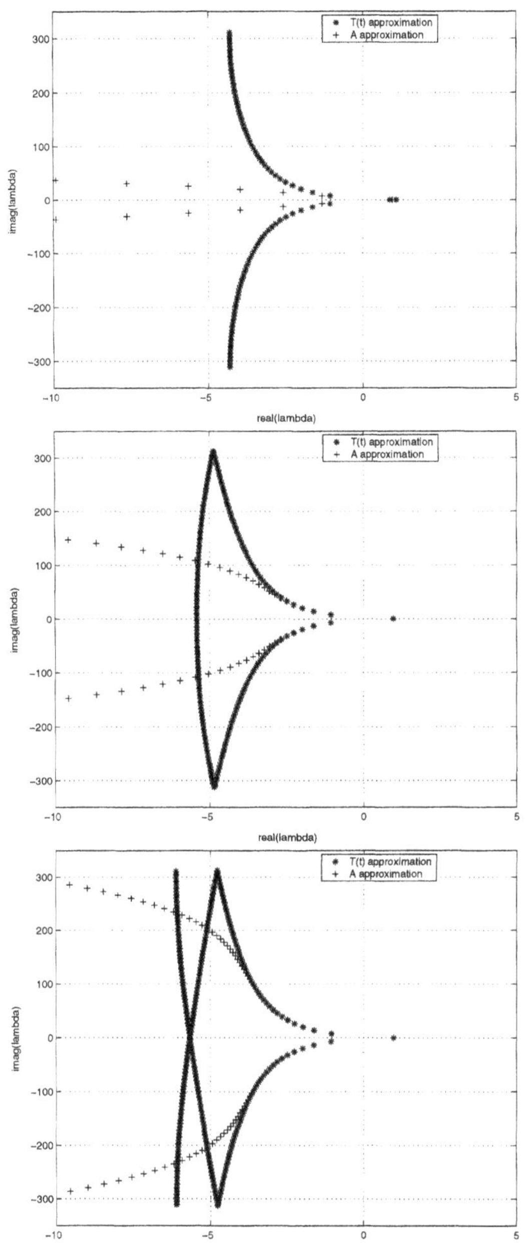

Fig. 2. Method comparison for $N = 100$ applied to (19): $s = 1$ (top), $s = 2$ (center) and $s = 3$ (bottom).

4. NUMERICAL RESULTS

The methods were implemented using *MATLAB*. Both algorithms are able to compute the rightmost approximation of the characteristic roots by setting a inferior limit for the real part.

Numerical tests were carried out varying the number N of discretization intervals (keeping $\tau = 1$ constant) and the order of the RK methods applied: in particualar RADAU-IIA methods of order $p = 1$ ($s = 1$, implicit Euler method), $p = 3$ ($s = 2$) and $p = 5$ ($s = 3$) were chosen.

The analyzed DDE ($m = 1$) was:

$$y'(t) = 2y(t) - ey(t-1) \tag{19}$$

which is known to have a double real root in 1. Distribution of the approximated characteristic roots of (19) is showed in Figure 1 for both methods. Complete spectrums are quite different wether in shape or in scale, but goodness of the approximation with regard to the rigthmost values can be viewed in detail in Figure 2.

Numerical tests shown that both methods are convergent to the rightmost characteristic roots of (19) and in particular convergence rapidity is accelerated increasing the convergence order p of the RK method applied (Figure 2) and decreasing the stepsize h (Figure 3).

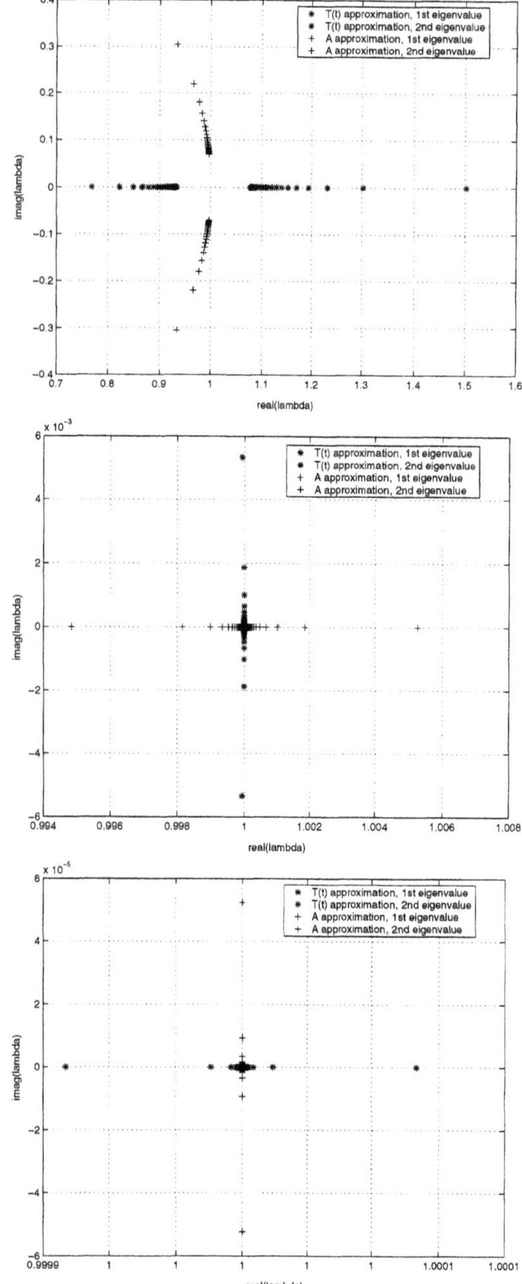

Fig. 3. Convergence to the double real root $\lambda = 1$ of (19), N varying from 10 to 200: $s = 1$ (top), $s = 2$ (center) and $s = 3$ (bottom).

Error analysis clearly confirmed the trend proved by Engelborghs and Roose (pre-print) for linear multistep methods (Figure 4):

$$e = |\lambda^* - \lambda| = Ch^{-p/\nu} \tag{20}$$

where C is a constant independent of N, p is the convergence order of the RK method and ν is the multiplicity of the eigenvalue λ. Relating (20) for two different values of the step h we obtained a relation for the convergence order of the method:

$$p/\nu = -\frac{ln(e_1/e_2)}{ln(h_1/h_2)} \tag{21}$$

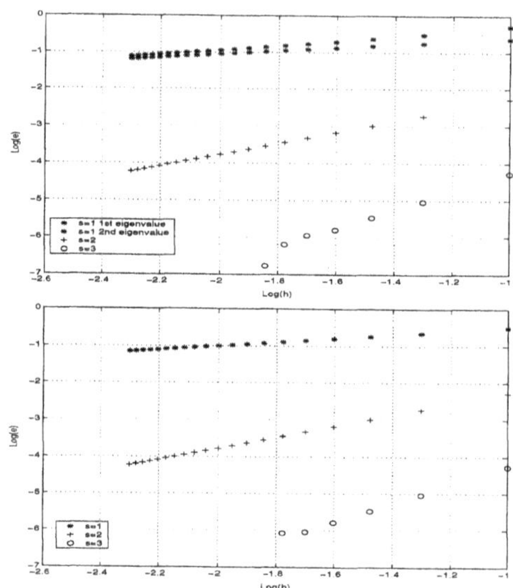

Fig. 4. Error $e = |\lambda^* - \lambda|$ analysis relative to the double real root $\lambda = 1$ of (19): $T(\tau)$ approximation method (top) and $\mathcal{A}$ approximation method (bottom) for $N = 100$.

which validity is asymptotically showed in Figure 5 and 6.

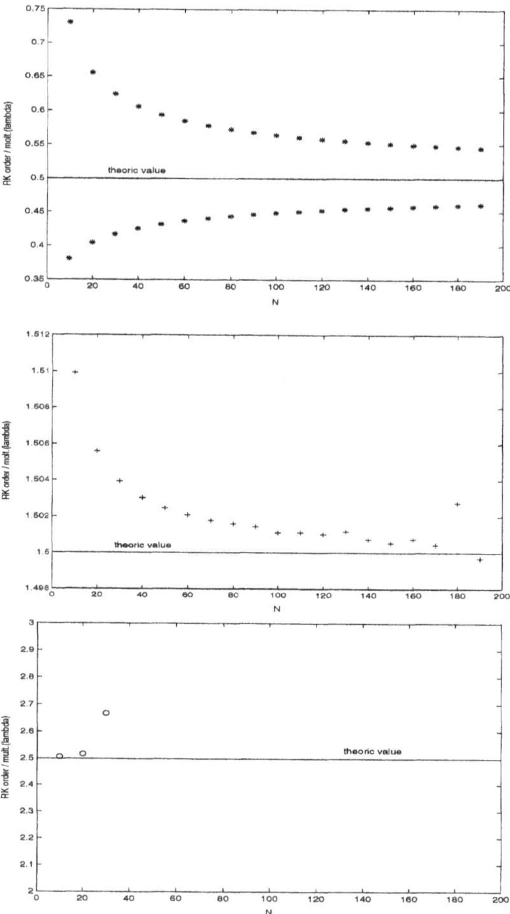

Fig. 5. Convergence order analysis relative to the double real root $\lambda = 1$ of (19) using $T(\tau)$ approximation method: $s = 1$ (top), $s = 2$ (center) and $s = 3$ (bottom).

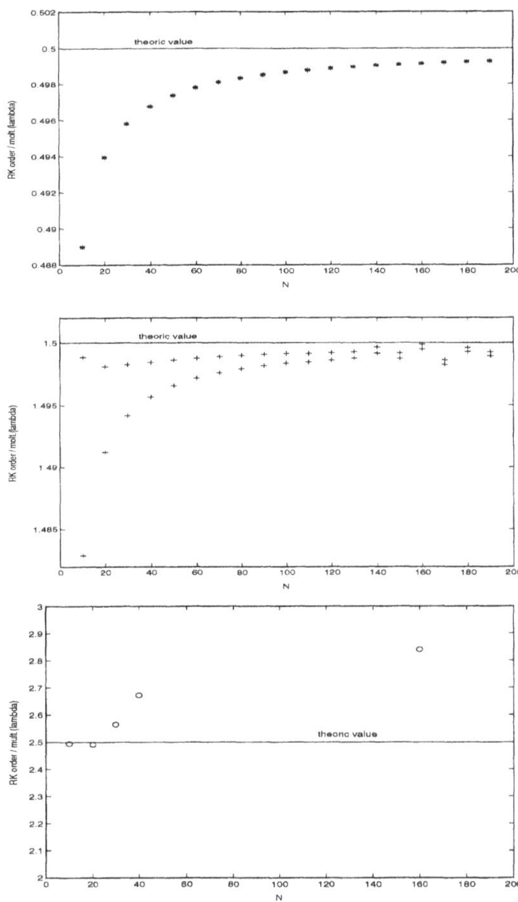

Fig. 6. Convergence order analysis relative to the double real root $\lambda = 1$ of (19) using $\mathcal{A}$ approximation method: $s = 1$ (top), $s = 2$ (center) and $s = 3$ (bottom).

REFERENCES

Diekmann, O., S.A. van Gils, S.M. Verduyn Lunel and H.O. Walther, (1995). *Delay Equations Functional, Complex and Nonlinear Analysis*, Springer Verlag, AMS series n. 110.

Maset, S., (2000). Numerical Solution of Retarded Functional Differential Equations as Abstract Cauchy Problems. *Numer. Math.*, **xxx**.

Engelborghs, K. and D. Roose, (pre-print). On Stability of LMS-methods and Characteristic Roots of Delay Differential Equations.

www.elsevier.com/locate/ifac

IMPLEMENTATION OF A DISTRIBUTED CONTROL LAW FOR A CLASS OF SYSTEMS WITH DELAY

V. Van Assche * **M. Dambrine** ** **J.-F. Lafay** *
J.-P. Richard **

* *IRCCyN, UMR CNRS 6597, E. C. Nantes, 1, rue de la Noë, B.P. 92101, 44321 Nantes Cedex 03, France*
** *LAIL, UPRESA CNRS 8021, E. C. Lille, B.P. 48, 59651 Villeneuve d'Ascq Cedex, France*

Abstract: The use of distributed delays in the control law of time-delay systems has been proposed by several authors. The implementation of such controllers is not trivial, and recent publication shown that replacing the distributed delay operator by an approximation computed through pointwise delay operators was unsafe with respect to the stability of the system. In this paper the use of digital controller is addressed, for the control of systems with an input delay. Practically, this means replacing an integral by a recurrent system. The first approach was to use a method of numerical approximation of an integral to build the recurrence law. In this paper it is shown that a such controller, built with the Simpson method, leads to an unstable closed-loop system. A second approach leads to the construction of a control law which realizes a sampled pole assignment, in the same way as the distributed control law realizes a pole assignment in continuous time. *Copyright © 2001 IFAC*

1. INTRODUCTION

This paper deals with some pole assignment methods for systems with delays. Throughout the paper, we will consider systems with commensurate delays, that is all delays are integer multiple of a real noted h. We define ∇, the delay operator, such that, for any integer k, $\nabla^k x(t) = x(t - kh)$. A linear system with commensurate delays will be written the following way:

$$\dot{x}(t) = A(\nabla)x(t) + B(\nabla)u(t) \quad (1)$$

$A(\nabla)$ and $B(\nabla)$ being two matrices of polynomials in ∇. The Laplace transform of $\nabla^k x(t)$ is $e^{-khs}X(s)$, where $X(s)$ is the Laplace transform of $x(t)$.

Precisely, we are interested in method of finite spectrum assignment, i.e. methods allowing to reduce the number of poles of a system with delays to a finite number and to assign arbitrarily the location of these poles in the complex plane. The control laws will have the form :

$$u(t) = h(u_t) + f(x_t), \quad (2)$$

where $u_t(.)$ and $x_t(.)$ are functions defined on the interval $[-\tau, 0]$ by $u_t(\theta) = u(t+\theta)$ and $x_t(\theta) = x(t+\theta)$, for $\theta \in [-h, 0]$.

Very often, the study and the control design of time-delay systems deal only with punctual delays, i.e. of the form $x(t-\tau)$, excluding integrals over the past of the command and state, this allows to work in a nice algebraic framework. In (Morse, 1976), it is shown that such a law allows a finite spectrum assignment only if the system fulfills the rather restrictive condition of strong controllability. Manitius and Olbrot (1979) show that spectral controllability is a sufficient condition for finite spectrum assignment with a control law of the form

$$u(t) = F(\nabla)x(t) + \int_{t-\tau}^{t} f(\theta)x(\theta)d\theta$$
$$+H(\nabla)u(t) + \int_{t-\tau}^{t} h(\theta)u(\theta)s\theta.$$

The system (1) is said to be spectrally controllable if, and only if, for any $s \in \mathbb{C}$, the matrix $(sI - A(e^{-s})\ B(e^{-s}))$ is of rank n.

In (Watanabe *et al.*, 1983), the necessity of this condition is proven for SISO systems, and in (Watanabe, 1986), for MIMO systems.

Recently, an algebraic background has been formalized in (Brethé and Loiseau, 1996) and (Brethé, 1997) for systems with distributed delays. The authors define the ring of the Laplace transform of both distributed delays and punctual delays. This formalism leads to a n-assignation algorithm which is much simpler than those of (Watanabe *et al.*, 1983) and (Watanabe, 1986).

The problem we want to deal with in this article is the implementation of the distributed delays. As the integral term in (2) cannot be exactly computed with a numerical computer, to use these command laws implies to approximate it. We will here show that the use of quadrature methods involves stability problems especially when delays are on the input variables.

As an illustration of this problem, let us consider the very simple unstable system:

$$\dot{x}(t) = x(t) + u(t-1). \tag{3}$$

With the control law

$$\begin{aligned} u(t) &= -(1+\lambda_d)\left(e^{1}x(t)\right. \\ &\left. + \int_0^1 e^{\theta}\, u(t-\theta)\, d\theta\right), \end{aligned} \tag{4}$$

which is a Volterra equation of second kind, the closed-loop system has one pole in $-\lambda_d$. With $\lambda_d > 0$, the system is stable, but, as we stated above, the problem we are faced with is the way of realizing the integral term

$$z(t) = \int_0^1 e^{\theta}u(t-\theta)d\theta.$$

A simple way of proceeding is to differentiate the previous expression. We obtain the differential equation

$$\dot{z}(t) = z(t) + u(t) - eu(t-1).$$

But, as explained in (Manitius and Olbrot, 1979), this realization is unsafe since unstable, and in this case the stabilization of the process is the result of a right half-plane pole-zero cancellation. In order to avoid such a phenomenon, Manitius and Olbrot recommend to compute directly $z(t)$ using a numerical quadrature method. Van Assche *et al.* (1999) for the example above, and Mondié *et al.* (August 2001) for the generalization to the case of systems with input delay, show that an approximation of the distributed delay with a finite sum of pointwise delays may give an unstable closed-loop system. Moreover, several examples of numerical regulators computed to approximate the continuous theoretical control law are given in (Van Assche *et al.*, 1999), with either the rectangular, trapezoidal and Simpson quadrature methods. The first two methods give a stable closed-loop system, although the last one, which is supposed to be the most precise, gives an unstable system, even when changing the sampling period.

In this paper, it is shown that a numerical regulator constructed through an approximation of the control law (4) will give an unstable closed-loop system regardless of the sampling period. Then, our contribution is to provide a safe way to implement a stabilizing control law of the type (2) using a numerical regulator without loss of the stability for the closed-loop system.

2. SYSTEM WITH INPUT DELAY

2.1 *Continuous system with an input delay*

In this paper, we consider systems with a delay on the input. The state space description of such systems is:

$$\dot{x}(t) = Ax(t) + \nabla Bu(t). \tag{5}$$

The dimension of the vectors $x(t)$ and $u(t)$ are noted respectively n and m.

Remark 1. A larger class of systems can be turned into a system with delay on the input. Indeed, if a system correspond to the scheme 1, it can be written in state space form:

$$\begin{aligned} \dot{x}_1(t) &= A_{11}x_1(t) + Bu(t), \\ \dot{x}_2(t) &= \nabla A_{21}x_1(t) + A_{22}x_2(t), \end{aligned} \tag{6}$$

where the vectors $x_1(t)$ and $x_2(t)$ are of dimensions n_1 and n_2, and $u(t)$ is a real valued function, and where A_{11}, A_{21} and A_{22}, and B are matrices of convenient dimensions over the field $\mathbb{R}$ of real numbers.

Fig. 1. System (6)

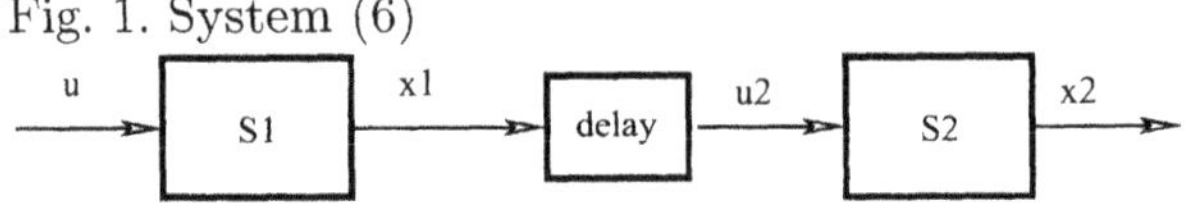

Such a system can be turned into a system with a delay only on the input with $x_1'(t) = \nabla x_1(t)$:

$$\begin{pmatrix} \dot{x}_1'(t) \\ \dot{x}_2(t) \end{pmatrix} = \begin{pmatrix} A_{11} & 0 \\ A_{21} & A_{22} \end{pmatrix} \begin{pmatrix} x_1'(t) \\ x_2(t) \end{pmatrix} + \begin{pmatrix} \nabla B \\ 0 \end{pmatrix} u(t). \quad (7)$$

One notes then

$$x(t) = \begin{pmatrix} x_1'(t) \\ x_2(t) \end{pmatrix}$$

and

$$A = \begin{pmatrix} A_{11} & 0 \\ A_{21} & A_{22} \end{pmatrix}.$$

2.2 *Sampling of the system (5)*

In order to implement a control law with a numerical device, the system has to be sampled. The sampling period is noted Δ, and the input $u(t)$ is constant between the sampling times. For a system without delay, it is well known that the equation of the sampled system is given by:

$$x(k\Delta + \Delta) = e^{A\Delta} x(k\Delta) + \int_{k\Delta}^{k\Delta+\Delta} e^{A(k\Delta+\Delta-\tau)} B d\tau u(k\Delta).$$

For a system with a delay of value h on the input, the expression above is directly adapted if the delay is a multiple of the sampling period, i.e. if there exists an integer p with $h = p\Delta$:

$$x(k\Delta + \Delta) = e^{A\Delta} x(k\Delta) + \int_0^{\Delta} e^{A(\Delta-\tau)} B d\tau u(k\Delta - p\Delta). \quad (8)$$

If the delay is not a multiple of the sampling period, we note $h = (p-1)\Delta + h'$ and, following Wittenmark (1985), one has

$$\begin{aligned} x((k+1)\Delta) &= e^{A\Delta} x(k\Delta) \\ &+ \int_0^{h'} e^{A(\Delta-h'+\tau)} B d\tau u((k-p)\Delta) \\ &+ \int_0^{\Delta-h'} e^{A\tau} d\tau u((k-p+1)\Delta). \end{aligned} \quad (9)$$

By writing

$$\Phi(t) = e^{At} \quad (10)$$

$$\Gamma(t) = \int_0^t e^{A\tau} d\tau, \quad (11)$$

(9) can be rewritten

$$\begin{aligned} x((k+1)\Delta) &= \Phi(\Delta) x(k\Delta) \\ &+ \Phi(\Delta - h')\Gamma(h') u((k-p)\Delta) \\ &+ \Gamma(\Delta - h') u((k-p+1)\Delta). \end{aligned} \quad (12)$$

3. CONTROL LAW FOR SPECTRUM ASSIGNMENT OF THE SYSTEM WITH AN INPUT DELAY

If the system (5) is spectrally controllable, one can apply the finite spectrum assignment methods. This gives the following control law:

$$u(t) = -F \int_0^h e^{A(-h-\tau)} B u(t-\tau) d\tau - Fx(t) + v(t), \quad (13)$$

where H is a real matrix of dimension $1 \times n$. The spectrum of the closed-loop system realized by (5) controlled by (13) is the spectrum of the matrix

$$A + e^{-Ah} BH, \quad (14)$$

see (Manitius and Olbrot, 1979).

We propose here a numerical computation of the integral, resulting in a discrete controller, which is clearly implementable. Moreover, the discrete system obtained with this method is similar to the system one gets when discretizing the theoretical continuous system obtained with the control law (13).

Before stating this original result, we first prove that the use of a quadrature method is not a good approach to solve this problem.

3.1 *Instability due to a Simpson's quadrature*

In a first approach, we compute the integral term through quadrature methods such as the rectangular method, the trapezoidal method and the Simpson's rule, see (Van Assche *et al.*, 1999). This last method is the one of higher order. Meanwhile, it gives the worst result in our simulations.

We show that the closed-loop system is unstable for the closed-loop system (3)-(4), with $h = 1 = p\Delta$, p being an even integer, and with $\lambda_d = 1$.

For this choice of h, Δ p and λ_d, the system (3) becomes, with (8),

$$x((k+1)\Delta) = e^{\Delta} x(k\Delta) + \left(e^{\Delta} - 1\right) u(k\Delta). \quad (15)$$

Its z-transform is :

$$z^p \left(z - e^{\Delta}\right) x = \left(e^{\Delta} - 1\right) u. \quad (16)$$

The discrete control law is :

$$u(k\Delta) = -2\frac{\Delta}{3}w(k\Delta) - 2e^1x(k\Delta) + v(k\Delta) \quad (17)$$

with

$$w(k\Delta) = u(k\Delta) + \sum_{l=1}^{p/2-1} 2e^{2l\Delta}u((k-2l)\Delta) + \sum_{l=1}^{p/2} 4e^{(2l-1)\Delta}u((k-2l+1)\Delta) + e^1u((k-p)\Delta).$$

The z-transform of (17) is:

$$d_c(z)u = 2e^1z^px + z^pv \quad (18)$$

with

$$d_c(z) = z^p(1+2\frac{\Delta}{3}) + 2\frac{\Delta}{3}\left(\sum_{l=1}^{p/2-1} 2e^{2l\Delta}z^{(p-2l)} + \sum_{l=1}^{p/2} 4e^{(2l-1)\Delta}z^{(p-2l+1)} + e^1\right).$$

The denominator of the closed-loop system is then :

$$z^p\left(z - e^{\Delta}\right)d_c(z) - \left(e^{\Delta} - 1\right)2e^1z^p,$$

and the closed-loop system is stable if, and only if, the polynomial

$$d(z) = \left(z - e^{\Delta}\right)d_c(z) - \left(e^{\Delta} - 1\right)2e^1 \quad (19)$$

has all its root inside the unit disc.

First, let us simplify the term $d_1(z) = \left(z - e^{\Delta}\right)d_c$, by remarking that

$$\left(z - e^{\Delta}\right)\sum_{k=1}^{p} e^{k\Delta}z^{p-k} = e^{\Delta}(z^p - e^1), \quad (20)$$

and that

$$d_c(z) = z^p(1+2\frac{\Delta}{3}) + 2\frac{\Delta}{3}\left(\sum_{l=1}^{d-1} 2e^{l\Delta}z^{(p-2l)} + \sum_{l=1}^{p/2} 2e^{(2l-1)\Delta}z^{(p-2l+1)} + e^1\right).$$

With some computation, this leads to

$$d_1(z) = \frac{3+2\Delta}{3}z^{p+1} - \frac{3-2\Delta}{3}e^{\Delta}z^p + \frac{4\Delta}{3}\sum_{k=0}^{p-1}\left((-1)^k e^{(k+1)\Delta}z^{n-k}\right) - \frac{2\Delta}{3}e^1z + \frac{2\Delta}{3}e^{1+\Delta}.$$

Therefore,

$$d(z) = \frac{3+2\Delta}{3}z^{p+1} - \frac{3-2\Delta}{3}e^{\Delta}z^p + \frac{4\Delta}{3}\sum_{k=0}^{p-1}\left((-1)^k e^{(k+1)\Delta}z^{p-k}\right) - \frac{2\Delta}{3}e^1z + \frac{6-2\Delta}{3}e^{1+\Delta} - 2e^1,$$

and, for $z = -1$,

$$d(-1) = \frac{-3-2}{3}\Delta - \frac{3+2\Delta}{3}e^{\Delta} + \frac{6-2\delta}{3}e^{1+\Delta} + \frac{2}{3}e^1\Delta - 2e^1 + \frac{4}{3}\Delta\sum_{k=0}^{p-1} e^{(k+1)\Delta}.$$

As

$$\lim_{\Delta\to 0}\Delta\sum_{k=0}^{p-1} e^{(k+1)\Delta} = \int_0^1 e^{\tau}d\tau = e^1 - 1,$$

One can check that

$$\lim_{\Delta\to 0} d(-1) = -2 + \frac{4}{3}(e^1 - 1) > 0.$$

As the degree of $d(z)$ is $p+1$, an odd number, one has for any value of Δ,

$$\lim_{z\to-\infty} d(z) = -\infty,$$

hence, for Δ sufficiently small, $d(z)$ has at least one root on the real axis, of magnitude strictly superior to 1. This show that the system with this controller is unstable, for any sufficiently small value of Δ, i.e. for the most precise approximation of the integral term of (4).

The computation above show that approximating the controller with a quadrature method is not always a safe approach to the problem. The next section of this work presents another approach, which leads to a safe implementation of the controller.

4. SAFE IMPLEMENTATION

This parts deals with systems of the form (7), sampled into (12). The control law is also sampled with the hypothesis that $u(t)$ is a bloc-pulse function, i.e. $u(t) = u(k\Delta)$ for $k\Delta \leqslant t < (k+1)\Delta$. This gives

$$u(k\Delta) = -F\sum_{l=1}^{p}\int_{(l-1)\Delta}^{l\Delta} e^{A(-h+\tau)}Bd\tau u((k-l)\Delta) + F\int_{(p-1)\Delta+h'}^{p\Delta} e^{A(-h+\tau)}Bd\tau u((k-p)\Delta) - Fx(k\Delta) + v(k\Delta). \tag{21}$$

Proposition 1. The system with an input delay (5), sampled with a period Δ, and controlled with the feedback (21) has the same characteristic polynomial as the linear discrete system without delay realized by the sampling of the system

$$\dot{x}(t) = Ax(t) + Bu(t) \tag{22}$$

sampled with the same sampling period and controlled by the feedback

$$u(t) = Fx(t),$$

up to a multiplication by a power of z.

Proof : With the definitions (10) and (11) the control law is rewritten:

$$\begin{aligned} u(k\Delta) = {} & F\Phi(-h)\sum_{l=1}^{p}(\Phi(l\Delta)\Gamma(-\Delta)u((k-l)\Delta)) \\ & -\Phi(-h+p\Delta)\Gamma(h'-\Delta)u((k-p)\Delta) \\ & -Fx(k\Delta)+v(k\Delta). \end{aligned}$$

Its z-transform is then:

$$d_c(z)u = -Fz^p x + z^p v \tag{23}$$

with

$$\begin{aligned} d_c(z) = {} & z^p - F\Phi(-h)\sum_{l=1}^{p}(\Phi(l\Delta)\Gamma(-\Delta)z^{p-l}) \\ & +F\Phi(-h)\Phi(p\Delta)\Gamma(h'-\Delta). \end{aligned}$$

The z-transform of the system (12) is :

$$x = \frac{(\Gamma(\Delta-h')z + \Phi(\Delta-h')\Gamma(h'))}{z^d(I_n z - \Phi(\Delta))}u. \tag{24}$$

The characteristic polynomial of the closed-loop system is the determinant of

$$P(z) = \begin{pmatrix} (I_n z - \Phi(\Delta))z^p & -B(z) \\ Fz^p & I_m z^p - HM \end{pmatrix}, \tag{25}$$

with

$$\begin{aligned} M(z) &= \sum_{l=1}^{p}\Phi(l\Delta)\Gamma(-\Delta)z^{p-l} \\ & \quad -\Phi(p\Delta)\Gamma(h'-\Delta) \\ B(z) &= \Gamma(\Delta-h')z + \Phi(\Delta-h')\Gamma(h'). \end{aligned}$$

Right multiplying the matrix $P(z)$ by $\begin{pmatrix} I_n & Mz^{-p} \\ 0 & I_m \end{pmatrix}$ $\begin{pmatrix} I_n & 0 \\ -F & I_m \end{pmatrix}$ gives

$$\det P(z) = \det(D(z))z^{m.p}$$

with

$$\begin{aligned} D(z) = {} & (I_n z - \Phi(\Delta))(I_n z^p - \Phi(-h))MF \\ & +B(z)F. \end{aligned}$$

Now, let us expand

$$\begin{aligned} D_1(z) &= (I_n z - \Phi(\Delta))(I_n z^p - \Phi(-h)MF) \\ &= I_n z^{p+1} - \Phi(\Delta)z^p \\ & \quad -(I_n z - \Phi(\Delta))\sum_{l=1}^{p}\Phi(l\Delta - h)\Gamma(-\Delta)Fz^{p-l} \\ & \quad +\Phi(p\Delta - h)\Gamma(h'-\Delta)Fz \\ & \quad -\Phi((p+1)\Delta - h)\Gamma(h'-\Delta)F. \end{aligned}$$

With a remark similar to (20), and remarking that $\Gamma(-\Delta) - \Gamma(h'-\Delta) = -\Phi(p\Delta - h)\Gamma(h')$, that $p\Delta - h = \Delta - h'$, and that $\Phi(\Delta - h')\Gamma(h'-\Delta) = -\Gamma(\Delta - h')$, we get

$$\begin{aligned} D_1(z) &= I_n z^{p+1} - \Phi(\Delta)z^p - \Phi(\Delta-h)\Gamma(-\Delta)Fz^p \\ & \quad -\Gamma(\Delta-h')Fz - \Phi(\Delta-h')\Gamma(h')F \\ &= I_n z^{p+1} - \Phi(\Delta)z^p - \Phi(\Delta-h)\Gamma(-\Delta)Fz^p \\ & \quad -B(z)F. \end{aligned}$$

Hence, we have

$$D(z) = (I_n z - \Phi(\Delta) - \Phi(\Delta-h)\Gamma(-\Delta)F)\,z^p.$$

The characteristic polynomial of the closed-loop system is then

$$z^{mp}\det(D(z)) = z^{mp}\det\left(z^p\left(I_n z - e^{A\Delta} - \int_0^{\Delta} e^{A(\Delta-\tau)}d\tau e^{-Ah}BF\right)\right).$$

As the sampling of the system (22) gives

$$x((k+1)\Delta) = e^{A\Delta}x(k\delta) + \int_0^{\Delta} e^{A(\Delta-\tau)}d\tau e^{-Ah}Bu(k\Delta),$$

the proof is achieved. □

The continuous control law (13) was designed so that the spectrum of the system with delay is shifted to be the spectrum of the matrix $A + e^{-Ah}BF$, hence, the spectrum of the system without delay (22) controlled by the feedback $u(t) = Fx(t)$. The discrete control law we present here allow to retain this spectrum assignment property when sampling the system.

Remark 2. In order to control a system (6) when only $x_1(t)$ is measured, and not $x_1'(t)$, one can remark that, as $x_1'(t) = x_1(t-h)$, one has for the sampled system

$$x_1'(k\Delta) = e^{A(\Delta-h')}x((k-p)\Delta) + \int_0^{\Delta-h'} e^{A(\Delta-h'-\tau)}d\tau Bu((k-p)\Delta).$$

This equation allows to deduce $x_1'(k\Delta)$ from the knowledge of $x(t)$ and $u(t)$ at the sampling times.

5. CONCLUSION

We have shown above that our method for the numerical implementation of the control law with distributed delay (13) for a system with an input delay is equivalent to the numerical implementation of a stabilizing control law for a linear system without delay.

One can remark that the numerical control law we obtain is similar to the control law one would get by using a spectrum assignment method for sampled system. Indeed, our approach allows to implement a spectrum assignment control law, but this can be done without computing the sampled system, and especially the exponential of the matrix A. This associated with the method of (Brethé and Loiseau, 1996) allow the design of a numerical regulator for system with input delay with a limited amount of computation.

6. REFERENCES

Brethé, D. (1997). Contribution à l'étude de la stabilisation des systèmes linéaires à retards. Thèse de doctorat. Ecole Centrale de Nantes.

Brethé, D. and J. J. Loiseau (1996). A result that could bear fruit for the control of delay-differential systems. *Proc. IEEE MSCA* pp. 168–172.

Manitius, A. Z. and W. Olbrot (1979). Finite spectrum assigment problem for systems with delays. *IEEE Trans. Automatic Control* **AC-24**(4), 541–553.

Mondié, S., M. Dambrine and O. Santos (August 2001). Approximations of control laws with distributed delays : a necessary condition for stability. *To appear in IFAC SSSC 2001 (First Symposium on System Structure and Control, Prague).*

Morse, A. S. (1976). Ring models for delay-differential systems. *Automatica* **12**, 529–531.

Van Assche, V., M. Dambrine, Lafay J.-F. and J.-P. Richard (1999). Some problems arising in the implementation of distributed-delay control laws. In: *Proc. 38th IEEE CDC.* p. 4668.

Watanabe, K. (1986). Finite spectrum assignment and observer for multivariable systems with commensurate delays. *IEEE Trans. Automatic Control* **AC-31**(6), 543–550.

Watanabe, K., M. Ito, M. Kaneko and T. Ouchi (1983). Finite spectrum assignment problem of systems with delay in state variables. *IEEE Trans. Automatic Control* **AC-28**(4), 506–508.

Wittenmark, B. (1985). Sampling of a system with a time delay. *IEEE Transactions on Automatic Control* **30**(5), 507–510.

www.elsevier.com/locate/ifac

FINITE SPECTRUM ASSIGNMENT FOR INPUT DELAY SYSTEMS [1]

S. Mondié [*,2] **and J.J. Loiseau** [**]

[*] *Departamento de Control Automático CINVESTAV-IPN, Av. IPN 2508, A.P. 14-740, 07300 México, D.F., México, smondie@ctrl.cinvestav.mx*

[**] *Institut de Recherche en Communications et Cybernétique de Nantes, UMR CNRS 6597, Ecole Centrale de Nantes, Université de Nantes, Ecole des Mines de Nantes, BP 92101, 44321 Nantes Cedex 03, France, loiseau@irccyn.ec-nantes.fr*

Abstract: The problem of polynomial invariant factors assignment of input delay systems with classical spectrum assigment control laws with distributed delays is adressed. The multiplicities of the invariant factors are shown to be restricted by specified Rosenbrock type inequalities. The results are proved with the help of an equivalent linear assignment problem with no delay, and within the Bezout domain $\mathcal{E}$. A bidimensional illustrative example is given.

Keywords: delay systems, finite spectrum assignment, invariant factors.

1. INTRODUCTION

Consider a linear system with delay described by

$$\dot{x}(t) = \sum_{i=0}^{K} A_i x(t-ih) + \sum_{i=0}^{K} B_i u(t-ih) \ , \quad (1)$$

where the control input $u(t) \in \mathbb{R}^m$ and the instantaneous state $x(t) \in \mathbb{R}^n$ for $t \geq 0$, and the family of control laws described by integral Volterra equations of the second kind,

$$u(t) = \int_0^{Kh} f(\tau)u(t-\tau)d\tau + \sum_{i=0}^{K} g_i x(t-ih) + \int_0^{Kh} g(\tau)x(t-\tau)d\tau \ . \quad (2)$$

Such a control law was introduced in Olbrot (1978), where the spectral controllability of system (1) is shown to be a necessary and sufficient condition for the control law (2) to freely assign a closed loop finite spectrum. However, not only the location of the roots, but also their multiplicities, or equivalently the invariant factors of the closed loop, are crucial for the closed loop dynamics. The question that arises is to characterize the freedom in assigning the invariant factors.

The answer to this query was given in the recent result in Loiseau (2001): the limitation is that the sum of the degrees of the invariant factors must be equal to n.

In this work, we restrict our attention to input delay systems. This subfamily of the systems introduced above includes the models of a wide class of applications where the delay is due to transport phenomenons, time consuming information processing, sensors design among others. These systems are described by

$$\dot{x}(t) = Ax(t) + \Sigma_{i=0}^{N} B_i u(t-ih). \quad (3)$$

As shown in Manitius and Olbrot (1979), Olbrot (1978), these systems are n-assignable by control laws that are simpler that those described by (2), those of the form

$$u(t) = K[x(t) + \Sigma_{i=1}^{N} \int_{t-h}^{t} e^{(t-\sigma-ih)A} B_i u(\sigma)d\sigma] \quad (4)$$

This is indeed a particular case of the finite spectrum assignment problem introduced above. The motivation for using such control laws is their simplicity: the state feedback is linear and static,

[1] Supported by Conacyt, Mexico 31951-A and by GDR-Automatique: systèmes à retards, France.
[2] On leave at Heudyasic, UMR CNRS 6599, France.

and the distributed component consists of linear combination of elementary distributed delays. As shown below, it permit not only toassign a finite spectrum, but also finite invariant factors. The motivation for assigning invariant factors with a finite number of roots is the same as the one for spectrum assignment: the invariant factors give and additional freedom in shaping the dynamics. If they have a finite number of roots, they can be readily analyzed. The question is then whether or not the freedom, that exists in the general case, is restricted when such simpler laws are used. The problem under consideration is then the following.

Problem Consider a spectrally controllable input time delay system described by (3). Under what conditions does static state feedback distributed control laws (4) exists, such that the closed loop has prescribed invariant factors with finite number of roots.

The solution to this problem is organized as follows. Some preliminary results are shown or recalled in Section 2. Then the main result is established in Section 3 and is discussed in the light of the Bezout domain $\mathcal{E}$ in Section 4. An illustrative example is presented in Section 5 and some concluding remarks end the paper.

Notation: $\mathbb{R}[s]$ denotes the ring of polynomials over $\mathbb{R}$, the field of reals. The degree of its elements, $\alpha(s)$, is denoted $\deg(\alpha(s))$. $\mathbb{R}(s)$ stands for the field of rational functions over $\mathbb{R}$, while the rings of proper and strictly proper rational functions are denoted, respectively, by $\mathbb{R}_p(s)$ and $\mathbb{R}_{sp}(s)$. Further, $\mathbb{R}^{m\times n}$and $\mathbb{R}^{m\times n}[s]$, ..., denote the sets of $m \times n$ matrices having elements in $\mathbb{R}$, $\mathbb{R}[s]$,..., respectively. Units of the ring $\mathbb{R}^{m\times m}[s]$ are called *unimodulars* and those of the ring $\mathbb{R}_p^{m\times m}(s)$ *bipropers*. The set $\mathcal{E}$ is a Bezout domain whose elements are fractions of the form $\alpha(s, e^{-hs}) = \frac{n(s,e^{-hs})}{d(s)}$, where all the zeros of $d(s) \in \mathbb{R}[s]$ are zeros of $n(s, e^{-hs}) \in \mathbb{R}[s, e^{-sh}]$ the set of quasipolynomials. Notice that $\mathbb{R}[s] \subset \mathcal{E}$. Units of the Bezout domain $\mathcal{E}$ are called unimodulars in $\mathcal{E}$. (for detailed information on these sets and their properties, see Gantmacher (1959), Kailath (1980), Brethé and Loiseau (1998), and Loiseau (2000, 2001)).

2. PRELIMINARY RESULTS

We now give a new interpretation of the result established in Manitius and Olbrot (1979), Artstein (1982) that has been extensively used in the literature, in the light of n-assignability, namely the ability to assign a finite spectrum closed loop system, provided that the system is spectrally controllable. Indeed, there is more to say: it is also possible to assign prescribed invariant factors with a finite number of roots to the closed loop characteristic matrix.

Lemma 1. Consider a spectrally controllable linear multivariable system with delay in the input described by

$$\dot{x}(t) = Ax(t) + \Sigma_{i=0}^{N} B_i u(t - ih) \qquad (5)$$

where $A \in \mathbb{R}^{n\times n}, B \in \mathbb{R}^{n\times m}$ and the delays in the input are commensurate to $h \geq 0$. Then the following problems are equivalent.
(i) The control law

$$u(t) = K[x(t) + \Sigma_{i=0}^{N} \int_{t-ih}^{t} e^{(t-\sigma-ih)A} B_i u(\sigma) d\sigma] \quad (6)$$

assigns to the closed loop system (5-6) a prescribed set of invariant factors

$$\alpha_1'(s), \alpha_2'(s), \cdots, \alpha_n'(s) .$$

(i) The control law $u(t) = Ky(t)$ assigns to the controllable system

$$\dot{y}(t) = Ay(t) + \Sigma_{i=0}^{N} e^{-ihA} B_i u(t) \qquad (7)$$

a closed loop system with a prescribed set of invariant factors $\alpha_1'(s), \alpha_2'(s), \cdots, \alpha_n'(s)$.

Proof. Consider the transformation

$$y(t) = x(t) + \Sigma_{i=0}^{N} \int_{t-ih}^{t} e^{(t-\sigma-ih)A} B_i u(\sigma) d\sigma \quad (8)$$

The derivative of (8) is

$$\begin{aligned}
\dot{y}(t) &= \dot{x}(t) + \Sigma_{i=0}^{N} e^{-ihA} B_i u(t) \\
&\quad - \Sigma_{i=0}^{N} e^{(h-ih)A} B_i u(t - ih) \\
&\quad + \Sigma_{i=0}^{N} A \int_{t-ih}^{t} e^{(t-\sigma-ih)A} B_i u(\sigma) d\sigma \\
&= Ax(t) + \Sigma_{i=0}^{N} B_i u(t - ih) \\
&\quad + \Sigma_{i=0}^{N} e^{-ihA} B_i u(t) - \Sigma_{i=0}^{N} B_i u(t - ih) \\
&\quad + \Sigma_{i=0}^{N} A \int_{t-ih}^{t} e^{(t-\sigma-ih)A} B_i u(\sigma) d\sigma \\
&= Ay(t) + \Sigma_{i=0}^{N} e^{-ihA} B_i u(t).
\end{aligned}$$

It follows that the control law (6) assigns the closed loop

$$\dot{y}(t) = (A + \Sigma_{i=0}^{N} e^{-ihA} B_i K) y(t).$$

In order to establish the equivalence with (ii), consider the description of the system (5) and the closed loop (6) in the frequency domain. They are respectively $(sI_n - A)x(s) = \Sigma_{i=0}^{N} B_i e^{-ihs} u(s)$ and $(I_m - K(\Sigma_{i=0}^{N} e^{-ihA}(sI - A)^{-1}[I - e^{-ih(sI-A)}]B_i)u(s) = Kx(s)$.

Let

$B := \Sigma_{i=0}^{N} B_i e^{-ihs}$,

$M := \Sigma_{i=0}^{N} e^{-ihA}(sI - A)^{-1}[I - e^{-ih(sI-A)}]B_i$.

The closed loop characteristic matrix is

$$\begin{pmatrix} sI_n - A & -B \\ -K & I_m - KM \end{pmatrix}.$$

Post multiplication by well defined unimodular matrices leads to

$$\begin{pmatrix} sI_n - A & -B \\ -K & I_m - KM \end{pmatrix} \begin{pmatrix} I & -M \\ 0 & I \end{pmatrix} \begin{pmatrix} I & 0 \\ K & I \end{pmatrix}$$
$$= \begin{pmatrix} (sI_n - A)(I_n - MK) - BK & \star \\ 0 & I_m \end{pmatrix},$$

the symbol $\star$ standing for a nonspecified matrix. Then, the invariant factors of the closed loop are those of $(sI_n - A)(I - MK) - BK$, that is

$$(sI_n - A).(I_n - \Sigma_{i=0}^{N} e^{-ihA}(sI - A)^{-1}.$$
$$(I - e^{-ih(sI-A)})B_i K) - \Sigma_{i=0}^{N} B_i e^{-ihs} K$$
$$= sI_n - A - \Sigma_{i=0}^{N} e^{-ihA}[I - e^{-ih(sI-A)}]B_i K)$$
$$-\Sigma_{i=0}^{N} B_i e^{-ihs} K$$
$$= sI_n - A - \Sigma_{i=0}^{N} e^{-ihA} B_i K$$

and the result follows. ■

Remark 1. The proof follows closely the steps of the proof in Manitius and Olbrot (1979) for n-assignability. We recall it here to show that it permits to conclude on the invariant factors of the closed loop matrix, not only on its spectrum.

Lemma 2. (Olbrot, 1978) The input delay system (5) is spectrally controllable if and only if the linear system (7) is controllable.

3. MAIN RESULT

We are now able to state our main result.

Theorem 3. Consider a spectrally controllable linear multivariable system with delay in the input described by

$$\dot{x}(t) = Ax(t) + \Sigma_{i=0}^{N} B_i u(t - ih), \qquad (9)$$

where $A \in \mathbb{R}^{n \times n}, B_i \in \mathbb{R}^{n \times m}$ and $h \geq 0$ is the delay. Let $c_1, c_2, ..., c_m$ be the controllability indices of the pair $(A, \Sigma_{i=0}^{N} e^{-ihA} B_i)$.

Then there exist a control law

$$u(t) = K[x(t) + \Sigma_{i=0}^{N} \int_{t-h}^{t} e^{(t-\sigma-ih)A} B_i u(\sigma) d\sigma] \quad (10)$$

that assigns a closed loop whose invariant factors are monic polynomials $\alpha'_1(s), \alpha'_2(s), \cdots, \alpha'_k(s)$ where $\alpha'_i(s)$ divides $\alpha'_{i-1}(s), i = 2, ..., m$, if and only if

$$\sum_{i=1}^{j} c_i \leq \sum_{i=1}^{j} \deg \alpha'_i \ , \ j = 1, ..., m \qquad (11)$$

with equality for $j = m$, and $\alpha'_i(s) = 1, i = m + 1, ..., n$.

Proof.
According to Lemma 2, the pair $(A, \Sigma_{i=0}^{N} e^{-ihA} B_i)$ is controllable. Let $c_1, c_2, ..., c_m$ denote its controllability indices. From the Rosenbrock Control Structure Theorem recalled in the appendix, there exists a control law (10) that assigns to the system (7) a closed loop with invariant factors $\alpha'_1(s), \alpha'_2(s), \cdots, \alpha'_n(s)$ where $\alpha'_i(s)$ divides $\alpha'_{i-1}(s), i = 2, ..., m$, if and only if conditions (11) hold. Finally, the result follows from Lemma 1. ■

Remark 2. Clearly, the use of simpler control laws has a cost: the degrees of the invariant factors cannot be assigned arbitrarily.

Remark 3. Lemma 1 implies that a simple static state feedback distributed control law described by (10) permits to assign to the input delay system (9) a set of desired invariant factors that satisfies the inequalities (11). The parameter K of the control law, can be calculated in a straightforward manner as the solution to the problem of a static state feedback invariant factors assignment for the linear system with no delay (7). Notice that toolboxes for the analysis and design of linear systems are available (The Polynomial Toolbox, 1998). If a set of desired invariant factors does not satisfies the inequalities (11), the results obtained in Loiseau (2001) guarantee that there exists a control law described by (2) that allows to assign them. This control law if fully determined by $h_i, i = 1, 2, ..., f(\tau)$ and $g(\tau)$.

4. INTERPRETATION OF THE RESULT IN THE BÉZOUT DOMAIN $\mathcal{E}$

The Bézout domain, which definition and basic properties are recalled in the notations, section 1, has proved to be a powerful tool for the study of commensurate time delay systems, by providing a solid algebraic framework that allows the generalization of many results established for linear systems without delays (Brethé, 1997). The following remarks enlightens some subtleties of the machinery in $\mathcal{E}$, in connexion with our problem. First of all, we establish the following clue fact.

Fact 1. The invariant factors in the Bezout domain $\mathcal{E}$ of a polynomial matrix coincide with its invariant factor over the ring $\mathbb{R}[s]$.

Consider indeed a polynomial matrix $D(s) \in \mathbb{R}^{n\times m}[s]$ of rank r. It is well known (Gantmacher, 1959) that there exist unimodular polynomial matrices $U(s) \in \mathbb{R}^{n\times n}[s]$ and $V(s) \in \mathbb{R}^{m\times m}[s]$ so that $D(s)$ can be factored as

$$D(s) = U(s)\begin{pmatrix} diag\{\alpha_i(s), i = 1,...,r\} & 0 \\ 0 & 0 \end{pmatrix} V(s). \quad (12)$$

where $\alpha_i(s), i = 1, ..., r$ are unique monic polynomials such that $\alpha_i(s)$ divides $\alpha_{i-1}(s)$ called the invariant factors of $D(s)$. Further, one can show (Brethé, 1987) that $\mathcal{E}$ is also a so-called invariant factor domain. Every matrix $D(s)$ over $\mathcal{E}$ can also be factored in the form (12), were now the matrices $U(s)$ and $V(s)$ and their inverses are over $\mathcal{E}$, and the $\alpha_i(s)$, $i = 1, ..., r$ are elements of $\mathcal{E}$, which are also called the invariant factors of $D(s)$. Now, since $\mathbb{R}[s] \subset \mathcal{E}$, it is clear that when $D(s)$ is polynomial, the polynomial factorisation (12) is also a factorisation over $\mathcal{E}$: $U(s)$ and $V(s)$ are indeed matrices over $\mathcal{E}$, which are unimodular over $\mathcal{E}$, the rank of $D(s)$ is r over $\mathcal{E}$, and, since such a factorisation over $\mathcal{E}$ is unique, the $\alpha_i(s)$, $i = 1, ..., r$, are also the uniquely defined invariant factors of $D(s)$ over $\mathcal{E}$.

Fact 2. If the input delay system (5) is spectrally controllable, then its transfer matrix can be factored in the form

$$(sI_n - A)^{-1}B = N(s, e^{-hs})D^{-1}(s) ,$$

where $D(s)$ and $N(s, e^{-hs})$ are right coprime matrices that are respectively polynomial in the variables s and s, e^{-hs}.

We can further state that if

$$\begin{aligned}(sI_n - A)^{-1}B &= \Sigma_{i=0}^{N}(sI_n - A)^{-1}B_i e^{-ihs} \\ &= \Sigma_{i=0}^{N} e^{-ihs} N_i(s)D^{-1}(s) ,\end{aligned}$$

whith

$$N_i(s) = (sI_n - A)^{-1}B_i D(s) ,$$

then the transfer of the system without delay (7) can be expressed as

$$\Sigma_{i=0}^{N}(sI_n - A)^{-1}e^{-ihA}B_i = \left(\Sigma_{i=0}^{N} e^{-ihA}N_i(s)\right) D^{-1}(s) .$$

Both systems have the same denominator, which is of course polynomial in s. As it is usual for systems without delays, we can assume that $D(s)$ is comumn reduced, with column degrees $c_1, c_2, \ldots, c_m$. We can notice in addition that the column degrees of such a column reduced polynomial denominator of the transfer of the system (5) are uniquely defined. Any other polynomial denominator, say $D'(s)$ is indeed so that $D(s)U(s, e^{-hs}) = D'(s)$, for some matrix $U(s, e^{-hs})$ which is unimodular over $\mathcal{E}$. We can see that $U(s, e^{-hs}) = D^{-1}(s)D'(s)$ is actually rational in the variable s. Since it is also a matrix over $\mathcal{E}$, an analytic function without pole, we conclude that it is actually polynomial in s, which shows the result. In the sequel, we shall call such a factorization with a polynomial denominator the *natural* factorization of the system.

We are now able to revisit the problem in the framework of $\mathcal{E}$. The system and the control law are respectively described in the frequency domain by

$$(sI_n - A)x(s) = \Sigma_{i=0}^{N} B_i e^{-ihs} u(s)$$

and

$$\begin{aligned}(I_m - K(\Sigma_{i=0}^{N} e^{-ihA}(sI - A)^{-1}. \\ (I - e^{-ih(sI-A)})B_i)u(s) &= Kx(s).\end{aligned}$$

We first observe that a natural factorization also exists for the closed loop system. Let indeed $(N_K(s, e^{-s}), D_K(s, e^{-s}))$ be a coprime factorization over $\mathcal{E}$ of the transfer of the closed loop system. The closed loop denominator is given by

$$\begin{aligned}&D_K(s, e^{-s}) \\ &= ((I_m - K(\Sigma_{i=1}^{N} e^{-ihA}(sI - A)^{-1} \\ &(I - e^{-ih(sI-A)})B_i))D(s) - KN(s) \\ &= D(s) - KN(s) \\ &-K(\Sigma_{i=1}^{N} e^{-ihA}(sI - A)^{-1}(I - e^{-ih(sI-A)})B_i)D(s) \\ &= D(s) - KN(s) - K(sI - A)^{-1}\Sigma_{i=1}^{N} e^{-ihA}B_i D(s) \\ &+K(sI - A)^{-1}\Sigma_{i=1}^{N} e^{-ihs}B_i D(s)\end{aligned}$$

and it follows from (??) that

$$\begin{aligned}&D_K(s, e^{-s}) \\ &= D(s) - K(sI - A)^{-1}\Sigma_{i=1}^{N} e^{-ihA}B_i D(s) \\ &= (I_m - K(sI - A)^{-1}\Sigma_{i=1}^{N} e^{-ihA}B_i)D(s). (13)\end{aligned}$$

Observe that the matrix

$$(I_m - K(sI - A)^{-1}\Sigma_{i=0}^{N} e^{-ihA}B_i)$$

is a biproper matrix in $\mathbb{R}(s)$, hence $D_K(s, e^{-s})$ is actually rational in the variable s, and, since it has no pole, the result follows.

One can further notice that a left product by a biproper matrix does not modify the column degrees. Thus the column degrees of $D_K(s, e^{-s})$ are those of $D(s)$.

The natural coprime factorization bases a design algorithm for the finite spectrum assignment, which consists of the following steps.

(i) Calculate a natural factorization $(N(s, e^{-sh}), D(s))$ of the transfer matrix of system (5). $D(s)$ is column reduced with column degrees $c_1, c_2, \ldots, c_m$.

(ii) Choose a set of monic polynomials $\alpha'_1(s)$, $\alpha'_2(s)$, $\cdots, \alpha'_m(s)$ where $\alpha'_i(s)$ divides $\alpha'_{i-1}(s)$ for $i = 2, ..., m,$, and take $\alpha'_{m+1}(s) = \ldots = \alpha'_n(s)$.

(iii) Providing that (11) hold, there exist a polynomial matrix $D_K(s)$, that is column reduced, with column degrees c_i and invariant factors $\alpha'_i(s)$, for $i = 1, \ldots, m$, and constant matrices X and Y of convenient dimensions, X being invertible, so that

$$XD(s) + YN(s, e^{-hs}) = D_K(s).$$

Algoritms to compute $D_K(s)$, X, and Y are provided by Kučera (1991). The feedback $K = -X^{-1}Y$ assigns to the closed loop system (3–4) the invariant factors $\alpha'_i(s)$, for $i = 1, \ldots, n$.

The situation is quite different when considering factorizations over $\mathcal{E}$. As shown in Loiseau (2001), contrarily to what happens in $\mathbb{R}(s)$, post multiplication of elements of $\mathcal{E}$ by unimodular over $\mathcal{E}$ modify the column degrees. An immediate consequence of this fact is the non unicity of the column degrees of the denominator of a left coprime factorization in $\mathcal{E}$. Moreover, there is complete freedom in choosing the column degrees of such a factorization.

If a set of invariant factors $\alpha'_i(s)$ does not satisfy the inequalities (11), exept for $j = m$, one can find a factorization of the transfer where $D(s, e^{-sh})$ is column reduced, with column degrees $\deg \alpha'_i(s)$, $i = 1, \ldots, m$, and there exist a unimodular matrix in e^{-hs}, say $Z(e^{-hs})$ and matrices $F(s, e^{-sh})$ and $G(s, e^{-sh})$ over $\mathcal{E}$, such that

$$(I_m - F(s, e^{-sh})D(s, e^{-sh}) + G(s, e^{-sh})N(s, e^{-sh}) = Z(e^{-hs})\mathrm{diag}\{\alpha'_i(s)\}.$$

This shows that a control law of the form (2) permits to assign the invariant factors $\alpha'_i(s)$ to the closed loop system (3–2). It should be clear that, in that case, a simpler control law of the form (4) cannot get this assignment.

5. ILLUSTRATIVE EXAMPLE

The above results and design procedures are illustrated with the two dimensional academic example

$$\dot{x}(t) = \begin{pmatrix} 0 & 0 \\ 1 & 0 \end{pmatrix} x(t) + \begin{pmatrix} 1 & 1 \\ 0 & 0 \end{pmatrix} u(t-h).$$

This system is spectrally controllable because the pair $(A, \exp(-hA)B)$ is controllable (See Lemma 2). Its controllability indices are $\{2, 0\}$.

The natural coprime factorization for this system is

$$N(s, e^{-s}) = \begin{pmatrix} se^{-s} & 0 \\ e^{-s} & 0 \end{pmatrix}, D(s) = \begin{pmatrix} s^2 & -1 \\ 0 & 1 \end{pmatrix}.$$

It is possible to design a control law that assigns prescribed invariant factors with finite roots, of degree $\{2, 0\}$. A solution is given by

$$(I_2 - F((s, e^{-s}))D(s, e^{-s}) - G(s, e^{-s})N(s, e^{-s}) = D_K(s, e^{-s}). \quad (14)$$

According to Theorem 3, a control law where $G(s, e^{-s}) = K$, and $F(s, e^{-s}) = K(sI - A)^{-1}(e^{-hA} - e^{-hs}I)B$ does the job, and substituting $D(s)$, $N(s, e^{-s})$, A, B, and

$$K = \begin{pmatrix} k_{11} & k_{12} \\ k_{21} & k_{22} \end{pmatrix}$$

into (13), gives

$$D_K(s, e^{-s}) = \begin{pmatrix} s^2 - k_{11}s + k_{12}s - k_{12} & -1 \\ -k_{21}s + k_{22}s - k_{22} & 1 \end{pmatrix}. \quad (15)$$

The invariant factors of degree $\{2, 0\}$, namely $\{s^2 + (-k_{11} + k_{12} - k_{21} + k_{22}) - (k_{12} + k_{22}), 1\}$, can indeed be assigned to arbitrary finite locations by choosing K. We have in this simple case obtained a parametrization of all the invariant factors that can be obtained with the control laws under consideration.

Now, if we want to assign invariant factors of degrees $\{1, 1\}$, for instance both invariant factors taking the value $s + 1$, let us consider according to the design procedure given in Loiseau (2001) a coprime factorization with column degrees $\{1, 1\}$ for the denominator. This new factorization is obtained as

$$D'(s, e^{-s}) = D(s)U(s, e^{-s}),$$
$$N'(s, e^{-s}) = N(s, e^{-s})U(s, e^{-s}),$$

where $U(s, e^{-s})$ is given by

$$U(s, e^{-s}) = \begin{pmatrix} \dfrac{2 - e^{-s}}{s + \ln 2} & -\dfrac{1 - e^{-s}}{s} \\ -s & s + \ln 2 \end{pmatrix}.$$

This leads to

$$(Z(e^{-s}) - F(s, e{-}s))D'(s, e^{-s}) + G(s, e^{-s})N'(s, e^{-s}) = \mathrm{diag}\{s + 1, s + 1\},$$

with

$$G(s, e^{-s}) = \begin{pmatrix} (1 + \ln 2)(2 - e^{-s}) & \ln 2(2 - e^{-s}) \\ 2 - e^{-s} & 0 \end{pmatrix}$$

$$Z(e^{-s}) = \begin{pmatrix} 1 & 2 - e^{-s} \\ 1 & 3 - e^{-s} \end{pmatrix},$$

$$F_{11}(s, e^{-s}) = (s(1 + \ln 2)) + \ln 2) \left(\frac{1 - e^{-s}}{s}\right)^2,$$

$$F_{12}(s, e^{-s}) = (1 + \ln 2)\frac{(1 - e^{-s})^2}{s} + \frac{1 - e^{-s}}{s},$$

$$F_{21}(s, e^{-s}) = (1 - e^{-s})^2,$$

and

$$F_{22}(s, e^{-s}) = \frac{(1 - e^{-s})^2}{s} + (1 - \ln 2)\frac{2 - e^{-s}}{s + \ln 2}.$$

Of course this calculation is uneasy, and so is the expression of this control law in the time domain (Brethé, 1997), (Brethé and Loiseau,1998), since there is no specialized toolbox to deal with $\mathcal{E}$.

6. CONCLUSION

It is shown that static state feedback distributed control laws for input delay systems that allow the assignment of a finite spectrum, can be used to assign as well invariant factors with a finite number of roots. This feature is indeed of interest in shaping the dynamic of the closed loop system. It is shown that the multiplicities of the invariant factors cannot be assigned freely: they are restricted by Rosenbrock inequalities by a set of well defined integers. The result is explained in the framework of the Bezout domain $\mathcal{E}$, where more general control laws allow a complete freedom in assigning the multiplicities of the invariant factors. Clearly, static state feedback distributed control laws can be designed in a straightforward manner, but at a cost: the freedom in the assignment of the multiplicities of the invariant factors is restricted.

7. APPENDIX

We recall here the Theorem of Rosenbrock (for instance from Kučera (1991)), which is a clue for our proof.

Lemma 4. Consider a linear controllable system described by

$$\dot{x}(t) = Ax(t) + Bu(t)$$

where $A \in \mathbb{R}^{n \times n}, B \in \mathbb{R}^{n \times m}$ and such that the controllability indices of the pair (A, B) are $c_1, c_2, \cdots, c_m$. There exists a static state feedback

$$u(t) = Kx(t)$$

such that the invariant factors of the closed loop system are $\alpha_1'(s), \alpha_2'(s), \cdots, \alpha_k'(s)$ where $\alpha_i'(s)$ divides $\alpha_{i-1}'(s), i = 2, ..., m$, if and only if

$$\sum_{i=1}^{j} c_i \leq \sum_{i=1}^{j} \deg(\alpha_i'), j = 1, ..., m$$

with equality for $j = m$, and $\alpha_i'(s) = 1, i = m + 1, ..., n$.

REFERENCES

Artstein, Z. (1982). Linear systems with delayed controls: a reduction. *IEEE Transation on Automatic Control*, Vol. AC-27, No. 4, 869-879.

Bellman, R. and K.L. Cooke (1963). *Differential Difference Equations*, Academic Press, London.

Brethé, D. (1997). Contribution à l'étude de la stabilisation des systèmes linéaires à retards, Thèse de Doctorat, IRCYN, Nantes, France.

Brethé D. and Loiseau J.J. (1998). An effective algorithm for finite spectrum assignment of single-input systems with delays, Mathematics in computers and simulation, 45, 339-348.

Gantmacher, F. R. (1959). *The theory of matrix*, Vol 1,AMS Chelsea Publishing, New York.

Kamen E.W, Khargonekar P.P. and A. Tannenbaum (1986). Proper stable bezout factorizations and feedback control of linear time delay systems, *Int. J. Contr.*, Vol. 43, No. 3, 837-857.

Kailath T. (1980). *Linear Systems*, Prentice Hall.

Kolmanovski, V B. and V.R. Nosov (1986). *Stability of functional differential equations*, Academic Press, New York.

Kučera, V. (1991). *Analysis and design of discrete linear control systems*, Prentice-Hall, London, and Academia, Prague.

Loiseau J.J. (2000). Algebraic tools for the control and stabilization of time-delay systems, *Annual Reviews in Control* 24, 135-149.

Loiseau J.J. (2001). Invariant factors assignment for a class of time-delay systems, *Kybernetika*, Volume 37, Number 3, Pages 265-275.

Manitius, A. Z. and A.W Olbrot (1979). Finite Spectrum Assignment problem for Systems with Delays, *IEEE Trans. Autom. Contr.*, Vol. AC-24, No. 4, 541-553.

Mondié S., Dambrine M. and Santos O. (2001), Approximation of control laws with distributed delays: a necessary condition for stability, IFAC Conference on Systems, Structure and Control, Prague, Czek Republic.

Niculescu S. (2001). *Delays effects on stability, A robust control approach, Springer, Heidelberg.*

Morse, A.S. (1976). Ring Models for Delay-Differential Systems, *Automatica*, Vol. 12, 529-531.

Olbrot, A. (1978). Stabilizability, Detectability, and spectrum assignment for linear autonomous systems with general time delays, IEEE Trans. Autom. Contr., Vol. AC-23, No. 5, 887-890.

The polynomial toolbox, Polynomial Methods for Systems, Signal and Control (1998).

Vidyasagar, M. (1985). *Control System Synthesis*, The MIT Press, USA.

www.elsevier.com/locate/ifac

DELAY ROBUSTNESS OF CLOSED LOOP FINITE ASSIGNMENT FOR INPUT DELAY SYSTEMS [1]

S. Mondié [*,2] **, S.I. Niculescu** [**] **and J.J. Loiseau** [***]

[*] *Departamento de Control Automático CINVESTAV-IPN, Av. IPN 2508, A.P. 14-740, 07300 México, D.F., México, smondie@ctrl.cinvestav.mx*

[**] *Heudiasyc, UTC, UMR CNRS 6599, BP 20529, 60205 , Compiègne, France, Silviu.Niculescu@hds.utc.fr*

[***] *Institut de Recherche en Communications et Cybernétique de Nantes, UMR CNRS 6597, Ecole Centrale de Nantes, Université de Nantes, Ecole des Mines de Nantes, BP 92101, 44 321 Nantes Cedex 03, France, loiseau@irccyn.ec-nantes*

Abstract: The problem of the robustness with respect to delay uncertainty for the finite spectrum closed loop assignment of input delay systems is addressed. Numerically exploitable conditions in terms of the maximal deviation of the design delay, and of the assigned closed loop are given. An analytic expression of a lower bound of this deviation guaranteeing stability in the monovariable and multivariable case are obtained. The Smith predictor and related schemes are also revisited.

Keywords: delay systems, robustness, spectrum assignment.

1. INTRODUCTION

The results concerning finite spectrum assignment of time delay linear systems with control laws that include distributed delays presented a decade ago in Manitius and Olbrot (1979), Artstein (1982) have been revisited recently. In particular, renewed attention is given to the implementation of the distributed delay (Van Assche *et al.*, 1999), (Mondié *et al.*, 2001a). and to the algebraic and structural properties that result from the adequate algebraic setting provided by the Bezout domain $\mathcal{E}$ (Brethé and Loiseau, 1998).

In this paper, we restrict our attention to spectrally controllable (Olbrot, 1978) systems with delay in the input described by

$$\dot{x}(t) = Ax(t) + Bu(t - h_0), \qquad (1)$$

where $A \in \mathbb{R}^{n\times n}, B \in \mathbb{R}^{n\times m}$ and $h_0 \geq 0$ is the delay of the system.

If the delay h_0 is known with certainty, following the ideas in Smith (1959), Manitius and Olbrot (1979), Artstein (1982) a prediction $x_p(t)$ for the variable $x(t + h_0)$ is given by

$$x_p(t) = e^{h_0 A}x(t) + \int_t^{t+h_0} e^{(t+h_0-\sigma)A}Bu(\sigma - h_0)d\sigma.$$

Defining $\tau := \sigma - h_0$, $x_p(t)$ can be rewritten as

$$x_p(t) = e^{h_0 A}x(t) + \int_{t-h_0}^{t} e^{(t-\tau)A}Bu(\tau)d\tau. \qquad (2)$$

It appears that $x_p(t)$ is available at time t. Therefore, it is natural to use a static state feedback law

$$u(t) = Kx_p(t) + v(t). \qquad (3)$$

that assigns the closed loop system

$$\dot{x}(t) = (A + BK)x(t) + Bv(t - h_0)$$

Its spectrum $\det(sI_n - A - BK)$ is indeed finite. Next, we explain why this property is so appealing. A fundamental result in the field of delayed systems is that a system is stable if and only if the roots of its characteristic equation, which is a quasipolynomial, have all strictly negative real part (Bellman and Cooke, 1963). Quasipolynomials have an infinite number of roots, of which a

[1] Supported by Conacyt, Mexico 31951-A and by GDR-Automatique: systèmes à retards, France.
[2] On leave at Heudyasic, UMR CNRS 6599, France.

finite number can be isolated and the remaining ones are located with regularity on a finite number of logarithmic lines. The task of finding the location of the roots of a quasipolynomial is involved, hence the interest in obtaining a closed loop system whose characteristic equation is a polynomial. Indeed, a polynomial has a finite number of roots and the study of their location is a standard problem.

However, when there is uncertainty in the parameters or in the delay value, the closed loop system stability is not guaranteed. Moreover, the spectrum is again a quasipolynomial whose analysis is involved. This concern was indeed exposed in seminal works as well as in recent publications and some numerical examples or analysis of simple examples were presented.

Due to the well known difficulties in identifying a delay, the uncertainty in the delay is a fundamental issue for practical applications. The effect of the delay size on the stability properties of delayed system has been extensively studied in the past, and is still intensively explored, giving rise to a vast amount of publications (see Kolmanovski and Nosov (1986), Niculescu (2001) and references therein).

In this paper the robustness of the stable placement with respect to delay uncertainty is analyzed, more precisely, the problem can be stated as follows.

Problem 1. Consider a spectrally controllable system with delay in the input described by (1) and a control law (2), (3) that assigns a stable closed loop finite spectrum. Assume that there is no full knowledge of the system delay h_0, and that the delay value used in the controller is h. Then what is the maximal value of the deviation $\Delta := h_0 - h$, named δ, such that the closed loop stability is insured for $|\Delta| \in [0, \delta)$?

The paper is organized as follows: The closed loop system characteristic polynomial when there is uncertainty in the delay is obtained in section 2. In section 3 the robustness of the assignment with respect to the size of the delay uncertainty is analyzed in the monovariable case, in the framework of the Smith predictor, and in the general case. The paper ends with some concluding remarks.

2. CLOSED LOOP QUASIPOLYNOMIAL

In this section, the closed loop quasipolynomial when the design delay and the system delay differ is obtained. The expressions of this quasipolynomial in terms of the control gain K and in terms of the closed loop stable assignment Σ are respectively given in Propositions 1 and 2.

Proposition 1. Consider the system with delay in the input described by (1). If the nominal delay h_0 is not known with certainty, and a delay h is employed in the design of the control law (2), (3) then the closed loop characteristic equation is

$$\det\{(sI_n - A - [I - e^{hA}e^{-hs} + e^{hA}e^{-h_0 s}I]BK\}.$$

Proof. The design parameter h is now used in the control law (2), (3)

$$u(t) = K[e^{hA}x(t) + \int_{t-h}^{t} e^{(t-\sigma)A}Bu(\sigma)d\sigma] + v(t).$$

The system and the control law are respectively described in the frequency domain by

$$(sI_n - A)x(s) = Be^{-h_0 s}u(s)$$

and

$$\begin{aligned}(I_m - K(sI - A)^{-1}[I - e^{-h(sI-A)}]B)u(s)\\ = Ke^{hA}x(s) + v(s).\end{aligned}$$

The characteristic matrix of the closed loop system is

$$\det\begin{pmatrix} (sI_n - A) & Be^{-h_0 s} \\ Ke^{hA} & (I_m - K(sI-A)^{-1}[I - e^{-h(sI-A)}]B) \end{pmatrix}.$$

Using standard determinant and matrices properties, we obtain that the above is

$$\begin{aligned}&\det\{sI_n - A\}\det\{(I_n - (sI - A)^{-1}.\\ &[I - e^{hA}e^{-hs} + e^{hA}e^{-h_0 s}]BK\}\\ &= \det\{(sI_n - A - [I - e^{hA}e^{-hs} + e^{hA}e^{-h_0 s}I]BK\}\end{aligned} \quad (4)$$

and the result follows. ■

Remark 1. If there is no uncertainty in the delay, $h = h_0$, and the closed loop finite spectrum is

$$\det\{sI_n - A - BK\}. \quad (5)$$

An important fact is that in the problem statement, it is assumed that when there is no uncertainty, the closed loop system is stable, or equivalently, that the control law is a stabilizing one. Since not all control laws have this property, it is better to get an expression of the closed loop system in terms of the closed loop Hurwitz matrix Σ. Observe in (5) that this matrix is not arbitrary. It must satisfy $\Sigma = A + BK$ for some K. The existence of such a gain K is indeed equivalent to the condition $A - \Sigma \subset \mathrm{Im}B$.

Notice also that an expression in terms of the delay deviation Δ suits better our purpose of finding the maximal deviation such that the stability of the closed loop is guaranteed for $\Delta \in [0, \delta)$.

These two requirements are met by the following expression of the closed loop quasipolynomial when uncertainty in the delay is present.

Proposition 2. Consider the system with delay h_0 in the input described by (1). If a design delay h is employed in the control law (2), (3) that assigns a closed loop characteristic matrix $sI - \Sigma$ where Σ is a Hurwitz matrix such that $A - \Sigma \subset \text{Im}B$. Then, the closed loop characteristic equation is

$$\det\{(sI_n - \Sigma + e^{hA}e^{-hs}[e^{-\Delta s} - 1](A - \Sigma)\}. \quad (6)$$

where $\Delta := h_0 - h$ is the deviation from the design delay value h.

Remark 2. The conditions on Σ are not restrictive with respect to spectrum assignment. The spectral controllability of the system implies that the spectrum can be chosen arbitrarily.

Remark 3. The closed loop characteristic equation (4) can be rearranged as

$$\det\{(sI_n - A - BK - e^{hA}e^{-hs}[e^{-\Delta s} - 1]BK\}. \quad (7)$$

This expression depends explicitly on the gain K, and one can observe that the "optimal" gain K results from a trade of between the maximal delay deviation for which the stability is insured and the assigned closed loop finite spectrum.

3. ROBUSTNESS ANALYSIS

In this section the robustness analysis with respect to the size of the delay deviation is addressed. First the monovariable case is considered and an analytical explicit bound that insures stability is given. Next, the Smith predictor is revisited and a numerically computable expression for the exact bound is obtained. Finally, a numerically computable bound for the multivariable case is proposed.

These analysis are all based on a continuity argument, also known as $\mathcal{D}$-partition analysis (Neimark ,1949). This method for the stability analysis of a quasipolynomial $p(s, e^{-s}, \Delta)$ which is stable for $\Delta = 0$, is essentially based on the fact that the location of the roots changes continuously as the parameter Δ increases continuously.

By hypothesis, when $\Delta = 0$, the roots of $det(p(s, e^{-s}, \Delta))$ have all negative real part.

As $|\Delta|$ grows the only way that the system can become unstable is when a root, or a set of roots, that moves continuously, hits for the first time the imaginary axis. There are two possible situations for this to occur.

The first one is when the crossing happens at $s = 0$. In all the cases studied below, this does not occur because when $s = 0$, $det(p(s, e^{-s}, \Delta))$ reduces to a non zero constant.

The other possibility is that the crossing of the imaginary axis occurs at a pair of roots $s = j\omega$, $s = -j\omega$ with $\omega > 0$. It follows that to determine the interval $[0, \delta)$ such that the stability is guaranteed when $\Delta \in [0, \delta)$, we need to find the smallest value of $\Delta > 0$ for which the equation $p(j\omega, e^{-j\omega}, \Delta) = 0$ admits a solution for ω and Δ.

3.1 *Monovariable case*

Due to the complexity of the closed loop quasipolynomial, we are not able to find an analytical explicit exact bound, even in the monovariable case. However, in the unstable case studied in this section, it is possible to find a lower bound on the delay deviation for which the stability is guaranteed. In this case, the delayed system is

$$\dot{x}(t) = ax(t) + bu(t - h_0),$$

with $a \geq 0$. And, according to Proposition 2, the closed loop quasipolynomial (6) is

$$s - \sigma + (a - \sigma)e^{ah}e^{-hs}(e^{-\Delta s} - 1) = 0.$$

In this case, the assigned spectrum is $s - \sigma$, where $\sigma < 0$, and the stabilizing control law $k = (\sigma - a)/b$ always exists.
According the $\mathcal{D}$-partition analysis explained above, the system is stable for $\Delta = 0$, and it reaches an instability region when the roots hit the imaginary axis for the first time as Δ grows continuously. When $s = 0$, the closed loop quasipolynomial is $-\sigma \neq 0$, hence no crossing of the imaginary axis occurs at $s = 0$.

In the following proposition implicit analytic conditions for crossing of the imaginary axis at $s = \pm j\omega$ are given.

Proposition 3. Consider the quasipolynomial

$$s - \sigma + (a - \sigma)e^{ah}e^{-hs}(e^{-\Delta s} - 1) = 0$$

where $\sigma < 0$, $h \geq 0$ and $a \geq 0$ are given. This quasipolynomial has roots on the imaginary axis if and only if the equations

$$\sqrt{\omega^2 + \sigma^2} = 2(a - \sigma)e^{ha}\left|\sin(\frac{\omega\Delta}{2})\right|, \quad (8)$$

$$\arctan(-\frac{\omega}{\sigma}) = \frac{\pi}{2} - h\omega - \frac{\omega\Delta}{2}, (\pi), \quad (9)$$

have a solution for ω and Δ.

Proof. Clearly, we want to determine the solutions Δ and ω such that

$$j\omega - \sigma + (a-\sigma)e^{ah}e^{-j\omega h}(e^{-j\omega\Delta} - 1) = 0. \quad (10)$$

Observe that

$$e^{-j\omega\Delta} - 1 = -2j\sin(\frac{\omega\Delta}{2})e^{-\frac{1}{2}j\omega\Delta}. \quad (11)$$

Substituting in (10) we obtain

$$j\omega - \sigma - 2j(a-\sigma)e^{ah}e^{-j\omega h}\sin(\frac{\omega\Delta}{2})e^{-\frac{1}{2}j\omega\Delta} = 0.$$

The module and argument of this expression are respectively given by (8) and (9) and the equivalence is established. ■

These conditions can be exploited to provide a lower bound δ_c for which the stability is insured for $|\Delta| \in [0, \delta_c)$.

Proposition 4. Consider the quasipolynomial

$$s - \sigma + (a-\sigma)e^{ah}e^{-hs}(e^{-\Delta s} - 1) = 0$$

where $\sigma < 0$, $h \geq 0$ and $a \geq 0$ are given. Then this quasipolynomial is stable for $\Delta \in [0, \delta_c)$, where δ_c is equal to

$$\min(2\left|\frac{\operatorname{arccot}(-\frac{\omega_0}{\sigma})}{\omega_0} - h\right|, 2\left|\frac{\operatorname{arccot}(\frac{\omega_0}{\sigma})}{-\omega_0} - h\right|, h) \quad (12)$$

with

$$\omega_0 = \sqrt{4(a-\sigma)^2e^{2ha} - \sigma^2} \quad (13)$$

Proof. Since $\left|\sin(\frac{\omega\Delta}{2})\right| \leq 1$ we have from (8) that

$$|\omega| \leq \omega_0 = \sqrt{4(a-\sigma)^2e^{2ha} - \sigma^2}$$

Now, we can rewrite (9) as

$$\omega(h + \frac{\Delta}{2}) = \frac{\pi}{2} - \arctan(-\frac{\omega}{\sigma}) = \operatorname{arccot}(-\frac{\omega}{\sigma}).$$

For $0 < \omega \leq \omega_0$,

$$\Delta \geq 2\left(\frac{\operatorname{arccot}(-\frac{\omega_0}{\sigma})}{\omega_0} - h\right),$$

and for $-\omega_0 \leq \omega < 0$,

$$\Delta \leq 2\left(\frac{\operatorname{arccot}(\frac{\omega_0}{\sigma})}{-\omega_0} - h\right) < 0.$$

Hence for $\Delta \in [0, \delta_c)$, where δ_c defined in (12) the system of equations (8), (9) have no solution for Δ and ω and we can conclude that the roots never cross the imaginary axis. Since σ is negative, when $\Delta = 0$, the closed loop quasipolynomial, which simplifies to $s - \sigma$, is stable, then for $|\Delta| \in [0, \delta_c)$, it remains stable. ■

An analytical expression of the exact bound is not available. However, it is possible to determine it with the help of numerical methods.

3.2 *Smith predictor revisited*

The analysis of the robustness of Smith predictor of a SISO system with respect to uncertain delays in the input, leads to the following stability problem (Niculescu, 2001):

Problem 2. Find conditions on δ such that the following complex equation:

$$A(s) + B(s)e^{-sh} - B(s)e^{-s(h+\Delta)} = 0, \mid \Delta \mid \leq \delta, (14)$$

where $A(s), B(s)$ polynomials $(deg(A(s) > deg(B(s)))$, $A(s)$ Hurwitz, has all the solutions in $\mathbb{C}^-$.

Using continuity arguments, let us define the quantities:

$$\alpha_1(\omega, h) = Re\left(A(j\omega) + B(j\omega)e^{-j\omega h}\right),$$
$$\alpha_2(\omega, h) = Im\left(A(j\omega) + B(j\omega)e^{-j\omega h}\right),$$
$$\beta_1(\omega, h) = -Re\left(B(j\omega)e^{-j\omega h}\right),$$
$$\beta_2(\omega, h) = -Im\left(B(j\omega)e^{-j\omega h}\right).$$

Then (14) may be rewritten as:

$$\alpha_1(\omega, h) + j\alpha_2(\omega, h) +$$
$$(\beta_1(\omega, h) + j\beta_2(\omega, h)) \cdot \left(cos(\omega\bar{\delta}) - jsin(\omega\bar{\delta})\right) = 0$$

(for some real ω and $\Delta = \bar{\delta}$), which is equivalent to the following real equations:

$$sin(\omega\bar{\delta}) = \frac{-\alpha_1(\omega, h)\beta_2(\omega, h) + \beta_1(\omega, h)\alpha_2(\omega, h)}{\beta_1^2 + \beta_2^2},$$
$$cos(\omega\bar{\delta}) = -\frac{\alpha_1(\omega, h)\alpha_2(\omega, h) + \beta_1(\omega, h)\beta_2(\omega, h)}{\beta_1^2 + \beta_2^2}.$$

Note that it is not possible to have simultaneously $\beta_1(\omega, h)$ and $\beta_2(\omega, h)$ both 0 for some ω_0, since this will imply $\alpha_1(\omega_0, h)$ and $\alpha_2(\omega_0, h)$ also 0, which leads to $A(j\omega_0) = 0$, condition which contradicts A Hurwitz stable polynomial.
For each root ω_i of the equation

$$\mid B(j\omega)e^{-j\omega h} \mid = \mid A(j\omega) + B(j\omega)e^{-j\omega h} \mid,$$

we shall compute the corresponding $\bar{\delta} = \delta_i$, given by:

$$\delta_i = \frac{arcsin\left(\frac{-\alpha_1(\omega_i,h)\beta_2(\omega_i,h) + \alpha_1(\omega_i,h)\beta_2(\omega_i,h)}{\beta_1(\omega_i,h)^2 + \beta_2(\omega_i,h)^2}\right)}{\omega_i}. (15)$$

Note that ω_i may be positive or negative. Based on such argument, define the sets $\Lambda_{+,0}$ and $\Lambda_{-,0}$ as follows (Niculescu, 2001): if there does not exist any *positive* δ_i, then $\Lambda_{+,0} = \{+\infty\}$, and

$$\Lambda_{+,0} = \{\delta_i > 0 \quad : \quad \delta_i \text{ given by } (15)\}, (16)$$

elsewhere. Similarly, if there does not exist any *negative* δ_h, then $\Lambda_{-,0} = \{-h\}$, and

$$\Lambda_{-,0} = \{\delta_i < 0 \quad : \quad \delta_i \text{ given by } (15)\}, (17)$$

elsewhere. Then, the *real bounds* δ_1 and δ_2 will be given by:

$$\begin{cases} \delta_1 = \max \Lambda_{-,0}, \\ \delta_2 = \min \Lambda_{+,0}, \end{cases},$$

and the derived condition is *necessary and sufficient.* The proposed results can be summarized as follows (Niculescu, 2001):

Proposition 5. Define the real function $\mathcal{F}(\omega) = | A(j\omega) + B(j\omega)e^{-j\omega h} |^2 - | B(j\omega)e^{-j\omega h} |^2$. Then the stability of the closed-loop system is guaranteed for any inaccurate modeling delay Δ, $| \Delta | \leq \delta$ if:
i) $\mathcal{F}(\omega)$ has no roots. In such case, the stability property is of *delay-independent* type, i.e. it may hold for any positive δ;
ii) $\mathcal{F}(\omega)$ has at least one root. In such case, the stability property is of *delay-dependent* type (with the delay-interval containing 0), and it holds for any $\Delta \in (\delta_1, \delta_2)$ $(\delta_1 < 0 < \delta_2)$, where δ_1, δ_2 are given by:

$$\begin{cases} \delta_1 = \max \Lambda_{-,0}, \\ \delta_2 = \min \Lambda_{+,0}, \end{cases},$$

where the sets $\Lambda_{\pm,0}$ are defined in (16) and (17). In this case, $\delta < \min\{\delta_1, \delta_2\}$, and is always a finite value.

Remark 4. The above procedure can be applied indeed to the monovariable system studied above by letting $A(s) = s - \sigma$, which is Hurwitz, and $B(s) = (\sigma - a)e^{ha}$.

Remark 5. If the following inequality holds

$$2\left|B(j\omega)\right| < \left|A(j\omega)\right|, \forall \omega \in \mathbb{R},$$

then the closed loop system stability is guaranteed for all inaccurate modeling of the delay.

3.3 *General case*

The multivariable case is obviously more complex. It is possible to obtain numerically exploitable sufficient conditions generalizing the ideas developed for the monovariable case.

Proposition 6. Consider the complex equation

$$\det\{D(s) - N(s)e^{-hs}[e^{-\Delta s} - 1]\} = 0 \quad (18)$$

where $D(s)$ and $N(s)$ are coprime polynomial matrices (Kailath, 1980) such that the Hurwitz polynomial matrix $D(s)$ is column reduced and such that the column degrees of $N(s)$ are smaller than those of the polynomial matrix $D(s)$. If for all $\omega \in \mathbb{R}^*$

$$\left\|N(j\omega)D(j\omega)^{-1}\right\|_\infty < \frac{1}{2} \quad (19)$$

then the stability of (18) is guaranteed for $\Delta \in [0, h]$.

Proof. It follows from (11) that

$$\left|e^{-j\omega\Delta} - 1\right| = \left|2\sin(\frac{\omega\Delta}{2})\right| \leq 2$$

hence if (19) holds, then

$$\left\|e^{-hs}(e^{-j\omega\Delta} - 1)N(j\omega)D(j\omega)^{-1}\right\|_\infty \leq \left|e^{-hs}\right|\left|e^{-j\omega\Delta} - 1\right|\left\|N(j\omega)D(j\omega)^{-1}\right\|_\infty < 1. \quad (20)$$

Therefore, $\forall \omega \in \mathbb{R}$

$$\det\{D(j\omega) - N(j\omega)e^{-hj\omega}[e^{-\Delta j\omega} - 1]\} \neq 0, \quad (21)$$

Moreover, for $\Delta = 0$ $\det\{D(s) - N(s)e^{-hs}[e^{-\Delta s} - 1]\} = \det\{D(s)\}$ has all its roots in the left half plane because the polynomial matrix $D(s)$ is Hurwitz by hypothesis. Then we can conclude that as Δ increases continuously, the roots of (18) move continuously, but never cross the imaginary axis to the right half plane, hence (18) is always Hurwitz.. ■

Proposition 7. Let the complex equation (18) be as in the previous Proposition. If there exist at least one frequency value such that

$$\left\|N(j\omega)D(j\omega)^{-1}\right\|_\infty \geq \frac{1}{2}, \quad (22)$$

let

$$\omega_m = \sup\{\omega_i : \left\|N(j\omega)D(j\omega)^{-1}\right\|_\infty < \frac{1}{2}, \forall \omega \in (0, \omega_i]\} \quad (23)$$

then the stability of (18) is guaranteed for $\Delta \in [0, \delta_c)$ where δ_c is equal to

$$\min\{ \inf_{\omega \in (0,\omega_m)} \frac{2}{\omega} \arcsin \frac{1}{2\left\|N(j\omega)D(j\omega)^{-1}\right\|_\infty}, h\}.$$

Proof. If there exist at least one frequency value such that (22) holds, then one can deduce from the continuity argument used in the previous result and from (20) that no crossing of the imaginary axis will occur if $\Delta \in [0, \delta_c)$ where δ_c is such that

$$\left\|e^{-hs}(e^{-j\omega\delta_c} - 1)N(j\omega)D(j\omega)^{-1}\right\|_\infty = 1,$$

or equivalently,

$$\left|2\sin(\frac{\omega\delta_c}{2})\right|\left\|N(j\omega)D(j\omega)^{-1}\right\|_\infty = 1$$

and the result follows. ■

Proposition 8. The delay robustness analysis of the quasipolynomial (7) reduces to the analysis of (18) with

$$D(s) := D_0(s) - KN_0(s)$$
$$N(s) := Ke^{hA}N_0(s)$$

where $(N_0(s), D_0(s))$ is a coprime factorization (Kailath, 1980) for the pair (A, B).

Proof. Let $(N_0(s), D_0(s))$ be a coprime factorization for the pair (A, B). In this case, we have that

$$(sI_n - A)^{-1}B = N_0(s)D_0(s)^{-1} \quad (24)$$

where $N_0(s) \in \mathbb{R}^{n\times m}[s]$ and $D_0(s) \in \mathbb{R}^{m\times m}[s]$. Pre multiplying the closed loop expression (7) by the Hurwitz matrix $(sI_n - A)^{-1}$ we obtain

$$\det\{I_n - [I + e^{hA}e^{-hs}[e^{-\Delta s} - 1]](sI_n - A)^{-1}BK\}.$$

Substituting (24) leads to

$$\det\{I_n - [I + e^{hA}e^{-hs}[e^{-\Delta s} - 1]]N_0(s)D_0(s)^{-1}K\}$$

or

$$\det\{I_m - K[I + e^{hA}e^{-hs}[e^{-\Delta s} - 1]]N_0(s)D_0(s)\}.$$

Finally, post multiplying by the Hurwitz matrix $D_0(s)$ gives

$$\det\{D_0(s) - KN_0(s) - Ke^{hA}N_0(s)e^{-hs}[e^{-\Delta s} - 1]\}$$

This expression has indeed the same unstable roots as (7). Moreover, $D(s) := D_0(s) - KN_0(s)$ is also column reduced, the column degrees of $N(s) := Ke^{hA}N_0(s)$ are strictly smaller than those of $D(s)$, and $N(s)$ and $D(s)$ are coprime and the result follows. ■

Remark 6. The solutions to (23) can be determined using a frequency sweeping of the imaginary axis.

Remark 7. The lower bound of Proposition 7 can be used indeed in the monovariable case. Notice that the resulting bound is more conservative than the one given in Proposition 4 because the information given by the argument condition is not used.

4. CONCLUDING REMARKS

The robustness of finite spectrum control laws with respect to uncertainty in the size of the delay is analyzed. Conditions for which stability of the closed loop system is insured are given in terms of the maximal allowed deviation from a given design value and in terms of the assigned closed loop. An analytical expression for a lower bound of this maximal deviation is obtained in the monovariable case and a numerically exploitable bound is given for the Smith predictor. For the multivariable case, a computable lower bound is obtained.

REFERENCES

Artstein, Z. (1982). Linear systems with delayed controls: a reduction. *IEEE Transation on Automatic Control*, Vol. AC-27, No. 4, 869-879.

Bellman, R. and K.L. Cooke (1963). *Differential Difference Equations*, Academic Press, London.

Brethé D. and Loiseau J.J. (1998). An effective algorithm for finite spectrum assignment of single-input systems with delays, Mathematics in computers and simulation, 45, 339-348.

Gantmacher, F. R. (1959). *The theory of matrix*, Vol 1,AMS Chelsea Publishing, New York.

Kailath T. (1980). *Linear Systems*, Prentice Hall.

Kamen E.W, Khargonekar P.P. and A. Tannenbaum (1986). Proper stable bezout factorizations and feedback control of linear time delay systems, *Int. J. Contr.*, Vol. 43, No. 3, 837-857.

Kolmanovski, V B. and V.R. Nosov (1986). *Stability of functional differential equations*, Academic Press, New York.

Manitius, A. Z. and A.W Olbrot (1979). Finite Spectrum Assignment problem for Systems with Delays, *IEEE Trans. Autom. Contr.*, Vol. AC-24, No. 4, 541-553.

Mondié S., Dambrine M. and Santos O. (2001). Approximation of control laws with distributed delays: a necessary condition for stability, IFAC Conference on Systems, Structure and Control, Prague, Czek Republic.

Neimark J. (1949). $\mathcal{D}$-subdivisions and spaces of quasipolynomials, *Prickl. Math. Mech.*, 13, 349-380.

Niculescu S. (2001). *Delays effects on stability, A robust control approach, Springer, Heidelberg.*

Olbrot, A. (1978). Stabilizability, Detectability, and spectrum assignment for linear autonomous systems with general time delays, IEEE Trans. Autom. Contr., Vol. AC-23, No. 5, 887-890.

Smith O.J.M. (1959). Closer Control of loops with dead time, *Chem. Eng. Prog.*,53, 217-219.

Van Assche V., Dambrine, M., Lafay, J.F. and Richard, J.P. (1999). Some problems arising in the implementation of distributed-delay control laws, 38th IEEE Conference on Decision and Control, Phoenix, Arizona, USA.

www.elsevier.com/locate/ifac

ROBUST EIGENVALUE ASSIGNMENT FOR UNCERTAIN DELAY CONTROL SYSTEMS

L. Fridman*, P. Acosta*, A. Polyakov **

* *Av. Tecnologico 2909, Chihuahua, Chih., 31310, Mexico*
e-mail:[lfridman,pacosta]@itch.edu.mx
** *Voronezh State University,*
Universitetskaja pl. 1, Voronezh, 394693, Russia

Abstract: Time delay does not allow to realize an ideal sliding mode, but implies oscillations in the space of state variables. Two types of predictors are proposed for time delay compensation in input and output delayed systems. Relay control algorithms are suggested ensuring the existence of sliding modes on the intersection of some surfaces in the space of predictor variables, and desired dynamics of systems into this mode. *Copyright © 2001 IFAC*

Keywords: Time delay, Delay Compensation, Stabilization, Relay control, Sliding modes

1. INTRODUCTION

The relay control systems are widely used thanks to the following main reasons:

- the relay control law is one of the simplest control algorithms;
- relay controllers are robust;
- sliding motions on a discontinuity surface, a special kind of motions in discontinuous systems, are quite useful for design of an efficient control;
- there are control systems in which only sign of variables is observable (Choi and Hedrick, 1996, Li and Yurkovich, 1999).

Time delay in control systems is usually present and must be taken into account. In practice, time delay is caused by the following:

- *Delay in the output* An example of such systems is the controllers of exhausted gas in the fuel injector automotive control systems (Choi and Hedrick, 1996, Li and Yurkovich, 1999). In such systems the λ sensors are using which can measure only the sign of error for ratio with delay.
- *Delay in the input* It could be for example any system with delayed actuator. An example of such a system is the controller for stabilization of the fingers of an underwater manipulator (Bartolini *et al*, 1997).

In such situation for control design it is possible to use the value of state vector without delay but the control is transferred to the plant via delayed actuator and system has input delay.

Time delay doesn't allow to design the sliding mode control in the space of state variables. That is why it may be singled out two main approaches to use relay control for delay systems:

- time delay compensation

The idea of delay compensation to realize the sliding mode control was formulated by Drakunov and Utkin (1992). Relay control algorithms for systems with delay in state and control based on delay compensation was suggested in Li and Yurkovich (1999). Roh and Oh (1999) designed the sliding mode control in the space of predic-

[1] This research was partially supported under the grant of Consejo Nacional de Ciencia y Tecnologia (CONACYT) N990704

tor variables for input delay systems using unit control. It is necessary to take into account the comments of Sing (2001).

- investigation of oscillations properties and control of oscillations amplitudes and frequencies.

The knowledge of oscillations and periodicity properties of first order systems with relay delay control was used by Fridman *et al* (1993, 2000, 2001) and for control of motions amplitudes. Strygin *et al* (2001) have generalized this results for multidimensional case.

Akian *et al* (1997) have suggested P.I. control algorithms for amplitude control for one dimensional relay system with delay in the input.

This paper is devoted to delay compensation approach for output and input delayed systems. For input and output delay systems the predictors are proposed allowing to design the sliding modes in the space of predictor variables. Relay control algorithms are suggested for eigenvalues assignment into the sliding modes in the space of predictors variables.

2. PROBLEM STATEMENT

2.1 *Plant Model*

Consider the uncertain delay system

$$\dot{x}(t) = Ax(t) + Bu+$$
$$+g_1(x(t),t) + g_2(x(t-h),t), \qquad (1)$$

where $x \in \mathbf{R}^n$ is the state vector, $u(t) \in \mathbf{R}^m$ is control, $h \in \mathbf{R}^+$ is time delay, g_1, g_2 are smooth uncertainties presenting perturbations and nonlinearities in system (1), the pair $\{A, B\}$ is controllable.

2.2 *Modified Matching Conditions*

Suppose for control system (1) the modified matching conditions defining the insensibility of system to bounded perturbations and uncertainties are given in form

$$e^{Ah}g_1(x(t),t) = B\gamma_1(x(t),t), \qquad (2)$$

$$e^{Ah}g_2(x(t-h),t) = B\gamma_2(x(t-h),t). \qquad (3)$$

Assume that trivial solution is an equilibrium point of system (1), which means that

$$\gamma_1(0,t) \equiv \gamma_2(0,t) \equiv 0. \qquad (4)$$

2.3 *Specific Features of Output Delay System*

Suppose that the system (1) is the system with output delay, when the system (1) output has the form

$$y(t) = x(t-h) \qquad (5)$$

and to design the control law one can use only the information about state vector $x(t-h)$.

For output delay systems it is necessary to complete the modified matching conditions (2),(3) with the following assumptions:

$$||\gamma_1(x(t),t)|| \leq p_1,\, p_1 > 0, \qquad (6)$$

$$||\gamma_2(x(t-h),t)|| \leq p_2,\, p_2 > 0. \qquad (7)$$

2.4 *Specific Features of Input Delay Systems*

System (1) is the system with input delay, when to design the control law for system (1) one can use the information about state vector of system (1) without time delay. But it is possible to transmit the control to the system only with time delay. That is why one can define the control in (1) only in form

$$u = u(t-h). \qquad (8)$$

In this case it is necessary to complete matching conditions (2),(3),(4),(7) with the following assumption:

$$||\gamma_1(x(t),t)|| \leq q_1||x(t)||,\, q_1 > 0, \qquad (9)$$

2.5 *Initial Conditions*

Delay systems are not controllable on time interval $[0, h]$ that is why it is sufficient to introduce the initial conditions for systems (1), (5) and (1),(8) in the form:

$$x(\theta) = \varphi(\theta), -h \leq \theta \leq 0,\, \varphi(\theta) \in L_1[-h, 0],$$
$$x(0) = x^0,$$
$$u(\theta) = \psi(\theta),\, -h \leq \theta \leq 0,\, \psi(\theta) \in L_1[-h, 0].$$

3. DELAY COMPENSATION

Time delay doesn't allow to design the sliding mode control in the space of state variables. That is why to compensate the time delay it is necessary to introduce predictors in systems (1), (5) and (1),(8), reduce (1), (5) and (1),(8) to the delay free systems, and design relay sliding mode control in the space of predictor's variables to guarantee desired dynamics of systems.

3.1 *Predictors for Output Delay Systems*

To compensate the time delay in system (1),(5) consider the state predictor in the form:

$$z(t) = e^{Ah}x(t-h) + \int_{-h}^{0} e^{-A\theta}Bu(t+\theta)d\theta. \quad (10)$$

is useful. In this situation one has

$$\dot{z} = Ae^{Ah}x(t-h)) + e^{Ah}Bu(t-h) - e^{Ah}Bu(t-h) + \\ + Bu(t) + A\int_{-h}^{0} e^{-A\theta}Bu(t+\theta)d\theta + \\ + e^{Ah}(g_1(x(t-h),t)) + \\ + g_2(x(t-2h),t)) = Az + Bu(t) + \\ + B(\gamma_1(x(t-h),t)) + \gamma_2(x(t-2h),t)). \quad (11)$$

3.2 *Predictors for Input Delay Systems*

For the input delay system (1),(8) the predictor is given by formula

$$z(t) = e^{Ah}x(t) + \int_{-h}^{0} e^{-A\theta}Bu(t+\theta)d\theta. \quad (12)$$

Then

$$\dot{z} = Ae^{Ah}x(t) + e^{Ah}Bu(t-h) - e^{Ah}Bu(t-h) + \\ + Bu(t) + A\int_{-h}^{0} e^{-A\theta}Bu(t+\theta)d\theta + e^{Ah}(g_1(x(t),t)) \\ + g_2(x(t-h),t)) = Az + Bu(t) + \\ + B(\gamma_1(x(t),t)) + \gamma_2(x(t-h),t)). \quad (13)$$

Fiagedzi and Pearson (1986) have shown that there exists the input/ output mapping which transforms the input/output delay systems (1),(5) and(1),(8) into the forms (11),(13) such that systems (1),(8) and (1),(8) and (11),(11) correspondingly have the same dynamics. Consequently, to ensure desired dynamics for systems (1),(5) and (1),(8) it is possible to deal with the systems (11) and (13).

4. EIGENVALUE ASSIGNMENT IN SLIDING MODE SYSTEMS VIA RELAY CONTROL

4.1 *Design of Sliding Surfaces*

It is well known (Drazenovic, 1969) that the sliding equations for systems under matching conditions do not depends on disturbances. That is why to design the sliding mode surface for systems (11) and (13) it is sufficient to consider the system without uncertainties and disturbances in form

$$\dot{\bar{x}}(t) = A\bar{x}(t) + Bu(t), \quad (14)$$

and to transform the system to the regular form, which was introduced by Loukjanov and Utkin (1981). Suppose that $rank(B) = m$ and matrix B has the following form

$$B = [B_1^T \, B_2^T]^T,$$

where $B_1 \in \mathbf{R}^{(n-m)\times m}$, $B_2 \in \mathbf{R}^{m\times m}$, $det\, B_2 \neq 0$. The nonsingular coordinate transformation

$$T\bar{x} = [\begin{matrix}\hat{x}_1 \\ \hat{x}_2\end{matrix}], T = (\begin{matrix} I_{n-m} & -B_1B_2^{-1} \\ 0 & B_2^{-1}\end{matrix}), \quad (15)$$

reduce the system (14) to the regular form

$$\dot{\hat{x}}_1 = A_{11}\hat{x}_1 + A_{12}\hat{x}_2$$
$$\dot{\hat{x}}_2 = A_{21}\hat{x}_1 + A_{22}\hat{x}_2 + u, \quad (16)$$

where $\hat{x}_1 \in \mathbf{R}^{(n-m)}$, $\hat{x}_2 \in \mathbf{R}^m$, $det\, B_2 \neq 0$. Loukjanov and Utkin (1981) was shown that, if the pair $\{A, B\}$ is controllable, the matrix pair $\{A_{11}, A_{12}\}$ is controllable too. Handling $\hat{x}_2$ as an m - dimensional control in the $(n-m)$ dimensional first subsystem of (16), all $(n-m)$ eigenvalues may be assigned arbitrary by a proper choice of matrix C in form $\hat{x}_2 = -C\hat{x}_1$. To provide desired dependence between coordinates x_2 and x_1 of the state vector, sliding surfaces should be designed in form

$$\hat{s} = \hat{S}\hat{x} = (C\, I_m)(\begin{matrix}\hat{x}_1 \\ \hat{x}_2\end{matrix}) = 0.$$

In that case the motions into the sliding domain for systems (15) and (16) are governed by $(n-m)$ dimensional system

$$\dot{\hat{x}}_1 = (A_{11} - A_{12}C)\hat{x}_1.$$

Returning to the variables of system (14) one has the sliding surfaces in form:

$$\bar{s} = S\bar{x} = 0, \quad S = \hat{S}T^{-1}, \quad \hat{S} = (C\, I_m). \quad (17)$$

Then the equations for the system (14) dynamics into the sliding mode on $Sx = 0$ have the form

$$\dot{\bar{x}} = (I_n - B(SB)^{-1}S)A\bar{x}. \quad (18)$$

Nonzero eigenvalues of matrices $A_{11} - A_{12}C$ and $(I_n - B(SB)^{-1}S)A$ coincide.

5. DESIGN OF RELAY CONTROL

5.1 *Relay Control Design for Output Delay systems*

To guarantee the presence of the sliding modes in the systems (11) on the intersections of the surfaces

$$s = Sz(t) = 0, \quad (19)$$

to have the sliding dynamics in form (17), and eliminate the presence of uncertainties, one has to define the relay control in form

$$u_r(x(t-h),t) = -(SB)^{-1}SAz- \\ -U_0(x(t-h),t)SIGN(\sigma(t)) \quad (20)$$

where $\sigma = [SB]^T Sz$, $SIGN(\sigma(t)) =$
$= (sign(\sigma_1(t)), sign(\sigma_2(t)), ..., sign(\sigma_m(t)))^T$.

Theorem 1. For control law in form (20) there exist $p, q > 0$, such that for $U_0(x(t-h),t) = q||x(t-h)|| + p$, the sliding manifold $s = Sz = 0$ for system (1),(5) solutions is reachable for the finite time and sliding dynamics is described by equation (18).

Proof. Consider the Lyapunov function candidate

$$V(s) = \frac{1}{2}s^T s.$$

Taking into account (11) one has

$$\dot{V} = s^T S[Az + Bu + B(\gamma_1(x(t-h),t)) + \\ +\gamma_2(x(t-2h),t))] = -||\sigma||(U_0(x(t-h),t) - \\ -\gamma_1(x(t-h),t)) - \gamma_2(x(t-2h),t))),$$

It is necessary to remark that

$$||\gamma_1(x(t-h),t)) + \gamma_2(x(t-2h),t))|| \leq \\ \leq p_1 + p_2.$$

Then, choosing $q > 0, p > p_1 + p_2$, and taking into account that matrix SB is nonsingular, one has

$$\dot{V} \leq -\beta||\sigma|| \leq -\beta_1||s||||SB|| \leq -\beta_2\sqrt{V},$$

$\beta_2, \beta_1, \beta > 0$. This inequality guarantees the existence and finite time convergence to the sliding modes on $s = 0$. To find the sliding mode dynamics one has:

$$S[Az + Bu_{eq} + B(\gamma_1(x(t-h),t)) + \\ +\gamma_2(x(t-2h),t))] = 0 \\ u_{eq} = -(SB)^{-1}SAz- \\ -(\gamma_1(x(t-h),t)) + \gamma_2(x(t-2h),t)) \quad (21)$$

Substituting (11) into (21) one has the sliding dynamics in the form (18). ■

5.2 *Relay Control Design for Input Delay Systems*

Analogously the relay control law for input delay system (13) has the form

$$u_r(x(t),t) = -(SB)^{-1}SAz \\ -U_0(x(t),t)SIGN(\sigma(t)). \quad (22)$$

and the following theorem is true:

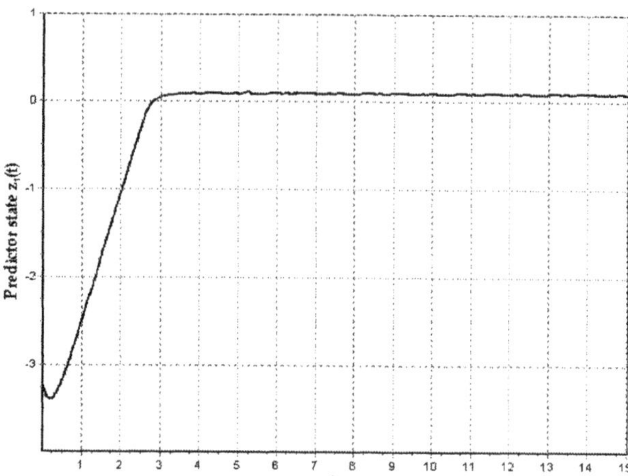

Fig. 1. Deviation of predictor variable z_1 when the modified matching conditions are not valid

Theorem 2. For control law in form (22) there exist $p, q > 0$ such that for

$$U_0(x(t),t) = q||x(t)|| + p, \; q > q_1, \; p > p_2,$$

the sliding manifold for system (13) solution is reachable for the finite time and sliding dynamics is described by equations (18).

6. SOME COMMENTS TO THE THEOREM 1

6.1 *Modified Matching Conditions*

On the one hand modified matching conditions (2), (3) are strong. On the other hand the standard matching conditions by Drazevovic (1969) in form:

$$g_1(x(t),t) = B\gamma_1(x(t),t), \quad (23)$$

$$g_2(x(t-h),t) = B\gamma_2(x(t-h),t). \quad (24)$$

could not guarantee the invariance of sliding mode equations for uncertainties (23), (24) for systems (11) and (13). Actually, suppose that one can guarantee the sliding mode existence for systems (11) on the manifold $s = Sz = 0$. In this case the sliding equations for system (11) have the form: $\dot{z} = (I_n - B(SB)^{-1}S)Az + (I_n - B(SB)^{-1}S)e^{Ah}$

$$\times B(\gamma_1(x(t),t)) + \gamma_2(x(t-h),t)). \quad (25)$$

It is obvious that in general case when B equation (6.1) could depend from bounded uncertainties g_1, g_2 (see subsection 7.1).

6.2 *Deviations of Equilibrium Point*

Fortunately even in the case when modified matching conditions hold, one can not guarantee the invariance of motions in original systems (1),(5) and (1),(8) with respect to uncertainties

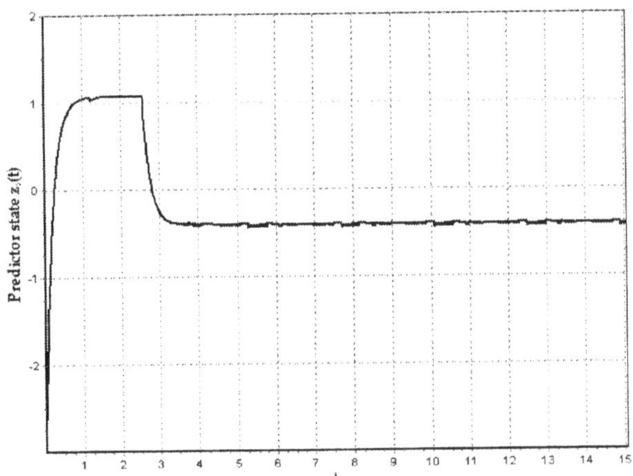

Fig. 2. Deviation of predictor variable z_2 when the modified matching conditions are not valid

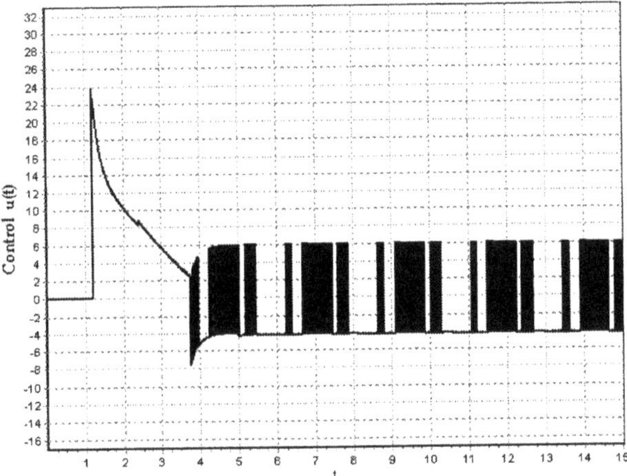

Fig. 3. Relay control u

g_1, g_2. For example in fact, from formula (10) it follows that if

$$lim_{t\to\infty} z(t) = 0,$$

then

$$lim_{t\to\infty}[e^{Ah}x(t-h) + \int_{-h}^{0} e^{-A\theta}Bu_{eq}(t+\theta)d\theta] = 0.$$

Taking into account (21), one can conclude that condition (4) is natural sufficient condition for absence of equilibrium point deviations in the original system (1) (see subsection 7.2).

6.3 *Razumikhin Approach and Sliding Modes*

Roh and Oh (1999) considered the case when delayed uncertainty in system (1) can be majorize by linear function using Razumikhin approach (see Krasovskii (1963)). It is questionable, because the main assumption of Razumikhin approach is that for all $t > 0$ one has $||x(t+\theta), t)|| \leq q^*||x(t)||$, $q^* > 1$, $-h \leq \theta \leq 0$. For systems with sliding modes it is not true at least for the case when B is square matrix and when $t = t^*$ is the point of entrance into the sliding mode, because $x(t^*) = 0$.

7. MUNERICAL EXAMPLES OF RELAY DELAY CONTROL UNDER CONSTANT PERTURBATIONS

7.1

Consider the control system

$$\dot{x} = \begin{pmatrix} 0 & 1 \\ 3 & -2 \end{pmatrix} x + Bu(t-h)$$

$$+ \begin{pmatrix} 0 \\ 1 \end{pmatrix} (g_0(x(t),t) + g_1(x(t-h),t) \quad (26)$$

with output delay $h = 1.2$, control matrix $B = [0,1]^T$, and initial conditions $\varphi(\theta) \equiv [-1.6, 1]^T$, $\theta \in [-1.2, 0]$. Assume that the constant perturbation in system (26) is given in the form:

$$g_0(x(t),t) = 0.5,\ g_1(x(t-h),t) = 0, \quad (27)$$

and the switching surface is designed in the form: $\sigma = Sz = [5,1]z$, and, consequently, desired sliding mode surface has the form $z_2 + 5z_1 = 0$. The matching conditions (23),(24) for system (26) are held BUT the modified matching conditions (2),(3) are not held. Figures 1 - 3 illustrate that the presence of constant perturbation in system leads to deviation of equilibrium point for system (11), describing the predictor variables. This means that in this case even the behavior of predictor variables is not invariant with respect to constant perturbations.

7.2 *Deviation of Equilibrium in the Original System*

Now assume that for system (26) the modified matching conditions (2),(3) hold but condition (4) is not valid. It is true for $S = I, B = I$. Figures 4,5 show that in this case the trivial solution of system (10) is stable, but there is the deviation of equilibrium point for variables of original system (1).

8. CONCLUSION

The time delay compensation approach for relay control of uncertain input and output delayed systems is developed. Two types of predictors are investigated for time delay compensation in input and output delayed systems. Relay control algorithms are proposed ensuring the existence of sliding mode on the intersection of some surfaces in the space of predictor variables, and desired dynamics of systems into this mode.

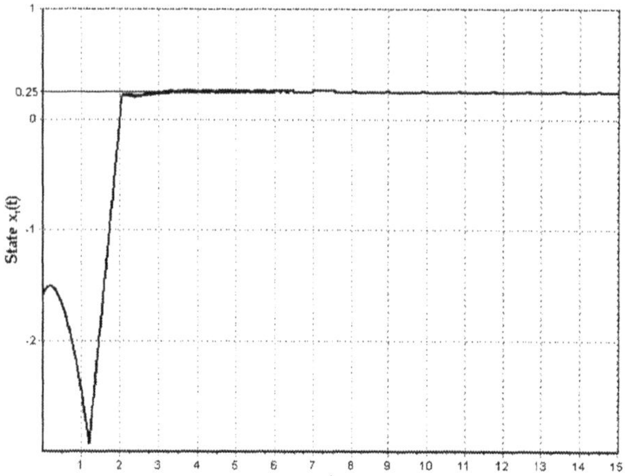

Fig. 4. Deviation of state variable x_1 when the condition (4) is not valid

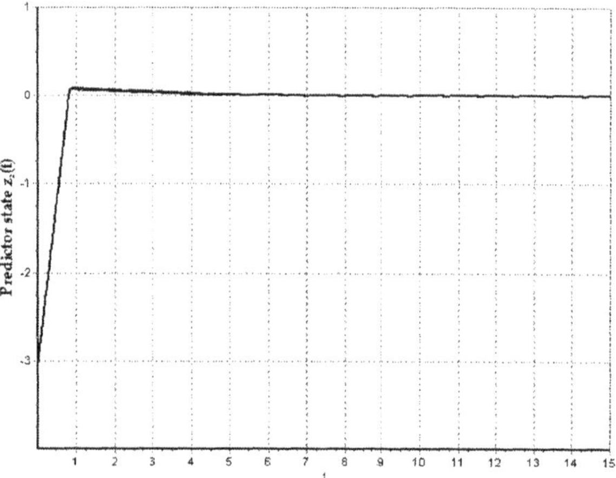

Fig. 5. Convergnece of predictor variable z_1 when the modified matching conditions are true

REFERENCES

Akian, M., P.-A. Bliman and M. Sorine (1997). P.I. control of periodic oscillations of relay systems." Proceedings of Conference on Control and Chaos, St-Petersburg, Russia.

Bartolini, G., W. Caputo, M. Cecchi, A. Ferrara and L. Fridman (1997). Vibration damping in elastic robotic structure via sliding modes. *Int. Journ. of Robotic Systems,* **14,** , 675-696.

Choi, S.-B. and J. K. Hedrick (1996). Robust Throttle Control of Automotive Engines. *ASME J. of Dynamic Systems, Measurement and Control,* **118,** 92-98.

Drakunov, S.V. and V.I. Utkin (1993). Sliding mode control in dynamic systems. *Int. Journ. of Control,* **55,** 1029-1037.

Drazenovic B. (1969). The time invariant conditions for invariant systems. *Automatica,* **5,** 287-295.

Fiagedzi Y.A. and A.E. Pearson (1986). Feedback stabilization of linear autonomous time lag systems. *IEEE Trans. on Automatic Control,* **31,** 847-855.

Fridman, L., E. Fridman and E. Shustin, (1993). Steady modes in an autonomous system with break and delay. *Differential Equations,* **29,** 1161-1166.

Fridman, L., E. Fridman and E. Shustin (2001). Steady modes and sliding modes in relay control systems with delay. in: *Sliding Mode Control in Engineering,* Barbot J.P., Perruquetti W. (Ed.), Marcel Dekker, New York, 261-292.

Fridman, E., L. Fridman and E. Shustin (2000). Steady modes in the relay control systems with delay and periodic disturbances. *ASME Journal of Dynamical Systems, Control and Measurement,* **122,** 4, 732-737.

Gouaisbalt, F., W. Perruquetti, Y. Orlov and J.-P. Richard (1999). Sliding Mode Controller Design for Linear Time-Delay Systems . Proceedings of European Control Conference ECC'1999, Carlsruhe, Germany, 1999.

Krasovksii N.N. (1963). *Stability of motions,* Stanford University Press.

Li, X. and S. Yurkovitch (1999). Sliding Mode Control of Systems with Delayed States and Controls, in: *Variable Structure Systems, Sliding Mode and Nonlinear Control,* K.D. Young, U. Ozguner (Eds.), Lecture Notes in Control and Information Sciences, **247,** Springer, Berlin, 93-108.

Lukjanov, A. and V. Utkin (1981). Methods of reducing equations for dynamic systems to a regular form, *Automation and Remote Control,* **42,** 413-420.

Nguang, S.K. (2001) Comments on "Robust stabilization of uncertain input delay systems by sliding mode control with delay compensation". *Automatica,* **37,** 1677.

Roh, Y.- H. and J.-H. Oh,(1999). Robust stabilization of uncertain input delay systems by sliding mode control with delay compensation. *Automatica,* **35,** 1861-1865.

Strygin, V., L.Fridman and A. Polyakov (2001). Local stabilization of relay systems with delay. *Doklady Mathematics,* **64,** 106 - 108.

Utkin, V., J. Guldner and J. Shi (1999). *Sliding Modes in Electromechanical Systems,* Taylor and Francis,London.

www.elsevier.com/locate/ifac

RELAY CONTROL OF OSCILLATIONS AMPLITUDES FOR SYSTEMS WITH DELAY

L. Fridman *, V. Strygin **, A. Polyakov **

*Division of Postgraduate Study and Investigation,
Chihuahua Institute of Technology,
Av. Tecnologico 2909, Chihuahua, Chih., 31310, Mexico
e-mail:lfridman@itch.edu.mx*
** *Department of Applied Mathematics, Voronezh State University,
Universitetskaja pl. 1, Voronezh, 394693, RUSSIA*

Abstract: Time delay does not allows to realize an ideal sliding mode, but implies oscillations in the space of state variables. The notations of ε stabilization and weak controllability are introduced. The algorithms for local stabilization of oscillation's amplitudes for weak controllable control systems with input and output delay via relay control was suggested. *Copyright © 2001 IFAC*

Keywords: Stabilization, Time delay, Relay control

1. INTRODUCTION

Time delay doesn't allow to design the sliding mode control in the space of state variables (Fridman *et al* 1993, 2000). That is why it may be singled out two main approaches to use relay control for delay systems:

- time delay compensation

Relay control algorithms for systems with delay in state and control based on delay compensation was suggested by Li and Yu (1999). Roh and Oh (1999) designed the sliding mode control in the space of predictor variables for input delay systems using unit control, but it is necessary to take into account the comments by Nguang (2001).

- control of oscillations amplitudes

The knowledge of oscillations and periodicity properties of first order systems with relay delay control Fridman *et al* (1993, 2000) used in for control of amplitudes of motions.

Akian *et al* (1997) suggested P.I. control algorithms for amplitudes control for one dimensional relay system with delay in the input.

Fridman *et al* (1993) have shown that any solution of equation

$$\dot{x}(t) = kx - p\,sign[x(t-1)],$$

with initial conditions

$$|\varphi(0)| < p\frac{2-e^k}{ke^k}$$

under conditions

$$0 < k < ln\,2 \qquad (1)$$

is situated in the band $|x| < p(e^k - 1)/k$ for all $t \in [0, \infty)$.

Strygin *et al* (2001) and Fridman *et al* (2001) was generalized the stabilization condition (1) for controllable systems for multidimensional case.

The main result of this paper is the generalization of results of Strygin *et al* (2001) and Fridman

[1] This research was partially supported under the grant of Consejo Nacional de Ciencia y Tecnologia (CONACYT) N990704

et al (2001) for the class of so called "weak controllable" system.

The notations of ε stabilization and weak controllability are introduced. The algorithm for local stabilization of oscillation's amplitudes for weak controllable systems with input and output delay via relay control is suggested.

2. ε STABILIZATION

Consider the system

$$\frac{dx}{dt} = f(x) + Bu, \qquad (2)$$

where $x \in R^n$, $f : R^n \to R^n$ is a smooth function, $f(0) = 0$, B is a $(n \times m)$ matrix, $u \in R^m$. Assume that the matrix $A = f_x(0)$ can have the eigenvalues with positive real part.

In this paper to stabilize the zero solution of system (2) we will try to find relay delayed m dimensional feedback in the form:
$u = F(signS_1(x(t-1)), ..., signS_k(x(t-1)))$, where the pair $\{S, F\}$ belongs to the class Q consisting of the pairs of smooth mappings, transforming

$$S : R^n \to R^k,\ S = (S_1, S_2, ..., S_k)^T.$$

$$F : R^k \to R^m.$$

Let us denote as $x(t)$ the solution to the system (2) with initial conditions

$$x(t) = \varphi(t) \quad (-1 \le t \le 0).$$

Definition 1. The trivial solution to the system (2) is said to be ε stabilizable, if for any $\varepsilon > 0$ there exist $\delta > 0$, integer $k > 0$, and the pair $\{S, F\} \in Q$, such that from the inequality $sup_{t\in[-1,0]}||\varphi(t)|| < \delta$ it follows that $sup_{t\in[-1,\infty]}||x(t)|| < \varepsilon$.

Remark 2. It is necessary to remark, that S, F usually depend on $\varepsilon > 0$.

3. WEAK CONTROLLABILITY

Assume that the spectrum $\sigma(A)$ of the matrix $A = f_x(0)$ consists of two parts

$$\sigma(A) = \sigma_+ \cup \sigma_-,$$

where σ_+ and σ_- are the sets of matrix A eigenvalues with the positive and negative real parts.

Then the state space $E = R^n$ could be represented in the following form $E = E_+ \oplus E_-$, where E_+, E_- are the invariant subspaces with respect to A.

Consider two projectors P and Q, transforming

$$P : E \to E_+, Q : E \to E_-.$$

For any $x \in E$ the following representation $x = Px + Qx$ is true. Then system (2) has the form

$$(\dot{Px}) + (\dot{Qx}) = A(Px + Qx) + Bu + \overline{f}(x). \qquad (3)$$

Denoting $y = Px$, $z = Qx$ one can rewrite system (3) in the form

$$\begin{cases} \dot{y} = A^+y + B^+u + f_1(y, z) \\ \dot{z} = A^-z + B^-u + f_2(y, z), \end{cases} \qquad (4)$$

where $y \in E_+$, $z \in E_-$, $A^+ = PA$, $A^- = QA$, $B^+ = PB = (b_1^+, b_2^+, ..., b_m^+)$, $B^- = QB$, $f_1(y, z) = P\overline{f}(y + z)$, $f_2 = Q\overline{f}(y + z)$.

Definition 3. The pair of the matrices $\{A, B\}$ is said to be weak controllable, if $rank(B^+) = dim(E_+)$

4. ε STABILIZATION OF MIMO SYSTEMS

Let us denote by $G_{n,k}$ the set of $(n \times n)$ matrices with the following spectrum's properties:

$$\sigma_+ = \{\lambda_i\}_{i=1}^l \cup \{\alpha_j \pm i\beta_j\}_{j=1}^\nu,\ l + 2\nu = k \qquad (5)$$

$$\lambda_i \in (0, L), L = \ln(2), \qquad (6)$$

$$\alpha_j \in (0, M), M = \max_{0\le t\le \frac{\pi}{4}}\{\frac{1}{2}t\cos(t + \frac{\pi}{4})\}, \qquad (7)$$

and all the eigenvalues from σ_+ are simple.

Theorem 1. Assume

1) $f_x(0) \in G_{n,k}$;

2) the pair $\{f_x(0), B\}$ is weak controllable.
Then the system (2) is ε stabilizable.

Let us describe the control algorithm allowing to realize the theorem 1.

$rank\, B^+ = dim\, E_+ = k = l + 2\nu$ that is why the vectors $\{b_j^+\}$ $(j = \overline{1, k})$ are linearly independent. This means that the following representation holds:

$$Ah_i = \lambda_i h_i, h_i = s_{1i}b_1^+ + s_{2i}b_2^+ + ... + s_{ki}b_k^+,$$

$$Af_{1j} = \alpha_j f_{1j} - \beta_j f_{2j}, Af_{2j} = \beta_j f_{1j} - \alpha_j f_{2j},$$

$$f_{1j} = s_{1l+2j-1}b_1^+ + s_{2l+2j-1}b_2^+ + ... + s_{kl+2j-1}b_k^+,$$

$$f_{2j} = s_{1l+2j}b_1^+ + s_{2l+2j}b_2^+ + ... + s_{kl+2j}b_k^+,$$

where $i = 1, ..., l, j = 1, , ..., \nu$. Consider the matrix S_0 consisting of $\{s_{ij}\}$ coefficients for representations of matrix A eigenvectors in the basis $b_1^+, ..., b_k^+$. Let us design the control u in the form:

$$u = \begin{pmatrix} S_0 & 0 \\ 0 & 0 \end{pmatrix} \sigma,$$

where the function $\sigma = (\sigma_1(y(t-1)), \sigma_2(y(t-1)), ..., \sigma_m(y(t-1)))^T$ we will define bellow. In this case the first equation of system (4) has the form:

$$\dot{y} = A^+ y + B^+ S_0 \sigma + f_1(y, z). \qquad (8)$$

Denote $B^+ S_0 = (h_1, h_2, ..., h_l, f_{11}, f_{21}, f_{12}, f_{22}, ..., f_{1\nu}, f_{2\nu}, 0, ..., 0) = (T\ 0)$. Matrix A^+ has only simple eigenvalues which, means that T is a nonsingular matrix. Substituting the variables $v = T^{-1}y$ into (8), we will have

$$\dot{v} = \begin{pmatrix} \lambda_1 & 0 & 0 & 0 & 0 & ... & 0 & 0 \\ 0 & \lambda_2 & 0 & 0 & 0 & ... & 0 & 0 \\ ... & ... & ... & ... & ... & ... & ... & ... \\ 0 & ... & 0 & 0 & \lambda_l & 0 & ... & 0 \\ 0 & ... & 0 & 0 & 0 & \alpha_1 & -\beta_1 & 0 \\ 0 & ... & 0 & 0 & 0 & \beta_1 & \alpha_1 & 0 \\ ... & ... & ... & ... & ... & ... & ... & ... \\ 0 & ... & 0 & 0 & 0 & 0 & \alpha_\nu & -\beta_\nu \\ 0 & ... & 0 & 0 & 0 & 0 & \beta_\nu & \alpha_\nu \end{pmatrix} v$$

$$+(I\ 0)\sigma + g(v, z), \qquad (9)$$

where $g(v, z) = T^{-1} f_1(Tv, z)$. Let us design $\sigma(v(t-1))$ in form

$$\sigma_i(v(t-1)) = -\alpha_i' \varepsilon sign(v_i(t-1)),$$
$$\sigma_{l+2j-1}(v(t-1)) = -\alpha_{l+j}' \varepsilon sign(\psi(t-1)),$$
$$\psi(t-1) = v_{l+2j-1}(t-1)\cos\beta_j - v_{l+2j}(t-1)\sin\beta_j,$$
$$\sigma_{l+2j}(v(t-1)) = -\alpha_{l+j}' \varepsilon sign(\bar{\psi}(t-1)),$$
$$\bar{\psi}(t-1) = v_{l+2j-1}(t-1)\sin\beta_j + v_{l+2j}(t-1)\cos\beta_j,$$
$$\sigma_r(v(t-1)) = 0, r = \overline{k+1, m},$$

where $i = 1, 2, ..., l$, $j = 1, 2, ..., \nu$. Here α_i' are some constants depending from λ_i, α_i and $\bar{f}(x)$. Returning to the original system (2) we will have

$$\dot{x} = f(x) + BS\sigma([PB_0S_0]^{-1}Px(t-1)), \qquad (10)$$

where $B_0 = (b_1, b_2, ..., b_k)$.

Now from Strygin *et al* (2001) and Fridman *et al* (2001) one can conclude that the system (10) is ε stabilizable.

5. EXAMPLES

5.1 *Invertible pendulum*

Consider the problem of an invertible pendulum stabilization with the help of relay delayed control. The model of the pendulum has the form:

$$\ddot{\theta} + k\dot{\theta} - p\sin(\theta) = u(t-1). \qquad (11)$$

where θ is an inclination angle, k is a friction coefficient, $p = g/l$, where l is a length of pendulum. Linearizing (11) we will have

$$\dot{x} = \begin{pmatrix} 0 & 1 \\ p & -k \end{pmatrix} x$$

$$+ \begin{pmatrix} 0 \\ 1 \end{pmatrix} u(t-1) + \begin{pmatrix} 0 \\ 1 \end{pmatrix} g(x_1(t)), \qquad (12)$$

where $g(x_1) = o(x_1)$. Assume that the eigenvalues λ_1, λ_2 of the matrix

$$A = \begin{pmatrix} 0 & 1 \\ p & -k \end{pmatrix}$$

have the different signs, and $0 < \lambda_1 < L$. Substituting

$$x = \begin{pmatrix} 1 & 1 \\ \lambda_1 & \lambda_2 \end{pmatrix} y, \qquad (13)$$

into the system (12) we will have

$$\dot{y} = \begin{pmatrix} \lambda_1 & 0 \\ 0 & \lambda_2 \end{pmatrix} y - \frac{1}{k} \begin{pmatrix} 1 \\ 1 \end{pmatrix} u(t-1)$$

$$-\frac{1}{k} \begin{pmatrix} 1 \\ 1 \end{pmatrix} g(y_1(t) + y_2(t)). \qquad (14)$$

Let us design $u(t-1)$ in form

$$u(t-1) = k\alpha' \varepsilon sign(y_1(t-1)). \qquad (15)$$

Returning to the system (11), we will have

$$\ddot{\theta} + k\dot{\theta} - p\sin(\theta) = k\alpha'\varepsilon sign(\dot{\theta}(t-1) - \lambda_2\theta(t-1)).$$

Consider the case when

$$\ddot{\theta} + 0.3\dot{\theta} - 0.04\sin(\theta) = u \qquad (16)$$

$$u = 0.3\alpha'\varepsilon sign(\dot{\theta}(t-1) - \lambda_2\theta(t-1)).$$

$$\theta(t) = 0.005\cos(2t), \qquad (17)$$

$$\dot{\theta}(t) = -0.01 sin(2t) \text{ for } t \in [-1, 0]. \qquad (18)$$

Figures 1- 3 shows the behavior of pendulum (16) - (18) for $\alpha' = 0.1$, $\varepsilon = 0.01$.

5.2 *ε stabilization of weak controllable system*

Consider the system

$$\begin{pmatrix} \dot{x_1} \\ \dot{x_2} \end{pmatrix} = \begin{pmatrix} \sin(0.7x_1 + 0.1x_2) + 0.3x_2 \\ -0.1\sin(x_1) + \ln(1 - x_1 - 1.7x_2) \end{pmatrix}$$

$$+ \begin{pmatrix} 2 \\ -1 \end{pmatrix} u(t-1) \qquad (19)$$

It is obvious that $f(0) = 0$ and

$$f_x(0) = \begin{pmatrix} 0.7 & 0.4 \\ -1.1 & -1.7 \end{pmatrix}$$

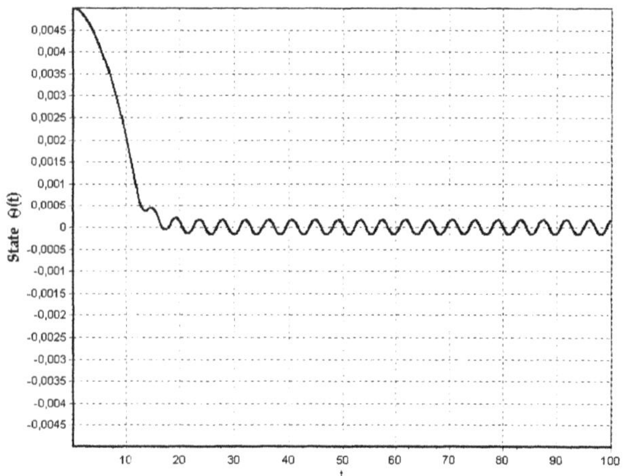

Fig. 1. Inclination angle θ

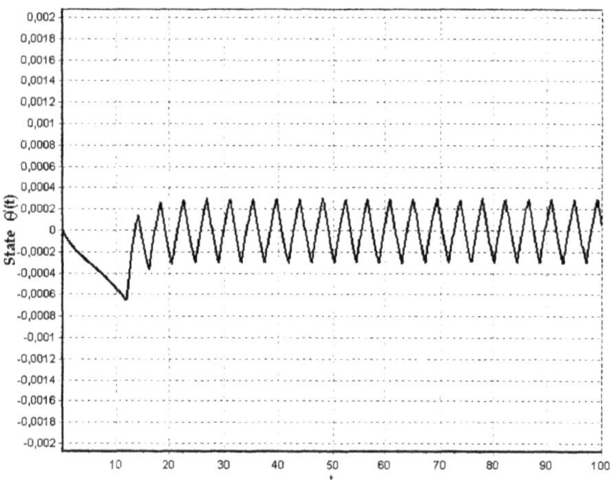

Fig. 2. Angular speed $\dot{\theta}$

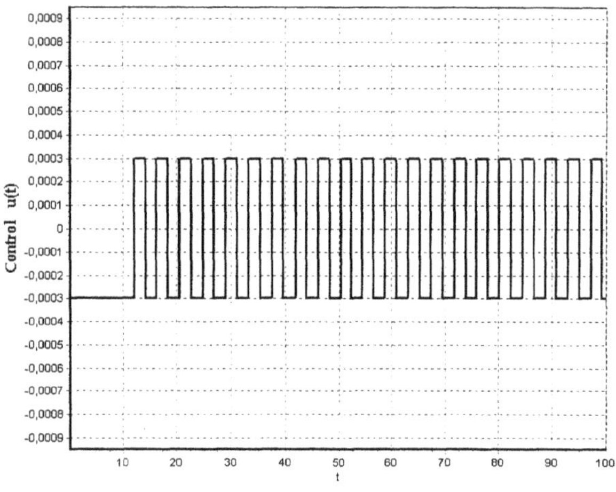

Fig. 3. Relay control law u

The pair $\{f_x(0), b\}$ is not controllable, but this pair is weak controllable. It allows us to design the control law in the following form:

$$u = -\alpha' \varepsilon sign[11x_1(t-1) + 2x_2(t-1)].$$

Figures 4 - 6 are shown the behavior of control system (19) for $\varepsilon = 0.02$, $\alpha' = 0.3$ and initial functions
$x_1^0(t) = 0.01e^t, x_2^0(t) = 0.2sin(t) \quad (-1 \leq t \leq 0)$.

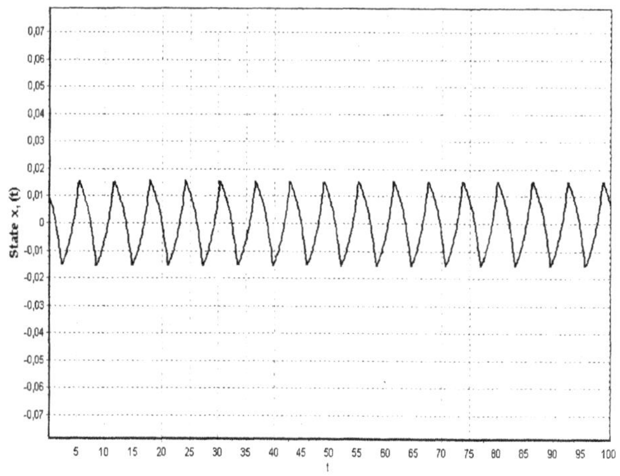

Fig. 4. x_1 coordinate

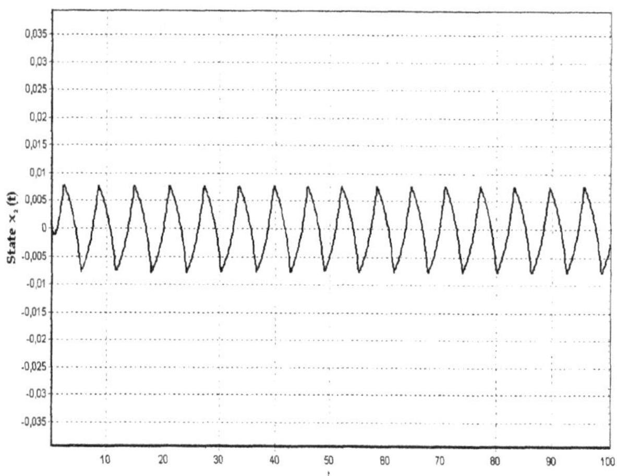

Fig. 5. x_2 coordinate

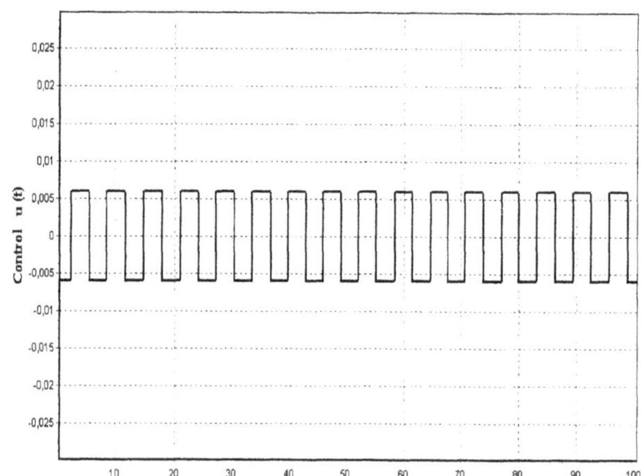

Fig. 6. Relay control law u

CONCLUSION

The sufficient conditions are found for stabilization of oscillation's amplitudes for weak controllable systems with input or output delay via relay control.

REFERENCES

Akian, M., P.-A. Bliman and M. Sorine (1997). P.I. control of periodic oscillations of relay systems. Proceedings of Conference on Control and Chaos, St-Petersburg, Russia.

Bartolini, G., W. Caputo, M. Cecchi, A. Ferrara and L. Fridman (1997). Vibration damping in elastic robotic structure via sliding modes. *Int. Journ. of Robotic Systems,* **14,** , 675-696.

Choi, S.-B. and J. K. Hedrick (1996). Robust Throttle Control of Automotive Engines. *ASME J. of Dynamic Systems, Measurement and Control* **118,** 92-98.

Drakunov, S.V. and V.I. Utkin (1993). Sliding mode control in dynamic systems, *Int. Journ. of Control,* **55,** 1029-1037.

Fridman, L., E. Fridman and E. Shustin (1993). Steady modes in an autonomous system with break and delay. *Differential Equations,* **29,** 1161-1166.

Fridman, E., L. Fridman and E. Shustin (2000). Steady modes in the relay control systems with delay and periodic disturbances. *ASME Journal of Dynamical Systems, Control and Measurement,* **122**, 4, 732-737.

Fridman, L.,V. Strygin and A. Polyakov (2001). Stabilization of Oscillations Amplitudes via Relay Delay Control. Proc. of 40th Conference on Decision in Control, Orlando, FL.

Gouaisbalt, F., W. Perruquetti, Y. Orlov and J.-P. Richard (1999). Sliding Mode Controller Design for Linear Time-Delay Systems . Proceedings of European Control Conference ECC'1999, Carlsruhe, Germany, 1999.

Li, X. and S. Yurkovitch (1999). Sliding Mode Control of Systems with Delayed States and Controls, in: *Variable Structure Systems, Sliding Mode and Nonlinear Control,* K.D. Young, U. Ozguner (Eds.), Lecture Notes in Control and Information Sciences, **247,** Springer, Berlin, 93-108.

Nguang, S. K. (2001). Comments on "Robust stabilization of uncertain input delay systems by sliding mode control with delay compensation". *Automatica,* **37,** 1677.

Roh, Y.- H. and J.-H. Oh (1999). Robust stabilization of uncertain input delay systems by sliding mode control with delay compensation. *Automatica,* **35,** 1861-1865.

Strygin, V., L.Fridman and A. Polyakov (2001). Local stabilization of relay systems with delay. *Doklady Mathematics,* **64,** 106 - 108.

www.elsevier.com/locate/ifac

SLIDING MODE CONTROL OF SYSTEMS WITH DISTRIBUTED DELAY

F. Gouaisbaut, M. Dambrine, J.P. Richard

LAIL cnrs upresa 8021, Ecole Centrale de Lille, BP 48, 59651 Villeneuve d'Ascq CEDEX - FRANCE.
e-mail : frederic.gouaisbaut@ec-lille.fr
FAX: (+33) 3 20 33 54 18

Abstract: The present article is devoted to the construction of sliding mode controllers for a class of distributed, possibly time-varying, delay systems. By the use of Lyapunov-Krasovskii functionnals or Lyapunov-Razumikhin functions, the reduced system is proven to be asymptotically stable for all delays less than an upperbound, which is optimized via LMIs. Finally, an example illustrates our approach.

Keywords: Distributed Delay system, Robust Control, Sliding Mode Control, LMI

1. INTRODUCTION

Sliding mode control (SMC), based on the use of switching control laws, belongs to the variable structure approaches. It is known to efficient in many problems of robust stabilization, as non-holonomic systems for instance or, more generally, systems which do not fit the Brocket's conditions (see (Perruquetti and Barbot, 2001 (to appear); Misawa and Utkin, 2001)). However, the presence of delays in combination with switching devices makes the design more complex (see for instance (Fridman *et al.*, 1996)). Some general frameworks were proposed (infinite dimensional systems (Orlov, 2000), differential inclusions for delay systems (Kolmanovskii and Myshkis, 1999)), but the concrete control results were mainly concerning systems with discrete delays (systems with point-wise delays). In (Choi, 1999; Shyu and Yan, 1993; Koshkouei and Zinober, 1996; Shyu and Yan, 1993) as here, results are considering delayed state variables, but the delays are pointwise and the control laws are independent of the delay value (and, consequently, may be more conservative than delay-dependent results (Gouaisbaut *et al.*, September 1999; Gouaisbaut *et al.*, 2001)). To our best knowledge, the only result concerned with SMC and distributed delays (with constant value) was provided in (Zheng *et al.*, 1995) : the controller was based on the predictor approach proposed in (Fiagbedzi and Pearson, 1987; Pearson and Pandiscio Jr, 1997). The resulting control law needs to solve a transcendental matrix equation.

The present paper considers uncertain systems with distributed possibly time-varying but unknown state-delay and additive perturbations which may be nonvanishing. It aims at designing the sliding surface in such a way that it maximizes the calculable set of admissible delays, where "admissible" means here those that don't destabilize the closed-loop, relay-delay system. The outline of the paper is as follows : after some notations, Section 3 is devoted to the transformation of the original system into a suitable form for sliding mode design called *regular form.* Then in Section 4, a sliding mode controller is developed by means of an LMI approach for distributed and constant time delays. This approach is extended to the case of a time-varying delay in Section 5. In both sections, the reduced system is proven to be asymptotically stable for any value of the delay less than a bound obtained by solving a convex optimization problem.

Lastly, in Section 6, an illustrative example shows the effectiveness of the method. In our opinion, this work has two original features:

(1) It considers distributed delays with weak hypotheses (except linearity).
(2) It can be implemented via classical LMI algorithms.

2. NOTATIONS AND ASSUMPTIONS

In this paper, the following system, with possibly time-varying delay $\tau = \tau(t) \geq 0$, will be considered:

$$\begin{cases} \dot{x}(t) = Ax(t) + A_d \int\limits_{t-\tau}^{t} x(w)dw + Bu(t) + f_1(t, x_t), \\ x(t) = \phi(t), \text{ for } t \in [-\tau, 0], \ \tau = \tau(t) \end{cases} \tag{1}$$

The following notations are used: $x(t) \in \mathbb{R}^n$; A and A_d are constant $n \times n$ matrices; B is a $n \times m$ matrix, $u \in \mathbb{R}^m$ is the input vector; f_1 is a term representing the neglected dynamics and external disturbances,τ is a uniformly bounded delay ($\tau \leq \tau_{sup}$).. x_t is the function associated with x and defined on $[-\tau, 0]$ by $x_t(\theta) = x(t+\theta)$. ϕ is the initial function defined on $[-\tau_{sup}, 0]$ and can be eventually discontinuous.

We will denote $||e||$ the Euclidean norm of the n-vector e, and $||M|| = \sup\limits_{||e||=1} ||Me||$ the spectral norm of the $n \times n$ matrix M. Finally $\widetilde{M}$ will denote an orthogonal complement of M.

We will use the following assumptions:

A1) The pair of matrix (A, B) is controllable.
A2) The perturbation term f_1 satisfies the classical *matching conditions*, *i.e.*

$$f_1(x(t), t) = Bf(x(t), t), \tag{2}$$

and is bounded as follows

$$||f|| < \Psi(x_t), \tag{3}$$

where Ψ is a known functional of x_t.
A3) B is full-rank: $rank(B) = m$.

3. REGULAR FORM

The aim is to design a sliding mode controller for (1). We first transform the original system into a special form, appropriate for sliding mode control (classically called *regular form* (Lukyanov and Utkin, 1981)). Let us choose the following sliding surface

$$s(x) = Sx = B^T X^{-1} x = 0,$$

where $S \in \mathbb{R}^{(n-m)\times n}$, and define a nonsingular transformation

$$z(t) = Mx(t), \quad \det M \neq 0, \tag{4}$$

where

$$M = \begin{bmatrix} \tilde{B}^T \\ S \end{bmatrix}. \tag{5}$$

Lemma 1. (Choi, 1997) The original system (1) is equivalent to

$$\dot{z}(t) = \hat{A}z(t) + \hat{A}_d \int_{t-\tau}^{t} z(w)dw + \hat{B}(u(t) + f), \tag{6}$$

where

$$\begin{aligned} \hat{A} &= \begin{pmatrix} \hat{A}_{11} & \hat{A}_{12} \\ \hat{A}_{21} & \hat{A}_{22} \end{pmatrix} \\ &= \begin{pmatrix} \tilde{B}^T AX\tilde{B}(\tilde{B}^T X\tilde{B})^{-1} & \tilde{B}^T AB(SB)^{-1} \\ SAX\tilde{B}(\tilde{B}^T X\tilde{B})^{-1} & SAB(SB)^{-1} \end{pmatrix}, \\ \hat{A}_d &= \begin{pmatrix} \hat{A}_{d11} & \hat{A}_{d12} \\ \hat{A}_{d21} & \hat{A}_{d22} \end{pmatrix} \\ &= \begin{pmatrix} \tilde{B}^T A_d X\tilde{B}(\tilde{B}^T X\tilde{B})^{-1} & \tilde{B}^T A_d B(SB)^{-1} \\ SA_d X\tilde{B}(\tilde{B}^T X\tilde{B})^{-1} & SA_d B(SB)^{-1} \end{pmatrix}, \\ \hat{B} &= \begin{pmatrix} 0 \\ SB \end{pmatrix}. \end{aligned} \tag{7}$$

In the following, we first present a sliding mode control for a system with one constant delay, then the result is extended to the case of systems with a time-varying delay.

4. SLIDING MODE CONTROL SYNTHESIS: CASE OF A CONSTANT DELAY

In this section, we consider that the delay τ is constant but unknown.

Theorem 2. Let $\Lambda \in \mathbb{R}^{n\times n}$ be an Hurwitz matrix, P is the solution of the Lyapunov equation

$$\Lambda^T P + P\Lambda = -I,$$

and $m_1 > 0$ a real number.

Assume that conditions $(A1) - (A2) - (A3)$ hold. Then, with the control law

$$u(t) = -(SB)^{-1}(-\Lambda s(x) + SAx(t) + m\frac{Ps}{||Ps||}), \tag{8}$$

where

$$m = m_1 + \sup_{\theta \in [-\tau_{\max}, 0]} \left(\begin{array}{c} ||SB|| \, \Psi(x_t(\theta)) \\ + \left\| SA_d \int_\theta^0 x(t+w)dw \right\| \end{array} \right)$$

, the surface $s(x) = 0$ is globally attractive in finite time, and the solution $x = 0$ of the system (6) is asymptotically stable for $\tau \in [0, \tau_{\max}]$, where $\tau_{\max}$ is the solution of the following optimization problem :

$$\tau_{max}^{-1} = min(\tau^{-1}), \tag{9}$$

under the constraints ($\sigma, \alpha \in \mathbb{R}$) :

$$\begin{bmatrix} \tau^{-1}\Theta + Y & A_d X \\ X A_d^T & -Y \end{bmatrix} - \sigma \begin{bmatrix} BB^T & 0 \\ 0 & BB^T \end{bmatrix} < 0,$$
$$X > 0$$
$$Y > 0$$
$$Y - \alpha BB^T > 0 \tag{10}$$

where $\Theta = AX + XA^T$ and $X, Y \in \mathbb{R}^n$

Remark 3. In practice, the stable matrix Λ allows to choose the dynamics of s when the solution of the system is far from the manifold $s(x) = 0$.

The proof is decomposed into two subproblems: firstly, prove the attractivity of the surface in finite time; secondly, prove the asymptotic stability of the reduced system (on the surface).

4.1 *Attractivity of the surface*

Proof. Let us consider the function

$$V(t) = s^T(x(t))Ps(x(t)), \tag{11}$$

Its derivative along the trajectories of (6) with (8) is

$$\dot{V}(t) = -s^T(x)s(x) + 2s^T(x)P(SAx(t) \tag{12}$$
$$+ SA_d \int_{t-\tau}^{t} x(w)dw + SBf - m\frac{Ps}{||Ps||}), \tag{13}$$

$$\dot{V}(t) < -s(x)^T s(x) - 2m_1\sqrt{\lambda_{min}(P)}\sqrt{V(t)} \tag{14}$$
$$< -2m_1\sqrt{\lambda_{min}(P)}\sqrt{V(t)}. \tag{15}$$

This last inequality is known to prove the finite time convergence of the system (6) towards the surface (Utkin, 1991). ∎

4.2 *Asymptotic stability of the reduced system*

The vector z appearing in the regular form (6) is partitioned into $z = [z_1 \ z_2]^T$, where $z_1 \in \mathbb{R}^{n-m}, z_2 \in \mathbb{R}^m$.

Once in sliding mode, the equations $s(x) = \dot{s}(x) = 0$ lead to the reduced system:

$$\dot{z}_1(t) = \hat{A}_{11}z_1(t) + \hat{A}_{d11}\int_{t-\tau}^{t} z_1(w). \tag{16}$$

Proof. Let us choose the following Lyapunov-Krasovskii's functional

$$V = z_1^T(t)Pz_1(t) + \int_{t-\tau}^{t}\int_{w}^{t} z_1(c)^T R z_1(c)dcdw, \tag{17}$$

where P, $R \in R^{(n-m)\times(n-m)}$ are positive definite matrices.

The derivative of V along the solutions of the reduced system (16) gives us :

$$\dot{V}(z_{1t}) = z_1^T(t)\left(\hat{A}_{11}^T P + P\hat{A}_{11}\right)z_1(t)$$
$$+2z_1^T(t)P\hat{A}_{d11}\int_{t-\tau}^{t} z_1(w)dw$$
$$-\int_{t-\tau}^{t} z_1^T(w)Rz_1(w)dw + \tau z_1^T(t)Rz_1(t)dw.$$

By using the following inequality :

$$2z_1(t)P\hat{A}_{d11}\int_{t-\tau}^{t} z_1(w)dw \leq \tau z_1(t)P\hat{A}_{d11}R^{-1}\hat{A}_{d11}^T P z_1(t)$$
$$+\int_{t-\tau}^{t} z_1^T(w)Rz_1(w)dw,$$

where R is a positive definite matrix, we obtain an estimate of $\dot{V}$:

$$\dot{V}(z_{1t}) \leq z_1^T(t)Nz_1(t),$$

with $N(\tau) = \hat{A}_{11}^T P + P\hat{A}_{11} + \tau R + \tau P\hat{A}_{d11}R^{-1}\hat{A}_{d11}^T P$. If there exists P and R two positive definite matrices and a real $\tau_{\max}$ such that $N(\tau_{\max})$ is a negative definite matrix, then the reduced system is asymptotically stable for all delays $\tau \in [0, \tau_{\max}]$. To optimize the upper bound $\tau_{\max}$, let post and pre-multiply N by $S = P^{-1}$ and take $P = (\tilde{B}^T X\tilde{B})^{-1}$. The inequality $N < 0$ becomes

$$S\hat{A}_{11}^T + \hat{A}_{11}S + \tau SRS + \tau P\hat{A}_{d11}R^{-1}\hat{A}_{d11}^T < 0,$$

or

$$\tilde{B}^T AX\tilde{B} + \tilde{B}^T XA^T\tilde{B} + \tau Q$$
$$+\tau\tilde{B}^T A_d X\tilde{B}Q^{-1}\tilde{B}^T XA_d^T\tilde{B} < 0,$$

with $Q = SRS$. By choosing $Q = (\tilde{B}^T Y\tilde{B})$, with Y a positive definite matrix, and by using schur's complement, the inequality is known to be equivalent to (10). ∎

5. SLIDING MODE CONTROL SYNTHESIS: CASE OF A TIME-VARYING DELAY.

In this section, we consider that the delay $\tau = \tau(t)$ is varying, but remains uniformly bounded:

Theorem 4. Let $\Lambda \in \mathbb{R}^{n\times n}$ be an Hurwitz matrix, P is the solution of the Lyapunov equation

$$\Lambda^T P + P\Lambda = -I,$$

and $m_1 > 0$ a real number.

Assume that conditions $(A1) - (A2) - (A3)$ hold. Then, with the control law (8) the surface $s(x) = 0$ is globally attractive in finite time, and the system is asymptotically stable for $\tau \in [0, \tau_{\max}]$, where $\tau_{\max}$ is the solution of the following optimization problem :

$$\tau_{max}^{-1} = min(\tau^{-1}), \tag{18}$$

under the constraints ($\sigma \in \mathbb{R}$) :

$$\begin{bmatrix} \Theta & A_d X \\ X A_d^T & -X \end{bmatrix} - \sigma \begin{bmatrix} BB^T & 0 \\ 0 & BB^T \end{bmatrix} < 0, \quad (19)$$
$$X > 0,$$

where $\Theta = \tau^{-1}\left(AX + XA^T\right) + X$, $X \in \mathbb{R}^{n \times n}$.

As previously, the proof is decomposed into two subproblems: firstly, prove the attractivity of the surface in finite time; secondly, prove the asymptotic stability of the reduced system (on the surface). As the proof of the attractivity of the surface remains the same, we do not repeat it. The proof of the second part is the following:

Proof. Once in sliding mode, the equations $s(x) = \dot{s}(x) = 0$ lead to the reduced system (16). Let us choose the following function of Lyapunov-Razumikhin :

$$V(t) = z_1^T(t) P z_1(t), \quad (20)$$

where $P \in \mathbb{R}^{(n-m)\times(n-m)}$ is a positive definite matrix.
following the approach proposed by Razumikhin, let us assume that the following inequality :

$$V(x(t+\theta)) < qV(x(t)), \quad (21)$$

with $q > 1$ is satisfied for $\theta \in [-\tau(t), 0]$.
The derivative of V along the solutions of the reduced system (16) give us :

$$\dot{V}(z_{1t}) = z_1^T(t)\left(\hat{A}_{11}^T P + P\hat{A}_{11}\right) z_1(t)$$
$$+2z_1^T(t) P\hat{A}_{d11}\int_{t-\tau}^{t} z_1(w)dw.$$

By using the following inequality :

$$2z_1(t)P\hat{A}_{d11}\int_{t-\tau}^{t} z_1(w)dw \le \tau z_1 P\hat{A}_{d11}P^{-1}\hat{A}_{d11}^T P z_1$$
$$+\int_{t-\tau}^{t} z_1^T(w) P z_1(w)dw,$$

we derive that :

$$\dot{V}(z_{1t}) \le z_1^T(t) N z_1(t),$$

where $N(\tau) = \begin{pmatrix} \hat{A}_{11}^T P + P\hat{A}_{11} + \tau q P \\ +\tau P\hat{A}_{d11}P^{-1}\hat{A}_{d11}^T P \end{pmatrix}$.
If there exists a positive definite matrix P and a real $\tau_{\max}$ such that $N(\tau_{\max})$ is negative definite, then the reduced system is asymptotically stable for $\tau \in [0, \tau_{\max}]$.
To optimize the upper bound $\tau_{\max}$, let us post and pre-multiply N by $S = P^{-1}$ and let us assume that $P = (\tilde{B}^T X \tilde{B})^{-1}$. The inequality $N < 0$ becomes

$$S\hat{A}_{11}^T + \hat{A}_{11}S + \tau q S + \tau P\hat{A}_{d11}SPS\hat{A}_{d11}^T < 0,$$

or, equivalently

$$\tilde{B}^T A X \tilde{B} + \tilde{B}^T X A^T \tilde{B} + \tau q S$$
$$+\tau \tilde{B}^T A_d X \tilde{B} P \tilde{B}^T X A_d^T \tilde{B} < 0. \quad (22)$$

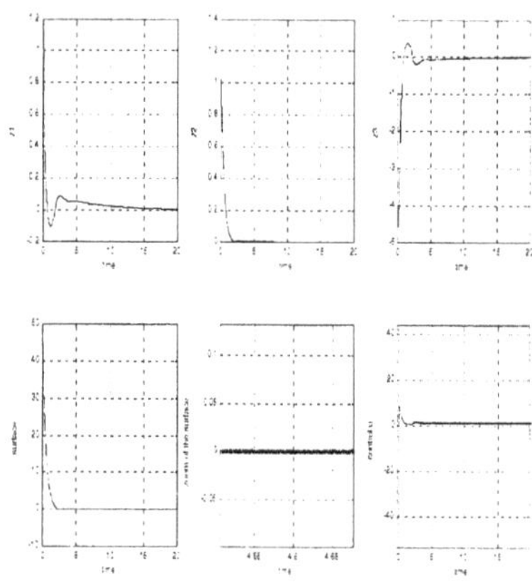

Fig. 1. Response of (24) with control (8) and $\tau = 1.99$.

By using the Schur's complement, we derive that (22) is equivalent to

$$\begin{bmatrix} \tau^{-1}\left(AX + XA^T\right) + qX & A_d X \\ X A_d^T & -X \end{bmatrix} - \sigma \begin{bmatrix} BB^T & 0 \\ 0 & BB^T \end{bmatrix} < 0. \quad (23)$$

For $q = 1$,the inequality (23) is exactly the constraint (19). By continuity, if we assume that there exists $X, \tau, \sigma, \alpha, \beta$ such that (19) is satisfied, then there exists $q_1 > 1$ such that (23) is satisfied too. ■

6. EXAMPLE

Consider system (1) with

$$A = \begin{pmatrix} 2 & 0 & 1 \\ 1.75 & 0.25 & 0.8 \\ 0 & 2 & 3 \end{pmatrix}, A_d = \begin{pmatrix} -1 & 0 & 0 \\ -0.1 & 0.25 & 0.2 \\ 1 & -2 & -1 \end{pmatrix}, \quad (24)$$

$$B, = \begin{pmatrix} 0 \\ 0 \\ 1 \end{pmatrix}, p = \begin{pmatrix} 0 \\ 0 \\ \sin(z_1(t-\tau)) \end{pmatrix}$$

and $z(t) = \begin{bmatrix} 1 \\ 1 \\ 1 \end{bmatrix}$ for $t \in [-\tau, 0]$. (25)

Constant delay: By using semi-definite programming and theorem 2, we find that the system (24) with control (8) is asymptotically stable for all constant delays $\tau < 1.99$. The simulation provided in Fig.1.was obtained with a first-order integration scheme of step 0.001.

Varying delay: Now, if we don't restrict to constant delays, then according to Theorem 4, the system is asymptotically stabilized for $\tau(t) < 0.83$ which is of course more constraining. The simulation leads to Fig.2-Fig.3.

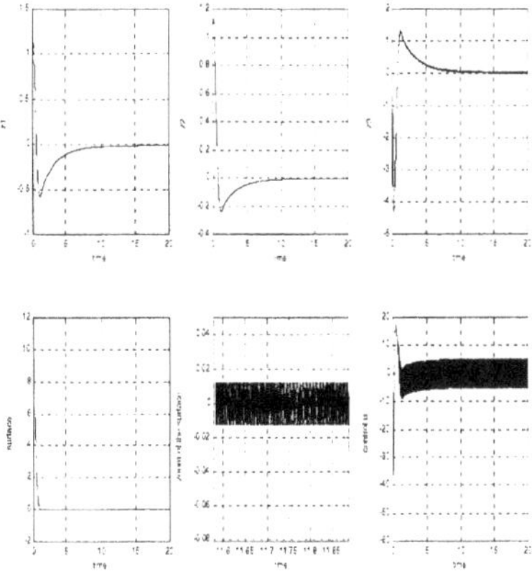

Fig. 2. Response of (24) with control (8) and a time varying delay $\tau_{max}(t) = 0.83$.

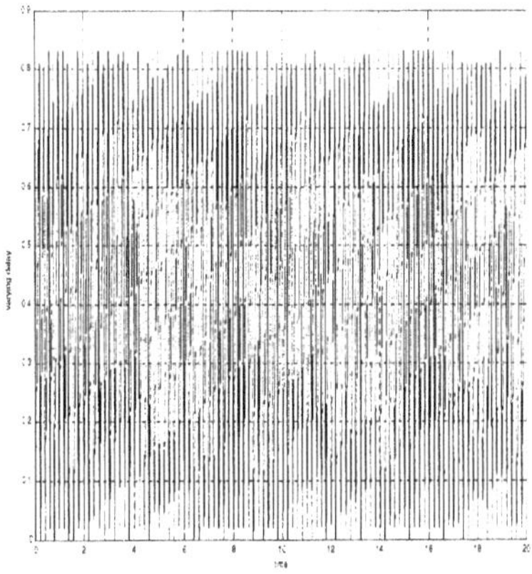

Fig. 3. Implemented delay (saw)

7. CONCLUSION

In this paper, we have developped a sliding mode control appoach to distributed time delay systems. Two cases were considered : the case of a constant distributed delay and the case of a time varying delay. Both approaches lead to LMI optimization problems, which can be solved very efficiently.

8. REFERENCES

Choi, H.H. (1997). A new method for variable structure control system design : A linear matrix inequality approach. *Automatica* **33**, 2089–2092.

Choi, H.H. (1999). An LMI approach to sliding mode control design for a class of uncertain time delay systems. In: *Proc. ECC'99*. Karlsruhe, Germany.

Fiagbedzi, Y.A. and A.E. Pearson (1987). A multistage reduction technique for feedback stabilizing distributed time-lag systems. *Automatica* **23**(3), 311–326.

Fridman, L.M., E. Fridman and E.I. Shustin (1996). Steady modes and sliding modes in the relay control systems with time delay. In: *35th IEEE CDC'96*. Kobe, Japan. pp. 4601–4606.

Gouaisbaut, F., M. Dambrine and J.P. Richard (2001). Sliding mode control of perturbed systems with time varying delays. In: *Proc. 1th IFAC Symposium on System Structure and Control, SSSC'01*. Prague.

Gouaisbaut, F., W. Perruquetti, Y. Orlov and J.P. Richard (September 1999). A sliding mode controller for linear time delay systems. In: *Proc. ECC'99 European Control Conference*. Karlsruhe.

Kolmanovskii, V.B. and A. Myshkis (1999). *Introduction to the theory and applications of functional differential equations*. Kluwer Acad.. Dordrecht.

Koshkouei, A.J. and A.S.I. Zinober (1996). Sliding mode time-delay systems.. In: *Proc. International Workshop on VSS*. Toskyo. pp. 97–101.

Lukyanov, A.G. and V.I. Utkin (1981). Methods of reducing equations of dynamics systems to regular form. *Automat. Remote Control* **42**, 413–420.

Misawa, E., A. and V. Utkin (2001). Special issue on sliding mode control. *Journal of Dynamic Systems, Measurement and Control.*

Orlov, Y.V. (2000). Discontinuous unit feedback control of uncertain infinite-dimensional systems. *IEEE Trans. Aut. Control* **45**(5), 834–843.

Pearson, A.E. and A. A. Pandiscio Jr (1997). Control of time lag systems via reducing transformations. In: *Imacs'97 World Congress*. Berlin. pp. 9–14.

Perruquetti, W. and J.P. Barbot (2001 (to appear)). *Sliding Mode Control in Engineering*. Marcel Dekker.

Shyu, K.K. and J.J. Yan (1993). Robust stability of uncertain time-delay systems and its stabilization by variable structure control. *Int. J. of Control* **57**, 237–246.

Utkin, V.I. (1991). *Sliding Modes in Control Optimization*. CCES Springer-Verlag.

Zheng, F., M. Cheng and W-B. Gao (1995). Variable structure control of time-delay systems with a simulation study on stabilizing combustion in liquid propellant rockect motors. *Automatica* **31**(7), 1031–1037.

www.elsevier.com/locate/ifac

STABILITY OF TIME VARYING NONLINEAR DIFFERENCE EQUATIONS WITH CONTINUOUS TIME

P. Pepe

Dipartimento di Ingegneria Elettrica, Università dell'Aquila, Monteluco di Roio, 67040 L'Aquila, Italy, e-mail: pepe@ing.univaq.it

ABSTRACT:

The stability of Time Varying Nonlinear Difference Equations with Continuous Time is here addressed. Lyapunov theorems are built up for the absolute stability of the origin. Among many other cases such as economic models, these systems are encountered as input output linearizable nonlinear delay systems internal dynamics. Indeed, when a nonlinear delay system admits full delay relative degree, the output and its derivatives until a suitable order can be driven to zero by a suitable control law, while the system variables depend on such output and derivatives and on their past values, on continuous time. This paper shows a Lyapunov methodology to check such internal dynamics. An example is reported of linearizable nonlinear delay system, which has a stable internal dynamics, proved by the methodology here shown.

Keywords: Difference Equations with Continuous Time, Nonlinear Delay Systems.

1. INTRODUCTION

In this paper the stability of a class of nonlinear time varying difference equations with continuous time is addressed, usefull for the study of the internal dynamics in linearizable nonlinear delay systems. Results concerning existence and uniqueness of the solution and stability of the origin are found in [10], in the linear case. In [14] the absolute stability, that is independent of the delay, of linear difference equations with continuous time is studied, and sufficient conditions are found, first of all the Schur stability of the dynamic matrix obtained if all delays were null. In [20] the asymptotic stability is studied of linear difference equations with continuous time and delay, under random perturbations of the coefficients. In [30] asymptotic properties of the solutions of difference equations with continuous argument are treated. Solutions are selected of relaxation and turbulence types.

In this paper time varying nonlinear difference equations with continuous time are addressed and Lyapunov theorems are built up for the absolute stability of the origin. Nonlinear difference equations are considered too with a parameter, which in nonlinear delay systems internal dynamics can be ragarded as the initial value of the normal form new state vector (euclidean part).

2. CONTINUOUS TIME DIFFERENCE EQUATIONS

Time varying nonlinear difference equations with continuous time and delay here considered are as follows

$$\begin{aligned} x(t) &= \phi(t, x(t-\Delta), x(t-2\Delta), \ldots, x(t-n\Delta)) \\ & \quad t \geq 0 \\ x(\tau) &= x_0(\tau), \qquad \tau \in [-n\Delta, 0] \end{aligned} \tag{1}$$

where $\Delta > 0$ is the delay, $t \in I\!R^+$, $x(t) \in I\!R^n$, ϕ is a continuous function from $I\!R^+ \times I\!R^{n^2}$ to $I\!R^n$, x_0 is

This work is supported by the Italian National Research Council CNR Grant 2030728, it was developed mainly at the Department of Electrical Engineering and Computer Sciences of the University of California at Berkeley, where the author was in 2000 visiting scholar.

an absolutely continuous function with square integrable derivative. Let us here report the stability definition (see [10] for a treatise). We suppose that

$$\phi(t, 0, \dots, 0) = 0 \qquad t \geq 0 \tag{2}$$

Definition 1. The null solution of the system (1) is said to be stable if for any $\epsilon > 0$ there exist $\delta > 0$ such that $||x_0||_\infty < \delta$ implies $||x(t)|| < \epsilon$, $t \geq 0$. ∎

In this paper we consider too a slightly different equation, by which the behaviour of internal dynamics for a delay input output linearized system can be studied, in the hypothesis of full delay relative degree [9].

$$\begin{aligned} x(t) &= \phi(t, \Theta, x(t-\Delta), x(t-2\Delta), \dots, x(t-n\Delta)) \\ & \qquad t \geq 0 \\ x(\tau) &= x_0(\tau), \qquad \tau \in [-n\Delta, 0] \end{aligned} \tag{3}$$

where, as in (1), $\Delta > 0$ is the delay, $t \in I\!R^+$, $x(t) \in I\!R^n$,x_0 is an absolutely continuous function with square integrable derivative and ϕ is now a continuous function from $I\!R^+ \times I\!R^n \times I\!R^{n^2}$ to $I\!R^n$. $\Theta \in I\!R^n$ is a vector of parameters. Such parameter vector can be regarded as the initial value in [9] of the normal form new euclidean state vector, composed of the output and its derivatives until $n-1$. Such initial value depends on initial values $x(0)$, $x(-\Delta), \dots, x(-(n-1)\Delta)$, so that any condition on Θ can be generally transformed in a condition on x_0 [9].

In the hypothesis that

$$\phi(t, 0, 0, \dots, 0) = 0 \qquad t \geq 0 \tag{4}$$

let us consider the following definition for system (3), usefull for studying nonlinear delay systems internal dynamics [9].

Definition 2. The null solution of the system (3) with $\Theta = 0$ is said to be stable if, for any $\epsilon > 0$, there exist $\delta > 0$ such that $||x_0||_\infty < \delta$ and $||\Theta|| < \delta$ imply, for the solution $x(t)$ of (3), $||x(t)|| < \epsilon$, $t \geq 0$. ∎

3. STABILITY BY LYAPUNOV THEORY

System (1) can be reduced, by using an extended state,

$$X(t) = [\, x(t-(n-1)\Delta) \quad \dots \quad x(t) \,]^T \tag{5}$$

to the following one

$$X(t) = A_B X(t-\Delta) + B_B \phi(t, X(t-\Delta)) \tag{6}$$

where $X(t) \in I\!R^N$, $N = n^2$, A_B and B_B are Brunowsky matrices.

So let us consider stability for the difference equation with continuous time

$$\begin{aligned} X(t) &= \Phi(t, X(t-\Delta)) \qquad t \geq 0 \\ X(\tau) &= X_0(\tau), \qquad \tau \in [-\Delta, 0] \end{aligned} \tag{7}$$

where $X(t) \in I\!R^N$, Φ is a continuous function from $I\!R^+ \times I\!R^N$ to $I\!R^N$, X_0 is an absolutely continuous function with square integrable derivative.

It can be easily recognized that stability of the null solution for system (1) implies and is implied by stability of the null solution of (7).

Let us here recall [28] that a function $\alpha(\cdot) : I\!R^+ \mapsto I\!R^+$ is said to be of class K if it is continuous, strictly increasing and $\alpha(0) = 0$.

Theorem 3. *Let there exists a continuous function $V(t, X) : [-\Delta, \infty) \times I\!R^N \to I\!R^+$ such that*

i] $V(t, 0) = 0 \qquad \forall\, t \geq -\Delta$;

ii] there exists a function α of class K, such that for a suitable $r_1 > 0$ it is

$$V(t, X) \geq \alpha(||X||), \qquad \forall\, ||X|| < r_1,\ t \geq -\Delta \tag{8}$$

iii] $\forall\, \epsilon > 0$, *there exists $\delta > 0$ such that*

$$\sup_{||X||<\delta, \tau\in[-\Delta,0]} V(\tau, X) < \epsilon \tag{9}$$

iiii] there exists a suitable $r_2 > 0$, such that

$$V(t, \Phi(t, X)) \leq V(t-\Delta, X) \qquad \forall\, ||X|| < r_2, \quad t \geq 0$$

Then, the null solution of the difference equation with continuous time (7) is stable. ∎

Let us consider now the system (3), which can be transformed in the following difference equation

$$\begin{aligned} X(t) &= \Phi(t, \Theta, X(t-\Delta)) \qquad t \geq 0 \\ X(\tau) &= X_0(\tau), \qquad \tau \in [-\Delta, 0] \end{aligned} \tag{10}$$

where $X(t) \in I\!R^N$, Φ is a continuous function from $I\!R^+ \times I\!R^n \times I\!R^N$ to $I\!R^N$, X_0 is an absolutely continuous function with square integrable derivative.

Theorem 4. *Let there exists a continuous function $V(t, \Theta, X) : [-\Delta, \infty) \times I\!R^n \times I\!R^N \to I\!R^+$ such that*

i] $V(t, 0, 0) = 0 \qquad \forall\, t \geq -\Delta$;

ii] there exists a function α of class K, such that for suitable $r_0 > 0,\ r_1 > 0$ it is

$$V(t,\Theta,X) \geq \alpha(\|X\|), \quad \forall\ \|\Theta\| < r_0,\ \|X\| < r_1,\ t \geq -\Delta \tag{11}$$

iii] $\forall\ \epsilon > 0$, there exists $\delta > 0$ such that

$$\sup_{\|X\|<\delta,\|\Theta\|<\delta,\tau\in[-\Delta,0]} V(\tau,\Theta,X) < \epsilon \tag{12}$$

[iiii] there exist suitable $R_0 > 0, R_1 > 0$, such that

$$V(t,\Theta,\Phi(t,\Theta,X)) \leq V(t-\Delta,\Theta,X) \quad \forall\ \|\Theta\| < R_0,\ \|X\| < R_1,\ t \geq 0 \tag{13}$$

Then, the null solution of the difference equation with continuous time (10) with $\Theta = 0$ is stable.

4. ON A BANACH SPACE SETTING

Let us consider (10). Let $C([-\Delta,0];I\!R^N)$ denote the set of continuous functions from $[-\Delta,0] \to I\!R^N$, endowed with the sup norm. Let $X_k \in C([-\Delta,0];I\!R^N)$, $k = 0,1,2,\ldots$. System (10) can be rewritten in discrete time form as follows

$$X_{k+1} = \Psi(k,\Theta,X_k) \tag{14}$$

where $\Psi : Z^+ \times I\!R^n \times C([-\Delta,0];I\!R^N) \to C([-\Delta,0];I\!R^N)$,

$$\begin{aligned} X_k(\tau) &= x(k\Delta+\tau) \\ \Psi(k,\Theta,X_k)(\tau) &= \Phi(k\Delta+\Delta+\tau,\Theta,X_k(\tau)) \end{aligned} \tag{15}$$

Standard Lyapunov theorems for discrete time systems evolving on Banach Spaces can be used to check stability of (10), easily adapted to take into account the parameter Θ.

Theorem 5. *Let in (10)*

$$\Phi(t,\Theta,X(t-\Delta)) = \bar{\Phi}(z(t,\Theta),X(t-\Delta)) \qquad t \geq 0 \tag{16}$$

where $\bar{\Phi} : I\!R^n \times I\!R^N \to I\!R^N$ and $z : I\!R^+ \times I\!R^n \to I\!R^n$ are continuous functions

Let

$$\begin{aligned} \xi_c(t,\Theta) &= [\,|z_{1,t,\Theta}|_\infty \quad |z_{2,t,\Theta}|_\infty \quad \ldots \quad |z_{n,t,\Theta}|_\infty\,]^T \\ & \qquad t \geq \Delta \\ \chi_c(t) &= [\,|X_{1,t}|_\infty \quad \ldots \quad |X_{N,t}|_\infty\,]^T \qquad t \geq 0 \end{aligned} \tag{17}$$

where, as usual, $z_{i,t,\Theta} \in C([-\Delta,0];I\!R)$, $i = 1,2,\ldots,n$, $X_{i,t} \in C([-\Delta,0];I\!R)$, $i = 1,2,\ldots,N$ and

$$\begin{aligned} z_{i,t,\Theta}(\tau) &= z_i(t+\tau,\Theta) \\ X_{i,t}(\tau) &= X_i(t+\tau) \end{aligned}$$

Let $\xi(k,\Theta) = \xi_c((k+1)\Delta,\Theta)$, $k = 0,1,\ldots$.
Let there exist a vector of functions

$$l = [\,l_1 \quad l_2 \quad \ldots \quad l_N\,]^T : I\!R^{n+} \times I\!R^{N+} \mapsto I\!R^{N+} \tag{18}$$

such that

a] for any

$$\begin{aligned} (v,\bar{v}) &\in I\!R^{n+} \times I\!R^{n+} \\ (w,\bar{w}) &\in I\!R^{N+} \times I\!R^{N+} \end{aligned} \tag{19}$$

such that

$$v_i \leq \bar{v}_i,\ i = 1,2,\ldots,n, \tag{20}$$

and

$$w_i \leq \bar{w}_i,\ i = 1,2,\ldots,N, \tag{21}$$

the inequality holds

$$l_i(v,w) \leq l_i(\bar{v},\bar{w}),\ i = 1,2,\ldots,N; \tag{22}$$

b] there exist a positive real R such that, if $\|\Theta\| < R$ and, for some $\bar{t} \geq \Delta$,

$$|X_{j,\bar{t}-\Delta}|_\infty < R, \qquad j = 1,2,\ldots,N, \tag{23}$$

then

$$\begin{aligned} &\sup_{\tau\in[-\Delta,0]} |\bar{\Phi}_i(z(\bar{t}+\tau,\Theta),X(\bar{t}-\Delta+\tau))| \\ &\leq l_i\begin{pmatrix} \xi_c(\bar{t},\Theta) \\ \chi_c(\bar{t}-\Delta) \end{pmatrix} \qquad i = 1,2,\ldots,N \end{aligned} \tag{24}$$

Consider the following discrete time system

$$\chi(k+1) = l\begin{pmatrix} \xi(k,\Theta) \\ \chi(k) \end{pmatrix} \qquad k = 0,1,\ldots \tag{25}$$

Let us suppose that for any $\epsilon > 0$ there exist δ such that $\|\chi(0)\| < \delta$ and $\|\Theta\| < \delta$ imply $\|\chi(k)\| < \epsilon,\ \forall\ k \geq 0$.

Then, the null solution in (10) with $\Theta = 0$ is stable.
■

5. ROBUST EASY CONDITIONS

The hypotesis of this section is that for equation (1) the following inequality holds

$$\|x(t)\| \leq a_{n-1}\|x(t-\Delta)\| + \ldots + a_0\|x(t-n\Delta)\| \tag{26}$$

where a_i, $i = 0, 1, \ldots, n-1$ are non negative unknown constants belonging to known intervals $[a_i^-, a_i^+]$, $0 \le a_i^- \le a_i^+$, $i = 0, 1, \ldots, n-1$.

In this case stability, indeed asympotic stability, and robust, that is independent of coefficients in their intervals, can be studied by the stability of the family of polynomials

$$\begin{aligned} &z^n - a_{n-1}z^{n-1} - \ldots - a_1 z - a_0 \\ &a_i \in [a_i^-, a_i^+] \quad 0 \le a_i^- \le a_i^+ \quad i = 0, 1, \ldots, n-1 \end{aligned} \tag{27}$$

Schur stability of (27) implies stability of the difference equation satisfying (26) (see [9] and references therein).

A methodology to study the Schur stability of the polynomial family (27) is to use the bilinear transformation such to transport the discrete time stability problem to the continuous time one and so solve it by the celebrated Kharitonov theorem [16]. This transportation however is obtained by a different parameterization which generates a new family of polynomials containing the original one, and so the stability of the new family is only sufficient for the stability of the original one [12].

In [12] it is proved that the Schur stability of the vertex polynomials of the family (27) is a necessary and sufficient condition for the Schur stability of the entire family, when roughly half the parameters, the half highest indexes ones, are exactly known. In [15] it is proved that, in the hypotheses of [12], the Schur stability of roughly $n+2$ vertex polynomials, clearly defined, is necessary and sufficient for the Schur stability of an entire diamond of polynomials.

A methodology for studying (27) is given by the Edge Theorem [3], which gives necessary and sufficient conditions. It requires the stability evaluation of $n * 2^{n-1}$ one parameter polynomials, the edge polynomials.

However, the polynomial family (27) is Schur asymptotically stable if and only if the condition

$$\sum_{i=0}^{n-1} a_i^+ < 1 \tag{28}$$

is verified.

A condition like (28), that is

$$\sum_{i=0}^{n-1} \sup(|a_i^-|, |a_i^+|) < 1 \tag{29}$$

may be usefull for studying stability of families of polynomials with any sign coefficients. Such condition (29) is sufficient, not necessary, for the stability of the whole family of polynomials (27). But it is immediate to be checked, and can avoid the edge theorem, while the bilinear transformation can fail and other extreme point results in literature cannot be applied because all coefficients are unknown.

Conditions (28) and (29) are the straigthforward trivial application to polynomial families like (27) of very well known results for polynomials (see [13],[22], see also [10], section 9.6) or can be regarded as the application of the zero exclusion principle. Maybe the necessity of condition (28) has not been pointed out in literature.

Here we just point out that there exist families of polynomials whose Schur stability can be easily proven by condition (29), while it cannot be proven by bilinear transformation, neither by extreme point results in literature. And the edge theorem computation time may well be much bigger than the checking time of condition (29). For instance, consider the following family

$$\begin{aligned} &z^5 + a_4 z^4 + a_3 z^3 + a_2 z^2 + a_1 z + a_0 \\ &a_0 \in [-0.5, 0.5] \quad a_1 \in [-0.15, 0.15] \\ &a_2 \in [-0.13, 0.13] \quad a_3 \in [-0.13, 0.13] \\ &a_4 \in [-0.07, 0.07] \end{aligned} \tag{30}$$

Other conditions can help for studying polynomial families stability, such as monotonic condition.

Condition (28) proves at sight the stability of the example given in [14].

The following known result (see [19], pp. 126, Thrm. 1.3, see also [9]) will be used in the example.

Proposition 6. *Let $zero(t)$, $t \ge 0$ be a non negative function going to zero, bounded by the constant M. Let a_i^+, $i = 0, 1, \ldots, n-1$ be non negative constants, verifying condition (28).*

Let, for $t \in [0, \infty)$,

$$\begin{aligned} \|x(t)\| \le &a_{n-1}(t)\|x(t-\Delta)\| + \ldots + a_0(t)\|x(t-n\Delta)\| \\ &+ zero(t) \end{aligned} \tag{31}$$

where functions $a_i(t)$, $i = 0, 1, \ldots, n-1$ are non negative and verify

$$a_i(t) \le a_i^+ \qquad i = 0, 1, \ldots, n-1 \tag{32}$$

Then

$$\begin{aligned} &1) \ \|x(t)\| \le \frac{1}{1 - \sum_{i=0}^{n-1} a_i^+} \cdot \\ &\qquad (M + \sum_{i=0}^{n-1} a_i^+ \sup_{-n\Delta \le \tau \le 0} \|x(\tau)\|); \\ &2) \ \lim_{t \to \infty} \|x(t)\| = 0. \end{aligned}$$

■

6. EXAMPLE

Let us consider the following continuous time nonlinear delay system

$$\begin{aligned}
&\dot{x}_1(t) = x_2(t) + 1 - (x_2(t-\Delta)+1)^{\frac{1}{3}} \qquad t \geq 0 \\
&\dot{x}_2(t) = u(t) \\
&y(t) = x_1(t) \\
&x(\tau) = x_0(\tau) \qquad \tau \in [-\Delta, 0]
\end{aligned} \tag{33}$$

Let

$$z(t) = \begin{pmatrix} x_1(t) \\ x_2(t) + 1 - (x_2(t-\Delta)+1)^{\frac{1}{3}} \end{pmatrix} \tag{34}$$

The following control law, obtained by methodologies in [9],

$$u(t) = \begin{cases} 0 & 0 \leq t < \Delta \\ \frac{u(t-\Delta)}{3(1+x_2(t-\Delta))^{\frac{2}{3}}} + Kz(t) & t \geq \Delta \end{cases} \tag{35}$$

where K is a suitable constant matrix, is such that $z(t) = e^{A(t-\Delta)}\Theta$, $t \geq \Delta$, with A Hurwitz and

$$\Theta = \begin{pmatrix} x_1(\Delta) \\ x_2(\Delta) + 1 - (1 + x_2(0))^{\frac{1}{3}} \end{pmatrix} \tag{36}$$

By

$$\begin{aligned}
&|x_2(\Delta)| = |x_2(0)|; \\
&\sup_{\tau \in [-\Delta,0]} |x_1(\tau+\Delta)| \leq |x_1(0)| + \Delta |x_2(0)| + \\
&\qquad \Delta \sup_{\tau \in [-\Delta,0]} |1 - (1+x_2(\tau))^{\frac{1}{3}}|
\end{aligned} \tag{37}$$

it follows that $||\Theta||$ can be less than an arbitrary fixed positive number, provided $||x_0||$ is sufficiently small. The following difference equation with continuous time describes the internal dynamics,

$$\bar{x}(t) = e^{At}\Theta + \begin{bmatrix} 0 \\ -1 + (\bar{x}_2(t-\Delta)+1)^{\frac{1}{3}} \end{bmatrix} \qquad t \geq 0 \tag{38}$$

where

$$\bar{x}(t) = [\bar{x}_1(t) \quad \bar{x}_2(t)]^T = [x_1(t+\Delta) \quad x_2(t+\Delta)]^T \tag{39}$$

Le us use the following Lyapunov function

$$V(t, \Theta, \bar{x}) = \alpha M e^{\lambda t} ||\Theta||^2 + \beta ||\bar{x}||^2 \tag{40}$$

for suitable positive constants α and β, with M and λ such that

$$||e^{At}|| \leq M e^{\lambda t}$$

By applying theorem (4) it can be easily proved that the given example shows a stable internal dynamics. That is, while the output and its derivative ($z(t)$ in (34)) are driven to zero by the control law, the system state can stay in any fixed ball with center in the origin, provided the initial conditions have sufficiently small norm. The following limit can be usefull for the above proof

$$\lim_{\bar{x}_2 \to 0} \frac{|(1+\bar{x}_2)^{1/3} - 1|^2}{\bar{x}_2^2} = \frac{1}{9} \tag{41}$$

By proposition (6), the input (35) is bounded and goes to zero.

7. CONCLUSIONS

Here Lyapunov theorems are given for the stability of time varying nonlinear difference equations with continuous time. Such theorems can be usefull for checking internal stability of input output linearizable nonlinear delay sytems.

8. Acknowledgements

I wish to thank the Professor Alfredo Germani for discussions, criticism, advice, suggestions.

References

[1] B.R. Barmish, R. Tempo, C.V. Hollot, H.I.Kang, An Extreme Point Result for Robust Stability of a Diamond of Polynomials, *IEEE Transactions on Automatic Control*, Vol. 37, No. 9, September 92, pp. 1460–1462.

[2] B. Ross Barmish, Zhicheng Shi, Robust Stability of a Class of Polynomials with Coefficients Depending Multilinearly on Perturbations, *IEEE Transactions on Automatic Control*, Vol. 35, No. 9, September 1990, pp. 1040–1043.

[3] A. C.Bartlett, C.V. Hollot, Huang Lin, Root Locations of an Entire Polytope of Polynomials: it Suffices to Check the Edges, *Mathematics of Control, Signals, and Systems*, No. 1, 1988, pp. 61–71.

[4] F. Blanchini, R. Tempo, F. Dabbene, Computation of the Minimum Destabilizing Volume for Interval and Affine Families of Polynomials, *IEEE Transactions on Automatic Control*, Vol. 43, No. 8, August 98, pp. 1159–1163.

[5] N. K. Bose, E. Zeheb, *Proc. Instn. elect. Engrs, Pt D, 133, 187*

[6] Y. Domshlak, Oscillatory Properties of Linear Difference Equations with Continuous Time, *Differential Equations Dynam. Systems*, ,No. 4, 1993, pp. 311–324

[7] L. Dugard and E. Verriest (Eds), Stability and Control of Time Delay Systems, *Springer*, 1998

[8] Minyue Fu, Andrzej W. Olbrot, Michael P. Polis, Robust Stability for Time Delay Systems: the Edge Theorem and Graphical Tests, *IEEE Transactions on Automatic Control*, Vol. 34, No. 8, August 89, pp. 813–820.

[9] A. Germani, C. Manes and P. Pepe, Local Asymptotic Stability for Nonlinear State Feedback Delay Systems, *Kybernetika,*, Vol.36, n. 1, pp. 31-42, 2000

[10] J. K. Hale and S. M. Verduyn Lunel, Introduction to Functional Differential Equations, *Springer Verlag*, 1993

[11] R. Hernàndez, S. Dormido, R. Dormido, On the Thirty-Two Virtual Polynomials to Stabilize an Interval Plant, *IEEE Transactions on Automatic Control*, Vol. 43, No. 10, October 98, pp. 1460–1465.

[12] C. V. Hollot and A. C. Bartlett, Some Discrete-Time Counterparts to Kharinov's Stability Criterion for Uncertain Systems, *IEEE Transactions on Automatic Control*, Vol. AC-31, No. 4, April 1986, pp. 355–356.

[13] E. I. Jury, Theory and Application of the z-Transform Method, *New York: Wiley*, 1964.

[14] K. Kaiser, D. G. Korenevski, Algebraic Coefficient Conditions for the Absolute Stability of Linear Difference Systems with Continuous Time and Delay,(Russian) *Avtomat. i Telemekh.*, No. 1, 1998, pp. 22–27, translation in *Automat. Remote Control*, No. 1, part.1, 1998, pp. 17–21

[15] Hwan Il Kang, Some Discrete-Time Counterparts to Extreme Point Results for Robust Stability of a Diamond of Polynomials, *IEEE Transactions on Automatic Control*, Vol. 44, No. 8, August 1999, pp. 1613–1615.

[16] V. L. Kharitonov, Asymptotic Stability of an Equilibrium Position of a Family of Systems of Linear Differential Equations, *Differential'nye Uravneniya*, Vol. 14, No. 11, 1978, pp. 2086–2088.

[17] Vladimir L. Kharitonov, Alexei P. Zhabko, Robust Stability of Time Delay Systems, *IEEE Transactions on Automatic Control*, Vol. 39, No. 12, December 94, pp. 2388–2397.

[18] J. Kogan, Schur Stability of Interval Polynomials, *Int. J. Systems Sci.*, Vol. 25, No. 9, 1994, pp. 1405–1415.

19] V. Kolmanovskii and A. Myshkis, Applied Theory of Functional Differential Equations, *Kluwer Academic Publishers*, 1992

[20] D. G. Korenevski, Asymptotic Stability of Solutions of Systems of Linear Difference Equations with Continuous Time and with Delay Under Random Perturbations of the Coefficients,(Russian) *Dokl. Akad. Nauk.*, No. 1, 1999, pp. 13–16.

[21] D. G. Korenevski, Criteria for the Mean-Square Asymptotic Stability of Solutions of Systems of Linear Stochastic Difference Equations with Continuous Time and with Delay,(Russian) *Ukrain. Mat. Zh,*, No. 8, 1998, pp. 1073–1081 translation in *Ukrainian Math. J.* No. 8, 1998, pp. 1224–1232.

[22] M. Marden, Geometry of Polynomials, *RI: Amer. Math. Soc.*, 1966.

[23] Nikos E. Mastorakis, Comments Concerning Multidimensional Polynomials' Properties, *IEEE Transactions on Automatic Control*, Vol. 41, No. 2, February 1996, pp. 260.

[24] Kanh T. Ngo and Kelvin T. Erickson, Stability of Discrete-Time Matrix Polynomials, *IEEE Transactions on Automatic Control*, Vol. 42, No. 4, April 1997, pp. 538–542.

[25] P. Pepe, Some Results on Nonlinear Delay Control Systems - Delayless Obeservers, Adaptive Control, Stability, Discretization, *Research Report CNR Grant 2030728*, 2000

[26] Ian R. Petersen, A Class of Stability Regions for which a Kharitonov-Like Theorem Holds, *IEEE Transactions on Automatic Control*, Vol. 34, No. 10, October 1989, pp. 1111–1115.

[27] Anders Rantzer, Stability Conditions for Polytopes of Polynomials, *IEEE Transactions on Automatic Control*, Vol. 37, No. 1, January 1992, pp. 79–89.

[28] S. Sastry, Nonlinear Systems, Analysis, Stability, and Control, *Springer*, 1999

[29] A. N. Sharkovsky, Ideal Turbulence in an Idealized Time-Delayed Chua's Circuit, *Internat. J. Bifur. Chaos Appl. Sci. Engrg.*, ,No. 2, 1994, pp. 303–309

[30] A. N. Sharkovsky,Yu. L. Mastrenko,E. Yu. Romanenko, Difference Equations and their Applications. Translated from the 1986 Russian original by D.V. Malyshev, P.V. Malyshev and Y.M. Pestryakov, *Mathematics and its Applications, Kluwer Academic Publishers Group,* , Dordrecht, 1993

[31] Roberto Tempo, A Dual Result to Kharitonov's Theorem, *IEEE Transactions on Automatic Control*, Vol. 35, No. 2, February 1990, pp. 195–198.

[32] Yu. P. Yatsenko, A Study of an Integro-Functional Equation that Arises in Optimization Problems,(Russian) *Kibernetika, Kiev*, No. 6, 1989, pp. 117–118

www.elsevier.com/locate/ifac

TRIANGULAR FORMS FOR NONLINEAR TIME-DELAY SYSTEMS

Luis A. Marquez [*] **Claude H. Moog** [**]
E. Aranda-Bricaire [***]

[*] *CICESE Research Center. Dept. of Electronics and Telecommunications. Km 107 Carretera Tijuana–Ensenada. 22800 Ensenada B.C. MEXICO. E-mail: lmarquez@cicese.mx*
[**] *IRCCyN - UMR CNRS 6597. 1, rue de la Noë, B.P 92101. 44321 Nantes* CEDEX *3. FRANCE. E-mail: moog@irccyn.ec-nantes.fr*
[***] *Sección de Mecatrónica. Depto. de Ing. Eléctrica, CINVESTAV-IPN. Apartado Postal 14-740. 07000 Mexico, D.F., MEXICO. e-mail: earanda@mail.cinvestav.mx*

Abstract: In this paper we study some structural properties of nonlinear systems with delays. A specially adapted mathematical setting is developed and used to characterize the equivalence to the so-called triangular form. We claim that this result opens several research lines in the study of the considered class of systems.

Keywords: Time-delay, nonlinear systems, state-space models, structural properties.

1. INTRODUCTION

In this paper, we present a mathematical framework adapted for a broad class of nonlinear time-delay systems. These mathematical tools have been used to display some structural properties of nonlinear time-delay systems (Márquez-Martínez *et al.*, 2000; Márquez-Martínez, 1999; Márquez-Martínez and Moog, 1999). Now, they will be used to characterize the equivalence of a time-delay nonlinear system to the so-called triangular form.

Our motivation for considering this particular issue is that the notion of triangular form has proved to be useful for studying basic properties of nonlinear systems without delays.

Triangular form equivalence is a first step for full feedback linearization. It has also been used to study stability issues (Celikovský and Nijmeijer, 1996). An extension to the case of time-delay systems opens new research lines for the study of these problems.

The goals of this paper are: first, to investigate the structural properties of nonlinear time-delay systems by extending the notion of triangular forms to this class of systems. Second, to characterize whether a system can be transformed into this form. We claim that this will open several research lines in the previously mentioned directions.

The rest of this paper is organized as follows: in section 2 the class of systems under study is defined and the mathematical tools are introduced. In section 3, a necessary and sufficient condition for the equivalence of a triangular form is presented. Finally, some conclusions and open issues are discussed in section 4.

2. NOTATIONS AND PRELIMINARY DEFINITIONS.

In this section, the class of considered systems will be defined, and the mathematical setting to be used in this paper, which was introduced in (Moog *et al.*, 2000), will be recalled and completed.

2.1 *Considered systems*

Our approach is valid for systems with non-commensurable delays. Even if all of the contributions set forth may be extended to this case, we will restrict ourselves to the commensurable case, where all the delays are multiples of an elementary delay T. Furthermore, it will be assumed that the time axis has been scaled so as $T = 1$. Under these assumptions, the considered nonlinear time–delay systems (without output) are described by

$$\Sigma : \begin{cases} \dot{x}(t) = f(x(t-\tau), u(t-\tau), \tau \in \underline{S}), \\ \qquad \underline{S} := \{0, 1, \ldots, s\} \\ x(\tau) = \varphi, \quad u(\tau) = u_0, \quad \forall \tau \in [t_0 - s, t_0] \end{cases} \tag{1}$$

where only a finite number of *constant time delays* appear. The state $x \in I\!R^n$ and the input $u \in I\!R$. The entries of f are meromorphic functions of their arguments. The notation $f(x(t-\tau), u(t-\tau), \tau \in \underline{S})$ stands for $f(x(t), u(t), x(t-1), u(t-1), \ldots, x(t-s), u(t-s))$, for some $s \in I\!N$. The initial conditions are a continuous function φ.

2.2 *Mathematical setting*

Let $\mathcal{K}$ be the field of meromorphic functions of a finite number of variables in

$$\{x(t-\tau), u^{(k)}(t-\tau), \quad \tau \in \underline{S}\}.$$

These variables are independent in the sense that they are not related by any algebraic or trascendental equation. Let also $\mathcal{E}$ be the formal vector space over $\mathcal{K}$ given by

$$\mathcal{E} = \mathrm{span}_{\mathcal{K}} \{\mathrm{d}\,\xi \mid \xi \in \mathcal{K}\}.$$

Define the *shift* operators $\delta : \mathcal{K} \to \mathcal{K}$ and $\nabla : \mathcal{E} \to \mathcal{E}$ as:

$$\begin{aligned} \delta f(t) &:= f(t-1) \\ \nabla f(t) &:= f(t-1)\nabla \\ \nabla \mathrm{d} &:= \mathrm{d}\,\delta \end{aligned} \tag{2}$$

Let $\mathcal{K}[\nabla]$ be the (left) ring of polynomials in ∇ with coefficients on $\mathcal{K}$. Every element of this ring may be written as

$$\begin{aligned} \alpha(\nabla) &= \alpha_0(t) + \alpha_1(t)\,\nabla + \cdots + \alpha_{r_\alpha}(t)\,\nabla^{r_\alpha} \\ & \alpha_i(t) \in \mathcal{K} \end{aligned}$$

where r_α is the polynomial degree of $\alpha(\nabla)$. Addition and product on this ring are defined by

$$\begin{aligned} \alpha(\nabla) + \beta(\nabla) &= \sum_{i=0}^{\max\{r_\alpha, r_\beta\}} (\alpha_i(t) + \beta_i(t))\nabla^i \\ \alpha(\nabla)\beta(\nabla) &= \sum_{i=0}^{r_\alpha}\sum_{j=0}^{r_\beta} \alpha_i(t)\beta_j(t-i)\nabla^{i+j} \end{aligned}$$

The following technical results have been proved in (Márquez-Martínez, 2000).

Lemma 1. For every $a(\nabla)$, $b(\nabla) \in \mathcal{K}[\nabla]$ there exist non-zero polynomials $\alpha(\nabla)$, $\beta(\nabla) \in \mathcal{K}[\nabla]$ such that

$$\alpha(\nabla)a(\nabla) + \beta(\nabla)b(\nabla) = 0 \tag{3}$$

Corollary 2. For every $a(\nabla)$, $b(\nabla) \in \mathcal{K}[\nabla]$ there exist $\alpha(\nabla)$, $\beta(\nabla) \in \mathcal{K}[\nabla]$ such that

$$\alpha(\nabla)a(\nabla)b(\nabla) = \beta(\nabla)b(\nabla)a(\nabla).$$

In the case of linear time delay systems, the notion of closure, introduced in (Conte and Perdon, 1995), has proved to be a powerful tool. In the nonlinear case, this notion is extended accordingly.

Definition 3. Let A be a submodule of $\mathcal{M}$ defined over a ring $\mathcal{R}$. Its closure over this ring, or $\mathcal{R}$-closure is defined as

$$cls_{\mathcal{R}}\{A\} = \{x \in \mathcal{M} \mid \exists P \in \mathcal{R}, Px \in A\}.$$

Definition 4. A submodule A is said to be $\mathcal{R}$-closed if it is equal to its $\mathcal{R}$-closure.

Remark 5. The $\mathcal{R}$-closure of a module A has the following properties:

- It is uniquely defined.
- It is the largest submodule of $\mathcal{M}$ containing A while having the same rank.

Definition 6. A finitely generated submodule with basis vectors $\{\omega_1, \ldots, \omega_k\}$ is said to be integrable if there exist k functions $\varphi_i \in \mathcal{K}$ such that, locally:

$$\mathrm{span}_{\mathcal{K}[\nabla]}\{\omega_1, \ldots, \omega_k\} = \mathrm{span}_{\mathcal{K}[\nabla]}\{\mathrm{d}\varphi_1, \ldots, \mathrm{d}\varphi_k\}$$

Proposition 7. (Change of basis). Consider a module $\mathcal{M}$. Let $\mathcal{A}$ be a finitely generated submodule of $\mathcal{M}$ with basis vectors $\omega = \{\omega_1, \ldots, \omega_s\}$. Then $\bar{\omega} = \{\bar{\omega}_1, \ldots, \bar{\omega}_s\}$ is also a basis of $\mathcal{A}$ if and only if there exists a unimodular matrix $T(\nabla) \in \mathcal{K}^{s \times s}[\nabla]$ such that

$$\begin{bmatrix} \omega_1 \\ \vdots \\ \omega_s \end{bmatrix} = T(\nabla) \begin{bmatrix} \bar{\omega}_1 \\ \vdots \\ \bar{\omega}_s \end{bmatrix}. \tag{4}$$

Definition 8. (Change of coordinates). Consider system (1). Then,

$$z(t) = \varphi(x(t-\tau), \tau \in \underline{S})$$

are said to be a set of state coordinates if every component of the original state $x(t)$ may be expressed as a causal function of $z(t)$. In other words, if there exists an inverse function such that, locally, $x(t) = \varphi^{-1}(z(t-\tau), \tau \in \underline{S})$.

Proposition 9. $z(t) = \varphi(x(t-\tau), \tau \in \underline{S})$ is a change of coordinates for system Σ if and only if

$$\mathrm{span}_{\mathcal{K}[\nabla]}\{\mathrm{d}\,z\} = \mathrm{span}_{\mathcal{K}[\nabla]}\{\mathrm{d}\,x\}$$

Let $\mathcal{M}$ be the formal left-module over the ring $\mathcal{K}[\nabla]$:

$$\mathcal{M} = \text{span}_{\mathcal{K}[\nabla]}\{\mathrm{d}\xi \mid \xi \in \mathcal{K}\}.$$

For a given set of vectors $\{\omega_1, \ldots, \omega_k\}$ over $\mathcal{E}$, $\text{span}_{\mathcal{K}[\nabla]}\{\omega_1, \ldots, \omega_k\}$ represents the submodule of $\mathcal{M}$ generated by $\{\omega_1, \ldots, \omega_k\}$.

Let $\omega(t) = \sum_i \alpha_i \mathrm{d}x_i(t)$ be a one-form, with $\alpha_i \in \mathcal{K}[\nabla]$. Its time-derivative $\dot{\omega}(t)$ is given by

$$\dot{\omega}(t) := \sum_i \dot{\alpha}_i \mathrm{d}x_i(t) + \sum_i \alpha_i \mathrm{d}(\dot{x}_i(t)).$$

The relative degree for a given one-form is defined as follows.

Definition 10. The relative degree r of a one-form $\omega \in \text{span}_{\mathcal{K}[\nabla]}\{\mathrm{d}x\}$ is given by

$$r = \min\left\{k \in I\!N \mid \omega^{(k)} \notin \text{span}_{\mathcal{K}[\nabla]}\{\mathrm{d}x\}\right\}.$$

If, for all $k \in I\!N$, $\omega^{(k)} \in \text{span}_{K[\nabla]}\{\mathrm{d}x\}$, one sets $r = \infty$

Finally, we recall the definition of relative shift.

Definition 11. Let ω be a one-form with finite relative degree r. It is said to have a relative shift μ if

$$\mu = \min\{k \in I\!N \mid \omega^{(r)} \in \text{span}_{\mathcal{K}[\nabla]}\{\mathrm{d}\,x, \nabla^k \mathrm{d}\,u\} \quad \omega^{(r)} \notin \text{span}_{\mathcal{K}[\nabla]}\{\mathrm{d}\,x, \nabla^{k-1}\mathrm{d}\,u\}\}$$

This notion has played a crucial role in the search of causal solutions to control problems for systems with time delays (Márquez-Martínez and Moog, 1998; Moog *et al.*, 2000).

2.3 *Filtration of $\mathcal{M}$*

The filtration $\mathcal{M} \supset \mathcal{H}_0 \supset \mathcal{H}_1 \supset \cdots \supset \mathcal{H}_k \supset \cdots$ has been introduced in the case of nonlinear systems without delays for the discrete-time case (Aranda-Bricaire *et al.*, 1996), and the continuous-time case (Aranda-Bricaire *et al.*, 1995). Its importance comes from the structural information that it provides. For example, it gives the structure between the input and the state, from which it easy to know whether the system is input-to-state linearizable (Hunt *et al.*, 1983; Jakubczyk, 1987; Jakubczyk and Respondek, 1980). It also allows to define the so-called "Brunovský's infinitesimal canonical form", which, in the linear case, is equivalent to the Brunovský's canonical form. As a last example, it gives an easy-to-check characterization of system's accessibility (Jakubczyk and Sontag, 1990; Sussmann and Jurdjevic, 1972).

The definition of this filtration has been extended to the case of nonlinear time-delay systems as follows:

$$\mathcal{H}_k := \left\{\omega \in \text{span}_{\mathcal{K}[\nabla]}\{\mathrm{d}x\} \mid \text{rel. deg.}(\omega) \geq k\right\}. \tag{5}$$

For nonlinear time-delay systems, the submodules $\mathcal{H}_k$ may be infinitely generated. In (Márquez-Martínez, 1999), it has been shown that

- they can be characterized in a finite number of steps;
- they are closed over $\mathcal{K}[\nabla]$, which means that they are unique and, if $\text{rank}_{\mathcal{K}[\nabla]}\,\mathcal{H}_k = \text{rank}_{\mathcal{K}[\nabla]}\,\mathcal{H}_{k+1}$, then $H_k = \mathcal{H}_{k+1}$.

3. TRIANGULAR FORMS

Triangular forms have been widely used for the analysis of nonlinear systems without delays (Celikovský and Nijmeijer, 1996). They have been used to study small time controllability and stability properties. Also, the triangular-form equivalence is a necessary condition for full input-to-state linearization.

In this section, the problem of whether a nonlinear time-delay system admits a triangular form is stated and solved.

Problem statement 12. (Triangular form). Given system (1), find, if possible, a change of coordinates such that the system may be written as:

$$\begin{aligned}\dot{z}_i &= f_i\left(z_1(\cdot), \ldots, z_{i+1}(\cdot)\right) \qquad i = 1, \ldots n-1 \\ \dot{z}_n &= f_n\left(z_1(\cdot), \ldots, z_n(\cdot), u(\cdot)\right)\end{aligned}$$

with

$$\frac{\partial f_i(z_1(\cdot), \ldots, z_{i+1}(\cdot))}{\partial z_{i+1}(t - \tau_i')} \neq 0 \qquad \forall i \in \{1, \ldots, n-1\}$$

and

$$\frac{\partial f_n(z_1(\cdot), \ldots, z_n(\cdot), u(\cdot))}{\partial u(t - \tau_n')} \neq 0,$$

for some $\tau_1', \ldots, \tau_n' \in I\!N$. $f(z_i(\cdot))$ stands for $f(z_i(t - \tau), \tau \in \underline{S})$.

Now, it is possible to check if a system is equivalent to a triangular one.

Theorem 13. System Σ is equivalent to a triangular form under change of coordinates if and only if

i) $\mathcal{H}_\infty = 0$

ii) $\exists\, \varphi_i$ s.t. $\mathcal{H}_k = \text{span}_{\mathcal{K}[\nabla]}\{\mathrm{d}\,\varphi_1, \ldots, \mathrm{d}\,\varphi_{n-k+1}\}$, $k = 1 \ldots n$.

Proof. Assume that conditions *i)* and *ii)* hold. First condition implies that there is no one-form having an infinite relative degree, so

$$\dot{\mathcal{H}}_k \not\subset \mathcal{H}_k, \qquad k = 1 \ldots n \tag{6}$$

From *ii)* we have

$$\mathcal{H}_1 = \text{span}_{K[\delta]}\{d\varphi_1,\ldots,d\varphi_n\} = \text{span}_{K[\delta]}\{dx_1,\ldots,dx_n\}, 1\leq k\leq n.$$

From previous equation and Proposition 9, $z_i = \varphi_i(t-\tau)$ is a change of coordinates for system Σ. Since $\dot{\mathcal{H}}_k \subset \mathcal{H}_{k-1}$ then $\dot{z}_i = f_i(z_1,\ldots,z_{i+1})$ for $i<n$. Equation (6) and condition *ii)* imply that $(\partial/\partial z_{i+1})f_i(\cdot)\neq 0$. Finally, from condition *i)* one has that $(\partial/\partial u)f_n(\cdot)\neq 0$, and the system is in triangular form under state coordinates z_i.

Now assume that system has a triangular form. Then, a basis for the submodule $\mathcal{H}_k$ is given by $\{dz_1,\ldots,dz_{n-k+1}\}$, which implies *ii)*. Since $\mathcal{H}_n = \text{span}_{K[\delta]}\{dz_1\}$ and rel. deg.$(dz_1)<\infty$ this yields condition *i)*. ∎

In the case of accessible nonlinear systems without delays, it is possible to choose a basis for the submodules $\mathcal{H}_k$ as

$$\mathcal{H}_k = \text{span}_{\mathcal{K}[\nabla]}\{d\varphi, d\dot{\varphi},\ldots,d\varphi^{(n-k)}\}$$

where $d\varphi$ is a basis for $\mathcal{H}_n$. For time-delay systems, even in the linear case, this is not true, as shown by the following example.

Example 14. Let consider the system

$$\dot{x}_1(t) = x_2(t-1)$$
$$\dot{x}_2(t) = u(t)$$

for which $\mathcal{H}_n = \text{span}_{\mathcal{K}[\nabla]}\{d x_1\}$, but

$$\text{span}_{\mathcal{K}[\nabla]}\{d x_1, d\dot{x}_1\} = \text{span}_{\mathcal{K}[\nabla]}\{d x_1(t), d x_2(t-1)\} \neq \mathcal{H}_1 = \text{span}_{\mathcal{K}[\nabla]}\{d x_1(t), d x_2(t)\}$$

However, for any triangular system, the following proposition holds.

Proposition 15. Consider a triangular system Σ, with φ_i such that

$$\mathcal{H}_k = \text{span}_{\mathcal{K}[\nabla]}\{d\varphi_1,\ldots,d\varphi_{n-k+1}\},$$

for $k=1\ldots n$. Then

$$\mathcal{H}_{k-1} = \text{cls}_{\mathcal{K}[\nabla]}\{d\varphi_1,\ldots,d\varphi_{n-k+1}, d\dot{\varphi}_{n-k+1}\}, \quad k=2\ldots n$$

Proof. From Theorem 13, any $\mathcal{H}_k$ and $\mathcal{H}_{k-1}$ may be written as

$$\mathcal{H}_k = \{d\varphi_1,\ldots,d\varphi_{n-k+1}\}$$
$$\mathcal{H}_{k-1} = \{d\varphi_1,\ldots,d\varphi_{n-k+2}\}$$

By definition, $\dot{\mathcal{H}}_k \subset \mathcal{H}_{k-1}$. Also, the relative degree of any one-form is not affected by any non-trivial multiplicative factor. From this, it is easy to show that

$$\text{cls}_{\mathcal{K}[\nabla]}\{d\varphi_1,\ldots,d\varphi_{n-k+1}, d\dot{\varphi}_{n-k+1}\} \subset \mathcal{H}_{k-1}$$

Recall that $d\dot{\varphi}_{n-k+1}$ also belongs to $\mathcal{H}_{k-1}$, so

$$d\dot{\varphi}_{n-k+1} = \sum_{i=1}^{n-k+2} a_i d\varphi_i$$

with $a_{n-k+2}\neq 0$. Any one-form $\omega\in\mathcal{H}_{k-1}$ may be written as

$$\omega = \sum_{i=1}^{n-k+2} b_i d\varphi_i$$

From Lemma 1 there exist α and β such that $\alpha a_{n-k+2} = \beta b_{n-k+2}$. Thus,

$$\beta\omega = \beta\sum_{i=1}^{n-k+1} b_i d\varphi_i + \alpha a_{n-k+2} d\varphi_{n-k+2}$$
$$= \beta\sum_{i=1}^{n-k+1} b_i d\varphi_i + \alpha(d\dot{\varphi}_{n-k+1} - \sum_{i=1}^{n-k+1} a_i d\varphi_i)$$
$$= \sum_{i=1}^{n-k+1} b_i' d\varphi_i + \alpha d\dot{\varphi}_{n-k+1}$$

which implies that

$$\mathcal{H}_{k-1} \subset \text{cls}_{\mathcal{K}[\nabla]}\{d\varphi_1,\ldots,d\varphi_{n-k+1}, d\dot{\varphi}_{n-k+1}\},$$

and this ends the proof. ∎

As a final remark, writing a system as its triangular-form equivalent, does not change the problems found while searching for causal solutions to control problems. This is because the relative shift (i.e., the time lag of one variable with respect to present time) of state variables and control inputs, is invariant under change of coordinates, as it can be easily proved from 9.

4. CONCLUSIONS AND OPEN ISSUES

In this paper, a new mathematical setting adapted to the study of a broad class of nonlinear time-delay systems has been presented. It has been used to display some structural properties of this class of systems. Under this approach, a definition of triangular forms has been proposed. We argue that this result opens new research lines in several control issues. One example is the problem of feedback linearization; that is, the transformation of a nonlinear time-delay systems into a linear one by means of a change of coordinates and a static or dynamic feedback.

Another example is the study of stability issues through the use of triangular forms is a completely unexplored approach for time-delay systems.

Finally, one should note that the causality of any solution to a particular control problem remains the same, independently if the system is in its original form, or it is written as its triangular-form equivalent, as discussed at the end of previous section.

5. REFERENCES

Aranda-Bricaire, E., C.H. Moog and J.-B. Pomet (1995). A linear algebraic framework for dynamic feedback linearization. *IEEE Trans. on Automatic Control* **40**, 127–132.

Aranda-Bricaire, E., Ü. Kotta and C.H. Moog (1996). Linearization of discrete-time systems. *SIAM J. Control & Opt.* **34**, 1999–2023.

Celikovský, S. and H. Nijmeijer (1996). Triangular forms, local controllability and stabilizability of single-input nonlinear systems. In: *13th IFAC World Congress*. San Fco., USA. pp. 227–233.

Conte, G. and A.M. Perdon (1995). The disturbance decoupling problem for systems over a ring. *SIAM J. Control & Opt.* **33**(3), 750–764.

Hunt, L.R., R. Su and G. Meyer (1983). *Design for multi-input nonlinear systems*. pp. 268–298. Birkhäuser.

Jakubczyk, B. (1987). Feedback linearization of nonlinear discrete-time systems. *Syst. & Ctrl. Lett.* pp. 411–416.

Jakubczyk, B. and E.D. Sontag (1990). Controllability of nonlinear discrete-time systems: a lie-algebraic approach. *SIAM J. Control & Opt.* **28**, 1–33.

Jakubczyk, B. and W. Respondek (1980). On linearization of control systems. *Bull. Acad. Pol. Sci., Ser. Sci. Math.* **18**, 517–522.

Márquez-Martínez, L.A. (1999). Note sur l'accessibilité des systèmes non linéaires à retards. *C.R. Acad. Sci. Paris* (329), 545–550.

Márquez-Martínez, L.A. (2000). Analyse et commande de systmes non linaires retards. PhD thesis. Universit de Nantes / Ecole Centrale de Nantes. Nantes, France.

Márquez-Martínez, L.A. and C.H. Moog (1998). On the input-output linearization of nonlinear time-delay systems. In: *IFAC Conf. Syst. Struct. & Control*. Vol. 3. Nantes. pp. 575–580.

Márquez-Martínez, L.A. and C.H. Moog (1999). New results on the analysis and control of nonlinear time-delay systems. In: *IEEE Conf. on Dec. Ctrl.*. Phoenix, USA.

Márquez-Martínez, L.A., C.H. Moog and M. Velasco-Villa (2000). The structure of nonlinear time-delay systems. *Kybernetika* **36**(1), 53–62.

Moog, C.H., R. Castro-Linares, M. Velasco-Villa and L.A. Márquez-Martínez (2000). Disturbance decoupling for time-delay nonlinear systems. *IEEE Trans. on Automatic Control* **48**(2), 305–309.

Sussmann, H. and V. Jurdjevic (1972). Controllability of nonlinear systems. *J. Differ. Eq.* **12**, 95–116.

www.elsevier.com/locate/ifac

GLOBAL STABILIZATION OF MULTIPLE INTEGRATORS WITH TIME-DELAY AND INPUT CONSTRAINTS [1]

W. Michiels * and **D. Roose** *

* *Katholieke Universiteit Leuven,Department of Computer Science,Celestijnenlaan 200A,B-3001 Leuven, Belgium*

Abstract: In this paper we construct parametrized controllers for the semi-global and global asymptotic stabilization of a multiple integrator with time-delay and constraints on the input variable. The stability results are semi-global in the delay.

Keywords: delay equations, stability, input constraints

1. INTRODUCTION

We study the global stabilization of linear systems with a time-delay and constraints on the input variable,

$$\begin{aligned}\dot{y}(t) = Ay(t) + B\text{sat}(u(t-\tau)),\\ y \in \mathbb{R}^n, \quad u \in \mathbb{R},\end{aligned} \tag{1}$$

by means of static state feedback, $u = F(y)$. Here sat(.) represents the classical saturation function defined as, $\text{sat}(y) = y$ when $|y| \leq 1$ and $\text{sat}(y) = \text{sign}(y)$ otherwise.

When $\tau = 0$, it is well known that the system (1) can only be globally asymptotically stabilized when the pair (A, B) is controllable and A has all its eigenvalues in the *closed* left half plane, see (Sussmann *et al.*, 1994) and the references therein. Moreover both in (Sussmann *et al.*, 1994), where a saturation design is used, and in (Lin, 1999), in the context of low-gain design, procedures were given for the explicit construction of a globally stabilizing controller.

In the literature on delay equations the (robust) stabilization of the system

$$\dot{y}(t) = Ay(t) + A_d y(t-\tau) + B\ \text{sat}(u(t)), \tag{2}$$

has been widely studied, see e.g. (Garcia and Tarbouriech, 1998; Niculescu *et al.*, 1996; Tarbouriech, 1998; Tarbouriech and Gomes da Silva Jr., 2000). Although the uncontrolled system contains a delayed term and is thus more general than in (1), the controller design problem is facilitated by the fact that the control input is not delayed. Further, research up to now has been focussed on the derivation of sufficient conditions for (robust) local or global stabilization in a Lyapunov framework. Typically, such conditions are expressed by the feasibility of a LMI or the solvability of an ARE, on which also the construction of a controller relies. However in the context of global stabilization, few attention has been paid to the structural requirements on the controller and to the study of the feasibility of such LMIs or the solvability of such AREs in the case where the uncontrolled system has its rightmost eigenvalues on the imaginary axis [2]. Note for instance that in this case, generally *nonlinear* feedback is needed for global stabilization, even when $\tau = 0$, see (Sussmann *et al.*, 1994).

[1] This research presents results of the project IUAP P4/02 funded by the programme on Interuniversity Poles of Attraction, initiated by the Belgian State, Prime Minister's Office for Science, Technology and Culture.

[2] This case is of particular interest because linear systems with eigenvalues in the open right half plane cannot be globally asymptotically stabilized with a bounded input, see e.g. (Sussmann *et al.*, 1994; Lin, 1999).

In this paper we assume that (1) has its rightmost eigenvalues on the imaginary axis and restrict ourselves to the simple, yet practically important case where the system (1) corresponds to the multiple integrator,

$$\begin{aligned} \dot{y}_1 &= y_2,\ \dot{y}_2 = y_3, \ldots, \dot{y}_{n-1} = y_n, \\ \dot{y}_n &= \text{sat}(u(t-\tau)), \end{aligned} \tag{3}$$

i.e. in (1) we make the assumption that

$$A = \begin{bmatrix} 0 & 1 & & \\ & \ddots & \ddots & \\ & & 0 & 1 \\ & & 0 & 0 \end{bmatrix}, \quad B = \begin{bmatrix} 0 \\ \vdots \\ 0 \\ 1 \end{bmatrix}, \tag{4}$$

and focus on the explicit construction of (semi-)globally stabilizing controllers.

The structure of the paper is as follows. First we study the local and semi-global stabilization of (3) with linear low-gain feedback and briefly discuss an alternative approach, where the time-delay is first compensated with a static prediction. Then we consider the global stabilization problem. We conclude with some preliminary remarks on the global stabilization of more general types of delay equations.

Preliminaries: The state of the delay equation

$$\dot{y}(t) = f(y, y(t-\tau)), \quad f: \mathbb{R}^n \times \mathbb{R}^n \to \mathbb{R}^n, \tag{5}$$

at time t can be described as a vector $y(t) \in \mathbb{R}^n$ or as a function segment y_t defined by

$$y_t(\theta) = y(t+\theta),\ \theta \in [-\tau,\ 0].$$

When the right hand side of (5) is continuous and Lipschitz in both of its arguments, a solution is uniquely defined by specifying as initial condition a function segment y_0, where $y_0 \in \mathcal{C}([-\tau,\ 0], \mathbb{R}^n)$, the Banach space of continuous functions mapping the delay-interval $[-\tau,\ 0]$ into $\mathbb{R}^n$ and equipped with the supremum-norm $\|.\|_s$. Stability definitions are analogous to the ODE-case, see e.g. (Hale and Verduyn Lunel, 1993, Section V).

The delay equations considered in this paper originate from a stabilization problem of the form

$$\dot{y} = f(y, u(t-\tau)), \quad u = g(y), \tag{6}$$

and we analyse the stability of the closed-loop system,

$$\dot{y} = f(y, g(y(t-\tau))), \quad y_0 \in \mathcal{C}([-\tau,\ 0], \mathbb{R}^n). \tag{7}$$

However for (6) it is more natural to take as initial conditions $u(t) = 0$ for $t \in [-\tau,\ 0]$ and $y(0) \in \mathbb{R}^n$. Then (6) is equivalent with

$$\begin{cases} \dot{y} = f(y,0) & y(0) \in \mathbb{R}^n \ \ t \in [0,\ \tau], \\ \dot{y} = f(y, g(y(t-\tau))) & t \geq \tau. \end{cases} \tag{8}$$

However, due to the time-invariance of (7), stability of (8) is implied by the stability of (7).

2. LINEAR LOW-GAIN FEEDBACK

In this section we consider the stabilization of (3) with the parametrized family of linear state feedback laws,

$$\begin{aligned} u &= K(\epsilon)^T y \\ &= -k_1 \epsilon^n y_1 - k_2 \epsilon^{n-1} y_2 - \ldots - k_n \epsilon y_n, \end{aligned} \tag{9}$$

where $\epsilon > 0$ and the following assumption of $K(\epsilon)$ is satisfied:

Assumption 1. The polynomial $\lambda^n + \sum_{i=1}^n k_i \lambda^{i-1}$ is Hurwitz.

Since $\|K(\epsilon)\| \to 0$ as $\epsilon \to 0$, the control laws (9) correspond to low-gain feedback.

Instrumental to the derivation of the stabilizability properties of (9) is the similarity transformation

$$z = T(\epsilon) y, \tag{10}$$

defined by

$$z_i = \epsilon^{n-i+1} y_i, \quad i = 1 \ldots n,$$

and the re-scaling of time

$$t_{(\text{new})} = \epsilon t_{(\text{old})}, \tag{11}$$

which allow to transform the closed-loop system (3)-(9) into

$$\dot{z} = Az + BK^T \text{sat}\left(z(t - \epsilon\tau)\right), \tag{12}$$

where $K = [-k_1 \ \ -k_2 \ldots -k_n]^T$ and A and B defined by (4). Note that (in)stability is not affected by the transformation (10)-(11) and therefore the stability of (3)-(9) is implied by the stability of (12). Note that the re-scaling of time (11) causes a scaling of the eigenvalues by $1/\epsilon$.

2.1 *Local stability and performance*

In this paragraph we consider local properties, and therefore we study the linearization of (12),

$$\dot{z} = Az + BK^T z(t - \epsilon\tau). \tag{13}$$

Because of Assumption 1, $\bar{A} \triangleq A + BK^T$ is Hurwitz and the system (13) is asymptotically stable for small values of $\epsilon\tau$. As a consequence we have

Theorem 1. The system (3) can be locally asymptotically stabilized, semi-globally in the delay with the control law (9), as $\epsilon \to 0$, i.e. $\forall \bar{\tau} > 0$, $\exists \bar{\epsilon} > 0$ such that the system (3)-(9) is locally asymptotically stable for any $0 \leq \tau \leq \bar{\tau}$ and $0 < \epsilon \leq \bar{\epsilon}$.

For given $\bar{\tau}$, a suitable value of $\bar{\epsilon}$ can be found by continuing the eigenvalues of (13) as a function of $\epsilon\tau$, see the example below.

Intuitively the semi-gobal stabilizability in the delay is explained by the fact that, whatever the value of the input delay, it becomes negligible compared to the system's dynamics as $\epsilon \to 0+$, a property exploited by the transformation (10)-(11). Note however that delay-independent stabilization is not possible, since A is not Hurwitz (Hale and Verduyn Lunel, 1993).

Despite of the good stabilizability properties of the controller (9), the input delay puts severe restrictions on its performance. At one hand, small values of ϵ lead to a slow convergence of the closed-loop solutions to zero. The convergence rate even tends to zero as $\epsilon \to 0+$. On the other hand, from (13) it follows that for a fixed value of τ, increasing ϵ ultimately leads to instability. Hence there exists an optimal value of ϵ, maximizing the exponential decay rate of the solutions (proportional to $1/\tau$!).

Example 2. In Figure 1 the rightmost eigenvalues of the (linearized) system (3)-(9) are shown for $n = 2$ and $K = [-2 \ \ -2]^T$, as a function of ϵ and τ. The numerical calculations were carried out using the publically available software package DDE-BIFTOOL (Engelborghs, 2000).

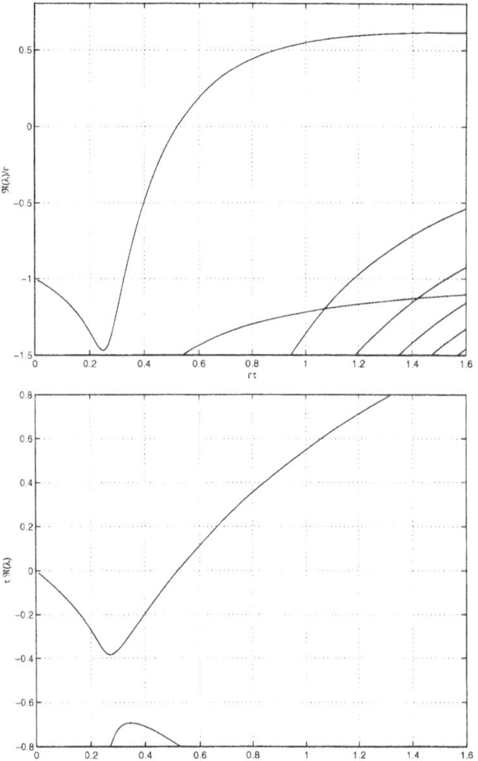

Fig. 1. Rightmost eigenvalues of the (linearized) system (3) controlled with (9) for $n = 2$ and $K = [-2 \ -2]^T$ as a function of ϵ and τ. The maximal achievable exponential decay of the closed-loop solutions, $\approx e^{-0.38/\tau}$, is obtained for $\epsilon\tau \approx 0.27$.

2.2 *Semi-global stabilization*

We return to the constrained problem (3) and show that the control law (9) also achieves semi-global asymptotic stability of (3) in the state variable y, i.e. that any bounded set of initial conditions is included in the attraction domain of the zero solution as $\epsilon \to 0+$. Because of Theorem 1, it is sufficient to prove that for any given (bounded) set of initial conditions Ω, the input constraint is not active along the solutions when ϵ is sufficiently small. For this, we consider the linearized system (13) and first construct a compact set in $\mathbb{R}^n$ which contains all solutions with initial conditions in Ω. Then we guarantee that inside this set, the input $u = K^T z = K(\epsilon)^T y$ is smaller than one and hence, these solutions of the linearized system coincide with the solutions of the original nonlinear problem. This approach is inspired by (Lin, 1999).

2.2.1. *Generalization of positively invariant sets*

In the ODE-case, bounds on the solutions can be derived based on the fact that sublevel sets of an appropriate Lyapunov function are positively invariant. We now outline how these ideas can be generalized.

Because in equation (13), $\bar{A} = A + BK^T$ is Hurwitz, we can find matrices $P > 0$ and $Q > 0$, satisfying the Riccati equation

$$\bar{A}^T P + P\bar{A} = -Q. \tag{14}$$

Then

$$V = z^T P z = y^T T(\epsilon)^T P T(\epsilon) y \triangleq y^T P(\epsilon) y \tag{15}$$

is a Lyapunov function for the linearized closed-loop system when $\tau = 0$. For $\tau \neq 0$, a sublevel set S of V,

$$S = \{z \in \mathbb{R}^n : V(z) = z^T P z \leq c\}, \tag{16}$$

with c an arbitrary constant, is not positively invariant since A is not Hurwitz and as a consequence, (15) cannot serve as a Lyapunov-Razumikhin function (Hale and Verduyn Lunel, 1993) for (13). Indeed, it is always possible to construct an initial condition $z_0 \subseteq S$ with $z(0) \in \partial S$ such that along the solution,

$$\lim_{t\to 0+} \dot V(t) = z(0)^T\underbrace{(A^TP+PA)}_{\text{indefinite}}z(0) \\ +2z(0)^TPBK^Tz(-\epsilon\tau) > 0.$$

Note that when we let $\epsilon\tau \to 0$, the initial condition of such 'escaping' solution must have arbitrarily large time-derivatives, but these do not occur for $t \in [0,\ \epsilon\tau]$, since in that interval the time-derivative is bounded by $|A||z|+|BK^T||z(t-\epsilon\tau)|$. This motivates us to extend the definition of a positively invariant set to

Definition 3. For the delay equation equation $\dot z = f(z(t), z(t-\tau))$ a set $D \subset \mathbb{R}^n$ is 1-positively invariant iff $z_0 \in C([-\tau,\ 0], \mathbb{R}^n)$ and $z(t) \in D$, $\forall t \in [-\tau,\ \tau]$ implies $z(t) \in D$, $\forall t \geq 0$.

An alternative consists of only requiring the invariance property for solutions with initial conditions satisfying a Lipschitz condition, as in (Bartholomeus-Goubet, 1996, Subsection V.6).

We now give conditions on ϵ under which the set (16) is 1-positively invariant for the system (13) or equivalently the set $S(\epsilon) = \{y \in \mathbb{R}^n :\ y^TP(\epsilon)y \leq c\}$ is 1-positively invariant for the linearization of the closed-loop system (3)-(9). Therefore we first rewrite (12) as

$$\dot z = \bar A z + BK^T(z(t-\epsilon\tau) - z(t)),$$

where the second term can be interpreted as a perturbation of the asymptotically stable ODE $\dot z = \bar A z$. When a solution reaches the boundary of the set S at time $t \geq \epsilon\tau$ we have,

$$\begin{aligned}\dot V(t) &\leq -z^TQz + 2|z||PBK^T||z(t-\epsilon\tau)-z(t)| \\ &\leq -z^TQz + 2|z||PBK^T|\epsilon\tau|\dot z(\theta)|, \\ &\qquad\qquad \theta \in [t-\epsilon\tau,\ t] \\ &\leq -z^TQz + 2|z||PBK^T|\epsilon\tau(|Az(\theta)|+ \\ &\quad |BK^Tz(\theta-\epsilon\tau)|) \\ &\leq -\lambda_{\min}(Q)|z|^2 + 2|PBK^T|\epsilon\tau(|A|+ \\ &\quad |BK^T|)\frac{\lambda_{\max}(P)}{\lambda_{\min}(P)}|z|^2.\end{aligned}$$

The set (16) is 1-positively invariant when $\dot V(t) \leq 0$. This is satisfied if

$$\epsilon\tau \leq \frac{\lambda_{\min}(Q)}{2|PBK^T|(|A|+|BK^T|)}\frac{\lambda_{\min}(P)}{\lambda_{\max}(P)}. \quad (17)$$

Under this condition, the sublevel sets of V can be used to derive bounds on solutions, as we illustrate below.

2.2.2. *Checking input constraints* In order to check whether the input satisfies $|u| = |K(\epsilon)^T)y| \leq 1$, we calculate the critical sublevel set of (15), i.e. the smallest sublevel set of V on which $u = K(\epsilon)^Ty$ reaches the value ± 1, by solving

$$\min_{K^TT(\epsilon)y=1} y^TP(\epsilon)y = \min_{K^Tz=1} z^TPz. \quad (18)$$

Denote the value of the minimum by α. Then the critical sublevel set is given by

$$y^TP(\epsilon)y \leq \alpha, \quad (19)$$

and since $\lim_{\epsilon\to 0} P(\epsilon) = 0$, any compact set in the state space is contained in the critical sublevel set when ϵ is sufficiently small.

We now return to the semi-global stabilization of (3). Take any bound on the delay $\bar\tau$ and any compact set $\Omega \subset \mathbb{R}^n$. Then there exists a compact set $\Omega_2 \subset \mathbb{R}^n$ such that $\forall \tau \leq \bar\tau$, $\forall y_0 \in C([-\tau,\ 0],\ \mathbb{R}^n)$ with $y_0 \subset \Omega$, the following holds: $y(t) \in \Omega_2$. $\forall t \in [-\tau,\ \tau]$. When taking ϵ sufficiently small, the critical sublevel set contains Ω_2, is 1-positively invariant because of (17), and thus contains all trajectories with $y_0 \subset \Omega$. Furthermore, the linearized system, describing the closed-loop dynamics on the critical sublevel set, is asymptotically stable for small ϵ because of Theorem 1. Hence we have proven the following result:

Theorem 4. The system (3) can be semi-globally asymptotically stabilized in both y and the delay with a feedback of the form (9), as $\epsilon \to 0$.

For Ω and $\bar\tau$ given, an suitable value of ϵ can be calculated from (17) and (19).

2.3 *An alternative approach*

From the above analysis, it follows that a small value of ϵ guarantees large regions of attraction of the zero solution, while larger values lead to improved (local) performance. However, as shown in the previous section and illustrated with Figure 1, the maximal achievable exponential decay rate of the solutions is limited. This restriction of (delayed) state feedback could at first sight be removed by first compensating the delay with a static prediction. With the low-gain distributed delay feedback law,

$$u = K(\epsilon)^T\hat y(t, t+\tau) = K(\epsilon)^T \left(e^{A\tau}y(t) + \int_0^\tau e^{A(\tau-\theta)}Bu(t+\theta-\tau)d\theta\right), \quad (20)$$

where $\hat y(t, t+T)$ is the prediction of the state y over one delay interval, the characteristic equation of the linearized closed-loop system is given by

$$\det\left(\lambda I - A - BK(\epsilon)^T\right) = 0.$$

Now there are only n closed-loop eigenvalues which can be assigned *arbitrarily* in the complex

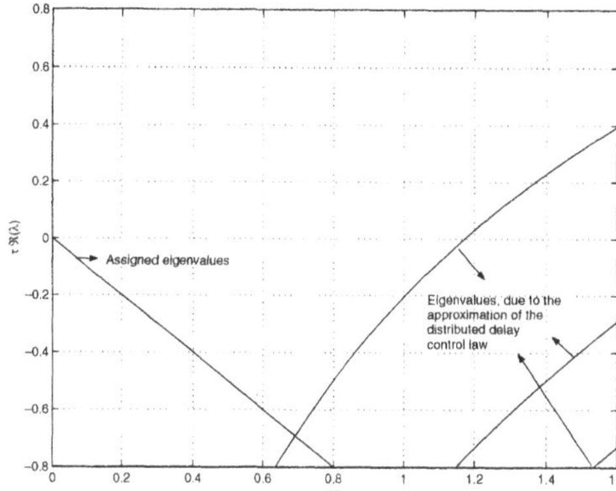

Fig. 2. Rightmost eigenvalues of the double integrator controlled with the distributed delay feedback law(20) with $K = [-2 \ -2]^T$, in function of ϵ and τ. The assigned eigenvalues are $(-1 \pm j)\epsilon$. The other eigenvalues are caused by an approximation of the integral term in with a finite sum. The maximal achievable exponential decay of the closed-loop solutions is approx. $e^{-0.69/\tau}$.

plane. However in order to implement the control law (20), a numerical approximation of the integral term is needed (e.g. using a quadrature rule), and in (Engelborghs *et al.*, 2001) it is shown that even an *arbitrarily* small approximation error introduces an essential spectrum which may affect both stability and performance. In (Engelborghs *et al.*, 2001) it is also explained how this essential spectrum can be numerically calculated. As we now illustrate with an example, the practical limitations on stability and performance are qualitatively the same as for pure state feedback.

Example 5. In Figure 2, the rightmost eigenvalues of the (linearized) system (3)-(20) with $n = 2$ and $K = [-2 \ -2]^T$ are shown in function of ϵ and τ. Theoretically only the assigned eigenvalues $(-1 \pm j)\epsilon$ occur. The other eigenvalues are caused by the approximation of the integral with a finite sum. They limit the maximal achievable exponential decay rate and cause instability for large values of ϵ, a situation qualitatively comparable to the state feedback case, treated in Example 2.

3. GLOBAL STABILIZATION

For global stability in y, nonlinear feedback is necessary when $n > 2$, even for $\tau = 0$ (Sussmann *et al.*, 1994). A first possibility consists of gain scheduling, i.e. using a control law of the form (9), where ϵ now depends on y and is thus continuously adapted along the solutions. This approach is not further worked out in this paper.

Alternatively an adaptation to the time-delay case can be made of the so-called saturation design, initiated by a paper of Teel (Teel, 1992), which motivated many researchers to consider the explicit construction of globally asymptotically stable feedback laws. In saturation design the control law typically consists of a (non)linear combination of saturated linear functions.

For the multiple integrator with time-delay, we have the following result:

Theorem 6. The feedback

$$u = -\epsilon \, \mathrm{sat} x_n - \epsilon^2 \mathrm{sat}\frac{x_{n-1}}{\epsilon} - \epsilon^3 \mathrm{sat}\frac{x_{n-2}}{\epsilon^2} - \cdots - \epsilon^n \mathrm{sat}\frac{x_1}{\epsilon^{n-1}}, \qquad (21)$$

where

$$x_k = \sum_{j=k}^{n} \epsilon^{n-j} \binom{n-k}{n-j} y_j, \qquad k = 1 \ldots n, (22)$$

stabilizes the system (3) globally in y and semi-globally in the delay τ, as $\epsilon \to 0$.

Due to space limitations, we only briefly sketch the proof, which can be found in (Michiels and Roose, 2001). Despite of some technicalities, its structure is analogous to the proof of Theorem 2.1 of (Teel, 1992). First we apply the transformation (22) to (3), which results in the system

$$\begin{cases} \dot{x}_1 = \epsilon x_2 + \cdots + \epsilon x_n + \mathrm{sat}(u(t-\tau)) \\ \dot{x}_2 = \epsilon x_3 + \cdots + \epsilon x_n + \mathrm{sat}(u(t-\tau)) \\ \vdots \\ \dot{x}_{n-1} = \epsilon x_n + \mathrm{sat}(u(t-\tau)) \\ \dot{x}_n = \mathrm{sat}(u(t-\tau)) \end{cases}, (23)$$

with the control-law (21). For an arbitrary solution of the closed-loop system, it follows from the last equation of (23) that after a finite time the first saturation function in (21) must act in its linear region when ϵ is sufficiently small, since this equation reads as $\dot{x}_n = -\epsilon \mathrm{sat}(x_n(t-\tau)) + O(\epsilon^2)$. Using the same type of arguments the existence is proven of a finite time $T > 0$ such that for $t \geq T$, all saturation functions in (21)-(23) are in their linear region and thus the closed-loop system is linear. Asymptotic stability of the linearized system follows from an application of Theorem 1. Typical for the time-delay case is that also the linearization of the control law (21) depends on ϵ, which is necessary for achieving semi-global asymptotic stability in the delay.

4. SOME PRELIMINARIES ON THE GENERAL CASE

So far our research focussed on the (semi-)global stabilization of the multiple integrator. For the general case we conjecture:

Conjecture 7. Consider the system (1) and assume that the pair (A, B) is controllable and that all eigenvalues of A are in the closed left half plane. Then for any delay values $0 \leq \tau_1 \leq \tau_2$, satisfying

$$\tau_2 - \tau_1 < \min \left\{ \frac{\pi}{|\Omega|} : j\Omega \in \sigma(A) \right\},$$

there exists a control law $u = F(x)$ which achieves global asymptotic stability for all $\tau \in [\tau_1, \tau_2]$.

Our motivation comes from the fact that global stabilizability is related with semi-global stabilizability with linear low-gain feedback. It is easy to show that when a low-gain control law $u = K(\epsilon)^T y$ moves an isolated eigenvalue at $j\Omega$ to $j\Omega + \epsilon\Delta\lambda$ for $\tau = 0$, the behavior of the eigenvalue as a function of the delay can be approximated by $\lambda \approx j\Omega + \epsilon\Delta\lambda e^{-j\Omega\tau}$ as $\epsilon \to 0$.

When there is also a delayed term in the uncontrolled system, as in

$$\dot{y} = Ay + A_d y(t - \tau_1) + B\text{sat}(u(t - \tau_2)), \quad (24)$$

the condition that all its eigenvalues are in the closed left half plane is obviously necessary for (semi-)global stabilization, but not sufficient. Note that, despite of the input delay, with linear low-gain control the stabilization problem of (1) is still of a finite-dimensional nature in the sense that for small ϵ, only the imaginary eigenvalues (at most n) need to be controlled with the n controller parameters. For (24) however, more than n rightmost eigenvalues on the imaginary axis are possible, which may be an obstruction to stabilizability. For instance the scalar system

$$\dot{y} = \frac{1}{\tau_1} y - \frac{1}{\tau_1} y(t - \tau_1) + \text{sat}(u(t - \tau_2)) \quad (25)$$

has, for $u = 0$, all its eigenvalues in the closed left half plane, yet a double eigenvalue at zero. Based on (Kolmanovskii and Myshkis, 1999), where the stability of equation $\dot{y} = ay + by(t - \tau)$ is investigated, it is easy to see that (25) can be (at least) locally asymptotically stabilized with the low-gain control law $u = -k\epsilon y$ for $\tau_2 = 0$, while this is impossible for $\tau_2 = \tau_1$.

5. REFERENCES

Bartholomeus-Goubet, A. (1996). Sur la stabilité et la stabilisation des systemes retardés: critères dépendant des retard. PhD thesis. L'Université des Sciences et Technologies de Lille.

Engelborghs, K. (2000). DDE-biftool: a Matlab package for bifurcation analysis of delay differential equations. TW Report 305. Department of Computer Science. Katholieke Universiteit Leuven, Belgium.

Engelborghs, K., M. Dambrine and D. Roose (2001). Limitations of a class of stabilization methods for delay equations. *IEEE Transactions on Automatic Control* **46**(2), 336–339.

Garcia, G. and S. Tarbouriech (1998). Preliminaries on nonlinear bounded control for time-delay systems. In: *Proceedings of the first IFAC workshop on linear time-delay systems.* Grenoble, France. pp. 93–98.

Hale, J.K. and S.M. Verduyn Lunel (1993). *Introduction to Functional Differential Equations.* Vol. 99 of *Applied Mathematical Sciences.* Springer-Verlag.

Kolmanovskii, V.B. and A. Myshkis (1999). *Introduction to the theory and application of functional differential equations.* Vol. 463 of *Mathematics and its applications.* Kluwer Academic Publishers.

Lin, Z. (1999). *Low gain design.* Vol. 240 of *Lecture notes in control and information sciences.* Springer-Verlag.

Michiels, W. and D. Roose (2001). Global stabilization of a multiple integrator with time-delay and input constraints. TW Report 325. Department of Computer Science. Katholieke Universiteit Leuven, Belgium. Available at `http://www.cs.kuleuven.ac.be/publicaties/rapporten/TW2001.html`.

Niculescu, S.-I., J.-M. Dion and L. Dugard (1996). Robust stabilization for uncertain time-delay systems containing saturating actuators. *IEEE Transactions on Automatic Control* **41**(5), 742–747.

Sussmann, H.J., E.D. Sontag and Y. Yang (1994). A general result on the stabilization of linear systems using bounded controls. *IEEE Transactions on Automatic Control* **39**(12), 2411–2425.

Tarbouriech, S. (1998). Local stabilization of continuous-time delay systems with bounded inputs. In: *Stability and control of time-delay systems* (L. Dugard and E.I. Verriest, Eds.). Vol. 228 of *Lecture notes in control and information sciences.* Chap. 14. Springer-Verlag.

Tarbouriech, S. and J.M. Gomes da Silva Jr. (2000). Synthesis of controllers for continuous time delay systems with saturating controls via LMIs. *IEEE Transactions on Automatic Control* **45**(1), 105–113.

Teel, A.R. (1992). Global stabilization and restricted tracking for multiple integrators with bounded controls. *Systems & Control Letters* (18), 165–171.

www.elsevier.com/locate/ifac

REPETITIVE LEARNING TIME-DELAYED CONTROL FOR CHAOTIC SYSTEMS

Yanxing Song*, Xinghuo Yu*, Guanrong Chen, Jian-Xin Xu*****

*Faculty of Informatics and Communication, Central Queensland University, Rockhampton QLD 4702 Australia
**Department of Electronic Engineering, City University of Hong Kong, Kowloon, Hong Kong
***Department of Electrical Engineering, National University of Singapore, Singapore

Abstract: In this paper, a time-delayed chaos control method based on repetitive learning is proposed. A general repetitive learning control structure based on the invariant manifold of chaotic system is given. The integration of the repetitive learning control principle and the time-delayed chaos control technique enables adaptive learning of appropriate control actions from learning cycles. In contrast to the conventional repetitive learning control, no exact knowledge (analytic representation) of the target periodic orbits is needed, except for the time delay constant, which can be identified via either experiments or adaptive learning. The controller effectively stabilizes the states of the continuous-time chaos on desired unstable periodic orbits. Simulations on Duffing and Lorenz chaos are provided to verify the design and analysis. *Copyright © 2001 IFAC*

Key words: Time-delay, Chaos control, Learning control.

1. INTRODUCTION

Recently, stabilizing unstable periodic orbits (UPOs) of chaotic systems has become an active and focusing direction in the field of chaos control (Yu, et al., 1999). The problem can be formulated as a (periodic target) tracking problem in classical control theory, yet for unstable targets and chaotic systems. The rich literature of conventional tracking control may be modified for the task, provided that the UPO as the reference signal is available for design. In practice, however, it is very difficult to obtain an exact and analytic formula for a UPO, except for the degenerate case of an unstable equilibrium. Moreover, it is extremely difficult, if not impossible, to implement a UPO as the reference signal by physical means such as circuitry due to the instability nature of such orbits. A time-delayed feedback control (TDFC) principle was proposed for chaos control (Pyragas, 1992), which can overcome the problems mentioned above. The novel idea in this method is to use the current as well as past system states in the feedback loop, thereby avoiding a direct use of the target UPO signals in the controller. This method has lately been extended and applied to various systems (see Yu, *et al.*, 1999 and references therein).

One existing problem with the TDFC methods is that they may require a priori knowledge of the dynamics of the underlying chaotic system, so that a crude control can be designed to suppress/eliminate the known nonlinearities. Tailored adaptive TDFC methods for discrete-time chaotic systems have been developed, which can be used to learn based on past and current states, for example, a neural network approach (Konishi and Kokame, 1997) and a learning control approach (Konishi and Kokame, 1998). These methods are based on the discretized time-delayed systems of finite dimension. However, the TDFC principle for continuous-time chaotic systems is yet to be fully explored. One difficulty is that TDFC for continuous-time chaotic systems essentially results in infinite-dimensional systems, and this remains to be a very challenging control problem from a mathematical viewpoint.

In this paper, the TDFC problem for continuous-time chaotic systems using the repetitive learning control

(RLC) technique is addressed. RLC is a control technique specifically tailored for control tasks that involve periodic reference signals (Moore, 2000; Xu, 2000). It is based on the conventional iterative learning control (ILC). By taking into account the special characteristics of periodic signals, repetitive learning integrated with a suitable controller can result in a better (less crude) control performance. Such control strategies have been used successfully for a number of control tasks (Bien and Xu, 1998). However, the conventional ILC cannot be used directly for TDFC as they require the resetting of initial conditions for each learning cycle and this is impractical as far as chaos control is concerned. The main reason is that the time evolution of a chaotic state is continuous and cannot be interrupted in general; otherwise the chaotic orbit may escape to another region of attraction and never come back. Nevertheless, the design principle of RLC is useful for developing a repetitive time delayed chaos control since the tracking to a UPO is similar to RLC, except that the reference signal is not completely known. Based on the invariant manifold method, a repetitive learning TDFC is proposed for chaos control in this paper, which integrates the RLC and TDFC principles. The advantage of this integration is that it enables the controller to adaptively learn from learning cycles an appropriate control force, so that crude control can be avoided. In contrast to the conventional RLC, knowledge (analytic representation) of the target periodic orbit is not necessarily to be known, except for the time delay constant, which can be easily identified via either experiments or adaptive learning (Yu, 1999). This advantage alone makes it attractive, especially for chaos control.

The organization of this paper is as follows. In Section 2, the descriptions of the background for the concerned problem and the designed controller of RLC based on the invariant manifold method are given. In Section 3, two simulations are presented for controlling the Duffing and Lorenz systems, respectively, to show the effectiveness of the proposed approach. In Section 4, a brief conclusion is given.

2.GENERAL DESCRIPTION OF THE PROBLEM AND MAIN RESULTS

2.1 General Description

Consider the controlled chaotic system described by

$$\dot{x} = f(x) + g(x)\gamma \ . \tag{1}$$

Where $x \subset R^n$ is the state vector, $\gamma \subset R^1$ is the control, and both f and g are smooth vector fields on R^n.

We now study the chaos control using only partial states. Without loss of generality, assume that the partial state is expressed on a one-dimensional (1-D) manifold as follows:

$$y = h(x) \ . \tag{2}$$

Where h is a scalar smooth field on R^n, which may contain only partial information about the states. Using the well-known input-output linearization approach, (1) can be reformulated as

$$\dot{\mu}_i = \mu_{i+1}, \ i = 1, ..., r-1 \ , \tag{3}$$

$$\dot{\mu}_r = a(\mu,\phi) + b(\mu,\phi)\gamma \ , \tag{4}$$

$$\dot{\phi} = w(\mu,\phi) \ . \tag{5}$$

Where $\mu_i = L_f^{i-1}h(x)$ for $i = 1, ..., r-1$, $a(\mu,\phi) = L_f^r h(x)$, and $b(\mu,\phi) = L_g L_f^{r-1} h(x)$, r being the relative degree of $h(x)$, with output defined as $y = \mu_1$ and state vectors

$$\mu = (\mu_1, \ ... \ ,\mu_r)^T, \quad \phi = (\phi_1, \ ... \ ,\phi_{n-r})^T \ .$$

In the UPO tracking control, the reference signal is the time delayed state with delay constant τ, where τ is the period of the target UPO. Choose the manifold as

$$s = \sum_{i=1}^{r} c_i e_i \ , c_r = 1 \ . \tag{6}$$

Where $e_i = \mu_i(t-\tau) - \mu_i(t)$, $i = 1, \dots, r$, c_i, $i = 1, \dots, r$, and (6) is Hurwitz. If $s = 0$, then $e_i \to 0$, $t \to \infty$, $i = 1, \dots, r$; thus $\mu_i(t) \to \mu_i(t-\tau)$, $t \to \infty$, $i = 1, \dots, r$. Assume that the zero dynamics, $\dot{\phi} = w(0,\phi)$, is asymptotically stable, that is $\phi_i(t) \to \phi_i(t-\tau)$, $t \to \infty$, when $e_i = 0$, $i = 1, \dots, n-r$.

The control task is to make s an invariant manifold, so that s=*0* becomes an attractor. When this is done, from the above assumptions, tracking a target UPO is done. Taking the time derivative of s leads to

$$\begin{aligned} \dot{s} &= \sum_{i=1}^{r-1} c_i \dot{e}_i + \dot{e}_r = \sum_{i=1}^{r-1} c_i \dot{e}_i - \dot{\mu}_r \\ &= \sum_{i=1}^{r-1} c_i \dot{e}_i - (a(\mu,\phi) + b(\mu,\phi)\gamma) \ . \end{aligned} \tag{7}$$

Using (3)-(5) in (7) yields

$$\begin{aligned} \gamma = {} & b^{-1}(\mu,\phi)(\sum_{i=1}^{r-1} c_i(\dot{\mu}_i(t-\tau) - \dot{\mu}_i(t)) \\ & + \dot{\mu}_r(t-\tau) - a(\mu,\phi)) - b^{-1}(\mu,\phi)\dot{s} \ . \end{aligned} \tag{8}$$

In the design, one necessary step is to adjust the controller to ensure that $b(\mu,\phi) = L_g L_f^{r-1} h(x) > 0$, $b^{-1}(\mu,\phi)$ exists, and $|b^{-1}(\mu,\phi)| \le \rho$, where ρ is an appropriate positive constant.

2.2 Repetitive Learning Control

The formulae (8) can be expressed as follows:

$$\begin{aligned} & \gamma = \mathbf{\Psi}^T(\mu,t)\mathbf{\theta}(\mu(t-\tau),t) \ , \\ & + \kappa(\mu, \mu(t-\tau), t) - b^{-1}(\mu,\phi)\dot{s} \ . \end{aligned} \tag{9}$$

Where, $\mathbf{\Psi} \subset R^{n_1}$ is a known vector of μ and t,

$\theta \in R^{n_1}$ is the learnable structured uncertainty, which will be the repetitive learning control part; κ is the unstructured uncertainty function vector with a known finite upper bound, that is, $\|\kappa\| \le l_\kappa$, $l_\kappa > 0$; $n_1 \in Z^+$. In this arrangement, there is an option in assigning a few of the partially known terms. The terms involving

$$b^{-1}(\mu,\phi)\sum_{i=1}^{r-1} c_i(\dot{\mu}_i(t-\tau) - \dot{\mu}_i(t)) + \dot{\mu}_r(t-\tau))$$

have a bounding function and can be included in κ. The same terms can also be factorized into known state dependent functions and unknown state-independent functions, and therefore can be put in $\Psi^T(\mu,t)\theta(\mu(t-\tau),t)$. The control structure to be chosen will contain two parts. The first part is to learn the structured uncertainty repetitively and then control it. The second part is the robust control, which suppresses the unstructured uncertainty.

In order to derive the repetitive learning control, firstly assume that the learning control is done repeatedly over time based on a finite time interval $[0,\tau]$. It consists of two parts:

$$\gamma_i(t) = \Psi_i^T \mathbf{v}_i(t) + \mathbf{w}_i(t)\,, \Psi_i = \Psi_i(\mu,t,\tau)\,. \quad (10)$$

Where i is the number of learning trials and $\mathbf{v}_i(t) \in R^{n_1}$ is the recursive learning control part updated as follows:

$$\mathbf{v}_i(t) = \mathbf{v}_{i-1}(t) + \alpha\Psi_{i-1}s_{i-1}(t)\,. \quad (11)$$

Where $\alpha > 0$ is a feed-forward gain, $\mathbf{w}_i(t) \in R^{n_1}$ is the robust control part, which can be determined by minimizing the difference of an evaluation function to be present later. That is, the control in the chaotic system to be controlled consists of two parts; one is the leaning part $\Psi_i^T \mathbf{v}_i(t)$, and another one is the robust control $\mathbf{w}_i(t)$, which is used to control the uncertainty of that includes the uncertainty caused by unknown parameters.

2.3 Controller Design

The evaluation function used for the learning control law as follows:

$$E_i(t) = \int_0^t \|\theta(\xi) - \mathbf{v}_i(\xi)\|^2 d\xi\,, i = 1,2,\ldots \quad (12)$$

It can be easily verified that

$$\Delta E_i = E_{i+1} - E_i$$

$$= \int_0^t [\mathbf{v}_{i+1} - \mathbf{v}_i]^T [\mathbf{v}_{i+1} + \mathbf{v}_i - 2\theta] d\xi$$

$$= \int_0^t [\alpha^2 s_i^2 \Psi_i^2 - 2\beta\Psi_i^T s_i(\theta - \mathbf{v}_i)] d\xi\,. \quad (13)$$

From equations (9) and (10), we have

$$\gamma_i(t) = \Psi_i^T\theta + \kappa_i - b_i^{-1}\dot{s}_i(t) = \psi_i^T \mathbf{v}_i + w_i\,.$$

Therefore,

$$\Psi_i^T(\theta - \mathbf{v}_i) = w_i - \kappa_i + b_i^{-1}\dot{s}_i(t). \quad (14)$$

Substituting (14) into (13) yields

$$\Delta E_i(t) = \int_0^t [\alpha^2 s_i^2 \Psi_i^2 - 2\alpha s_i(w_i - \kappa_i + b_i^{-1}\dot{s}_i)] d\xi$$

$$= -\alpha b_i^{-1} s_i^2\Big|_0^t + \int_0^t \left[\alpha^2 s_i^2 \Psi_i^2 - 2\alpha s_i(w_i - \kappa_i - \frac{1}{2}\dot{b}_i^{-1} s_i)\right] d\xi$$

$$\le \int_0^t [\alpha^2 s_i^2 \Psi_i^2 - 2\alpha s_i w_i + 2\alpha|s_i\kappa_i| + \alpha|\dot{b}_i^{-1}| s_i^2] d\xi$$

$$- \alpha b_i^{-1} s_i^2\Big|_0^t\,. \quad (15)$$

In order to ensure the decay of the evaluation function to zero and, therefore, ensure the convergence of the learning algorithm, the robust control part is chosen as

$$w_i = \frac{1}{2}\alpha\Psi_i^2 s_i + (\frac{1}{2}\rho + \delta)s_i + l_\kappa \operatorname{sgn}(s_i)\,, \quad (16)$$

so that we can make $\Delta E_i(t) < 0$, where $\delta > 0$ is a constant feedback gain. Putting (16) in (15) yields

$$\Delta E_i(t) \le -\alpha s_i^2(\xi) b_i^{-1}(\xi)\Big|_0^t - 2\int_0^t \alpha\delta s_i^2 d\xi\,. \quad (17)$$

Therefore, at the i th iteration, for the final time of the period instant $t = \tau$, we have

$$\Delta E_i = \Delta E_i(\tau) \le -\alpha b_i^{-1} s_i^2(\xi)\Big|_0^\tau - 2\int_0^\tau \alpha\delta s_i^2 d\xi\,. \quad (18)$$

Theorem: For the controlled chaotic system (1), if the learning control is designed as

$$\mathbf{v}(t) = \begin{cases} 0 & if\ t < \tau \\ \mathbf{v}(t-\tau) + \alpha\Psi^T(\mu,t-\tau,\tau)s(t-\tau) & otherwise. \end{cases} \quad (19)$$

Where

$$w(t) = \frac{\alpha}{2}\Psi^2(\mu,t,\tau)s(t) + l_\kappa \operatorname{sgn}(s(t)) + \beta s(t)\,,$$

$$\beta = \frac{1}{2}(\rho + \delta)\,, \quad (20)$$

$$\gamma(t) = \Psi^T(t)\mathbf{v}(t) + w(t)\,, \quad (21)$$

then $s(t) \to 0$ asymptotically, that is, *s=0* becomes an attractor.

Proof. For the chosen evaluation function

$$E(t) = \int_0^\infty \|\mathbf{v}(s) - \theta\|^2 d\xi\,, \quad (22)$$

by segmenting the time axis $[0, \infty)$ into a series of even time intervals of the form $[i\tau, (i+1)\tau]$, $i = 0,1,\ldots$, equation (22) can be rewritten as

$$E(t) = \sum_{i=0}^\infty E_i = \sum_{i=0}^\infty \int_{i\tau}^{(i+1)\tau} \|\theta - \mathbf{v}(s)\|^2 d\xi$$

$$= \sum_{i=0}^\infty \int_0^\tau \|\theta - \mathbf{v}_i(s)\|^2 ds\,. \quad (23)$$

Where

$$\mathbf{v}_i(t) = \mathbf{v}(i\tau + t), \quad t \in [0, \tau]\,, \quad (24)$$

is the learning control part. Using the above definition, the learning control $\mathbf{v}(t)$ in (19) can be expressed in the following recursive form:

$$\mathbf{v}_i(t) = \mathbf{v}_{i-1}(t) + \alpha \mathbf{\Psi}_{i-1}^T(\mu(t), t, \tau) s_{i-1}(t) . \quad (25)$$

In this case, both E_i and $\mathbf{v}_i(t)$ can be treated as being defined over the finite periodic interval $[0, \tau]$. Notice that the evaluation functions (12), (24) and the learning control formulas (25) and (19) now have the same forms. If we treat a RLC process as an iterative learning control process, based on the ILC principle, the state starting point of current iteration must be the finishing point of the previous iteration, that is $\mu_i(\tau) = \mu_{i+1}(0)$. Therefore, $s_i(\tau) = s_{i+1}(0)$ that holds for all trials. It thus follows from (18) that

$$\begin{aligned} &\sum_{i=0}^{k} \Delta E_i = E_{k+1} - E_0 \\ &\le -\sum_{i=0}^{k} \alpha b_i^{-1}(\tau) s_i^2(\tau) + \sum_{i=0}^{k} \alpha b_i^{-1}(0) s_i^2(0) - \sum_{i=0}^{k} \alpha \delta \int_0^\tau s_i^2 d\xi \\ &= -\sum_{i=1}^{k+1} \alpha b_i^{-1}(0) s_i^2(0) + \sum_{i=0}^{k} \alpha b_i^{-1}(0) s_i^2(0) - \sum_{i=0}^{k} \alpha \delta \int_0^\tau s_i^2 d\xi \\ &= -\alpha b_{k+1}^{-1}(0) s_{k+1}^2(0) + \alpha b_0^{-1}(0) s_0^2(0) - \sum_{i=0}^{k} \alpha \delta \int_0^\tau s_i^2 d\zeta . \quad (26) \end{aligned}$$

Consequently,

$$\sum_{i=0}^{k} \alpha \delta \int_0^\tau s_i^2 d\xi \le E_0 + \delta s_0^2(0) ,$$

Furthermore,

$$\sum_{i=0}^{k} \int_0^\tau s_i^2 d\xi \le \frac{1}{\alpha\delta}(E_0 + \delta s_0^2(0)) . \quad (27)$$

As k increases, the right-hand side of (27) maintains as a constant. So $\int_0^\tau s_i^2 d\xi \to 0$, $i \to \infty$ thus resulting in $s_i^2 \to 0$, $i \to \infty$. Consequently, there exists an instant, $i_0 > 0$, such that whenever $i > i_0$, $|s_i(\tau)| < |s_i(0)|$. That is, from the relation $s_i(\tau) = s_{i+1}(0)$, $|s_{i+1}(0)| < |s_i(0)|$. This contraction ensures the asymptotical convergence of the system state to the UPO, that is, $s(t) \to 0$ asymptotically. This completes the proof.

Note that when *s=0* is reached, because of (6), there exist $e_i \to 0$, $t \to \infty$, $i = 1, \dots, r$; thus $\mu_i(t) \to \mu_i(t-\tau)$, $t \to \infty$, $i = 1, \dots, r$. Based on the assumptions that the zero dynamics $\dot\phi = w(0, \phi)$ is asymptotically stable and that $\phi_i(t) \to \phi_i(t-\tau)$, for $t \to \infty$, $i = 1, \dots, n-r$ when $e_i = 0$, then the tracking UPO task is completed.

3 SIMULATION RESULTS

In order to demonstrate the effectiveness of the learning control scheme for tracking UPOs, we have done simulations on controlling the Duffing and Lorenz systems.

3.1 Duffing System

Firstly consider the tracking of UPOs in the Duffing system. Assume that only the knowledge of the value of the target period is known. The controlled Duffing equation is introduced as follows

$$\begin{aligned} &\dot\mu_1 = \mu_2 \\ &\dot\mu_2 + p_1\mu_1 + p_2\mu_2 + \mu_1^3 - q\cos(\omega t) = \gamma(t), \end{aligned} \quad (28)$$

with $p_1 = -1.1$, $p_2 = 0.4$, $q = 2.3$, $\omega = 1.8$. The Duffing system exhibits chaos [7] and its trajectories are uniformly bounded, see Figure 5.1.

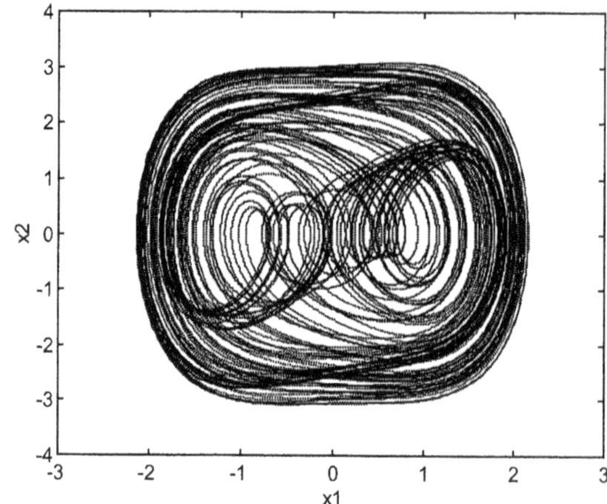

Figure1. Duffing Chaos

For convenience, reformulate the Duffing system as follows:

$$\begin{aligned} &\dot\mu_1 = \mu_2 \\ &\dot\mu_2 = -p_1\mu_1 - p_2\mu_2 - \mu_1^3 + q\cos(\omega t) + \gamma(t). \end{aligned} \quad (29)$$

Formulae (29) has the normal form, in this case, with $b = 1, \dot b^{-1} = 0, \rho = 0$. Let $\mu = [\mu_1 \ \mu_2]^T$. The control objective is to ensure the following condition to be satisfied: Given a small tolerance $\varepsilon > 0$, there exists a $t_1 > 0$, such that $t > t_1$, $\|\mu(t-\tau) - \mu(t)\| < \varepsilon$, where τ is an inherent period of target UPO of the Duffing system. The ultimate goal is to enable $\mu(t) \to \mu(t-\tau)$ for $t \to \infty$, which indicates that the controlled state $\mu(t)$ exhibits a periodic orbit of period τ. To design the repetitive learning control, as in formulae (9), an invariant manifold into which we want to drive the Duffing system is designed as follows:

$$s(t) = c_1(\mu_1(t-\tau) - \mu_1(t)) + (\mu_2(t-\tau) - \mu_2(t)), c_1 > 0 . (30)$$

It is apparent that when $s = 0$ is realized, $\mu(t) \to \mu(t-\tau)$. From (22), we have

$$\mu_2(t-\tau) - \mu_2(t) = -c_1(\mu_1(t-\tau) - \mu_1(t)) . \, (31)$$

Since $\dot\mu_1 = \mu_2$ from (29), one can see that (31) becomes $\dot\mu_1(t-\tau) - \dot\mu_1(t) = -c_1(\mu_1(t-\tau) - \mu_1(t))$. Hence $\mu_1(t) \to \mu_1(t-\tau)$ so that $\mu(t) \to \mu(t-\tau)$. Taking the derivative of $s(t)$ with respect to t, one have

$$\begin{aligned} \dot s(t) &= c_1(\mu_2(t-\tau) - \mu_2(t)) + \dot\mu_2(t-\tau) \\ &+ p_1\mu_1 + p_2\mu_2 + \mu_1^3 - q\cos(\omega t) - \gamma(t) . \end{aligned}$$

Therefore, the control $\mu(t)$ is

$$\gamma(t) = -c_1\mu_2(t) + c_1\mu_2(t-\tau)$$

$$+\dot{\mu}_2(t-\tau)+p_1\mu_1+p_2\mu_2+\mu_1^3-\dot{s}(t)$$

$$=\Psi^T(x,t)\theta(t)+\kappa(x,t,\tau)-\dot{s}(t),$$

$$\Psi(\mu,t,\tau)=[-c_1\mu_2+c_1\mu_2(t-\tau)+\dot{\mu}_2(t-\tau)\quad \mu_1\quad \mu_2\quad \mu_1^3]^T$$

$$\theta(t)=[1\quad p_1\quad p_2\quad 1]^T,\ \kappa(\mu,t,\tau)=-q\cos(\omega t).$$

The RLC algorithm presented in the Theorem is readily applicable. In the simulation, the computation step size was 0.001 and it is well known that the period one UPO of the Duffing chaos is 3.491. The parameters for the controller were given as

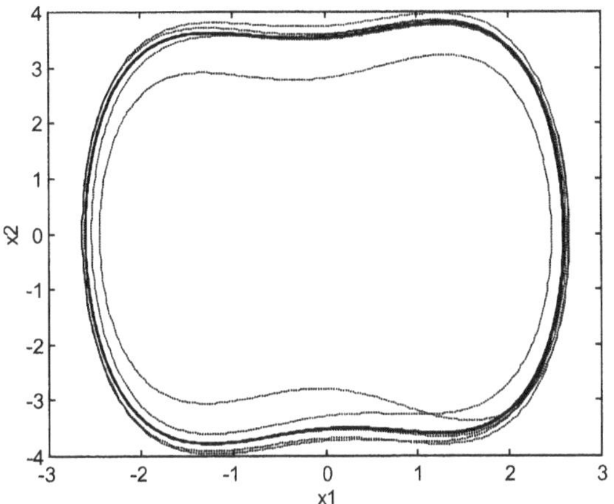

Figure 2. Controlled Duffing system

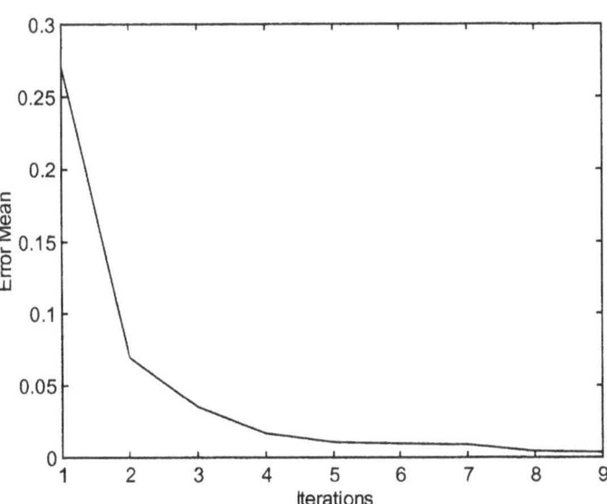

Figure 3. Error versus iteration

$\alpha=0.001$, $c_1=2, \beta=0.01$, $l_\kappa=0.05$. To make sure the controlled system can track the inherent UPO, we designed a criterion as

$$\frac{1}{N}\sum_{j=1}^{N}\|\mu_1(j-\tau)-\mu_1(j)\|$$

to monitor the chaotic dynamics in real time. When this is less than ε (in our simulation, $\varepsilon=0.06$, $N=3491$ j is the iteration number), starts the learning control. Figure 2 illustrates the controlled Duffing chaos, the initial value is (-1.5, 2). Figure 3 depicts the error versus iteration, showing clearly the convergence of the error mean of $\mu_1(t-\tau)-\mu_1(t)$. One can see that chaos is well under control.

3.2 Lorenz system

Consider control of the Lorenz system is as follows:

$$\begin{aligned}\dot{x}&=c(y-x)\\ \dot{y}&=rx-y-xz+\gamma.\\ \dot{z}&=xy-bz\end{aligned}\qquad(30)$$

Where c,r,b are 10, 28 and 8/3, respectively. Without any control, $\gamma=0$, the system shows the well-known butterfly chaos (see Figure 4).

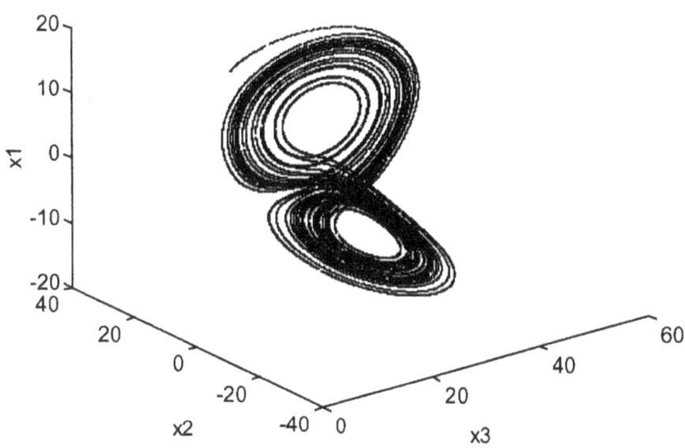

Figure 4. Lorenz Chaos

Here, we add a control in the second equation of (32) and choose output as $h=y$, then the output has relative degree one. Rewrite (32) as follows:

$$\begin{aligned}\dot{\mu}&=r\phi_1-\mu-\phi_1\phi_2+\gamma\\ \dot{\phi}_1&=c(\mu-\phi_1)\\ \dot{\phi}_2&=\mu\phi_1-b\phi_2.\end{aligned}\qquad(33)$$

Let $X_\tau=X(t-\tau),\ X=\gamma,\mu,\phi_1,\phi_2$. We then have

$$\begin{aligned}\dot{\mu}_\tau&=r\phi_{1\tau}-\mu_\tau-\phi_{1\tau}\phi_{2\tau}+\gamma_\tau\\ \dot{\phi}_{1\tau}&=c(\mu_\tau-\phi_{1\tau})\\ \dot{\phi}_{2\tau}&=\mu_\tau\phi_{1\tau}-b\phi_{2\tau}.\end{aligned}\qquad(34)$$

Let $e_{\phi 1}=\phi_{1\tau}-\phi_1; e_\mu=\mu_\tau-\mu; e_{\phi 2}=\phi_{2\tau}-\phi_2$. Subtracting formula (34) from formula (33) yields

$$\begin{aligned}\dot{e}_\mu&=re_{\phi 1}-e_\mu-(\phi_{1\tau}\phi_{2\tau}-\phi_1\phi_2)+\gamma_\tau-\gamma\\ \dot{e}_{\phi 1}&=c(e_\mu-e_{\phi 1})\\ \dot{e}_{\phi 2}&=\mu_\tau\phi_{1\tau}-\mu\phi_1-be_{\phi 2}.\end{aligned}\qquad(35)$$

The objective of control is to make a chosen properly manifold invariant, so as to have $e_\mu\to 0$, $e_{\phi 1}\to 0$, $e_{\phi 2}\to 0$. One can see, from the first equation in (35), that if choose the manifold as $s = e_\mu$ then when $e_\mu \to 0$ there will be $e_{\phi 1} \to 0$, so that $\mu_\tau \phi_{1\tau} \to \mu\phi_1$. Furthermore, from third equation in (35), one knows $e_{\phi 2} \to 0$. From (35), there is $\dot{e}_\mu = re_{\phi 1} - e_\mu - (\phi_{1\tau}\phi_{2\tau} - \phi_1\phi_2) + \gamma_\tau - \gamma$; therefore,

$$\gamma_\tau - \gamma = re_{\phi 1} - e_\mu - (\phi_{1\tau}\phi_{2\tau} - \phi_1\phi_2) - \dot{s} = \mathbf{\Psi}^T(t)\mathbf{\theta} - \dot{s}, \quad (36)$$

where $\mathbf{\Psi} = (e_{\phi 1} \quad -e_\mu - \phi_{1\tau}\phi_{2\tau} + \phi_1\phi_2)^T$, $\mathbf{\theta} = (r\ 1)^T$. Comparing (36) with formula (8) and letting $\Delta\gamma = \gamma - \gamma_\tau$, one can see that the augmented control, $\Delta\gamma$ satisfy the learning control design structure. The iterative learning algorithm can be given as follows:

$$\mathbf{v}_i = \mathbf{v}_{i-1} + \alpha\mathbf{\Psi}_{i-1}^T s_{i-1}$$

$$\Delta\gamma_i = \mathbf{\Psi}_i^T \mathbf{v}_i + w_i$$

$$\gamma_i = \gamma_{i-1} + \Delta\gamma_i$$

Where i represents the i th iteration. Moreover, the repetitive control algorithm for Lorenz chaos can be implemented as follows:

$$\mathbf{v}(t) = 0,\ t < \tau$$

$$v(t) = \mathbf{v}(t-\tau) + \alpha\mathbf{\Psi}(t-\tau)s(t-\tau),\ t \geq \tau$$

$$\Delta\gamma(t) = \mathbf{\Psi}^T(t)v(t) + w(t)$$

$$\gamma(t) = \gamma(t-\tau) + \Delta\gamma(t)$$

In simulation, the step size was chosen as 0.001. The dynamics of chaos was shown in figure 4 with the initial point *(16.7447, 12.0664, 13.1038).* The period of UPO we tried to track was *1.717,* which was for a period 2 UPO. We also designed a criterion as

$$\frac{1}{N}\sum_{j=1}^{N}\left\| y(j-\tau) - y(j) \right\|.$$

When this is less than a small positive $\varepsilon = 0.05$, the learning control switches on. The parameters used were $\alpha = 0.05$, $l_\kappa = 0.03$, $\beta = 0.05$. When we started the chaotic system controlled with the initial value (16.7447, 12.0664, 13.1038), after *1143.52* seconds, system began entering the neighborhood of the target UPO, the learning control was then started. From Figure 5, one can see that the system states never move out of its UPO after that moment.

4. CONCLUSION

In this paper, a time-delayed chaos control method based on the repetitive learning principle has been developed; two simulations, on Duffing system and Lorenz system, were presented in order to illustrate the effectiveness of the proposed approach. The method combines RLC and TDFC, so is very different from that of [Konishi, K. and H. Kokame, 1997, 1998]. The main advantage of the integration of RLC and TDFC is that the controller is able to adaptively learn from learning cycles an appropriate control force. Unlike conventional RLC, the new controller does not require exact knowledge (analytic representation) of the target UPO except the time delay constant, which however can be easily identified via either experiments or adaptive learning techniques. All these together make the new design a valuable technique for chaos control to UPOs.

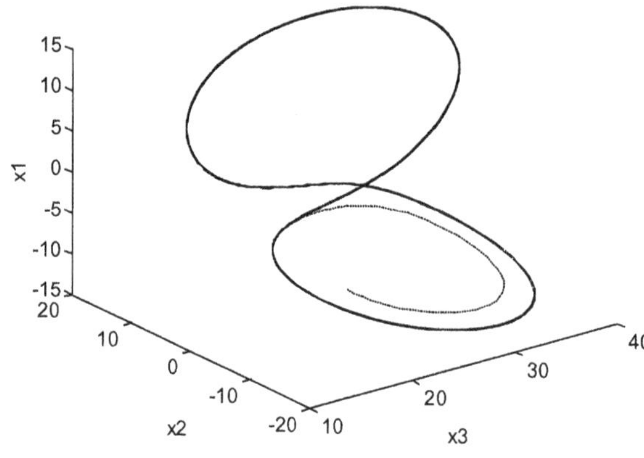

Figure 5. The controlled Lorenz system to Period-2

REFERENCE

Bien, Z. and J.-X. Xu, (1998). Iterative Learning Control - Analysis, Design, Integration and Applications, Kluwer Academic Press, Boston, USA.

Chen, G. and X. Dong, (1993). "On feedback control of chaotic continuous time systems," *IEEE Trans. Circuits and Systems - Part I,* vol. 40, pp. 591-600.

Konishi, K. and H. Kokame, (1997). "Control of chaotic systems using an on-line trained linear controller", *Physica D,* vol. D100, pp. 423-438.

Konishi, K. and H. Kokame, (1998). "Learning control of time-delay chaotic systems and its applications," *Int. J. Bifurcation Chaos,* vol.8, no.12, pp. 2457-2465.

Moore, K. L. (2000). "A non-standard iterative learning control approach to tracking periodic signals in discrete-time nonlinear systems," *Int. J. Control,* vol. 73, no.10, pp. 955-967.

Pyragas, K. (1992). "Continuous control of chaos by self-controlling feedback," *Phys. Lett. A,* vol. A170, pp. 421-428.

Xu, J.-X., B. Viswanathan and Z. Qu, (2000). "Robust learning control for robotic manipulators with an extension to a class of nonlinear systems," *Int. J. Control,* vol. 73, no.10, pp. 858-870.

Yu, X., Y.-P. Tian and G. Chen, (1999). "Time delayed feedback control of chaos," in Controlling Chaos and Bifurcations in Engineering Systems, G. Chen (Ed), pp. 255-274, CRC Press.

Yu, X. (1999). "Tracking inherent periodic signals in chaotic systems using adaptive variable structure time delayed control," *IEEE Trans. Circuits and Systems - Part I,* vol. 46, no. 11, pp. 1408-1411.

www.elsevier.com/locate/ifac

OPTIMAL ESTIMATION AND CONTROL OF CONTINUOUS SYSTEMS WITH TIME-VARYING DELAYS

Michael V. Basin * Mikhail Skliar **

* *Department of Physical and Mathematical Sciences, Autonomous University of Nuevo Leon, Mexico*
** *Department of Chemical and Fuels Engineering, University of Utah, USA*

Abstract: This paper provides the solution of optimal filtering problems for the broad class of continuous linear systems with discrete and continuous measurements with time-varying sampling intervals and time-varying measurement delays caused by communication network, human in the loop and a priori unknown triggering of data acquisition. Using the duality principle, the dual control problem with delays in actuation is then solved. The solution of filtering and control problems is obtained using the integral model of linear dynamic systems in the form of Ito-Volterra integrals with discontinuous measures and later reduced to state space systems. *Copyright © 2001 IFAC*

Keywords: Time delay systems, measurement delays, actuation delays, optimal filtering, optimal control

1. INTRODUCTION AND MOTIVATION

The importance of the posed problem is most easily demonstrated by the example of the networked control systems. The decentralized and distributed control networks are becoming the reality in a variety of application areas. The practical interest in systems with a communication network as part of the information path for data acquisition and control is driven by the need for integration and coordination of multiple spatially remote processes. It has been recognized for some time that a network in the loop presents some unusual challenges. For example, the nondeterministic nature of network traffic leads to time-varying delays in the delivery of streaming measurements (which we treat in this paper as continuous [1]). Discrete (or packet) measurements are also delayed by a priori unknown and varying time. Furthermore, the loss of data during network delivery will effectively result in the time-varying sampling rate even if the sampling at the remote site is uniform and deterministic. From the control perspective, based on delayed information, the remote controller calculates commands (setpoints), which, because of the network properties, will be delayed in their arrival to the local controllers. Therefore, from the remote controller "perspective" we have delayed actuation of generated commands, and these delays are random and a priori unknown. Further examples of systems with time-varying and a priori unknown delays in discrete and continuous measurements and controls include systems with human in the loop making sampling and control decisions, and systems in which measurement and control are triggered by exogenous events.

[1] In reality, streaming measurements may represent discrete measurements sampled at a relatively high rate.

Our approach to the solution of the filtering and control problems with delayed discrete and continuous measurements and time-varying sampling rates is based on the integral Volterra description of deterministic linear systems and the integral Ito-Volterra description of stochastic linear systems, and the mathematical theory of optimal control of systems with discontinuities. We refer to this approach as an *integral* approach. This approach is applicable to systems with discontinuities in measurements, controls and states, – increasingly important theoretical and practical problems – and systems with time-varying and a priori unknown delays.

This paper has the following organization. We begin by introducing Ito-Volterra systems, first, for continuous systems, followed by the systems with discontinuities. We then present the optimal filtering results for these systems and their generalization on the case of systems with time-varying and a priori unknown time delays. After introducing the duality principle, we give the solution of the dual control problem with delays in actuation.

2. ITO-VOLTERRA DESCRIPTION OF DYNAMIC SYSTEMS

Let (Ω, F, P) be a complete probability space with an increasing right-continuous family of σ-algebras $F_t, t \geq 0$, and let $(W_1(t), F_t, t \geq 0)$ and $(W_2(t), F_t, t \geq 0)$ be independent unit-intensity Wiener processes. Here Ω is the sample space, F is a set of subsets on which the probability measure (or, simply, probability) is defined, and P is the probability defined on F. All subsets of F form a σ-algebra, and F_t denotes a family of subsets (σ-algebra) for each t such that for $t_1 < t_2$, $F_{t_1} \subset F_{t_2}$. The partially observed F_t-measurable random process $(x(t), y(t))$ can be described using the Ito-Volterra equations:

$$x(t) = \int_0^t a_0(t,s) + a(t,s)x(s)ds + \int_0^t b(t,s)dW_1(s) \quad (1)$$

$$y(t) = \int_0^t A_0(t,s) + A(t,s)x(s)ds + \int_0^t B(t,s)dW_2(s) \quad (2)$$

where $x(t) \in R^n$ is the state vector, and $y(t) \in R^m$ is a vector of measurements [2] *integrated* over the time interval $[0, t]$. The vector-valued function $a_0(t, s)$ describes the effect of system inputs (controls and disturbances). Matrix functions $a(t, s)$ and $b(t, s)$ and vector-function $a_0(t, s)$ are smooth functions of t uniformly in s, i.e. for each s these functions are smooth in t independently of s. Functions $A_0(t, s)$, $A(t, s)$, and $B(t, s)$ are continuous in t and s. Both t and s are independent (time) variables, and can be used, among other things, to assign a variable number of time-varying delays in both states and measurements to adequately describe the particular application at hand. We also assume that $A(t, s)$ is a nonzero matrix and $B(t,s)B^T(t,s)$ is a positive definite matrix. All coefficients in (1)–(2) are deterministic functions of appropriate dimensions.

The estimation problem is to find the estimate of the system state $x(t)$ described by the Ito–Volterra model (1) based on the observation process $Y(t) = \{y(s), 0 \leq s \leq t\}$, which minimizes the Euclidean 2-norm $J = E[(x(t) - \hat{x}(t))^T(x(t) - \hat{x}(t))]$ at each time moment t. In other words, our objective is to find the conditional expectation $m(t) = \hat{x}(t) = E(x(t) \mid F_t^Y)$. As usual, the matrix function $P(t) = E[(x(t) - m(t))(x(t) - m(t))^T \mid F_t^Y]$ is the estimate variance.

Our formulation is, in fact, the Kalman filtering problem for the integral Ito-Volterra system. The standard state space formulation is recovered by making all functional parameters in (1) and (2) dependent on s only.

The solution of the optimal filtering problem for the system (1)–(2) was recently reported in Basin and Villanueva Llanes (1999) by generalizing results Kleptsina and Veretennikov (1985); Shaikhet (1987) for systems with Ito-Volterra dynamics and standard differential measurements. It can be shown that the variance $P(t)$ alone is not sufficient to completely characterize the state estimation process and to obtain a closed form of filtering equations for dynamic systems in the integral form. However, the explicit solution can be obtained in terms of the integral cross-correlation function $f(t, s)$, which characterizes the deviation of the optimal estimate $m(t)$ from an unknown true state $x(t)$, and defined as: $f(t,s) = E[(x(s,t) - m(s,t))(x(s) - m(s))^T \mid F_{t,s}^Y]$ where $x(s, t)$ can be viewed as a state with independent (time) variable s and parameter t, and is equal

$$x(s,t) = \int_0^s a_0(t,r) + a(t,r)x(r)dr + \int_0^s b(t,r)dW_1(r) \quad (3)$$

The governing equation for $x(s, t)$ can be differentiated with respect to s to yield the state space form of equation (3). $F_{t,s}^Y$ is the σ-algebra generated by the stochastic process $y(s, t)$:

$$y(s,t) = \int_0^s A_0(t,s) + A(t,s)x(s)ds + \int_0^s B(t,s)dW_2(s) \quad (4)$$

[2] In integral formulation, $y(t)$ is a vector of integrated measurements. Differential measurements, $dy(t)$, is the vector of actual physical measurements. Note, however, that this interpretation may change during, as is the case with equation (18), where $y(t)$ is treated as the vector of actual measurements.

and $m(s,t) = E[x(s,t) \mid F^Y_{t,s}]$, where we again treat t as a parameter. Note that function f is a generalization of the variance P since $f(t,t) = P(t)$. Furthermore for $s = t$, $x(s,t) = x(t)$ and $y(s,t) = y(t)$.

Next, we outline the integral model for systems with discontinuities in measurements.

3. OPTIMAL FILTERING FOR SYSTEMS WITH BOUNDED DISCONTINUITIES

Consider a nondecreasing vector-valued function of bounded variation: $g(t) = (g_1(t), \ldots, g_m(t)) \in R^m$. In essence, $g(t)$ is an arbitrary function, and it is only required that it remains bounded on each finite subinterval of its definition, and $g(t_1) \leq g(t_2)$ if $t_1 \leq t_2$. Continuity of $g(t)$ is not required. In fact, any bounded variation function (including vector-valued case) can be written as $g(t) = \{g^c_k(t) + \sum_{i=1}^N \Delta g_{ki}\chi(t - t_{ki}),\ k = 1 \ldots m\}$ where $g^c_k(t)$ is a continuous nondecreasing function and the second term describes bounded jumps in k-th components of $g(t)$ at times t_{ki}, and where χ is the Heaviside unit step function and Δg_{ki} is the size of the jump.

The discontinuous measure g can be used to describe discontinuities in states and measurements. The Ito-Volterra model with discontinuous measure in the observation equation generated by a bounded variation function has the following form:

$$\begin{aligned} y_k(t) = & \int_0^t (A_{0k}(t,s) + (A_k(t,s)x(s))dg_k(s) \\ & + \int_0^t B_k(t,s)dW_{2k}(g_k(s)),\ k = 1 \ldots m \quad (5) \end{aligned}$$

where the notation is analogous to the one used in (2), k identifies k-th component of the measurement vector, and g_k is the k-th component of g.

The measurement model given by equation (5) is in the Stieltjes integral form. Obviously, if $g(t) = t$, the description of the system is reduced to the initial model (1)–(2). If $g^c \equiv 0$, the observation model given by equation (5) describes the case of a continuous system with discrete measurements. The general case of $g(t)$ describes the dynamic system with an arbitrary combination of discrete and continuous measurements.

The general result for the optimal filtering of integral systems with measurement discontinuities is given by the following theorem.

Theorem 1. The optimal in Kalman sense estimate $m(t)$ of the states of system (1) based on discontinuous integral measurements (5) satisfies the filter equation

$$\begin{aligned} m(t) = & \int_0^t (a_0(t,s) + a(t,s)m(s))ds \\ & + \int_0^t K(t,s)[dy(s) \\ & - (A_0(t,s) + A(t,s)m(s))dg(s)] \qquad (6) \end{aligned}$$

where $K(t,s) = f(t,s)A^T(t,s)(B(t,s)B^T(t,s))^{-1}$, and function $f(t,s)$ satisfies the Riccati-like equation

$$\begin{aligned} f(t,s) = & \int_0^s [a(s,r)f^T(t,r) + f(s,r)a^T(t,r) \\ & + \frac{1}{2}(b(t,r)b^T(s,r) + b(s,r)b^T(t,r))]dr \\ & - \int_0^s [K_{tsss}A(s,r)f^T(s,r) + K_{sttt}A(t,r)f^T(t,r) \\ & - \frac{1}{2}K_{ttts}A(s,r)f^T(s,r) \\ & - \frac{1}{2}K_{ssst}A(t,r)f^T(t,r)]dg(r) \qquad (7) \end{aligned}$$

$K_{tsss}(t,s,r) = f(t,r)A^T(s,r)[B(s,r)B^T(s,r)]^{-1}$, etc., and multiplication by an m-dimensional measure $dg(t)$ is in the componentwise sense, as in (5).

This result is identical in *form* to the optimal filter for Ito-Volterra systems without discontinuous measurements Basin and Villanueva Llanes (1999). The theoretical basis for obtaining the optimal Kalman filter for integral systems with discontinuous measurements is the theory of vibro-solutions of differential and integral equations Krasnoselskii and Pokrovskii (1989), which states that the solution of the integral equation with discontinuous measure $g(t)$ can be expressed as a sum of the continuous component in the form of the integral equation with continuous measure and the term that describes jumps in the solution at the points of discontinuity.

In the case of filtering with continuous and discrete measurements, the optimal estimate will include a continuous part due to continuous measurements plus the contribution from discrete measurements, which lead to discontinuities in state estimates at the arrival time of a discrete measurement. The jumps in the state estimates due to discrete measurements can be obtained *explicitly* for any combination of discrete and continuous measurements. However, the explicit expressions of the jumps are rather complicated in the general case. Therefore, in this paper we limit our discussion to continuous dynamic systems in differential formulation. In this case, the explicit expressions for jumps in optimal estimates and variance P have a transparent form.

3.1 Case of State Space Systems

For the state space description of the system dynamics

$$dx(t) = (a_0(t) + a(t)x(t))dt + b(t)dW_1(t) \quad (8)$$

and the measurement model (5), the optimal filter given by Theorem 1 has a particularly simple form:

$$m(t) = \int_0^t (a_0(s) + a(s)m(s))ds + \int_0^t P(s-)[I + A^T(t,s)\big(B(t,s)B^T(t,s)\big)^{-1}A(t,s) \times P(s-)\Delta g(s)]^{-1} \times A^T(t,s)\big(B(t,s)B^T(t,s)\big)^{-1}[dy(s) - (A_0(t,s) + A(t,s)m(s-))]dg(s), \quad (9)$$

where $\Delta g(s)$ is a jump of $g(s)$ at s, and the standard notation for the value of the function at discontinuity is used. The variance $P(t)$ is found from the integral Riccati-like equation

$$P(t) = \int_0^t [a(s)P(s) + P(s)a^T(s) + b(s)b^T(s)]ds - \int_0^t P(s-)[I + A^T(t,s) \times \big(B(t,s)B^T(t,s)\big)^{-1}A(t,s)P(s-)\Delta g(s)]^{-1} \times A^T(t,s)\big(B(t,s)B^T(t,s)\big)^{-1}A(t,s)P(s-)dg(s) \quad (10)$$

where multiplication by m-dimensional measure $dg(s)$ is understood in a componentwise sense. If the bounded variation function $g(s)$ has continuous and discontinuous components, then the corresponding observation process $y(t) = \{y_k(t)\}$ also has continuous and discontinuous components, and can be written as $y_k(t) = y_k^c(t) + y_k^d(t)$. Physically, the continuous component of y corresponds to the integral of continuous measurements with discontinuous (discrete) measurements superimposed over them. The discontinuity in y leads to discontinuity in estimate $m(t)$ and the variance function $P(t)$. At the point of discontinuity t_{ki} when a new discrete measurement becomes available in k-th measurement channel, the optimal value of $m(t)$ and $P(t)$ can be directly calculated from (9)–(10), where $\Delta g(t_i) = \{\Delta g_{ki}\}$, and typically $\Delta g_{ki} = 1$.

If only discrete measurements are present, then between the time of $t = t_i$ of the last available measurement in any of the measurement channels and the next measurement, the estimate of the state is given by the following integral equation:

$$\hat{x}(t) = m(t) = m(t_i+) + \int_{t_i+}^t (a_0(s) + a(s)m(s))ds, \quad (11)$$

and the variance of the estimation process is found from

$$P(t) = P(t_i+) + \int_{t_i+}^t [a(s)P(s) + P(s)a^T(s) + b(s)b^T(s)]ds. \quad (12)$$

The state estimate and variance at the time of arrival of the new discrete measurement are equal

$$m(t_i+) = m(t_i-) + \Delta m(t_i), \quad (13)$$

$$P(t_i+) = P(t_i-) + \Delta P(t_i), \quad (14)$$

where jumps are explicitly given as

$$\Delta m(t_i) = K(t_i)\{dy(t_i) - [A_0(t_i) + A(t_i)m(t_i-)]\} \quad (15)$$

$$\Delta P(t_i) = -K(t_i)A(t_i)P(t_i-) \quad (16)$$

where the shorthand notation $A(t_i) = A(t_i, t_i)$, $A_0(t_i) = A_0(t_i, t_i)$ and $B(t_i) = B(t_i, t_i)$ was used, and

$$K(t_i) = P(t_i-)\{I + A^T(t_i)\left[B(t_i)B^T(t_i)\right]^{-1} \times A(t_i)P(t_i-)\}^{-1}A^T(t_i)\left[B(t_i)B^T(t_i)\right]^{-1} \quad (17)$$

The second limiting case (only continuous measurements are present) is obtained by setting $g(t) = t$. The resulting filter for the system in the differential form are obtained from equations (9)–(10), and is equivalent to the traditional Kalman filter. However, if the system is described in the integral form (1), the optimal filter for continuous systems with continuous measurements is not reducible to the Kalman filter, and is given by Theorem 1.

4. FILTERING FOR SYSTEMS WITH TIME DELAYS

Let us consider how time delays in discrete and continuous measurements can be handled within the framework of the integral approach. We first note that the integral approach to this problem appears to be substantively different from infinite-dimensional, algebraic, and functional differential systems approach, to systems with time delay. In addition to giving an alternative approach, which is important in itself, the advantages of the integral formulation is in its consistent theoretical foundation, and its ability to address systems with discontinuities of different origin, as discussed earlier.

An immediate observation is that when both continuous and discrete time-delayed measurements are present, we have a problem with discontinuous observations considered above without time delays. Now define a bounded variation function $g(t,s) = \sum_{i=1}^{N} \Delta g_i \chi(s - [t - d_i(t)])$, where s is an independent time variable, and t is treated as a parameter; $d_i(t)$ is the i-th time delay at time t. After substitution of $g(t,s)$ into the integral measurement model with discontinuous measure, the equation (5) takes the form

$$\begin{aligned} y(t) &= \textstyle\sum_i A_0(t, t - d_i(t)) \\ &+ \textstyle\sum_i A(t, t - d_i(t)) x(t - d_i(t)) \\ &+ \textstyle\sum_i B(t, t - d_i(t)) dW_2(t), \end{aligned} \tag{18}$$

where y is interpreted as a vector of differential continuous measurement. We took into account that for the white Gaussian noise $dW_2(t)$, $dW_2(t - d_i(t)) = dW_2(t)$, and assumed that $\Delta g_i = 1$. Equation (18) is the model of the measurement system, describing continuous measurements $y(t)$ as a linear function of states delayed by different and time-varying delays $d_i(t)$. It is in a form that allows the direct application of Theorem 1, giving the expression for the optimal filter with an arbitrary combination of delayed continuous measurements.

The model (18) can also be used to describe the discrete measurement system with time-varying delays. If we set $t = t_j$, where t_j's are discrete time instants when discrete measurements become available (*a priori* knowledge of the arrival time and the corresponding time delay of a discrete measurement is not required) and treat vector y as an integral of discrete measurements, then (18) describes the case of a continuous system with delayed discrete measurements. Delays and sampling rates in different channels can be different and time varying. Theorem 1 is again directly applicable, yielding an optimal filter for the continuous system with delayed discrete measurements.

The case of an arbitrary combination of delayed continuous and discrete measurements is obtained when different components of vector y describe continuous differential measurements, and discrete integral measurements. Theorem 1 is still applicable giving an optimal filter for this most general case.

5. DUAL CONTROL PROBLEM

Using the duality principle for integral systems Basin and Valadez Guzman (2000), the optimal state estimation problem for system with discrete and continuous measurements, equations (1) and (5), is dual to the optimal control problem of finding control u, which minimizes the quadratic cost function

$$\begin{aligned} J &= \frac{1}{2} [x(T)]^T \Psi^{-1} [x(T)] \\ &+ \frac{1}{2} \int_{t_0}^{T} u^T(t,s) B^T(s) B(s) u(t,s) dg(s) \\ &+ \frac{1}{2} \int_{t_0}^{T} x^T(s) b^T(s) b(s) x(s) ds], \end{aligned} \tag{19}$$

subject to the following Ito-Volterra system with states z and disturbances a_o^T:

$$\begin{aligned} z(t) &= \int_{t_0}^{t} (a_0^T(t,s) - a^T(t,s) z(s)) ds \\ &+ \int_{t_0}^{t} A^T(t,s) u(t,s) dg(s), \end{aligned} \tag{20}$$

and where $P(t_0) = \Psi > 0$. From the solution of the dual filtering problem, we obtain that the optimal control is given as

$$u^*(t,s) = (B^T B)^{-1}(s) A^T(t,s) f^T(t,s) z(s) \tag{21}$$

and $f(t,s)$ is the solution of the integral Riccati equation

$$\begin{aligned} f(t,s) &= f(t_0,t_0) + \int_s^T [a(s,r) f^T(t,r) \\ &+ f(s,r) a^T(t,r) + bb^T(t,r)] dr \\ &- \int_s^T [f(t,r) A^T(s,r) (B^T B)^{-1} A(s,r) f^T(s,r) \\ &+ f(s,r) A^T(t,r) (B^T B)^{-1} A(t,r) f^T(t,r) \\ &- \frac{1}{2} f(t,r) A^T(t,r) (B^T B)^{-1} A(s,r) f^T(s,r) \\ &- \frac{1}{2} f(s,r) A^T(s,r) (B^T B)^{-1} A(t,r) f^T(t,r)] dg(r), \end{aligned} \tag{22}$$

with terminal condition $f(T,T) = \Psi^{-1}$, and where $f(t_0,t_0)$ is found as a result of the solution of the integral equation and is included to indicate generally nonzero initial conditions of function f. After substituting the optimal control into the state equation (20), the state trajectory under optimal control can be determined.

Calculation of u requires the solution of the integral equation (22) with discontinuous measure generated by $g(t)$. It can be shown that jumps in the solution of the Riccati equation and in the optimal trajectory of states $z(t)$ can be computed from the solution of the following system of differential equations:

$$\frac{dz(g)}{dg} = A^T(t,t)(B^T B)^{-1} b^T(t,t) f(t,g) z(g) \tag{23}$$

$$z(0) = z(t-), \quad g \in [0, \Delta g(t)],$$

$$\frac{df(t,g)}{dg} = \\ -f(t,g)A^T(s,s)(B^TB)^{-1}A(s,s)f^T(s,g) \\ -f(s,g)A^T(t,s)(B^TB)^{-1}A(t,s)f^T(t,g) \\ +\frac{1}{2}f(t,g)A^T(t,s)(B^TB)^{-1}A(s,s)f^T(s,g) \\ +\frac{1}{2}f(s,g)A^T(s,s)(B^TB)^{-1}A(t,s)f^T(t,g) \quad (24)$$

$$f(t,0) = f(t,s-), \quad g \in [0, \Delta g(s)],$$

which allow for the analytical solution. Given the analytical solution, jumps $\Delta z(t) = z(\Delta g(t)) - z(t-)$ and $\Delta f(t,s) = f(t, \Delta g(s)) - f(t, s-)$ can be explicitly found. The closed form solution has a particularly simple form for the case of the state space systems (a_0, $-a^T$ and A^T in (20) are functions of s only). The optimal control in this case is

$$u^*(t,s) = (B^T(s)B(s))^{-1}(s)A^T(t,s)P(s)z(s),$$

with $P(s)$ satisfies the integral Riccati equation

$$P(t) = P(t_0) + \textstyle\int_t^T [a(s)P(s) \\ + P(s)a^T(s) + b^2(s)]ds \\ - \textstyle\int_t^T [P(s)A^T(t,s)(B^T(s)B(s))^{-1} \\ \times A(t,s)P(s)]dg(s) \quad (25)$$

with terminal condition $P(T) = \Psi^{-1}$, and where $P(t_0)$ is obtained from the solution of the above integral equation, and is included to indicate generally nonzero initial condition for $P(t)$. The jumps in z and P at the points of discontinuity of $g(t)$ have the following explicit form:

$$\Delta z(t) = A^T(t)(B^TB)^{-1}A(t)P(t-)\Delta g(t), \quad (26)$$

$$\Delta P(t) = -P(t-)[I + A^T(t)(B^TB)^{-1}A(t)P(t-) \\ \times \Delta g(t)]^{-1}A^T(t)(B^TB)^{-1}A(t)P(t-)\Delta g(t). \quad (27)$$

The assumptions that we need to make about the particular form of the discontinuous function g corresponds to different practical cases. If both continuous and discontinuous control commands are present, then point of discontinuity of g will specify the instances when discontinuous controls are applied. The assumption of the known g would then correspond to the case when the time of application of discontinuous controls is externally specified. If g is not known but can be manipulated, then the selection of points of discontinuity can be considered as part of the controller design, i.e. $\min_{u,g} J$, and it appears that $\min_{u,g} J = \min_g J(u^*(g))$, where u^* is determined in this section. Finally, if points of discontinuity of g are a priori unknown and determined by exogenous events, then some estimates of the instances of discontinuity are required to apply the developed results.

6. DELAYED ACTUATION

The introduction of time-varying delays in actuation can be accomplished similarly to the way we introduced time-varying delays into the measurements. The developed theory is directly applicable if actuation is delayed in an arbitrary and time-varying fashion, as long as the time delays are actually known. If delays are not known, as in the case of the networked control system, then measurement delays – a priori unknown, but available once the time-labelled measurement arrives – can be used as an estimation of the actuation delays. The performance degradation and stability bounds for uncertain actuation delays can be analyzed using minimax methods.

7. CONCLUSIONS

The combination of the Ito-Volterra formalism and the results on the necessary conditions for optimality of control systems with impulsive discontinuities allowed us to solve a broad range of optimal filtering problems for continuous linear systems with discrete and continuous measurements, including the case of time-delayed measurements. The optimal control problem with continuous and discontinuous controls is dual to the corresponding filtering problems. An important difference between filtering and control problems is that optimal estimates can be obtained without prior knowledge of the measurement delays, while the control problem requires the knowledge of the actuation delays to guarantee stability and optimality.

REFERENCES

M. V. Basin and I. R. Valadez Guzman. Optimal minimax filtering and control in discontinuous Volterra systems. In *Proc. Amer. Contr. Conf.*, pages 904–908, Chicago, IL, 2000.

M. V. Basin and M. A. Villanueva Llanes. On filtering problems over Ito-Volterra observations. In *Proc. Amer. Contr. Conf.*, pages 3407–3412, 1999.

M. L. Kleptsina and A. Yu. Veretennikov. On filtering and properties of conditional laws of Ito-Volterra processes. In *Statistics and Control of Stochastic Processes*, Steklov Seminar. 1984, pages 179–196. Optimization Software Inc., Publication Division, New York, 1985.

M. A. Krasnoselskii and A. V. Pokrovskii. *Systems with hysteresis*. Springer, 1989.

L. E. Shaikhet. On an optimal control problem of partly observable stochastic Volterra processes. *Problems of Control and Inform. Theory*, 16: 439–448, 1987.

RAPID MULTI-MODEL IDENTIFICATION IN SYSTEMS WITH DELAYS

S. L. Campbell [*,1] **K. Drake** [**] **R. Nikoukhah** [***]
F. Delebecque [****]

[*] *Dept. of Mathematics, North Carolina State University, Raleigh, NC 27695-8205. USA.*
[**] *Dept. of Mathematics, North Carolina State University, Raleigh, NC 27695-8205. USA.*
[***] *INRIA, Rocquencourt BP 105, 78153 Le Chesnay Cedex, France.*
[****] *INRIA, Rocquencourt BP 105, 78153 Le Chesnay Cedex, France.*

Abstract: Recently an approach for multi-model identification and failure detection in the presence of bounded energy noise over short time intervals has been introduced. In this paper it is shown how this approach can be extended to some problems with delays. Computational issues are carefully addressed and an example is worked. *Copyright © 2001 IFAC*

Keywords: Delay, Model Identification, Linear, Optimization, Failure Detection

1. INTRODUCTION

In (Campbell *et al.*, 2000; Campbell *et al.*, 2001; Nikoukhah *et al.*, 2000 Nikoukhah *et al.*, 2002) an approach for model identification and failure detection has been developed. This paper shows how to extend this approach to a class of problems which include delays. The inclusion of general variable delays or multiple delays in the models being identified results in a problem which is intrinsically infinite dimensional and requires additional theoretical and numerical results. Here, as a first but still important step, the focus is on a special class of delay problems which can be reformulated so that they can be solved using modifications of the existing theory. The reformulation is not new. However, the reformulation poises some new technical issues in terms of prior information. In addition, reformulation has important implications for the choice of software. An example will be carefully worked. The results of this paper are of interest in their own right. They are also a first step in fully extending the theory of (Campbell *et al.*, 2001; Nikoukhah *et al.*, 2000; Nikoukhah *et al.*, 2002) to systems with general and multiple delays.

The basic problem formulation is as follows. Assume there are m possible models of the form

$$x_i'(t) = A_i x_i(t) + G_i x_i(t-h) + B_i v(t) + M_i \nu_i(t) \quad \text{(1a)}$$

$$y(t) = C_i x_i(t) + N_i \nu_i(t) \quad \text{(1b)}$$

$$x_i(t) = \phi_i(t), \quad -h \le t \le 0 \quad \text{(1c)}$$

for $i = 1, \ldots, m$ where v is an auxiliary signal to be determined and y is the observed output. Note that v, y are independent of i and are considered known. For simplicity of discussion the coefficient matrices are constant. However, the discussion goes through with only the obvious changes if the coefficients are continuous functions of t. ϕ_i is not known and thus is a type of disturbance or noise as are the ν_i.

[1] Research supported in part by the National Science Foundation under DMS-0101802, ECS-0114095 and INT-9605114.

Detection is to be carried out over a finite interval $[0,\omega]$. Suppose that the disturbances ϕ_i, ν_i satisfy

$$S(\phi_i,\nu_i) = \int_{-h}^{0} \|\phi_i\|^2 + \int_{0}^{\omega} \|\nu_i\|^2 dt \leq 1 \quad (2)$$

Let $\mathcal{A}_i(v)$ be the set of outputs y for (1) given (2) holds. Then an auxiliary signal v is proper if it leads to disjoint output sets $\mathcal{A}_i(v)$. That is, $\mathcal{A}_i(v) \cap \mathcal{A}_j(v) = \emptyset$ if $i \neq j$. The first step is to calculate a minimal energy proper auxiliary signal. Once this auxiliary signal is determined one needs a test to determine which model is correct. This is done by using separating hyperplanes in (Campbell *et al.*, 2001; Nikoukhah *et al.*, 2000). Rrapid detection is again desired so that the time intervals are finite and "small." Also the computation of v and the functions needed in the hyperplane test may be computed offline. Only the application of the hyperplane test needs to be done online. This paper focuses on determining the auxiliary signal.

There are a number of variations one can take of this problem including having other inputs. For problems without delays these are discussed in (Campbell *et al.*, 2001). This paper focuses on the case with one delay and two models. The handling of the m-model case is then straightforward.

2. PROBLEM FORMULATION

If $\omega < h$, then alter (2) to

$$S(\phi_i,\nu_i) = \int_{-h}^{\omega-h} \|\phi_i\|^2 + \int_{0}^{\omega} \|\nu_i\|^2 dt \leq 1 \quad (3)$$

and treat ϕ_i on $[-h, \omega - h]$ as part of the initial disturbance. That is, $[\phi, \nu_i]$ is the new ν_i and $[G_i, M_i]$ is the new M_i. This is now the case considered before in (Campbell *et al.*, 2001) and no special consideration is needed. Thus for the rest of this paper assume that $\omega > h$. For convenience, assume that $\omega = \bar{\kappa}h$ for an integer $\bar{\kappa} \geq 1$. This is not restrictive for h small relative to ω. For h large relative to the desired ω this becomes somewhat restrictive. It can be handled but it requires some modification of the problem formulation and the use of phases. This will not be discussed here. Let $\kappa = \bar{\kappa} - 1$. For each model (1) there is an initial condition $x_i(0)$ and an initial function ϕ_i which is defined on $[-h, 0]$. As will become clear shortly, there are some optimization or numerical integrations that have to be performed and the choice of software to do this has several implications.

The theory for the approach developed in (Nikoukhah *et al.*, 2000; Nikoukhah *et al.*, 2002) has not yet been extended to general delay systems. Here it is shown how to solve the model identification problem for some delay systems by converting them to non-delay systems. The conversion is not new since it is based on the well known method of steps. What is new is the use of it in this multi-model setting and the discussion of numerical issues in making this a practical approach.

Assume (1) holds for $i = 1, 2$ to give the combined system

$$x_1'(t) = A_1x_1(t) + G_1x_1(t-h) + B_1v(t) + M_1\nu_1(t) \quad (4a)$$
$$y(t) = C_1x_1(t) + N_1\nu_1(t) \quad (4b)$$
$$x_2'(t) = A_2x_2(t) + G_2x_2(t-h) + B_2v(t) + M_2\nu_2(t) \quad (4c)$$
$$y(t) = C_2x_2(t) + N_2\nu_2(t) \quad (4d)$$

Elimination of y gives the delay descriptor system

$$x_1'(t) = A_1x_1(t) + G_1x_1(t-h) \quad (5a)$$
$$+B_1v(t) + M_1\nu_1(t) \quad (5b)$$
$$x_2'(t) = A_2x_2(t) + G_2x_2(t-h) + B_2v(t) + M_2\nu_2(t) \quad (5c)$$
$$0 = C_1x_1(t) - C_2x_2(t) + N_1\nu_1(t) - N_2\nu_2(t) \quad (5d)$$

Now let $z_{i,j}$ on $[0,h]$ be given by $z_{i,j}(t) = x_i(t+jh)$ for $i = 1, 2$ and $\kappa \geq j \geq 0$. $\mu_{i,j}$ is defined similarily from ν_i. For $0 \leq t \leq h$, let $\gamma_i(t) = \phi_i(t-h)$. Let $v_j(t) = v(t+jh)$. Then (5) can be rewritten without delay on $[0,h]$ as shown in (6),

$$z_{1,0}' = A_1z_{1,0} + G_1\gamma_1 + B_1v_0 + M_1\mu_{1,0} \quad (6a)$$
$$z_{1,1}' = A_1z_{1,1} + G_1z_{1,0} + B_1v_1 + M_1\mu_{1,1} \quad (6b)$$
$$\vdots \quad (6c)$$
$$z_{1,\kappa} = A_1z_{1,\kappa} + G_1z_{1,\kappa-1} + B_1v_\kappa + M_1\mu_{1,\kappa} \quad (6d)$$

$$z_{2,0}' = A_2z_{2,0} + G_2\gamma_2 + B_2v_0 + M_2\mu_{2,0} \quad (6e)$$
$$z_{2,1}' = A_2z_{2,1} + G_2z_{2,0} + B_2v_1 + M_2\mu_{2,1} \quad (6f)$$
$$\vdots \quad (6g)$$
$$z_{2,\kappa} = A_2z_{2,\kappa} + G_2z_{2,\kappa-1} + B_2v_\kappa + M_2\mu_{2,\kappa} \quad (6h)$$

$$0 = C_1z_{1,0} - C_2z_{2,0} + N_1\mu_{1,0} - N_2\mu_{2,0} \quad (6i)$$
$$\vdots \quad (6j)$$
$$0 = C_1z_{1,\kappa} - C_2z_{2,\kappa} + N_1\mu_{1,\kappa} - N_2\mu_{2,\kappa} \quad (6k)$$

Along with the initial and boundary conditions

$$z_{i,0}(0) = x_i(0) \quad i = 1, 2 \quad (7a)$$
$$z_{i,j}(0) = z_{i,j-1}(h), \quad i = 1, 2, \quad j = 1, \ldots, \kappa \quad (7b)$$

When working without delays there were two options on how to handle the initial conditions. In (Campbell *et al.*, 2001) they were allowed to be arbitrary and identification was made robust to any perturbation of the initial conditions. In (Nikoukhah *et al.*, 2002) the initial conditions were viewed as bounded noise. Other options are discussed in (Campbell *et al.*, 2001). Here the situation is somewhat different.

If G is invertible, then for bounded noise, and any fixed $\overline{v}$, (6a), (6e) show that any smooth $z_{i,0}$ is possible. But then (6b) allows any $z_{i,1}$ since $z_{i,0}$ is arbitrary. Continuing in this manner shows that if G is invertible and there is no information on ϕ_i, then model identification is not possible. Thus ϕ_i has to be included in (2).

What about $x_i(0)$? There are now three possibilities. The first option is that $x_i(0)$ is free and (2) is used to measure the noise. A second option is that $x_i(0)$ is viewed as another perturbation treated like ν_i and the noise measure used is

$$\mathcal{S}(\phi_i,\nu_i,x_i(0)) = x_i(0)^T Q x_i(0) + \int_{-h}^{0} \|\phi_i\|^2 + \int_{0}^{\omega} \|\nu_i\|^2 dt \quad (8)$$

for some $Q > 0$. These are analogous to the two prior options. However, there is now a third option. In this option the system has been operating prior to the test. Thus while ϕ_i is now known, there are the additional continuity boundary conditions

$$z_{i,0}(0) = \phi_i(0) \quad (9)$$

This boundary condition is somewhat unusual. In the algorithms that follow the noise measure is minimized subject to some constraints. In so doing the noises will be treated as controls. But then (9) is a boundary condition that links the state to the control. This is not a standard restriction on the control and will not be discussed further. In addition because of page limitations only the first option of $x_i(0)$ being free and using (2) is discussed in this paper.

For $q = z_1, z_2, v$, let $\overline{q}$ or $\overline{q}_i$ be the vector with components $q_0, q_1, \ldots, q_\kappa$, respectively $q_{i,0}, \ldots, q_{i,\kappa}$. $\overline{\mu}$ is the same except that γ_i is added to the first component of $\overline{\mu}_i$. Then (6) is the system (10) on $[0, h]$

$$\overline{z}_1' = \overline{A}_1\overline{z}_1 + \overline{B}_1\overline{v} + \overline{M}_1\overline{\mu}_1 \quad (10a)$$
$$\overline{z}_2' = \overline{A}_2\overline{z}_2 + \overline{B}_2\overline{v} + \overline{M}_2\overline{\mu}_2 \quad (10b)$$
$$0 = \overline{C}_1\overline{z}_1 - \overline{C}_2\overline{z}_2 + \overline{N}_1\overline{\mu}_1 - \overline{N}_2\overline{\mu}_2 \quad (10c)$$

where

$$\overline{A}_i = \begin{bmatrix} A_i & 0 & 0 & 0 \\ G_i & A_i & 0 & \ddots \\ 0 & \ddots & \ddots & \ddots \\ 0 & 0 & G_i & A_i \end{bmatrix}$$

$$\overline{B}_i = \mathrm{Diag}\{B_i, B_i, \ldots, B_i\}$$

$$\overline{z}_i = \begin{bmatrix} z_{i,0} \\ \vdots \\ z_{i,\kappa} \end{bmatrix}, \overline{\mu}_i = \begin{bmatrix} \gamma_i \\ \mu_{i,0} \\ \vdots \\ \mu_{i,\kappa} \end{bmatrix}, \overline{v} = \begin{bmatrix} v_0 \\ \vdots \\ v_\kappa \end{bmatrix}$$

$$\overline{M}_i = \begin{bmatrix} G_i & M_i & 0 & \cdot \\ 0 & 0 & \ddots & \cdot \\ 0 & 0 & 0 & M_i \end{bmatrix} \quad (11)$$

$$\overline{N}_i = \begin{bmatrix} 0 & N_i & 0 & \cdot \\ 0 & 0 & \ddots & \cdot \\ 0 & 0 & 0 & N_i \end{bmatrix}$$

$\overline{C}_i = \mathrm{Diag}\{C_i, C_i, \ldots, C_i\}$. The system (10) will be written more succinctly as

$$\overline{z}' = \overline{A}\overline{z} + \overline{B}\overline{v} + \overline{M}\overline{\mu} \quad (12a)$$
$$0 = \overline{C}\overline{z} + \overline{N}\overline{\mu} \quad (12b)$$

In (Campbell *et al.*, 2001) the constraint was used to eliminate part of the noise.

Note that weights in (2) can be accommodated by changing $\overline{M}_i, \overline{N}_i$ so as not to include any weights. The new disturbance measure is

$$\overline{\mathcal{S}}(\overline{\mu}_i) = \int_0^h \|\overline{\mu}_i\|^2 dt = \mathcal{S}(\phi_i, \nu_i) \quad (13)$$

It is advantageous to allow for more general boundary conditions. Thus suppose that (12) has the general boundary or initial conditions

$$\mathcal{B}_0\overline{z}(0) + \mathcal{B}_1\overline{z}(h) = 0 \quad (14)$$

These boundary conditions include (7b) as well as possibly other boundary conditions. Having additional types of boundary conditions allows the handling of more general models as shown in Section 2.2.3.

The advantage of formulation (10) is that with the exception of the additional boundary conditions (14), (10) is the same system as that studied in (Campbell *et al.*, 2001). In fact, the norms of the noises are even the same since they are L^2 norms. However, the price appears to be looking at much larger matrices. If Riccati equations are used as in (Nikoukhah *et al.*, 2000), there is a severe computational penalty. The Riccati equation is a matrix equation so that the number of variables will increase as κ^2. Then one has to integrate these Riccati equations. Depending on whether explicit or implicit integrators are used the computational effort is proportional to the dimension or a power of the dimension. Since the work also, loosely, scales with the interval length, at best computation effort scales with κ^2. For even modestly sized systems, and for reasonable values of τ, this can become excessive as the delay gets small.

In (Campbell *et al.*, 2001) there is a Riccati equation free formulation which performs an optimization subject to a boundary value problem constraint. If the optimizer integrates the dynamics in order to reduce problem size, then the previous problems of scale hold. However, there is one class of methods for which the numerical work for a given ω and set of models is essentially independent of h for a wide range of h. These are sparse direct transcription algorithms. Codes using this approach discretize the problem and then use sparse techniques in carrying out the optimization algorithm.

One such direct transcription code, SOCS (Sparse Optimal Control Software) developed at Boeing (Betts and Frank, 1994; Betts and Huffman, 1997), is used here. Thus for moderate values of κ, the optimizer does not do significantly more work. Using the formulation of the models and applying the algorithm below the optimizer finds one mesh on the interval $[0, h]$. This corresponds to a mesh on $[0, \omega]$ which is h periodic. The method of steps formulation requires significantly more computational work only when h is so small that the h periodic mesh on $[0, \omega]$ is significantly finer than the mesh needed on $[0, \omega]$ to resolve the dynamics. At this time SOCS does not have the ability to directly handle general delay problems and some reformulation, such as that given here, is necessary. The one exception to these comments about efficiency is that the delay can impact on the bandwidth of the sparsity patterns in the linear algebra routines used by the optimizer.

2.1 *The algorithm*

Independent of the particular choice of what is noise, the algorithm then takes the following form. Since the proofs are similar to those of (Campbell *et al.*, 2001) here it is only pointed out how the approach works and what specific form it takes.

Since the total noise is $\max_i \mathcal{S}_i$, the underlying norm is not an L^2 norm as in many other approaches. To get around this, the auxiliary cost function is defined to be

$$\overline{J}_{\overline{v}}(\beta) = \min \beta\overline{\mathcal{S}}(\overline{\mu}_1) + (1-\beta)\overline{\mathcal{S}}(\overline{\mu}_2) \quad (15)$$

where the minimum is taken over all $\overline{z}_i, \overline{\mu}_i$ satisfying (6), (7b). The auxiliary cost function is important since it turns out that $\overline{v}$ is proper if and only if $\overline{J}_{\overline{v}}(\beta) \geq 1$. Then the minimal proper v is the solution of

$$\min_{\overline{J}_{\overline{v}}(\beta)\geq 1} \|\overline{v}\|^2 \quad (16)$$

This leads to the optimization problem of minimizing (16) subject to

(1) The necessary conditions for the minimization for the right hand side of (15).
(2) An expression whose value is $J_v(\beta)$
(3) The constraint $J_v(\beta) \geq 1$.
(4) The boundary conditions (14).

To be more speciific, suppose that (14) is the boundary conditions (7b). The actual optimization problem for finding the auxiliary signal v can be formulated two ways. Let $\overline{\lambda}$ be the same size as $\overline{x}$ with block elements λ_i. Then one optimization formulation is

$$\min_{\overline{v}} \int_0^h \|\overline{v}\|^2 dt \quad (17a)$$

Subject to the constraints

$$\begin{aligned}
\overline{z}' &= \overline{A}\overline{z} + \overline{B}\overline{v} + \overline{M}\overline{\mu} && (17b)\\
\overline{\lambda}' &= -\overline{A}^T\overline{\lambda} + \overline{C}^T\overline{\eta} && (17c)\\
0 &= \overline{C}\overline{z} + \overline{N}\overline{\mu} && (17d)\\
0 &= V_\beta\overline{\mu} - \overline{M}^T\overline{\lambda} + \overline{N}^T\overline{\eta} && (17e)\\
Z' &= \frac{1}{2}\overline{u}^T V_\beta \overline{u} && (17f)\\
Z(h) &\geq 1 && (17g)\\
\lambda_0(0) &= \lambda_\kappa(h) = 0 && (17h)\\
z_j(0) &= z_{j-1}(h), \quad j = 1,\ldots,\kappa && (17i)\\
\lambda_j(0) &= \lambda_{j-1}(h), \quad j = 1,\ldots,\kappa && (17j)\\
Z(0) &= 0 && (17k)\\
0.01 &\leq \beta \leq 0.99 && (17\ell)
\end{aligned}$$

Here $\frac{1}{2}\overline{u}^T V_\beta \overline{u} = \beta\|\overline{\mu}_1\|^2 + (1-\beta)\|\overline{\mu}_2\|^2$.

2.2 *Problem Variations*

Sometimes the control or reference trajectory u acts along the same channels as the auxiliary signal

$$\begin{aligned}
x_i'(t) &= A_i x_i(t) + G_i x_i(t-h) + B_i u(t) && (18a)\\
&\quad + B_i v(t) + M_i \mu_i(t) && (18b)\\
y(t) &= C_i x_i(t) + N_i \mu_i(t) && (18c)
\end{aligned}$$

Defining $\overline{u}$ in the same way as $\overline{v}$ gives

$$\begin{aligned}
\overline{z}_1' &= \overline{A}_1\overline{z}_1 + \overline{B}_1\overline{v} + \overline{B}_1\overline{u} + \overline{M}_1\overline{\mu}_1 && (19a)\\
\overline{z}_2' &= \overline{A}_2\overline{z}_2 + \overline{B}_2\overline{v} + \overline{B}_2\overline{u} + \overline{M}_2\overline{\mu}_2 && (19b)\\
0 &= \overline{C}_1\overline{z}_1 - \overline{C}_2\overline{z}_2 + \overline{N}_1\overline{\mu}_1 - \overline{N}_2\overline{\mu}_2 && (19c)
\end{aligned}$$

The minimization problem is then to find the smallest $\overline{v}$ such that $\overline{u} + \overline{v}$ is proper. That is,

$$\min_{\overline{J}_{\overline{u}+\overline{v}}(\beta)\geq 1} \|\overline{v}\|^2 \quad (20)$$

Sometimes, however, the reference signal does not come in through the same channels as the auxiliary signal. This can easily happen, for example, in complex systems where some subsystems are exposed to a reference signal and the auxiliary signal is acting on another part of the system. Then

$$x_i'(t) = A_i x_i(t) + G_i x_i(t-h) + H_i u(t) + B_i v(t) + M_i \nu_i(t) \tag{21a}$$

$$y(t) = C_i x_i(t) + N_i \nu_i(t) \tag{21b}$$

Often the columns of H_i and B_i are linearily independent. However, sometimes they are not. Then in order to insure that the optimization software does not experience problems the following can be done. Assume that there is a matrix R so that $H_i + B_i R = [0\ K_i]$ where K_i is full column rank of minimum size for $i = 1, 2$. That is, $[K_i, B_i]$ has full column rank. There always exists such a matrix for each i. Note that this assumption is only on the differences in the input channels between the two models. It makes no restrictions on the dynamics, delays, or outputs. Then define $v = \widehat{v} + Ru$, or equivalently, $\widehat{v} = v - Ru$. Let $\widehat{B}_i = [K_i, B_i]$ and $\widehat{w} = [u^T, \widehat{v}^T]^T$. With these substitutions (21) becomes

$$x_i'(t) = A_i x_i(t) + G_i x_i(t-h) + \widehat{B}_i \widehat{w} + M_i \nu_i(t) \tag{22a}$$

$$y(t) = C_i x_i(t) + N_i \nu_i(t) \tag{22b}$$

(22) looks like the original problem (1) with auxiliary signal $\widehat{w}$. Let $\widehat{J}_{\widehat{w}}(\beta)$ be the auxiliary cost function for this problem. Then the minimization problem is

$$\min_{\substack{\widehat{J}_{\widehat{w}}(\beta) \geq 1 \\ \widehat{w}_1 = u}} \|\widehat{w}_2 + Ru\|^2 \tag{23}$$

Given (22) is the same as (1) it can be rewriten as a single system in the same way. However, in this case there are additional constraints given by $\widehat{w}_1 = u$ on each interval.

2.2.1. Delays in the control From a formulation point of view, delays in the control are easily handled. The major change is that if some elements of the output set are not immediately affected by the auxiliary signal, then there may be a minimum value of ω below which the model identification is not possible.

2.2.2. Other costs Bounds other than the one given can be gotten by rescaling the noise weights. For example a noise weighting

$$S(\phi_i, \nu_i) = \int_{-h}^{0} \phi_i^T Q \phi_i \, dt + \int_{0}^{\omega} \|\nu_i\|^2 dt$$

is accommodated by modifying $\overline{M}_i$ in (11) to be

$$\overline{M}_i = \begin{bmatrix} G_i Q^{-1/2} & M_i & 0 & \cdot \\ 0 & 0 & \ddots & \cdot \\ 0 & 0 & 0 & M_i \end{bmatrix}$$

Another variation occurs if it is known that the prior history differs from some trajectory $r(t)$. Then the disturbance measure is

$$S(\phi_i, \nu_i) = \int_{-h}^{0} \|\phi_i - r_i(t)\|^2 \, dt + \int_{0}^{\omega} \|\nu_i\|^2 dt$$

Let $\delta_i = \phi_i - r_i$. Then

$$\overline{z}_1' = \overline{A}_1 \overline{z}_1 + \overline{B}_1 \overline{v} + \overline{H}_1 \overline{u} + \overline{M}_1 \overline{\mu}_1 \tag{24a}$$

$$\overline{z}_2' = \overline{A}_2 \overline{z}_2 + \overline{B}_2 \overline{v} + \overline{B}_2 \overline{u} + \overline{M}_2 \overline{\mu}_2 \tag{24b}$$

$$0 = \overline{C}_1 \overline{z}_1 - \overline{C}_2 \overline{z}_2 + \overline{N}_1 \nu_1 - \overline{N}_2 \overline{\mu}_2 \tag{24c}$$

where

$$\overline{\mu} = \begin{bmatrix} \delta_1 \\ \mu_{i,0} \\ \vdots \end{bmatrix}, \overline{H}_i = \begin{bmatrix} G_i & 0 & \dots & 0 \\ 0 & 0 & \ddots & \vdots \\ \vdots & \ddots & \ddots & \vdots \\ 0 & \dots & \dots & 0 \end{bmatrix}, \overline{u} = \begin{bmatrix} r_i \\ 0 \\ \vdots \\ 0 \end{bmatrix}$$

which is now written so that the same technique as (21) can be applied.

2.2.3. Integral Terms In some control problems there are both point delays and distributed delays. In these cases the models (4) and (5) would have their first equation replaced by

$$x'(t) = Ax(t) + Gx(t-h) + \int_{-h}^{0} Ex(t+\tau)d\tau + Bv(t) + \int_{-h}^{0} Fv(t+\tau)d\tau + M\nu(t)$$

The method of steps can include these types of systems by the introduction of additional variables. For $i > 0$,

$$x_i' = Ax_i + Gx_{i-1} + E(z_i + w_{i-1}) + Bv_i + F(u_i + r_{i-1}) + M\nu_i \tag{25a}$$

$$z_i' = x_i, \quad z_i(0) = 0 \tag{25b}$$

$$w_i' = x_i, \quad w_i(h) = 0 \tag{25c}$$

$$u_i' = v_i, \quad u_i(0) = 0 \tag{25d}$$

$$r_i' = v_i, \quad r_i(h) = 0 \tag{25e}$$

This reformulation uses for $ih < t < (i+1)h$ that

$$\int_{-h}^{0} g(t+\tau)d\tau = \int_{ih}^{t} g(s)ds + \int_{t-h}^{ih} g(s)ds$$

Each of the integrals on the right can be written as a differential equation and a boundary condition. For $i = 0$ modify (25c) and (25e) as

$$w_0' = \phi, \quad w_0(h) = 0 \tag{26}$$
$$r_0' = \psi \tag{27}$$

The ϕ is not a problem. It is the initial condition on the state. The ψ is more problematic. That is the auxiliary signal prior to time 0 which does not exist. When putting the auxiliary signal down the same channel as a preexisting control some modifications to the theory would be needed. But in the case discussed here it makes the most sense to just use $\psi = 0$.

If (25a) is used back in the original model a new system results which has the same form as before. However, there is one difference. The states x_i are now made up of three parts. One part has an initial condition. One part has a final condition. One part has a continuity condition linking it to the same part of x_{i+1} and x_{i-1}. Thus these types of terms lead to the more general type of boundary conditions covered by (14). The multipliers that correspond to these states will have value zero at the opposite end of the interval.

3. COMPUTATIONAL EXAMPLE

To illustrate how this technique works consider an example taken from the control of the mach number in a wind tunnel (Manitius and Tran, 1986).

$$x_1'(t) = -ax_1(t) + akx_2(t-h) + \nu_1 \tag{28a}$$
$$x_2'(t) = x_3(t) + \nu_2 \tag{28b}$$
$$x_3'(t) = -\delta^2 x_2(t) - 2\eta\delta x_3(t) \tag{28c}$$
$$+\delta^2(u(t) + v(t)) + \nu_3 \tag{28d}$$
$$y = x_1 + x_2 + \nu_4 \tag{28e}$$

where $a = (1.964)^{-1}, k = -0.0117, \delta = 6, \eta = 0.8$, and $h = 1$ seconds. Note that the scaling of variables is set up so that for a constant control u there is a set point, equivalently a constant solution, $x_1 = ku, x_2 = u, x_3 = 0$. Assume the detection horizon is $[0, \kappa h]$. The auxiliary signal is sent down the same channel as u. The failure mode is the same as (28) except that $\delta = 4$ and $a = 1/2.892$. Neither δ nor a appear in the equations defining the set point so that if u is constant and the state is at the set point, then a change in δ or a would not be detectable no matter how long the system was observed. Thus an auxiliary signal is needed to notice changes in δ or a.

Figure 1 gives the minimal energy v for $\kappa = 1$ when u=2. The cost is $\|v\|^2 = 41.9591$. Since a constant u gives the same set point in both models, the minimal energy auxiliary signal for this problem is independent of u for constant u. In computing this answer it was found that limiting β to a smaller range was helpful in attaining a feasible solution on the initial coarse grids as was eliminating some of the constraints and thereby reducing the number of noise variables.

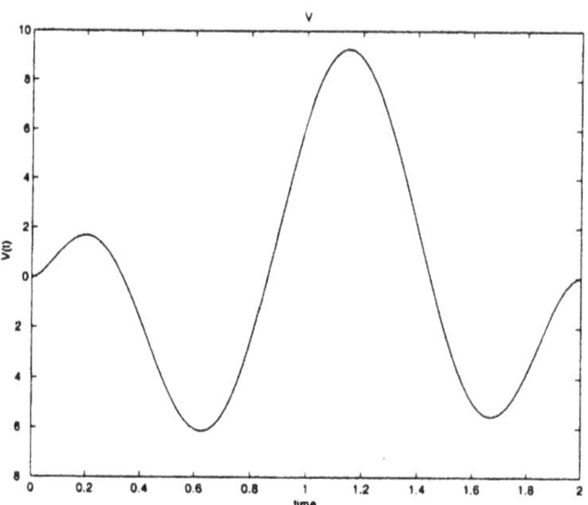

Fig. 1. Minimal Auxiliary Signal for (28).

4. DISCUSSION

When used in conjunction with direct transcription software, the method of steps can be used to solve a number of auxiliary signal design problems for systems with a single delay. This formulation, however, impacts on how the uncertainty in the initial conditions are handled. These results will also be helpful when examining the more complex case of more than two models and more than one delay.

REFERENCES

Betts, J. T. and P. D. Frank (1994). A sparse nonlinear optimization algorithm. *Journal of Optimization Theory and Applications* **82**, 519–541.

Betts, J. T. and W. P. Huffman (1997). Sparse optimal control software socs. *Mathematics and Engineering Analysis Technical Document MEA-LR-085.*

Campbell, S. L., K. Horton, R. Nikoukhah and F. Delebecque (2000). Rapid model selection and the separability index. *Proc. SAFEPROCESS.*

Campbell, S. L., K. Horton, R. Nikoukhah and F. Delebecque (2001). Auxiliary signal design for rapid multimodel identification using optimization. *Preprint.*

Manitius, A. and H. Tran (1986). Numerical simulation of a non-linear feedback controller for a wind tunnel model involving a time delay. *Optimal Control Applications and Methods* **7**, 19–39.

Nikoukhah, R., S. L. Campbell, and F. Delebecque (2000). Detection signal design for failure detection: a robust approach. *Int. J. Adaptive Control and Signal Processing* **14**, 701–724.

Nikoukhah, R., S. L. Campbell, K. Horton and F. Delebecque (2002). Auxiliary signal design for robust multi-model identification. *IEEE Trans. Aut. Control* p. to appear.

www.elsevier.com/locate/ifac

STABILITY CRITERIA FOR LINEAR TIME-DELAY SYSTEMS WITH MULTIPLE PASSIVE UNCERTAINTIES*

Qing-Long Han and Keqin Gu†

Department of Mechanical and Industrial Engineering
Southern Illinois University at Edwardsville
Edwardsville, Illinois 62026-1805, U.S.A.
E-mail: {qhan, kgu}@siue.edu

Abstract: This paper investigates the robust stability problem of time-delay systems with multiple passive uncertainties. A stability criterion is derived using the refined discretized Lyapunov functional method. The criterion is written in the form of a linear matrix inequality. Numerical examples are presented to illustrate the effectiveness of the method. *Copyright © 2001 IFAC*

Keywords: Stability; time-delay; Lyapunov functional; passive uncertainty; linear matrix inequality.

1. INTRODUCTION

Time-delays are frequently encountered in practical systems such as engineering, communications and biological systems (Hale and Lunel 1993). Their existence may induce instability, oscillation and poor performance (Malek-Zavarei and Jamshidi 1987). Therefore, the stability and control of time-delay systems have received significant attentions in the last two decades. Current efforts on this topic can be divided into two categories (Mori 1985), namely *delay-independent* stability criteria (Kokame *et al.* 1998, Han and Mehdi 1999a) and *delay-dependent* stability criteria (Li and de Souza 1997, Han and Mehdi 1999b, Park 1999, Han and Gu 2001a).

* This work was partially supported by the U.S. National Science Foundation under Grant INT-9818312.

† Corresponding author.

For a linear system with a constant time-delay, it has been proven that the existence of a more general quadratic form Lyapunov-Krasovskii functional is necessary and sufficient for the stability of an uncertainty-free time-delay system (Huang 1989). The related weight matrices have to satisfy a two point boundary value problem of ordinary differential equations. A solution construction for this system and the following analysis of this system is a very difficult task. However, a piecewise linear discretization scheme has been proposed to enable one to write the stability criterion in an LMI form (Gu 1997, 1999a, 1999b). The criteria have shown significant improvements over the existing results even under very coarse discretization. For uncertainty-free systems, the analytical results can be approached with fine discretization. Recently, a refinement has been proposed in Gu (2001) that significantly reduces conservatism for uncertain system and greatly accelerates the convergence to the analytical solution for uncertainty-free systems.

Since the notion of passivity was introduced by Willems (1971), it has played an important role in systems, circuits, and control. Passivity is an important tool in stability analysis. The passivity theorem (e.g. Desoer and Vidyasagar 1975) can be used to establish the stability of the feedback system, see for example, Willems (1971), Hill and Moylan (1976), Moylan and Hill (1978), Boyd and Yang (1989). Recently, there has been an increasing interest of applying passivity to time-delay system. In Niculescu and Lozano (2001), sufficient conditions are derived for time-delay systems to be passive and the passivity-based stabilization is also addressed. In Verriest and Aggoune (1998), some robust stability conditions for a class of time-varying delay system are obtained using the passivity of an appropriate linear system.

In this paper, the stability problem of linear time-delay systems with multiple passive uncertainties is investigated using the refined discretized Lyapunov functional approach (Gu 2001). A stability criterion is obtained. Numerical examples show the effectiveness of the method.

2. PROBLEM STATEMENT

Consider the linear time-delay system with feedback uncertainty

$$\dot{x}(t) = Ax(t) + A_d x(t-r) + Bu(t) \qquad (1)$$

$$y(t) = Cx(t) + C_d x(t-r) + Du(t) \qquad (2)$$

where $x(t)\in \mathbb{R}^n$ is the state, r is a constant time delay, $A,\ A_d \in \mathbb{R}^{n\times n}$, $B\in \mathbb{R}^{n\times p}$, $C,\ C_d \in \mathbb{R}^{q\times n}$ and $D\in \mathbb{R}^{q\times p}$ are known real constant matrices, $u(t)\in \mathbb{R}^p$ and $y(t)\in \mathbb{R}^p$. The system is subject to uncertain feedback

$$u = \Delta(y) \qquad (3)$$

where Δ is a possibly nonlinear causal operator mapping $[L^2_{loc}(\mathbb{R}_+)]^q$ into $[L^2_{loc}(\mathbb{R}_+)]^p$. It is assumed the characteristics of the uncertainty such that

$$\int_0^t u^T(\xi)Ky(\xi)d\xi \geq 0 \qquad (4)$$

for any $t \geq 0$, and for any

$$K \in \mathcal{K} \subset \mathbb{R}^{p\times p} \qquad (5)$$

where $\mathcal{K}$ is a known convex set. Typically, the set $\mathcal{K}$ describes block-diagonal dynamic or memoryless, linear or nonlinear uncertainty. The details will be described later.

Define $\mathcal{C}$ as the set of continuous $\mathbb{R}^n$ valued function on the interval $[-r,\ 0]$, and let $x_t \in \mathcal{C}$ be a segment of system trajectory defined as

$$x_t(\theta) = x(t+\theta)\ ,\ -r \leq \theta \leq 0\ . \qquad (6)$$

In this paper, we will attempt to formulate a practically computable criterion for robust stability of the above system.

3. MAIN RESULTS

Choose a Lyapunov functional V of a quadratic form

$$V: \mathcal{C} \mapsto \mathbb{R}$$

$$V(x_t) = \frac{1}{2}x(t)^T Px(t) + x^T(t)\int_{-r}^0 Q(\xi)x(t+\xi)d\xi$$

$$+\frac{1}{2}\int_{-r}^0 [\int_{-r}^0 x^T(t+\xi)R(\xi,\eta)x(t+\eta)d\eta]d\xi \quad (7)$$

where

$$P\in \mathbb{R}^{n\times n},\ P = P^T$$

$$Q:[-r,0] \to \mathbb{R}^{n\times n}$$

$$S:[-r,0] \to \mathbb{R}^{n\times n},\ S^T(\xi) = S(\xi)$$

$$R:[-r,0]\times[-r,0] \to \mathbb{R}^{n\times n},\ R(\eta,\xi) = R^T(\xi,\eta).$$

Then we have the following theorem.

Theorem 1: *The system described above is asymptotically stable if there exists a quadratic Lyapunov functional V of form (7) such that for some $\varepsilon > 0$, and arbitrary $x_t \in \mathcal{C}$*

$$V(x_t) \geq \varepsilon x^T(t)x(t) \qquad (8)$$

and its derivative along the solution of (1) satisfies

$$\dot{V}(x_t) + u^T Ky \leq -\varepsilon x^T(t)x(t)\ . \qquad (9)$$

Proof. See the full version of the paper (Han and Gu, 2001b).

Choose Q, R and S to be continuous piecewise linear, i.e.,

$$Q^i(\alpha)\overset{\Delta}{=}Q(\delta_{i-1}+\alpha h)=(1-\alpha)Q_{i-1}+\alpha Q_i \quad (10)$$

$$S^i(\alpha)\overset{\Delta}{=}S(\delta_{i-1}+\alpha h)=(1-\alpha)S_{i-1}+\alpha S_i \quad (11)$$

$$\begin{aligned} &R(\delta_{i-1}+\alpha h,\delta_{j-1}+\eta h)=R^{ij}(\alpha,\eta) \\ &\overset{\Delta}{=}\begin{cases}(1-\alpha)R_{i-1,j-1}+\eta R_{ij}+(\alpha-\eta)R_{i,j-1}, & \alpha\ge\eta \\ (1-\eta)R_{i-1,j-1}+\alpha R_{ij}+(\eta-\alpha)R_{i-1,j}, & \alpha<\eta\end{cases} \end{aligned} \quad (12)$$

for $0\le\alpha\le 1$, $0\le\eta\le 1$, where

$$\delta_i=-r+ih,\ i=0,\ 1,\ 2,\ \cdots,\ N$$
$$h=r/N$$

i.e., N is the number of divisions of the interval $[-r,\ 0]$, and h is the length of each division.

For the Lypanov functional condition, the following lemma was proven in Gu (2001).

Lemma 1 (Gu, 2001): *For piecewise linear Q, S and R as described by (10)-(12), equation (8) is satisfied for some $\varepsilon>0$, and arbitrary $x_t\in\mathcal{C}$ if*

$$\tilde{S}>0 \quad (13)$$

and

$$\begin{bmatrix} P & \tilde{Q} \\ \tilde{Q}^T & \frac{1}{h}\tilde{S}+\tilde{R}\end{bmatrix}>0 \quad (14)$$

where

$$S^c=diag(S_1\ S_2\ \cdots\ S_{N-1})$$
$$\tilde{S}=diag(S_0\ S^c\ S_N)$$
$$\tilde{R}=\begin{bmatrix} R_{00} & R_{01} & \cdots & R_{0N} \\ R_{10} & R_{11} & \cdots & R_{1N} \\ \vdots & \vdots & \ddots & \vdots \\ R_{N0} & R_{N1} & \cdots & R_{NN}\end{bmatrix}$$
$$\tilde{Q}=[Q_0\ Q_1\ \cdots\ Q_N].$$

Concerning condition (9), we have the following proposition.

Proposition 1: *For piecewise linear Q, S and R as described by (10)-(12), equation (9) is satisfied for some $\varepsilon>0$, and arbitrary $x_t\in\mathcal{C}$ if there exists $K\in\mathcal{K}$ such that*

$$\begin{bmatrix} G_{00} & -G_{01} & -G_{02} & H_0^s & H_0^a \\ -G_{01}^T & G_{11} & -G_{12} & H_1^s & H_1^a \\ -G_{02}^T & -G_{12}^T & G_{22} & H_2^s & H_2^a \\ H_0^{sT} & H_1^{sT} & H_2^{sT} & \frac{1}{h}S_d+R_d & 0 \\ H_0^{aT} & H_1^{aT} & H_2^{aT} & 0 & \frac{3}{h}S_d\end{bmatrix}>0 \quad (15)$$

where

$$G_{00}=-KD-D^TK$$
$$G_{01}=B^TP+KC$$
$$G_{02}=KC_d$$
$$G_{11}=-PA-A^TP-S_N-Q_N-Q_N^T$$
$$G_{12}=PA_d-Q_0$$
$$G_{22}=S_0$$
$$S_d=diag(S_{d1}\ S_{d2}\ \cdots\ S_{dN})$$
$$S_{di}=\frac{1}{h}(S_i-S_{i-1})$$
$$R_d=\begin{bmatrix} R_{d11} & R_{d12} & \cdots & R_{d1N} \\ R_{d21} & R_{d22} & \cdots & R_{d2N} \\ \vdots & \vdots & \ddots & \vdots \\ R_{dN1} & R_{dN2} & \cdots & R_{dNN}\end{bmatrix}$$
$$R_{dij}=\frac{1}{h}(R_{ij}-R_{i-1,j-1});\ i,j=1,\ 2,\ \cdots,\ N$$
$$H_j^s=[H_{j1}^s\ H_{j2}^s\ \cdots\ H_{jN}^s];\ j=0,\ 1,\ 2;$$
$$H_{0i}^s=-\frac{1}{2}B^T(Q_i+Q_{i-1})$$
$$H_{1i}^s=-\frac{1}{2}A^T(Q_i+Q_{i-1})+\frac{1}{h}(Q_i-Q_{i-1}) -\frac{1}{2}(R_{i,N}^T+R_{i-1,N}^T)$$
$$H_{2i}^s=-\frac{1}{2}A_d^T(Q_i+Q_{i-1})+\frac{1}{2}(R_{i,0}^T+R_{i-1,0}^T)$$
$$H_j^a=[H_{j1}^a\ H_{j2}^a\ \cdots\ H_{jN}^a];\ j=0,\ 1,\ 2;$$
$$H_{0i}^a=\frac{1}{2}B^T(Q_i-Q_{i-1})$$
$$H_{1i}^a=\frac{1}{2}A^T(Q_i-Q_{i-1})+\frac{1}{2}(R_{i,N}^T-R_{i-1,N}^T)$$
$$H_{2i}^a=\frac{1}{2}A_d^T(Q_i-Q_{i-1})-\frac{1}{2}(R_{i,0}^T-R_{i-1,0}^T).$$

Proof. See the full version of the paper (Han and Gu, 2001b).

Based on Lemma 1 and Proposition 1, we now state and establish the following theorem.

Theorem 2: *An uncertain time-delay system as described is asymptotically stable if there exist real matrices* $P=P^T$, Q_i, S_i, $(i=0, 1, 2, \cdots, N)$ *and* R_{ij} $(i,j=0, 1, 2, \cdots, N)$ *and* $K \in \mathcal{K}$ *such that (14) and (15) are satisfied.*

Proof: It is easy to see that (13) is implied by (15). Q. E. D.

In the rest of this section we will discuss the uncertainty characterization. A very general uncertainty can be modelled as feedback block-diagonal uncertainty as described in, for example, Boyd and Yang (1989), Doyle *et al.* (1982), Packard and Doyle (1993)

$$u=\begin{pmatrix} u_1^T & u_2^T & \cdots & u_m^T \end{pmatrix}^T$$
$$y=\begin{pmatrix} y_1^T & y_2^T & \cdots & y_m^T \end{pmatrix}^T$$

where

$$u_i \in \mathbb{R}^{p_i},\ y_i \in \mathbb{R}^{p_i}$$

and

$$u_i = \Delta_i(y_i)$$

We will assume that Δ_i all have unit gain. Otherwise, a standard scaling procedure can be used. Correspondingly, similar to Boyd and Yang (1989), $\mathcal{K}$ consists of all the diagonal matrices

$$diag\begin{pmatrix} K_1 & K_2 & \cdots & K_m \end{pmatrix}$$

where

$$K_i \in \mathcal{K}_i \subset \mathbb{R}^{p_i \times p_i}$$

Correspondingly

$$V_1 = \sum_{i=1}^{m} V_{1i}$$

with each quadratic V_{1i} corresponds to one block of uncertainty. The structure of $\mathcal{K}_i$ depends on the characteristics of the uncertainty. For the uncertainty to be known as passive, we can choose

$$\mathcal{K}_i = \left\{ -k_i I_{p_i} \;\middle|\; k_i \in \mathbb{R}^+ \right\}$$

4. EXAMPLES

Example 1: Consider the uncertain time-delay system (1)-(2) with

$$A=\begin{bmatrix} -2 & 0 \\ 0 & -0.9 \end{bmatrix},\ A_d=\begin{bmatrix} -1 & 0 \\ -1 & -1 \end{bmatrix}$$

$$B=\alpha\begin{bmatrix} 1 & 0 \\ 0 & 1 \end{bmatrix}\ \alpha \ge 0,\ C=C_d=D=\begin{bmatrix} 1 & 0 \\ 0 & 1 \end{bmatrix}$$

and

$$u=\Delta(y)$$

where $\alpha \ge 0$ is an uncertainty parameter.

In the following we consider two different structure of Δ.

(1) $\Delta=\begin{bmatrix} \Delta_{11} & \Delta_{12} \\ \Delta_{21} & \Delta_{22} \end{bmatrix}$ with $\sigma_{\max}(\Delta) \le 1$, where $\sigma_{\max}(\Delta)$ denotes the maximum singular value of the matrix Δ. We choose

$$\mathcal{K} = \left\{ -kI_2 \;\middle|\; k \in \mathbb{R}^+ \right\}$$

For $\alpha = 0.2$, the maximum values of $r_{\max}$ for the system to have guaranteed robust stability is estimated for different *N*. The results are listed in the following table

N	1	2	3
$r_{\max}$	4.1646	4.2264	4.2297

Now we consider the effect of the uncertainty bound α on the maximum time-delay for stability $r_{\max}$. The numerical results are estimated in the following table for $N=3$ and different α. It shows that as $\alpha \to 0$, the stability limit for delay approaches the uncertainty-free case (Gu 1999a). As α increases, $r_{\max}$ decreases.

α	0.00	0.05	0.10	0.15
r_{max}	6.17	5.58	5.06	4.62
α	0.20	0.25	0.30	0.35
r_{max}	4.22	3.88	3.58	3.32

Example 2: Consider another uncertain time-delay system (1)-(2) with

$$A = \begin{bmatrix} -3 & -2.5 \\ 1 & 0.5 \end{bmatrix}, \quad A_d = \begin{bmatrix} 1.5 & 2.5 \\ -0.5 & -1.5 \end{bmatrix}$$

$$B = \alpha \begin{bmatrix} 1 & 0 \\ 0 & 1 \end{bmatrix} \quad \alpha \geq 0, \; C = C_d = D = \begin{bmatrix} 1 & 0 \\ 0 & 1 \end{bmatrix}$$

and

$$u = \Delta(y)$$

where $\alpha \geq 0$ is an uncertainty parameter.

We also consider two different structure of Δ.

(1) $\Delta = \begin{bmatrix} \Delta_{11} & \Delta_{12} \\ \Delta_{21} & \Delta_{22} \end{bmatrix}$ with $\sigma_{max}(\Delta) \leq 1$, where $\sigma_{max}(\Delta)$ denotes the maximum singular value of the matrix Δ. We choose

$$\mathcal{K} = \left\{ -kI_2 \;\middle|\; k \in \mathbb{R}^+ \right\}$$

For $\alpha = 0.2$, the maximum values of r_{max} is estimated and listed in the following table for different N.

N	1	2	3
r_{max}	1.7032	1.7117	1.7121

The effect of the uncertainty bound α on the maximum time-delay for stability r_{max} is also studied. The numerical results are listed in the following table for $N = 3$ and different α. It shows again that as $\alpha \to 0$, the stability limit for delay approaches the uncertainty-free case (Chen *et al.* 1994). As α increases, r_{max} decreases.

α	0.00	0.05	0.10	0.15
r_{max}	2.41	2.19	2.01	1.85
α	0.20	0.25	0.30	0.35
r_{max}	1.71	1.59	1.48	1.38

(2) $\Delta = \begin{bmatrix} \Delta_1 & 0 \\ 0 & \Delta_2 \end{bmatrix}$ with $|\Delta_i| \leq 1$ (for $i = 1,2$). In this case we choose

$$\mathcal{K} = \left\{ -diag(k_1 \; k_2) \;\middle|\; k_i \in \mathbb{R}^+, \; i = 1,2 \right\}$$

For $\alpha = 0.2$, the results are in the following table for different N.

N	1	2	3
r_{max}	1.7597	1.7689	1.7694

It shows again that if the structure of Δ is more clearly known, less conservative results can be obtained.

5. CONCLUSION

The problem of robust stability of linear time-delay systems with multiple passive uncertainty has been addressed. A stability criterion has been obtained. Numerical examples have shown the effectiveness of the method.

REFERENCES

Boyd, S., L. El Ghaoui, E. Feron and V. Balakrishnan, (1994). *Linear Matrix Inequalities in Systems and Control Theory.* SIAM, Philadelphia.

Boyd, S. and Q. Yang (1989). Structured and simultaneous Lyapunov functions for system stability problems. *International Journal of Control*, **49**, 2215-2240.

Chen, J., G. Gu, and C. N. Nett (1994). A new method for computing delay margins for stability of linear delay systems. In: *Proc. 33rd IEEE Conf. Dec. Contr.*, Lake Buena Vista, FL, pp. 433-437.

Doyle, J. C., J. Wall and G. Stein (1982). Performance and robustness analysis for structured uncertainty. In: *Proceedings of 21st IEEE Conference on Decision and Control*, Florida, 629-636.

Gu, K. (1997). Discretized LMI set in the stability problem of linear uncertain time-delay systems. *International Journal of Control*, **68**, 923-934.

Gu, K. (1999a). A generalized discretization scheme of Lyapunov functional in the stability problem of linear uncertain time-delay systems. *International Journal of Robust and Nonlinear Control*, **9**, 1-14.

Gu, K. (1999b). Partial solution of LMI in stability problem of time-delay systems. In: *Proceedings of 38th IEEE Conference on Decision and Control*, Phoenix, Arizona, USA, vol. 1, December 1999, pp. 227-232.

Gu, K. (2001). A further refinement of discretized Lyapunov functional method for the stability of time-delay systems *International Journal of Control*, **74**, 967-976.

Hale, J. K. and S. M. Verduyn Lunel (1993). *Introduction to Functional Differential Equations.* Springer-Verlag, New York.

Han, Q.-L. and K. Gu (2001a). On robust stability of time-delay systems with norm-bounded uncertainty. In: *Proc the 2001 American Contr. Conf.*, Arlington, VA, June 25-27, 2001, pp. 4644-4649; Also to appear in: *IEEE Trans. Automat. Contr.*, **AC-46**, No. 8.

Han, Q.-L. and K. Gu (2001b). Stability criteria for linear time-delay systems with multiple passive uncertainties. Internal Technical Report (full version of the conference paper), Southern Illinois University Edwardsville, 2001.

Han, Q.-L. and D. Mehdi (1999a). Robust H_∞ controller synthesis for uncertain systems with multiple time-varying delays: An LMI approach. In *Proceedings of 14th IFAC World Congress*, Beijing, P. R. China, **Vol. C**, 271-276.

Han, Q.-L. and D. Mehdi (1999b). Robust stabilization for uncertain time-varying delay constrained systems with delay-dependence. *International Journal of Applied Mathematics and Computer Science*, **9**, 293-311.

Hill, D. and P. Moylan (1976). The stability of nonlinear dissipative systems. *IEEE Transaction on Automatic Control*, **21**, 708-711.

Huang, W. (1989). Generalization of Liapunov's theorem in a linear delay system. *Journal of Mathematical Analysis and Applications*, **142**, 83-94.

Kokame, H., H. Kobayashi and Mori, T. (1998). Robust H_∞ performance for linear delay-differential systems with time-varying uncertainties. *IEEE Transaction on Automatic Control*, **43**, 223-226.

Li, X. and C. E. de Souza (1997). Criteria for robust stability and stabilization of uncertain linear systems with state delay. *Automatica*, **33**, 1657-1662.

Malek-Zavarei, M. and M. Jamshidi (1987). *Time Delay Systems: Analysis, Optimization and Applications.* North-Holland, Amsterdam.

Mori, T. (1985). Criteria for asymptotic stability of linear time-delay systems, *IEEE Transactions on Automatic Control,* **30**, 158-161.

Moylan, P. and D. Hill (1978). Stability criteria for large-scale systems. *IEEE Transaction on Automatic Control*, **23**, 143-149.

Niculescu, S.-I. and R. Lozano (2001). On the passivity of linear delay systems. *IEEE Transaction on Automatic Control*, **46**, 460-464.

Packard, A. and J. C. Doyle (1993). The complex structured singular value. *Automatica*, **29**, 71-109.

Park, P. (1999). A delay-dependent stability criterion for systems with uncertain time-invariant delays. *IEEE Transaction on Automatic Control*, **44**, 876-877.

Verriest E. I. and W. Aggoune (1998). Stability of nonlinear differential delay systems. *Mathematics and Computers in Simulation*, 45, 257-269.

Willems, J. C. (1971). The generation of Lyapunov functions for input-output stable systems. *SIAM Journal of Control*, 9, 105-134.

AUTHOR INDEX

www.ingramcontent.com/pod-product-compliance
Ingram Content Group UK Ltd.
Pitfield, Milton Keynes, MK11 3LW, UK
UKHW050615260726
13967UKWH00008B/2879